高等院校卓越计划系列丛书

Fundamentals of Fluid Mechanics

工程流体力学

邵卫云 编著

中国建筑工业出版社

图书在版编目(CIP)数据

工程流体力学/邵卫云编著. —北京：中国建筑工业出版社，2015.5

（高等院校卓越计划系列丛书）

ISBN 978-7-112-17888-9

Ⅰ.①工… Ⅱ.①邵… Ⅲ.①工程力学-流体力学-高等学校-教材 Ⅳ.①TB126

中国版本图书馆 CIP 数据核字(2015)第 045382 号

本教材是高等院校卓越计划系列丛书中浙江大学建筑工程学院卓越计划系列丛书之一，主要根据教育部高等学校力学基础课程教学指导委员会流体力学及水力学课程教学指导小组编制的土建类专业的流体力学教学大纲编写。内容包括流体的物理特性、流体静力学、流体运动学、恒定总流基本方程、相似原理与量纲分析、流体阻力与能量损失、孔口与管嘴出流、管流、明渠流、堰流、渗流及可压缩一元流等。教材内容侧重于基本概念与原理的阐述及其在日常生活与工程中的应用。

本教材是普通高等学校土木、水利、海洋、市政工程等专业的专业基础课程流体力学的双语教学教材，也可作为环境、机械、能源工程等专业的流体力学课程双语教学用书，亦可供有兴趣了解流体力学英文术语词汇及其日常应用的学生与专业人员的参考。

* * *

责任编辑：赵梦梅　李东禧
责任设计：董建平
责任校对：姜小莲　赵　颖

高等院校卓越计划系列丛书
Fundamentals of Fluid Mechanics
工程流体力学
邵卫云　编著

*

中国建筑工业出版社出版、发行(北京西郊百万庄)
各地新华书店、建筑书店经销
北京红光制版公司制版
北京市安泰印刷厂印刷

*

开本：787×1092 毫米 1/16 印张：33¼ 字数：350 千字
2015 年 10 月第一版　2015 年 10 月第一次印刷
定价：**75.00** 元
ISBN 978-7-112-17888-9
(27145)

版权所有　翻印必究
如有印装质量问题，可寄本社退换
(邮政编码　100037)

浙江大学建筑工程学院卓越计划系列教材
丛 书 序 言

随着时代进步，国家大力提倡绿色节能建筑，推进城镇化建设和建筑产业现代化，我国基础设施建设得到快速发展。在新型建筑材料、信息技术、制造技术、大型施工装备等新材料、新技术、新工艺广泛应用新的形势下，建筑工程无论在建筑结构体系、设计理论和方法以及施工与管理等各个方面都需要不断创新和知识更新。简而言之，建筑业正迎来新的机遇和挑战。

为了紧跟建筑行业的发展步伐，为了呈现更多的新知识、新技术，为了启发更多学生的创新能力，同时，也能更好地推动教材建设，适应建筑工程技术的发展和落实卓越工程师计划的实施，浙江大学建筑工程学院与中国建筑工程出版社诚意合作，精心组织、共同编纂了"高等院校卓越计划系列丛书"之"浙江大学建筑工程学院卓越计划系列教材"。

本丛书编写的指导思想是：理论联系实际，编写上强调系统性、实用性，符合现行行业规范。同时，推动基于问题、基于项目、基于案例多种研究性学习方法，加强理论知识与工程实践紧密结合，重视实训实习，实现工程实践能力、工程设计能力与工程创新能力的提升。

丛书凝聚着浙江大学建筑工程学院教师们长期的教学积累、科研实践和教学改革与探索，具有了鲜明的特色：

（1）重视理论与工程的结合，充实大量实际工程案例，注重基本概念的阐述和基本原理的工程实际应用，充分体现了专业性、指导性和实用性；

（2）重视教学与科研的结合，融进各位教师长期研究积累和科研成果，使学生及时了解最新的工程技术知识，紧跟时代，反映了科技进步和创新；

（3）重视编写的逻辑性、系统性，图文相映，相得益彰，强调动手作图和做题能力，培养学生的空间想象能力、思考能力、解决问题能力，形成以工科思维为主体并融合部分人性化思想的特色和风格。

本丛书目前计划列入的有：《土力学》、《基础工程》、《结构力学》、《混凝土结构设计原理》、《混凝土结构设计》、《钢结构原理》、《钢结构设计》、《工程流体力学》、《结构力学》、《土木工程设计导论》、《土木工程试验与检测》、《土木工程制图》、《画法几何》等。丛书分册列入是开放的，今后将根据情况，做出调整和补充。

本丛书面向土木、水利、建筑、园林、道路、市政等专业学生，同时也可以作为土木工程注册工程师考试及土建类其他相关专业教学的参考资料。

浙江大学建筑工程学院卓越计划系列教材编委会
2014.10

前　　言

本教材是普通高等学校土木、水利、海洋、市政工程等专业的专业基础课程流体力学的双语教学教材，也可作为环境、机械、能源工程等专业的流体力学课程双语教学用书，亦可供有兴趣了解流体力学英文术语词汇及其日常应用的学生与专业人员的参考。

本教材的编写依据是全国高等学校土建类专业的《流体力学课程教学基本要求》，沿用国内中文流体力学课程教材的内容体系和符号系统，在一定程度上做了拓展，主要表现为：

1. 在有关章节中简单概述了常用的实验室及工程中的流体力学最新测量方法。流速（流场）与流量的测量是流体力学在工程实践中的一个重要环节。而在国内流体力学教材中，除了静水力学中的测压计、毕托管、文丘里流量计、明渠流量的堰流测量有所阐述外，其他的甚少涉及。为此，本教材除了在 2.3.3 节阐述了测压计、第 10 章阐述了明渠堰流测量原理外，在 6.8.4 节简单介绍了风洞与水洞的结构与应用，在 8.7 节介绍了管道中的点流速（热膜流速计、电磁流速计、LDV 等）、流场（PIV）、流量（机械类如涡轮流量计，水头类如文丘里、孔板与管嘴流量计）测量技术，在 9.9 节介绍了明渠中的流速测量仪器（旋桨流速仪、电磁流速仪、多普勒流速仪 ADV 和 ADCP、光学流速仪）与流量测量方法（水工结构法如堰、Parshall 量水槽和 Palmer-Bowlus 量水槽；流速面积法；底坡面积法；示踪剂法）。

2. 在例题与习题的选择上尽可能选用贴近生活与工程的题目，如例 3-9 的龙卷风、例 4-3 的计算机散热、例 4-10 的射流泵、例 8-1 的马桶冲水对淋浴头出水的影响等。同时精选思考题、习题，并对所有习题附有答案，利于学生自学。

3. 每章开篇附有相关的流体力学照片与导读，以引导学生；每章最后对本章内容做了详尽的概括，便于学生复习。

4. 书后附有流体力学学术词汇索引，本书中出现的流体力学专家索引与符号表。

本书编写过程中得到了浙江大学市政工程研究所、浙江大学建筑工程学院、中国建筑工业出版社的大力支持，同时顾建农教授提供了水洞原理图与照片、浙江大学市政工程研究所提供了管网、PIV、ADCP 等照片、浙江大学土木水利实验室提供了风洞照片、姜利杰博士绘制了书中图的 CAD 原图，在此一并表示衷心的感谢。另外，对参考文献作者、引用的网上相关照片的作者与网站表示衷心感谢，是你们为我提供了诸多宝贵的素材。最后，特别感谢我的先生和女儿，在我编写的过程中，始终给予我支持与鼓励。

限于作者水平，书中不妥之处敬请读者批评指正。

邵卫云
2014 年 12 月于浙江大学

Contents

1 Introduction ··· 1
   1.1 Brief Look at Fluid Mechanics ··· 1
   1.2 Forces Acting on Fluid ··· 7
   1.3 Density and Specific Weight ·· 9
   1.4 Fluid Properties ··· 10
   Summary ·· 22
   Word Problems ··· 23
   Problems ·· 24

2 Fluid Statics ··· 29
   2.1 Pressure at a Point ·· 29
   2.2 Equilibrium of a Fluid Element ··· 31
   2.3 Hydrostatic Pressure Distribution in Gravitational Field ··· 35
   2.4 Pressure Distribution in Rigid-Body Motion ·· 42
   2.5 Hydrostatic Force on a Submerged Plane Surface ·· 47
   2.6 Hydrostatic Force on a Curved Surface ·· 55
   2.7 Buoyancy, Flotation, and Stability ··· 58
   Summary ·· 64
   Word Problems ··· 65
   Problems ·· 68

3 Fluid Kinematics ··· 79
   3.1 Description Approaches of the Flow Field ·· 80
   3.2 Classification of Fluid Flows ·· 84
   3.3 Eulerian Concepts for Flow Description ·· 89
   3.4 Kinematic Description of Fluid Element ··· 99
   3.5 Differential Equations of Motion ··· 103
   3.6 The Bernoulli Equation ·· 109
   3.7 Steady Plane Potential Flows ·· 117
   Summary ·· 131
   Word Problems ··· 133
   Problems ·· 135

4 Fundamental Equations for Steady Total Flow ·· 145
   4.1 Average of Sectional Parameters ··· 145
   4.2 Continuity Equation ··· 147

4.3	Energy Equation	149
4.4	Linear Momentum Equation	169
	Summary	176
	Word Problems	178
	Problems	179

5 Similitude, Dimensional Analysis, and Modeling ... 191

5.1	Introduction	192
5.2	Dimensional Analysis	194
5.3	Similitude	200
5.4	Common Dimensionless Groups in Fluid Mechanics	203
5.5	Modelling	208
	Summary	213
	Word Problems	214
	Problems	215

6 Fluid Resistance and Energy Losses ... 220

6.1	Introduction to Head Losses	221
6.2	Laminar and Turbulent Flows	223
6.3	Fully Developed Flow and Its Shear Stress	227
6.4	Fully Developed Laminar Flow in Pipes	231
6.5	Fully Developed Turbulent Flow in Pipes	235
6.6	Frictional Losses in Turbulent Flow	245
6.7	Minor Losses	254
6.8	Boundary Layer and Drag Force	262
	Summary	277
	Word Problems	278
	Problems	281

7 Orifice Flow and Nozzle Flow ... 290

7.1	Introduction	290
7.2	Steady Flow through Sharp-edged Orifice	292
7.3	Steady Nozzle Flow	296
7.4	Unsteady Flow through Orifice	298
	Summary	302
	Word Problems	302
	Problems	303

8 Pipe Flow ... 307

8.1	Introduction	307
8.2	Steady Flow in Simple Short Pipeline	309
8.3	Steady Flow in Simple Long Pipeline	318
8.4	Steady Flow in Complex Long Pipeline	320

	8.5	Steady Flow in Pipe Networks	327
	8.6	Unsteady Flow in Pipeline	333
	8.7	Flow Measurements in Pipes	339
		Summary	346
		Word Problems	347
		Problems	350
9	**Open Channel Flow**		**363**
	9.1	Introduction	364
	9.2	Uniform Flow in Open Channel	368
	9.3	Uniform Flow in a Partially Full Circular Pipe	377
	9.4	Wave Speed and Froude Number	379
	9.5	Fundamentals of Nonuniform Flow	382
	9.6	Rapidly Varied Flow; The Hydraulic Jump	389
	9.7	Water Surface Profiles of GVF in Prismatic Channel	397
	9.8	Numerical Solution of Surface Profile	406
	9.9	Flow Measurements in Open Channel	409
		Summary	414
		Word Problems	416
		Problems	419
10	**Weir Flow**		**428**
	10.1	Introduction	428
	10.2	Sharp-crested Weirs	431
	10.3	Ogee-crested Weirs	434
	10.4	Broad-crested Weirs	435
	10.5	Flow under a Sluice Gate	441
	10.6	Hydraulic Design of Small Bridge Aperture	442
		Summary	446
		Word Problems	447
		Problems	447
11	**Seepage Flow**		**452**
	11.1	Introduction	453
	11.2	Darcy Law	455
	11.3	Unconfined Steady Gradually Varied Seepage Flow	458
	11.4	Seepage Flow around Wells and Clustered-Wells	463
		Summary	469
		Word Problems	470
		Problems	471
12	**One-dimensional Compressible Gas Flow**		**474**
	12.1	Fundamental of Compressible Flow	475

12.2　One-Dimensional Adiabatic and Isentropic Steady Flow ········· 478
12.3　Compressible Flow with Friction in Constant-area Duct ········· 484
　　Summary ·· 491
　　Word Problems ·· 491
　　Problems ··· 492
References ·· 495
Appendix A　Property Tables ·· 497
　　Table A-1 Properties of water at 1 atm pressure ··················· 497
　　Table A-2 Properties of air at 1 atm pressure ······················· 498
　　Table A-3 Properties of some common fluids at 1 atm pressure ··· 499
Appendix B　Bibliography ··· 500
Index of Vocabulary ··· 502
Nomenclature ··· 513
Answers to Problems ·· 517

1 Introduction

CHAPTER OPENING PHOTO: Water is a transparent fluid which forms the world's streams, lakes, oceans and rain, and covers 71% of the Earth's surface. On Earth, 96.5% of the planet's water is founded in seas and oceans, only 2.5% is freshwater, whereas 98.8% of which is in ice and groundwater, and less than 0.3% is in rivers, lakes, and the atmosphere. In civil, hydraulic and ocean engineering, water has the massive effects on the water related structures. We begin from its physical properties as in the present chapter. (© 2013 For Wallpaper.com)

In this introductory chapter, we start this chapter with a concept of fluid and continuum, the study methods and a brief history of the development of fluid mechanics. It follows the definition of surface and mass forces acting on the fluid that will be used in the fluid mechanics. Finally, we discuss properties that are encountered in the analysis of fluid flow: the density and specific gravity; the viscosity playing a dominant role in most aspects of fluid flow and the no-slip condition at solid-fluid interfaces; the coefficient of compressibility; the vapor pressure; and the surface tension determining the capillary rise under static equilibrium conditions.

1.1 Brief Look at Fluid Mechanics

1.1.1 The Concept of a Fluid

Fluid mechanics is the study of fluids either in motion (*fluid dynamics*) or at rest (*fluid statics*) and the subsequent effects of fluid upon the boundaries, which may be either solid surfaces or interfaces with other fluids. Fluid mechanics is one of the most important of all areas of physics. Its field varies from the study of blood flow in the capillaries (which are only a few microns in diameter) to the flow of the Three Gorges hydropower plant with a dam of 181m-high, i. e., breathing, blood flow, swimming, pumps, fans, turbines, airplanes, ships, rivers, windmills, pipes, missiles, icebergs, engines, filters, and sprinklers. Fluid mechanics principles are needed to explain the phenomena in this area, i. e., why airplanes are made streamlined with smooth surfaces for the most efficient flight, whereas golf balls are made with rough surfaces (dimpled) to increase their efficiency (see Section 6.8.3).

A *fluid* is any substance that deforms continuously when subjected to a shear stress, no matter how small. Thus, from the point of view of fluid mechanics, all matter can be classified as two states, *fluid* and *solid*. Their difference in reaction to an applied shear or tangential stress is perfectly obvious. A solid can resist a shear stress by a static deformation without flow; a fluid cannot. The fluid moves and deforms continuously as long as the shear stress is applied, whereas a fluid at rest must be in a state of zero shear stress.

Given the definition of a fluid above, both *liquids* and *gases* are classified as fluids. The technical distinction lies in the effect of cohesive forces. A *liquid*, being composed of relatively close-packed molecules with strong cohesive forces, tends to retain its volume and will form a free surface in a gravitational field if it is unconfined from above (Fig. 1-1a). A *gas*, being composed of widely spaced molecules with negligible cohesive forces, is free to expand to fill the entire available space and has no definite volume and free surface (Fig. 1-1b).

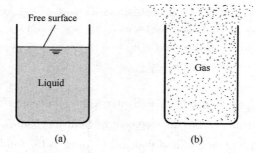

Fig. 1-1 (a) Liquid with a free surface and a definite volume;
(b) Gas which expands freely with no free surface

1.1.2 The Fluid as a Continuum

As far as we know, fluids are aggregations of molecules, widely spaced for a gas, closely spaced for a liquid. The distance between molecules is very large compared with the molecular diameter. For example, 1cm^3 of water at standard conditions (at 1atm and 20℃) contains approximately 3.3×10^{22} molecules, the distance between molecules is approximate 3×10^{-8} cm; 1cm^3 of air at standard conditions contains approximately 2.7×10^{19} molecules, the distance between molecules is approximate 3×10^{-7} cm. Thus, there must be many molecules even in a rather small volume of fluid, the statistical mean characteristics of flow could be obtained. The fluid properties of a fluid, such as pressure, velocity or density, can be assumed to be varied continually in space, that is, the variation in properties is so smooth that the differential calculus can be used to analyze the substance. For a fluid in such a case is called a *continuum*. The assumption is called the *continuum hypothesis*, the most fundamental idea for solving fluid mechanics problem.

Consider for example the density at a specific point in a fluid. It is defined as the ratio of the mass of molecules in a small volume surrounding that point to this given volume. However, the size of this small volume, δV, has to meet with certain criteria. It must be smaller than the physical dimensions of the region under consideration like the wing of an aircraft or the pipe in a hydraulic system. At the same time it must be sufficiently large to accommodate a large number of molecules to make the density meaningful. Too small a δV, the value of calculated density fluctuates because the number of molecules within δV is varying significantly with time. Too big a δV might mean that density itself is varying significantly within the region of interest. It is clear that there is a limit δV_0 below which molecular variations assume importance and above which one finds a macroscopic variation of density within the region. Therefore, the *density* ρ of a fluid is best defined as

$$\rho = \lim_{\delta V \to \delta V_0} \frac{\delta m}{\delta V} \tag{1.1}$$

where δm is the molecular mass within the given volume δV.

Equation 1.1 is plotted in Fig. 1-2. The limiting volume δV_0 is about 10^{-9} mm^3 for all liquids and for gases at atmospheric pressure. For example, 10^{-9} mm^3 of air at standard conditions contains approximately 3×10^7 molecules, which is sufficient to define a nearly constant density according to Eq. 1.1. In engineering practice, most problems are concerned with physical dimensions much larger than this limiting volume. Thus, *continuum hypothesis* is valid for most engineering problems.

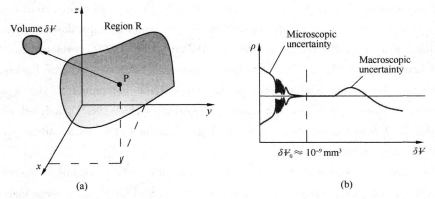

Fig. 1-2 The definition of continuum fluid density: (a) an elemental volume in a fluid region; (b) calculated density versus size of the elemental volume

1.1.3 Study Approaches

It should be noted that the basic practical understanding of the behavior of fluids dates back to the ancient civilizations, at least by the time of the ancient Egyptians. Through necessity there was a practical concern about the manner in which spears and arrows could be propelled through the air, in the development of water supply and irrigation systems, and in the design of boats and ships. In fact, the homes of well-to-do Romans had flushing toilets not very different from those in modern 21st-Century houses, and the Roman aqueducts are still considered a tremendous engineering feature. Some of the earliest writings that pertain to modern fluid mechanics are those of Archimedes (287-212 B. C.), a Greek mathematician and inventor who first expressed the principles of hydrostatics and flotation. Then, little has been added to further understanding of fluid in the next 1000 years behavior till about fifteenth century. Beginning with the Renaissance period, continuous series of contributions began to form the basis of fluid mechanics. Leonardo da Vinci (1452-1519) derived the equation of conservation of mass in one-dimensional steady flow and experimented with waves, jets, hydraulic jumps, eddy formation, etc. Galileo Galilei (1564-1642) marked the beginning of experimental mechanics, and Edme Mariotte (1620-1684), a Frenchman, built the first wind tunnel and tested models in it. Problems involving the momentum of fluids could finally be analyzed after Isaac Newton (1642-1727) postulating his laws of motion and the law of viscosity of the linear fluids now called Newtonian.

Following the numerous significant contributions of the theoretical and mathematical advances have been made, the theoretical and mathematical study of idealized, frictionless fluid behavior was termed *hydrodynamics* contributed by the eighteenth-century mathematicians (Daniel Bernoulli, Leonhard Euler, Jean d'Alembert, Joseph-Louis Lagrange, and Pierre-Simon Laplace). Among them, Leonhard Euler (1707-1783) developed both the differential equations of motion and their integral form, now called *Bernoulli equation*.

The applied or experimental aspects of real fluid behavior, particularly the behavior of water relying almost entirely on experiment, was termed *hydraulics* contributed by the experimentalists (Chézy, Pitot, Borda, Weber, Francis, Hagen, Poiseuille, Darcy, Manning, Bazin, and Weisbach) on a variety of flows such as open channels, ship resistance, pipe flows, waves, and turbines. William Froude (1810-1879) and his son Robert (1846-1924) developed laws of model testing and Lord Rayleigh (1842-1919) proposed dimensional analysis. Osborne Reynolds (1842-1912) published the classic pipe experiment and showed the importance of the dimensionless *Reynolds number*, named after him. Theodore von Kármán (1881-1963) analyzed what is now known as the *von Kármán vortex street*. Geoffrey Ingram Taylor (1886-1975) advanced statistical theory of turbulence and the Taylor microscale.

For the theoretical aspects of real fluid, the general differential equations of viscous flow theory, *Navier-Stokes equations*, named after Lowis Navier (1785-1836) and George Gabriel Stokes (1819-1903), was available but unexploited from nineteen century. Then, in 1904, a German engineer, Ludwig Prandtl (1875-1953) pointed out that fluid flows with small viscosity, e. g., water flows and air flows, can be divided into a thin *viscous layer*, or *boundary layer*, near solid surfaces and interfaces, patched onto a nearly inviscid outer layer, where the Euler and Bernoulli equations apply. The concept of "fluid boundary layer" laid the foundation for the unification of the theoretical and experimental aspects of fluid mechanics. Thus, Prandtl is generally accepted as the founder of modern fluid mechanics.

Besides the theoretical and experimental methods, today, because of the power of modern digital computers, there is yet a third way to study fluid dynamics: *computational fluid dynamics*, or *CFD* for short. In modern industrial practice CFD is used more for fluid flow analyses than either theory or experiment. But it is also important to understand that in order to do CFD one must have a fundamental understanding of fluid flow itself, from both the theoretical, mathematical side and from the practical, sometimes experimental, side. We will provide a brief introduction to each of these ways of studying fluid dynamics in the following subsections.

THE THEORETICAL APPROACH

For the theoretical approach, we employ the mathematical equations that govern the flow and try to capture the fluid behavior within a closed form solution i. e., formulas that can be readily used. Theoretical/analytical studies of fluid dynamics generally require considerable simplifications of the equations of fluid motion, the Navier-Stokes equations. This is perhaps the simplest of the approaches, but its scope is somewhat limited. Not every fluid flow renders itself to such an approach. The resulting equations may be too complicated to be solved easily.

In most practical engineering applications, various assumptions and simplifications need to be made to enable the analytical solution of the differential equations representing the physical situation. This at one hand limits the applicability of the methods to simple type problems, or limits the validity of the solutions if too many assumptions and simplifications are made.

Despite that, analytical methods played significant role in the past and they still play an important role. They have helped engineers and scientists in the understanding of the fundamental rules controlling the behavior of many engineering systems. In addition, they are used to help understand and interpret experimental results. Furthermore, they can be used as a first stage in the validation of CFD models.

THE EXPERIMENTAL APPROACH

Experimental approach is the oldest approach, perhaps also employed by Archimedes when he was to investigate a fraud. It is a very popular approach by which you will make measurements with a wind tunnel or a similar equipment. But this is a costly venture and is becoming costlier day by day.

Rather obviously, fluid experiments performed today in first class fluids laboratories are far more sophisticated. Nevertheless, until only very recently the outcomes of most fluid experiments were mainly a qualitative (and not quantitative) understanding of fluid motion. An indication of this is provided by the adjacent pictures of wind tunnel experiments. In each of these we are able to discern quite detailed qualitative aspects of the flow over different prolate spheroids. Basic flow patterns are evident from colored streaks, even to the point of indications of flow "separation" and transition to turbulence. However, such diagnostics provide no information on actual flow velocity or pressure—the main quantities appearing in the theoretical equations, and needed for engineering analyses.

There have methods for measuring pressure in a flow field for a long time, and these could be used simultaneously with the flow visualization to gain some quantitative data. On the other hand, it is till recently to be possible to accurately measure flow velocity simultaneously over large areas of a flow field. If point measurements are sufficient, *hot-wire anemometry* (HWA) or *laser-doppler velocimetry* (LDV) can be used; but for field measurements it is necessary to employ some form of *particle image velocimetry* (PIV). It is clear that the quantitative detail by PIV is far superior to the simple visualizations in the experiments, and as a consequence PIV is rapidly becoming the preferred diagnostic in many flow situations.

THE COMPUTATIONAL APPROACH

Computational Fluid Dynamics, usually abbreviated as CFD, is a branch of fluid mechanics that uses numerical methods and algorithms to solve the mathematical equations which govern the processes of fluid flow, heat transfer, mass transfer, chemical reactions, and related phenomena and analyze these problems that involve fluid flows. The fundamental basis of almost all CFD problems are the Navier-Stokes equations, which define any single-phase (gas or liquid, but not both) fluid flow. These equations can be simplified by removing terms describing viscous actions to yield the Euler equations. Further simplification, by removing terms describing vorticity yields the full potential equations. Finally, for small perturbations in subsonic and supersonic flows (not transonic or hypersonic) these equations can be linearized to yield the linearized potential equations.

CFD analysis complementing testing and experimentation can reduce the total effort required in the laboratory. In recent years with the development of computer technology, CFD has played an ever-increasing role in many areas of sports and athletics: from study and design of Olympic swimware to the design of a new type of golf ball providing significantly longer flight times, and thus driving distance (and currently banned by the PGA). The example of a race car also reflects current heavy use of CFD in numerous areas of automobile production ranging from the design of modern internal combustion engines exhibiting improved efficiency and reduced emissions to various aspects of the manufacturing process including, for example, spray painting of the completed vehicles.

It is essential to recognize that it is the CFD computer code that solves the Navier-Stokes equations. The user of such codes must understand the mathematics of these equations sufficiently well to be able to supply all required auxiliary data for any given problem, and he/she must have sufficient grasp of the basic physics of fluid flow to be able to assess the outcome of a calculation and determine, among other things, whether it is "physically reasonable", and if not, decide what to do next, or how to validate the outcome with scaled laboratory experiments.

1.1.4 Fluid Mechanics in Civil Engineering

As what discussed above fluid dynamics is one of the most important of all areas of physics.

The applications in fluid engineering are enormous. When you think about it, almost everything on this planet either is a fluid or moves within or near a fluid.

For the fluid mechanics on a Civil Engineering course, the provisions of adequate water services such as the potable water supply, drainage and sewerage are essential for the development of industrial society. It is these services civil engineers provide.

Fluid mechanics is involved in nearly all areas of Civil Engineering either directly or indirectly. Some examples of direct involvement are those where we are concerned with manipulating the fluid: sea and river (flood) defenses; water distribution/sewerage (sanitation) networks; hydraulic design of water/sewage treatment works; dams; pumps and turbines; water retaining structures. For other examples where the primary object is construction the analysis of fluid mechanics is essential: flow of air in/around buildings; bridge piers in rivers and ground-water flow.

Notice how nearly all of these involve water. The following course, although introducing general fluid flow ideas and principles, will demonstrate many of these principles through examples where the fluid is water.

1.2 Forces Acting on Fluid

The forces acting on a fluid element drawn from the fluid field usually can be classified as: surface force and mass force.

1.2.1 Surface Force

Surface force, designated by $\mathbf{F}_s$, is the force acting directly on the internal or external surface of the fluid element in a fluid field, which has a unit of N. Surface forces are due to only two sources: (a) the pressure distribution acting on the surface, imposed by the outside fluid surrounding the fluid element, and (b) the shear and normal stress distributions acting on the surface, also imposed by the outside fluid 'tugging' or 'pushing' on the surface by means of friction. Thus, surface force is directly proportional to the area of the surface with which fluid is in contact.

In Fig. 1-3 fluids exert both normal and tangential forces on surfaces with which they are in contact (e.g., surfaces of containers and 'surfaces' of adjacent fluid elements). Both of those two perpendicular forces, normal and tangential forces, are the components of surface force acting on the finite area, ΔA.

Shear stress, designated by τ, is the tangential force per unit area on the surface of fluid and defined as

$$\bar{\tau} = \frac{\Delta F_\tau}{\Delta A} \qquad (1.2)$$

where ΔF_τ is the tangential component of the surface force ΔF_s applied on the finite area ΔA.

Equation 1.2 indicates that it is the averaged shear stress acting on the finite area ΔA. The shear stress at point D is somewhat different from this averaged shear stress, which could be ob-

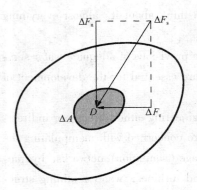

Fig. 1-3 Forces on the surface of a fluid element

tained while finite area ΔA tends to a limit, an infinitesimal limit which is sufficiently small to be negligible in comparison with macroscopic length scales squared, but still sufficiently large to contain enough molecules to permit calculation of averaged properties and 'construction' of fluid parcels. In mathematic expression this limit is replaced by zero, thus the shear stress at a point is defined as

$$\tau = \lim_{\Delta A \to 0} \frac{\Delta F_\tau}{\Delta A} \tag{1.3}$$

The shear stress has units of Pa, $1\text{Pa} = 1\text{N}/\text{m}^2$.

Pressure, designated by p, is the normal force per unit area on the surface of fluid. As we have done with the shear stress, we can define average pressure acting on a finite area ΔA as

$$\bar{p} = \frac{\Delta F_n}{\Delta A} \tag{1.4}$$

where ΔF_n is the normal component of the surface force ΔF_s applied over the finite area ΔA. Then the pressure at a point is given as

$$p = \lim_{\Delta A \to 0} \frac{\Delta F_n}{\Delta A} \tag{1.5}$$

where, as usual, the limit process is viewed within the confines of the *continuum hypothesis*. The pressure also has units of Pa.

1.2.2 Mass Force

Mass force, designated by $\mathbf{F_b}$, is the force acting directly on the volumetric mass of the fluid element such as gravitational, electric and magnetic forces. The value of mass force is directly proportional to the mass of fluid body. For homogeneous fluid, it is also directly proportional to the fluid volume, which gets it another name, *body force*. Mass force has units of N.

Unit mass force, designated by $\mathbf{f}$, is the mass force per unit mass of a fluid and defined as

$$\mathbf{f} = \frac{\mathbf{F_b}}{m} = f_x \mathbf{i} + f_y \mathbf{j} + f_z \mathbf{k} \tag{1.6}$$

where f_x, f_y and f_z are the components of the unit mass force in the x-, y- and z-direction, m is the mass of the fluid.

The unit mass force has units of m/s^2, same as the units of acceleration.

For a fluid at rest under gravity, the unit mass forces are $f_x = f_y = 0$, $f_z = -g$ (z- coordinate is upwards), and for a freely falling fluid, $f_x = f_y = f_z = 0$.

1.3 Density and Specific Weight

1.3.1 Density

The *density* of a fluid, designated by ρ, is defined as its mass per unit volume, that is

$$\rho = \frac{m}{V} \tag{1.7}$$

where m is the mass of fluid, kg; V is the volume of fluid, m^3; ρ has units of kg/m^3.

The density of a substance, in general, depends on temperature and pressure. The densities of most gases are proportional to pressure and inversely proportional to temperature and vary widely. For liquids, on the other hand, the variations of their densities with pressure are usually negligible, but depend more strongly on temperature than they do on pressure. For example, at 20℃, the density of water changes from $998kg/m^3$ at 1 atm to $1003kg/m^3$ at 100 atm, a change of just 0.5 percent. However, at 1 atm, the density of water changes from $998kg/m^3$ at 20℃ to $971.8kg/m^3$ at 80℃, a change of 2.6 percent, which can still be neglected in many engineering analyses. That is, density in liquids is nearly constant, for example, the density of water is about $1000kg/m^3$, and for mercury is about $13600kg/m^3$. Tables A.1 to A.3 in Appendix A list the values of densities for several common liquids at 1 atm.

The reciprocal of density is the *specific volume*, v, which is defined as volume per unit mass

$$v = \frac{1}{\rho} \tag{1.8}$$

and has units of m^3/kg.

This property is not commonly used in fluid mechanics but is used in thermodynamics.

1.3.2 Specific Weight

The *specific weight* of a fluid, designated by γ, is defined as its weight per unit volume and has units of N/m^3. Just as a mass has a weight of $G = mg$, density and specific weight are simply related by gravity

$$\gamma = \rho g \tag{1.9}$$

where g is the local acceleration of gravity, $g = 9.807 m/s^2$ under conditions of standard gravity.

Just as density is used to characterize the mass of a fluid system, the specific weight is used to characterize the weight of the system.

1.3.3 Specific Gravity

The *specific gravity* of a liquid, designated by S_G, is the ratio of the density of a liquid to the density of water at some specified temperature (usually at 4℃ with $\rho = 1000kg/m^3$)

$$S_G = \frac{\rho}{\rho_{H_2O@4℃}} \tag{1.10}$$

For example, the specific gravity of mercury at 20℃ is $S_{GHg} = \dfrac{13600\,\text{kg/m}^3}{1000\,\text{kg/m}^3} = 13.6$, which is dimensionless.

It is clear that density, specific weight, and specific gravity are all interrelated, and from a knowledge of any one of the three the others can be calculated.

1.4 Fluid Properties

1.4.1 Viscosity

NEWTON'S LAW OF VISCOSITY

When a glass ball is dropped into oil, the slower downward motion of the ball can be observed than in the water, which indicates that besides the density, there must be an additional property that represents the internal resistance of a fluid to motion or the 'fluidity': the viscosity.

Viscosity is that fluid property by virtue of which a fluid offers resistance to shear stresses. To obtain a relation for viscosity, consider a hypothetical experiment in which a fluid layer is placed between two very large parallel plates separated by a thin fluid layer of distance h as shown in Fig. 1-4. The bottom plate is rigidly fixed, but the upper plate moves continuously under the influence of the force F at a constant velocity U. A closer inspection of the fluid motion between the two plates would reveal that the fluid in contact with the upper plate moves with the plate velocity, U, and the fluid in contact with the bottom fixed plate has a zero velocity. The fluid 'sticks' to the solid boundaries is a very important one in fluid mechanics and is usually referred to as the *no-slip condition*, which is the characteristic of all viscous-fluid flows, both liquids and gases.

In the limit thickness of fluid layer, h, the fluid velocity between the plates varies linearly between 0 at the lower fixed plate and U at the upper moving plate, and thus the *velocity profile* and the *velocity gradient* are

$$u(y) = \dfrac{y}{h} U$$

$$\dfrac{du}{dy} = \dfrac{U}{h} \tag{1.11}$$

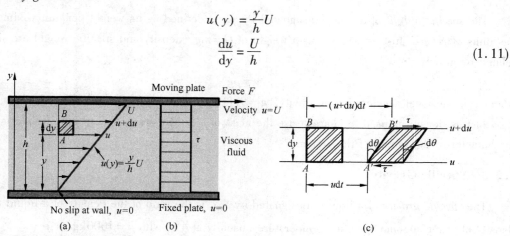

Fig. 1-4 Viscous flow induced by relative motion between two parallel plates. (a) linear velocity distribution in thin fluid layer; (b) uniform distribution of shear stress; (c) a fluid element straining at a rate of $d\theta/dt$

It indicates that in this particular case the velocity gradient is a constant, but in more complex flow situations this would not be true.

Consider a fluid element sheared drawn from Fig. 1-4a in one plane by a single shear stress τ, as in Fig. 1-4c. Since the upper surface moving at speed du larger than the lower, the shear strain angle dθ will continuously grow with time as long as the stress τ is maintained. In a small differential time increment, dt, the imaginary vertical line AB of the fluid element would rotate through a differential angle, dθ, so that

$$d\theta \approx \tan d\theta = \frac{du\,dt}{dy}$$

Thus

$$\dot{\gamma} = \frac{d\theta}{dt} = \frac{du}{dy} \qquad (1.12)$$

where $\dot{\gamma}$ is the *rate of shear strain*.

Equation 1.12 indicates that the rate of shear strain of a fluid element is equivalent to the velocity gradient, du/dy.

It was verified experimentally that for most fluids the force, F, needed to produce motion of the upper plate with constant speed U is proportional to the area of this plate, A, and to the speed U; furthermore, it is inversely proportional to the spacing between the plates, h. That is

$$F \propto \frac{AU}{h} \propto A\frac{du}{dy}$$

or

$$\tau \propto \frac{du}{dy} = \frac{d\theta}{dt} \qquad (1.13)$$

Note that this inverse proportionality with distance h between the plates further reflects physical viscous behavior arising from zero velocity at the lower plate. In particular, the upper plate motion acts, through viscosity, to attempt to 'drag' the lower plate.

In one-dimensional shear flow of Newtonian fluids, shear stress in Eq. 1.13 can be expressed by the linear relationship

$$\tau = \mu \frac{du}{dy} \qquad (1.14)$$

where the constant of proportionality μ is called the *dynamic viscosity, absolute viscosity* or simply the *viscosity* of the fluid, whose unit is kg/m·s or equivalently, N·s/m^2 (or Pa·s).

Equation 1.14 is also known as *Newton's law of viscosity*, which indicates that for a given rate of shear strain of a fluid, or velocity gradient, shear stress is directly proportional to viscosity.

KINEMATIC VISCOSITY

In engineering practice it is convenient to employ the combination of viscosity and density, designated by ν

$$\nu = \frac{\mu}{\rho} \qquad (1.15)$$

where ν is termed the *kinematic viscosity* and has units of m²/s, its variation on the temperature can be determined by the fitted empirical formula

$$\nu = \frac{0.01775 \times 10^{-4}}{1 + 0.0337T + 0.000221\,T^2} \qquad (1.16)$$

where T is the temperature with units of ℃.

Viscosity for several common liquids and gases are listed in Tables A.1 through A.3 in Appendix A. The viscosity of a fluid is a measure of its 'resistance to deformation', and caused by the cohesive forces between the molecules in liquids and by the molecular collisions in gases. The actual value of viscosity depends on the particular fluid, and for a particular fluid the viscosity is highly dependent on temperature, such as the viscosity of water and air listed in Tables A.1 and A.2 in Appendix A, and weakly dependent on the pressure. The viscosities of liquids decrease with the increasing temperature, whereas the viscosities of gases increase with temperature. It is because the liquid molecules are closely spaced with strong cohesive forces between molecules. As the temperature increases, these cohesive forces are reduced with a corresponding reduction in resistance to motion, so does the viscosity. In gases, however, the molecules are widely spaced and intermolecular forces negligible. In this case resistance to relative motion arises due to the exchange of momentum of gas molecules between adjacent layers. As molecules are transported by random motion from a region of low bulk velocity to mix with molecules in a region of higher bulk velocity (and vice versa), there is an effective momentum exchange which resists the relative motion between the layers. As the temperature of the gas increases, the random molecular activity increases with a corresponding increase in viscosity.

The dependence of viscosity on pressure is rather weak. For liquids, both the dynamics and kinematic viscosities are practically independent of pressure, and any small variation with pressure is usually disregarded, except at extremely high pressure. For gases, this is also the case for dynamic viscosity (at low to moderate pressure), but not for kinematic viscosity since the density of a gas is proportional to its pressure. For example, the dynamic and kinematic viscosities of air at 20℃ and 1 atm are 1.83×10^{-5} kg/m·s and 1.516×10^{-5} m²/s, respectively, but at 20℃ and 1 atm, they are 1.83×10^{-5} kg/m·s and 0.38×10^{-5} m²/s, respectively.

EXAMPLE 1-1 Shear stress in the oil between two plates

Suppose that the fluid being sheared in Fig. 1-4 is SAE 30 oil at 20℃. Compute the shear stress in the oil if $U = 3$ m/s and $h = 2$ cm.

Solution:

The shear stress is found from Eqs. 1.14 and 1.11:

$$\tau = \mu \frac{du}{dy} = \mu \frac{U}{h}$$

From Table A.3 in Appendix A for SAE 30 oil, $\mu = 0.29$ Pa·s. Then, for the given values of U and h, we have

$$\tau = \mu \frac{U}{h} = (0.29 \text{Pa·s}) \frac{3 \text{m/s}}{0.02 \text{m}} = 43.5 \text{Pa}$$

DISCUSSION   Although oil is very viscous, this is a modest shear stress, about 2400 times less than atmospheric pressure. Viscous stresses in gases and thin liquids are even smaller.

EXAMPLE 1-2 Shear stress in the oil layer sandwiched between two plates

A thin 20cm×20cm flat plate is pulled at a velocity of 1m/s horizontally through a 3.6-mm-thick oil layer sandwiched between two plates, one stationary and the other moving at a constant velocity of 0.3m/s, as shown in Fig. E1-2. The dynamic viscosity of oil is 0.027Pa·s. Assuming the velocity in each oil layer to vary linearly, (a) plot the velocity profile and find the location where the oil velocity is zero and (b) determine the force that needs to be applied on the plate to maintain this motion.

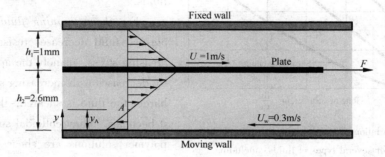

Fig. E1-2

Solution:

(a) Considering the linear distribution of the velocity in each oil layer and the no-slip condition, the velocity profile in each oil layer relative to the fixed wall can be plotted in the figure. The point of zero velocity is indicated by point A, and its distance from the lower plate is determined from geometric considerations (the similarity of the two triangles in the lower oil layer) to be

$$\frac{h_2 - y_A}{y_A} = \frac{U}{U_w}$$

Thus

$$y_A = \frac{h_2 U_w}{U + U_w} = \frac{(2.6\text{mm})(0.3\text{m/s})}{(1\text{m/s}) + (0.3\text{m/s})} = 0.6\text{mm}$$

(b) The magnitudes of shear forces acting on the upper and lower surfaces of the plate are

$$F_{upper} = \mu A \frac{U - 0}{h_1} = (0.027\text{Pa·s})(0.2\text{m})(0.2\text{m})\frac{1\text{m/s}}{0.001\text{m}} = 1.08\text{N}$$

$$F_{lower} = \mu A \frac{U - 0}{h_2 - y_A} = (0.027\text{Pa·s})(0.2\text{m})(0.2\text{m})\frac{1\text{m/s}}{0.0026\text{m} - 0.0006\text{m}} = 0.54\text{N}$$

Noting that both shear forces are in the opposite direction of motion of the plate, the force F is determined from a force balance on the plate to be

$$F = F_{upper} + F_{lower} = 1.08\text{N} + 0.54\text{N} = 1.62\text{N}$$

DISCUSSION Note that wall shear is a friction force between a solid and a liquid, and it acts in the opposite direction of motion.

NEWTONIAN AND NON-NEWTONIAN FLUIDS

Fluids for which the rate of shear strain is proportional to the shear stress are called *Newtonian fluids* after Sir Isaac Newton (1642-1727), who expressed it first in 1687. Most common fluids such as water, air, gasoline, and oils are Newtonian fluids. Fluids for which the shear stress is not

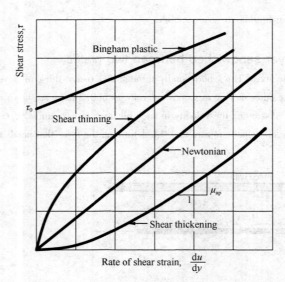

Fig. 1-5 Variation of shear stress with rate of shear strain for several types of fluids, including common non-Newtonian fluids

linearly related to the rate of shear strain are designated as *non-Newtonian fluids*. In this case, the slope of the curve on the τ versus du/dy is referred to the *apparent viscosity* of the fluid, μ_{ap} as shown in Fig. 1-5, in which the simplest and most common non-Newtonian fluids shown compared to the Newtonian fluids.

For *shear thinning fluids*, or *pseudoplastic*, fluid decreases resistance with increasing stress, namely the apparent viscosity decreases with increasing shear rate: the harder the fluid is sheared, the less viscous it becomes. Many colloidal suspensions and polymer solutions are shear thinning. For example, latex paint does not drip from the brush because the shear rate is small and the apparent viscosity is large. However, it flows smoothly onto the wall because the thin layer of paint between the wall and the brush causes a large shear rate (large du/dy) and a small apparent viscosity.

For *shear thickening fluids*, or *dilatant*, fluid increases resistance with increasing applied stress, namely the apparent viscosity increases with increasing shear rate: the harder the fluid is sheared, the more viscous it becomes. Common examples of this type of fluid include water-corn starch mixture and water-sand mixture ('quicksand'). Thus, the difficulty in removing an object from quicksand increases dramatically as the speed of removal increases.

The *Bingham plastic* is neither a fluid nor a solid. Such material can withstand a finite shear stress without motion (therefore, it is not a fluid), but once the *yield stress*, τ_0, is exceeded it flows like a fluid (hence, it is not a solid). An example of a yielding fluid is toothpaste, which will not flow out of the tube until a finite stress is applied by squeezing. Mayonnaise is also a common example of Bingham plastic materials.

IDEAL FLUID

All *real fluid*, or viscous fluid, including liquids and gases, have viscosity which causes the actual flow pattern within a fluid is usually complex and difficult to model mathematically. However, theory can be simplified considerably by the assumption that the fluid is *ideal*.

An *ideal fluid* (liquid or gas) is one which has the following properties: zero viscosity; incompressibility; zero surface tension; does not change phase.

Gases and vapors are compressible and can only be classed as ideal fluids when flow velocites are low. Gases can often be treated as perfect, in which case the perfect gas equations apply.

1.4.2 Compressibility of Fluids

Fluids usually expand as they are heated or depressurized and contract as they are cooled or pressurized. But the amount of volume change is different for different fluids. In fluid mechanics, there are two properties, the bulk modulus of elasticity K and the coefficient of volume expansion κ defined to relate volume changes to the changes in pressure and temperature.

BULK MODULUS OF ELASTICITY

To describe the phenomena that a fluid contracts when more pressure is applied on it and expands when the pressure acting on it is reduced, a *coefficient of volume compressibility*, also called the *isothermal compressibility*, β, for a fluid is introduced, which is defined as the fractional decrease in volume of a fluid corresponding to the increment in pressure while the temperature remains constant

$$\beta = -\frac{d\forall/\forall}{dp} = -\frac{1}{\forall}\frac{d\forall}{dp} \tag{1.17}$$

where dp is the change in pressure needed to create a change in volume $d\forall$ out of a volume $\forall$; β must be in the units of m^2/N since $d\forall/\forall$ is dimensionless; the negative sign '$-$' in Eq. 1.17 ensures β is a positive quantity.

Even though the fluid is compressed, the mass of fluid has not been changed, that is, $dm = 0$, which yields

$$dm = d(\rho\forall) = \rho d\forall + \forall d\rho = 0$$

or

$$-\frac{d\forall}{\forall} = \frac{d\rho}{\rho}$$

That is, the fractional changes in the specific volume and in the density of a fluid are equal in magnitude but opposite in sign.

Thus, Eq. 1.17 can also be expressed as

$$\beta = \frac{1}{\rho}\frac{d\rho}{dp} \tag{1.18}$$

A small value of β indicates that a large change in pressure is needed to cause a small fractional change in volume, and thus a fluid with a small β is essentially incompressible. This is typical for liquids, and explains why liquids are usually considered to be incompressible. For example, the pressure of water at normal atmospheric conditions must be raised to 210 atm (in which 'atm' is the *atmospheric pressure*, $1\,\text{atm} = 1.013 \times 10^5\,\text{N/m}^2$) to get a decrease of 1 percent in volume, corresponding to a value of $\beta = 4.70 \times 10^{-10}\,\text{m}^2/\text{N}$.

For a specific fluid, the coefficient of volume compressibility is dependent on the variation of temperature and pressure. For water at $0\,^\circ\text{C}$, the values of coefficient of volume compressibility at different pressure is tabulated in Table 1-1, in which 'at' is the *engineering atmospheric pressure*, $1\,\text{at} = 98\,\text{kN/m}^2$.

Table 1-1 The coefficient of volume compressibility, β, of water at 0℃

Pressure, at	5	10	20	40	80
$\beta \times 10^9$, m²/N	0.538	0.536	0.531	0.528	0.515

Source: Data of β are from 刘鹤年(2004).

The reciprocal of coefficient of volume compressibility is called the *bulk modulus of compressibility*, or *bulk modulus of elasticity*, K

$$K = \frac{1}{\beta} = -V\frac{dp}{dV} = \rho\frac{dp}{d\rho} \qquad (1.19)$$

where K is in the units of N/m².

Large values for the bulk modulus indicate that the fluid is relatively incompressible, that is, it takes a large pressure change to create a small change in volume. The use of bulk modulus as a property describing compressibility is most prevalent when dealing with liquids, although the bulk modulus can also be determined for gases.

COEFFICIENT OF VOLUME EXPANSION

The density of a fluid, in general, depends more strongly on temperature than it does on pressure, and the variation of density with temperature is responsible for numerous natural phenomena such as winds, currents in oceans, the operation of hot-air balloons, heat transfer by natural convection, and even the rise of hot air. To quantify these effects, a property that represents the variation of the density of a fluid with temperature at constant pressure, the *coefficient of volume expansion* (or *volume expansivity*) κ, is introduced and defined as

$$\kappa = \frac{1}{V}\frac{dV}{dT} = -\frac{1}{\rho}\frac{d\rho}{dT} \qquad (1.20)$$

where κ is in the units of 1/K or 1/℃, T is the temperature, K or ℃.

For a specific fluid, the coefficient of volume expansion is also dependent on the variation of temperature and pressure. Table 1-2 tabulates the values of κ of water at 1 atm with different temperature.

Table 1-2 The coefficient of volume expansion, κ, of water at 1atm

Temperature, ℃	1~10	10~20	40~50	60~70	90~100
$\kappa \times 10^4$, 1/℃	0.14	0.15	0.42	0.55	0.72

Source: Data of κ are from 刘鹤年(2004).

A large value of κ for a fluid means a large change in density with temperature, and the product κdT represents the fraction of volume change of a fluid that corresponds to a temperature change of dT at constant pressure.

Both Tables 1-1 and 1-2 show that water has negligible compressibility and expansivity. It would be reasonable to disregard the effects of compressibility and expansivity of water under general conditions. However, for some special cases, such as the water hammer in pressurized pipe and the propagation of explosive wave in water, the compressibility of water plays a key rule and should be considered seriously, so do the effects of expansivity in the hydraulic closed system and hot water heating system, in which the temperatures of the working fluids vary widely.

COMPRESSIBILITY AND EXPANSION OF GASES

Gases are highly compressible in comparison to liquids, with changes in gas density directly related to changes in pressure and temperature through the equation

$$\frac{p}{\rho} = RT \qquad (1.21)$$

where p is the absolute pressure, N/m^2; ρ the density, kg/m^3; T the absolute temperature, K; and R is a gas constant, J/kg·K, $R = 287$ J/kg·K for air under standard condition.

Equation 1.21 is commonly termed the *ideal or perfect gas law*, or the *equation of state* for an ideal gas. It is known to closely approximate the behavior of real gases under normal conditions when the gases are not approaching liquefactions.

When gases are compressed (or expanded) the relationship between pressure and density depends on the nature of the process. If the compression or expansion takes place under constant temperature conditions (*isothermal process*), then from Eq. 1.21 to give

$$\frac{p}{\rho} = \text{constant} \qquad (1.22)$$

If the compression or expansion is frictionless and no heat is exchanged with the surroundings (*isentropic process*), then

$$\frac{p}{\rho^k} = \text{constant} \qquad (1.23)$$

where k is the ratio of *specific heat* of gas at constant pressure.

For an ideal gas in an isothermal process, $p = \rho RT$, thus $\frac{dp}{d\rho} = RT = \frac{p}{\rho}$. Thus the coefficient of volume compressibility for an ideal gas in an isothermal process can be obtained from Eq. 1.19

$$K = \rho RT = p \qquad (1.24)$$

For an ideal gas with constant pressure, from *perfect gas law* to give $\frac{d\rho}{dT} = -\frac{\rho}{T}$. Thus the coefficient of volume expansion for an ideal gas can be obtained from Eq. 1.20

$$\kappa = \frac{1}{T} \qquad (1.25)$$

INCOMPRESSIBLE AND COMPRESSIBLE FLUIDS

As expected, values of K and κ for common liquids are very large as indicated in Tables 1-1 and 1-2 for water. For example, at normal atmospheric pressure and at 20°C it would require a pressure of 210 atm to compress water 1 percent, it follows $K = 21000$ atm. This result is representative of the compressibility of liquids. Since such large pressures are required to effect a change in volume, we conclude that liquids can be considered as *incompressible* for most practical engineering applications. However, small density changes in liquids can still cause interesting phenomena in piping systems and liquids should be considered as *compressible*, such as the *water hammer*, which occurs when a liquid in a piping network encounters an abrupt flow restriction (such as a closing valve) and is locally compressed.

As mentioned earlier the fact that gases are, in general, quite compressible and strongly depend on pressure and temperature; yet flows of gases can often be treated as incompressible flows. A simple, and quite important, example of this is flow of air in air-conditioning ducts at normal atmospheric pressure. For our purposes in this course, a flow will be considered as incompressible if its density is constant. This will often be the case in the problems treated here.

1.4.3 Vapor Pressure and Cavitation

For liquids, such as water and gasoline, placed in a closed container with a small vacuum air space left above the surface, a pressure will develop in the space as a result of the vapor that is formed by the escaping liquid molecules from the liquid surface into the air space. When a phase equilibrium condition is reached so that the number of molecules leaving the surface equals to the number entering, the vapor of a pure substance is said to be *saturated* and the pressure that the vapor exerts on the liquid surface at this given temperature is termed the *vapor pressure*, p_v. The value of vapor pressure for a particular liquid turns out to be identical to the *saturation pressure* of the liquid, p_{sat}, the pressure at which a pure substance changes phase at a given temperature.

Boiling, which is the formation of *vapor bubbles* within a fluid mass, is initiated when the absolute pressure in the fluid reaches its vapor pressure. At standard atmospheric pressure, water will boil when the temperature reaches 100 ℃, but it boils at 93 ℃ in an ordinary pan at a 2000-m elevation, where the atmospheric pressure is 0.8 atm. Boiling can be induced at a given pressure acting on the fluid by raising the temperature, or at a given fluid temperature by lowering the pressure.

If the liquid pressure is greater than the vapor pressure, the only exchange between liquid and its vapor is evaporation at the interface. However, if the pressure at some locations is lowered to the vapor pressure, vapor bubbles begin to appear in the liquid (Fig. 1-6a) and the unplanned vaporization can be observed. The *vapor bubbles*, also called *cavitation bubbles* since they form "cavities" in the liquid, collapse as they are swept away from the low-pressure regions into the higher-pressure regions, with highly destructive, extremely high-pressure waves and sufficient intensity. This process, the liquid pressure is dropped below the vapor pressure due to a flow phenomenon, is called *cavitation*.

Cavitation must be avoided (or at least minimized) in flow systems since it reduces performance, generates annoying vibrations and noise, and causes damage to equipment (Fig. 1-6b). The pressure spikes resulting from the large number of bubbles collapsing near a solid surface over a long period of time may cause erosion, surface pitting, fatigue failure, and the eventual destruction of the components or machinery. Thus, it is the important consideration in the engineering design such as hydraulic turbines and pumps.

Fig. 1-6 Cavitation bubble formation in liquid flows: (a) Spiral bubble sheets form from the surface of a marine propeller[1]; (b) Collapsing bubbles erode a propeller surface (*Courtesy of Wikipedia*)

1.4.4 Surface Tension and Capillarity

It is often observed that a drop of blood forms a hump on a horizontal glass; a drop of mercury forms a near-perfect sphere and can be rolled just like a steel ball over a smooth surface; water droplets from rain or dew hang from branches or leaves of trees; a liquid fuel injected into an engine forms a mist of spherical droplets; water dripping from a leaky faucet falls as spherical droplets; a soap bubble released into the air forms a spherical shape; a water striker walks on water surface (Fig. 1-7a); water surface in a glass cup is higher than the glass wall without overflow, with a coin floating on the water surface despite the fact that metal in the coin has a larger density than water (Fig. 1-7b).

Fig. 1-7 Phenomena of surface tension. (a) Water striders' leg impact on water surface (*Courtesy of ISAKA Yoji at Wikipedia*); (b) Coin floats on water surface[2]

---

[1] http://www.thehulltruth.com/boating-forum/173520-prop-cavitation-burn-marks.html#b
[2] http://blog.sina.com.cn/s/blog_604f73e70100p07p.html

It is the unbalanced cohesive forces acting on the liquid molecules at the liquid surface induce these consequences. Molecules deeply submerged in the interior of the liquid mass are surrounded by molecules that are attracted to each other equally as shown in Fig. 1-8. However, molecules at the surface of liquid are subjected to a net force toward the interior due to the asymmetric attractive forces applied by liquid molecules around and the gas molecules above as shown in Fig. 1-8, among which the force induced by the latter is usually very small and negligible. The apparent physical consequence of this unbalanced force along the surface is the surface acts like a stretched elastic membrane under tension. The pulling force that causes this tension acts parallel to the surface and is due to the attractive forces between the molecules of the liquid. Such a force on the surface of the liquid is called the *surface tension* and can be designated by the *coefficient of surface tension*, σ_s, the surface tension per unit length in the units of N/m. For a given liquid, the surface tension depends on the temperature and the fluid it is in contact with at the interface, and the magnitude of the surface tension decreases as the temperature increases.

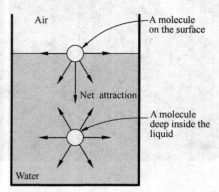

Fig. 1-8 Attractive forces acting on a water molecule at the surface and deep inside the water

The surface tension can arise at both liquid-solid and liquid-gas interfaces, and in general at the interface between any two immiscible (that is, non-mixable) fluids. If the interface is curved, say an idealized spherical droplet of water in air, a mechanical balance shows that there is a pressure difference across the interface, the pressure being higher on the concave side, as illustrated in Fig. 1-9.

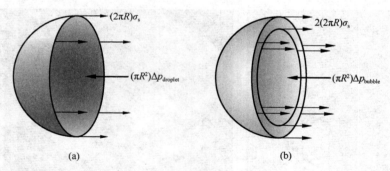

Fig. 1-9 Pressure change across a curved interface due to surface tension:
(a) interior of a spherical droplet; (b) interior of a bubble

If the spherical drop of liquid is cut in half (as shown in Fig. 1.9a) the force developed around the edge due to surface tension is $2\pi R\sigma_s$, where R is the radius of spherical drop. This force must be balanced by the pressure difference, $\Delta p_{droplet}$, between the internal pressure, p_i, and the external pressure, p_e, acting over the circular area, πR^2. Thus,

$$2\pi R \sigma_s = \pi R^2 \Delta p_{droplet}$$

or
$$\Delta p_{droplet} = p_i - p_e = \frac{2\sigma_s}{R} \quad (1.26)$$

which proves that the pressure inside the droplet is greater than the pressure surrounding the droplet.

We can use this result to predict the pressure increase inside a soap bubble, which has two interfaces with air as shown in Fig. 1-9b, an inner and outer surface of nearly the same radius R:

$$p_{bubble} \approx 2\Delta p_{droplet} = \frac{4\sigma_s}{R} \quad (1.27)$$

It indicates the pressure on the inside of a bubble of water would be two times of that on the inside of a drop of water of the same diameter and at the same temperature.

Another consequence of surface tension effect is the *capillarity*, the rise or fall of a liquid in a small-diameter tube inserted into the liquid. If the diameter of tube is small enough to make surface tension force non-negligible, the liquid in the tube curves up or down slightly at the edges where it touches the tube wall surface. Such a curved surface is named the *meniscus*, the angle the liquid interface intersecting with the solid surface is the *contact angle* θ (Fig. 1-10). If the contact angle is less than $90°$, the attraction (adhesive force) between the molecules of solid (tube wall) and liquid is strong enough to overcome the mutual attraction (cohesive force) of the liquid molecules and pull them up the wall, that is, the surface tension force tends to pull the liquid up in the tube along the wall (assume for simplicity that the tube is open to the atmosphere), such a liquid is said to be *wetting* the solid, i.e. water in a glass tube (Fig. 1-10a). If $\theta > 90°$, the surface tension force acts downward, the liquid surface curves down, i.e. mercury in a glass tube (Fig. 1-10c), such a liquid is termed *nonwetting*. For examples, water wets soap but does not wet wax. In natural world, the capillary effect is partially responsible for the rise of water to the top of tall trees.

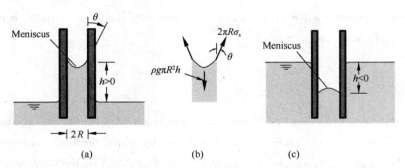

Fig. 1-10 Capillaries in small glass tubes. (a) Rise of column for a wetting liquid, i.e. water ($\theta < 90°$); (b) Free-body diagram for calculating column height; (c) Depression of column for a nonwetting liquid, i.e. mercury ($\theta > 90°$)

As illustrated in Fig. 1-10b, the *capillary rise* or *drop*, h, in a circular tube depends on the value of the surface tension, σ_s, the tube radius, R, the density of the liquid, ρ, and the contact angle, θ, between the fluid and tube. It can be determined by the force balance between the upward vertical force due to the surface tension, $2\pi R\sigma_s\cos\theta$, and the downward weight, $\rho g\pi R^2 h$

$$\rho g\pi R^2 h = 2\pi R\sigma_s\cos\theta$$

Solving for h gives the capillary rise

$$h = \frac{2\sigma_s\cos\theta}{\rho g R} \qquad (1.28)$$

This relation is also valid for the capillarity drop of the nonwetting liquids.

There are two most common capillarity in fluid mechanics occurring on the surfaces between water-air-glass and mercury-air-glass. For a clean surface of glass and a temperature of 20℃, the coefficient of surface tension is 0.0728N/m for air-water and 0.465N/m for air-mercury with a contact angle about 0℃ and 140℃, respectively. Take the densities of water and mercury at 20℃ to be 998.2kg/m^3 and 13550kg/m^3. Thus, the capillary rise of water and drop of mercury are

water-air-glass interface $\qquad h = \dfrac{29.8}{d}\quad$ (mm) $\qquad (1.29)$

mercury-air-glass interface $\qquad h = \dfrac{10.5}{d}\quad$ (mm) $\qquad (1.30)$

where the diameter of glass tube d has units of mm.

Note that the capillary rise is inversely proportional to the radius of the tube. Therefore, the thinner the tube is, the greater the rise (or fall) of the liquid in the tube. In practice, the capillary effect is usually negligible in tubes whose diameter is greater than 1 cm. When pressure measurements are made using manometers and barometers, it is important to use sufficiently large tubes to minimize the capillary effect. The capillary rise is also inversely proportional to the density of the liquid, as expected. Therefore, lighter liquids experience greater capillary rises.

An interesting example of the effects of surface tension can be found in wet sand. It is common experience that we cannot walk on water. But it is also difficult to walk on dry sand – think beach volleyball. However, when we wet sand with water sufficiently to 'activate' surface tension effects we can easily walk on the mixture.

Summary

In engineering practical, most problems have much larger physical dimensions than the limiting volume required by the *continuum hypothesis*, based on which the fluid properties, such as density, specific weight and pressure, can be defined and analyzed by the differential calculus.

Viscosity, by virtue of which a fluid offers resistance to shear stresses and satisfies the *no-slip condition* at solid-fluid interfaces, can be measured by the *dynamic viscosity* or *absolute viscosity*. The *shear stress* due to viscous effects in a simple flow with a linear distribution of velocity, $u = u(y)$, can be determined by the *Newtonian's law of viscosity*

$$\tau = \mu\frac{\mathrm{d}u}{\mathrm{d}y}$$

Fluids satisfying the Newtonian's law of viscosity are named as *Newtonian fluids*, otherwise *non-Newtonian fluids*.

Although all fluids have viscosity, they can still be considered as *ideal fluid* with the assumption of no viscosity. For an ideal gas, *perfect gas law* is used to relate pressure, temperature, and density in common gases.

Liquids are always considered as *incompressible* under standard atmospheric condition due to their large *coefficient of volume compressibility*, while many gases are assumed to be incompressible at low speeds, i. e. speed under 100 m/s for atmospheric air. *Compressible* is more discussed for gases (see Chapter 12), some times for *water hammer* of liquid in pipe system.

In engineering practical with large physical dimensions, *surface tension* is usually negligible. However, *vapor pressure* should be taken into serious consideration in some cases with high flow velocity, which may lead to *cavitation* and serious damage to the construction.

The forces on a liquid could be *surface forces* acting on the surface of the liquid, i. e. *shear stress* and *pressure*, and *mass forces* throughout the fluid body, i. e. inertia force and weight.

Word Problems

W1-1 What is continuum hypothesis of fluid? Can a fluid with bubbles in it be considered as a continuum?

W1-2 The mass force acting on a fluid includes (a) pressure force, (b) frictional force, (c) gravity, or (d) surface tension.

W1-3 What are the unit mass forces of a freely falling fluid under gravity? How about a fluid at rest under gravity? The z-coordinate is upward.

W1-4 What is specific gravity of a fluid? How is it related to density?

W1-5 What is the viscosity of a fluid? What are the cause of viscosity in liquids and in gases? Do liquids or gases have higher dynamic viscosities?

W1-6 What is the Newton's law of viscosity? It is the direct relation of (a) shear stress and pressure; (b) shear stress and rate of shear strain; (c) shear stress and shear strain; or (d) shear stress and velocity.

W1-7 How does the dynamic viscosity of (a) liquids and (b) gases vary with temperature?

W1-8 How does the kinematic viscosity of (a) liquids and (b) gases vary with temperature?

W1-9 What factors is the dynamic viscosity related to? How does the viscous affect the fluid behavior?

W1-10 What is a Newtonian fluid? Is water a Newtonian fluid?

W1-11 Consider two identical small glass balls dropped into two identical containers, one filled with water and the other with oil. Which ball will reach the bottom of the container first? Why?

W1-12 What is an ideal fluid?

W1-13 What does the coefficient of volume compressibility of a fluid represent?

W1-14 What does the coefficient of volume expansion of a fluid represent? How does it differ from the coefficient of volume compressibility?

W1-15 Can the coefficient of compressibility of a fluid be negative? How about the coefficient of volume expansion?

W1-16 What is vapor pressure? Does water boil at higher temperatures at higher pressures? Explain.

W1-17 What is cavitation? What causes it?

W1-18 What is surface tension? What is it caused by?

W1-19 Is the pressure inside the soap bubble higher or lower than the pressure outside?

W1-20 A small-diameter tube is inserted into a liquid whose contact angle is 70°. Will the level of liquid in the tube rise or drop? Explain.

W1-21 Is the capillary rise in a small-diameter tube larger than the one in a larger-diameter tube? Explain.

Problems

Density and Gravity

1-1 The specific gravity of mercury at 80 °C is 13.4. Determine its density and specific weight at this temperature.

1-2 The weight of $0.5 m^3$ oil is 4410N. Determine the density of the oil.

Viscosity

1-3 A fluid has a density of $850 kg/m^3$, dynamic viscosity $0.005 Pa \cdot s$. Determine the kinematic viscosity of the fluid.

1-4 Let a fluid layer between two plates be dragged along by the motion of an upper plate at a velocity of $U = 0.25 m/s$, while the bottom plate is stationary as in Fig. P1-4. It yields a shear stress of 2 Pa in the fluid layer. Determine the dynamic viscosity of this fluid.

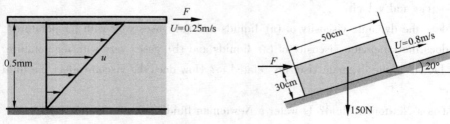

Fig. P1-4   Fig. P1-5

1-5 A force is applied horizontally on a 50cm × 30cm × 20cm block (weighing 150N) to make it move uphill at a constant velocity of 0.8m/s on an inclined surface with a friction coefficient of 0.27 as in Fig. P1-5. (a) Determine the magnitude of the required force F on the block.

(b) If a 0.4-mm-thick oil film, which has a dynamic viscosity of 0.012 Pa·s, is applied between the block and the inclined surface, determine the percent reduction in the required force.

1-6 A block of weight G slides down an inclined plane while lubricated by a thin film of oil with a thickness of h and a contact area of A as in Fig. P1-6. Assuming a linear velocity distribution in the film, derive an expression for the 'terminal' (zero-acceleration) velocity U of the block.

1-7 In Fig. P1-7 a thin plate is pulled at a velocity of U and separated from two fixed plates at distances of h_1 and h_2 by very viscous liquids μ_1 and μ_2, respectively. The contact area of the central plate with each fluid is A. (a) Assuming a linear velocity distribution in each fluid, derive the force F required to pull the plate at velocity U. (b) Is there a necessary relation between the two viscosities, μ_1 and μ_2?

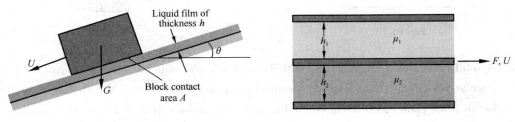

Fig. P1-6   Fig. P1-7

1-8 The 'no-slip' condition means that a fluid 'sticks' to a solid surface. This is true for both fixed and moving surfaces. Let two layers of fluid be dragged along by the motion of an upper plate, as in Fig. P1-8. The bottom plate is stationary. The top fluid puts a shear stress on the upper plate, and the lower fluid puts a shear stress on the bottom plate. Determine the ratio of these two shear stresses.

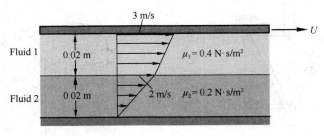

Fig. P1-8

1-9 A 6.00-cm-diameter shaft is being pushed axially through a 6.02-cm-diameter and 40-cm-long bearing sleeve. The clearance, assumed uniform, is filled with oil with a kinematic viscosity of 0.003 m²/s and a specify gravity of 0.88. Estimate the force required to pull the shaft at a steady velocity of 0.4 m/s.

1-10 A 120-mm-diameter shaft is pulled by a force of 8.43 N through a cylindrical bearing at a velocity of 0.493m/s, as in Fig. P1-10. Determine the viscosity of the lubricant oil filling the 0.2-mm gap between the shaft and the bearing.

1-11 The clutch system shown in Fig. P1-11 is used to transmit torque through a 3-mm-thick SAE 30W oil film of $\mu = 0.38$ Pa·s between two identical 30-cm-diameter disks. When the driving shaft rotates at a speed of 1450rpm, the driven shaft is observed to rotate at 1398rpm. Assuming a linear velocity profile in the oil film, determine the transmitted torque.

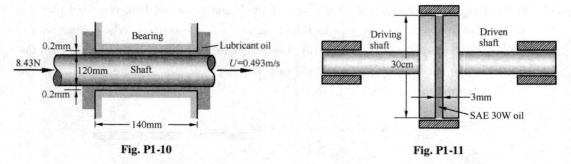

Fig. P1-10   Fig. P1-11

1-12 A solid cone with a circular base of 0.3m and a height of 0.5m is rotating with an initial angular velocity of 16rad/s inside a conical seat, as in Fig. P1-12. The clearance of 1 mm between the cone and the seat is filled with oil with a viscosity of 0.1Pa·s. Disregarding air drag, determine the resistant torque acting on the cone.

1-13 The viscosity of a fluid is to be measured by a viscometer constructed of two 75-cm-long concentric cylinders with a gap of 0.12cm, as in Fig. P1-13. The outer diameter of the inner cylinder is 15 cm. When the inner cylinder is rotated at 200rpm, the measured torque is 0.8 N·m, determine the viscosity of the fluid.

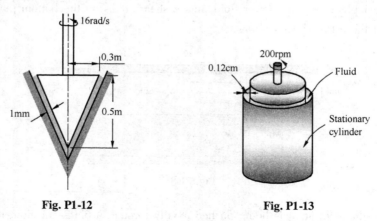

Fig. P1-12   Fig. P1-13

Compressibility

1-14 A 1-m³ volume of water is contained in a rigid container. Estimate the change in the volume of the water when a piston applies a pressure of 35 MPa on it.

1-15 A 1000m³ volume of pressurized liquid contained in a rigid cylinder is compressed by a piston to a pressure of 0.1 MPa. When the piston applies a pressure of 10 MPa, the volume of liquid is 995m³. Determine the bulk modulus of the liquid.

1-16 Take the temperature and the absolute pressure in the tires of car to be 20℃ and 350 kPa respectively when the car just runs. Determine the corresponding pressure in the tire after the temperature in the tires rise to 50℃.

1-17 It is observed that the density of an ideal gas decreases by 10 percent when compressed isothermally from 10 atm to 11 atm. Determine the percent decrease in density of the gas if it is compressed isothermally from 100 atm to 101 atm.

Vapor Pressure and Cavitation

1-18 Oil, with a vapor pressure of 20kPa, is delivered through a pipeline by equally spaced pumps in flow direction. Each pump can raise oil pressure by 1.3 MPa. Assume the friction loss in the pipeline is 150 Pa per meter of pipe. What is the maximum possible pump spacing to avoid cavitation of the oil?

1-19 A propeller is tested in a water tunnel at 20°C as in Fig. 1-6a. The lowest pressure on the blade can be estimated by a form of Bernoulli's equation (see Section 3.6):

$$p_{min} = p_0 + \frac{1}{2}\rho v^2$$

where $p_0 = 1.5$ atm and v is the average velocity in tunnel. If we run the tunnel at $v = 18$ m/s, can we be sure that there will be no cavitation? If not, can we change the water temperature and avoid cavitation?

Surface Tension

1-20 A 1.9-mm-diameter tube is inserted into an unknown liquid with a density of 960kg/m³. The observed liquid rise in the tube is 5mm and the contact angle is 15°. Determine the surface tension of the liquid.

1-21 The system in Fig. P1-21 is used to calculate the pressure p_1 at the liquid surface in the tank. If the height of liquid in the 1-mm-diameter tube is 15cm and the fluid is at 60℃. Calculate the true liquid height in the tube and the percent error due to capillarity if the liquid is (a) water and (b) mercury.

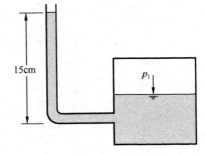

Fig. P1-21

1-22 The water striker bug shown in Fig. 1.7a is supported on the surface of a pond by surface tension acting along the interface between the water and the bug's legs. (a) Determine the minimum length of this interface needed to support the bug. Assume the bug weighs 10^{-4} N and the surface tension force acts vertically upwards. (b) Repeat part (a) if surface tension were to support a person weighing 750N.

1-23 A water droplet in the air has a diameter of 0.3mm. Determine the pressure difference between the interior and outside of the droplet.

1-24 The clear glass tubes are usually used as the piezometric tubes in the laboratory. Calculate the capillarity rise in the tube if the diameter of tube is (a) 0.6mm and (b) 1.2mm.

1-25 Assume the clear distance between two parallel vertical glass plate is 1.0mm and is filled with water. Determine the capillary rise of water within the clearance.

2 Fluid Statics

CHAPTER OPENING PHOTO[❶]: The Xin'anjiang Hydropower Station, the country's first large-scale power plant with the first reinforced concrete dam and the first dam higher than 100 m designed and built by Chinese in the 1950s, is located at the Xin'an River in Jiande city of Zhejiang Province, China. It forms a famous huge reservoir (Qiandao Lake) with 1,078 islands by the 105-m-tall, 466.5-m-long-crest Xin'an dam, on which the hydrostatic pressure, due to the weight of the standing water, can cause enormous forces and moments analyzed in the present chapter.

In this chapter we will discuss the fluid either at rest or moving in such a manner that there is no relative motion between adjacent particles, that is, the fluid is free of shear stress and only has pressure. We start this chapter with a detailed discussion of the pressure at a point, the pressure on a fluid element in the fluid body and its variation with the depth in a gravitational field and its measurements, followed by a discussion of pressure variation in a rigid-body fluid in motion either undergoing linear acceleration or in rotating containers. We then determine the hydrostatic forces applied on submerged bodies with plane or curved surfaces. Finally, we consider the buoyant force applied by fluids on submerged or floating bodies, and discuss their stabilities.

2.1 Pressure at a Point

2.1.1 Pressure at a Point in Static Fluid

Pressure at a point in a fluid at rest is independent of orientation. That is, it has magnitude but not a specific direction, and thus it is a scalar quantity. This can be demonstrated by considering a small tetrahedral fluid element in equilibrium with a size of dx by dy by dz, as in Fig. 2-1.

The fluid is at rest, so there are no shearing forces by definition, but we postulate that the

---

[❶] http://wl.jiande.gov.cn/Zpxd/201206/t20120608_148578.htm

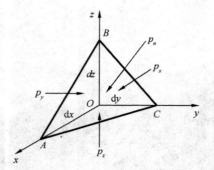

Fig. 2-1 Equilibrium of a small tetrahedral fluid element at rest

pressures acting at right angles to the surfaces, p_x, p_y and p_z in the x-, y- and z-direction and p_n in the direction normal to the inclined face, may be different on each face. The surface forces acting on the element in all directions are:

$$\Delta F_x = p_x \cdot \frac{1}{2}dydz, \quad \Delta F_y = p_y \cdot \frac{1}{2}dxdz,$$

$$\Delta F_z = p_z \cdot \frac{1}{2}dxdy, \quad \Delta F_n = p_n \cdot \Delta A_n$$

where ΔA_n is the area of the inclined face of the tetrahedral element.

The only mass force acting on the fluid element is the gravity. The components related to the coordinate system in Fig. 2-1 are

$$\Delta F_{bx} = \Delta F_{by} = 0, \quad \Delta F_{bz} = \rho g \cdot \frac{1}{6}dxdydz$$

As the fluid is at rest, or in equilibrium, the summation of the forces in any direction must equal zero (no acceleration): $\Sigma F_x = 0$, $\Sigma F_y = 0$ and $\Sigma F_z = 0$. Thus, in the x-direction, we have

$$\Sigma F_x = 0 = \Delta F_x - \Delta F_n \cos(\theta, x) + \Delta F_{bx} = p_x \cdot \frac{1}{2}dydz - p_n \cdot \Delta A_n \cos(\theta, x)$$

where (θ, x) is the angle between the outward normal line of the inclined face ABC and the x-axes, the geometric relation is such that

$$\Delta A_n \cos(\theta, x) = \frac{1}{2}dydz$$

Substituting back into the equilibrium equation in the x-direction yields

$$p_x = p_n$$

With similar manner from $\Sigma F_y = 0$ we have $p_y = p_n$; from $\Sigma F_z = 0$ and with $\Delta A_n \cos(\theta, z) = \frac{1}{2}dxdy$ to give

$$p_z \cdot \frac{1}{2}dxdy - p_n \cdot \Delta A_n \cos(\theta, z) - \rho g \cdot \frac{1}{6}dxdydz = (p_z - p_n) \cdot \frac{1}{2}dxdy - \rho g \cdot \frac{1}{6}dxdydz = 0$$

which yields

$$p_z - p_n - \rho g \cdot \frac{1}{3}dz = 0 \tag{2.1}$$

These relations illustrate two important principles of the hydrostatic, or shear-free, condition: (1) There is no pressure change in the horizontal direction, and (2) there is a vertical change in pressure proportional to the density, gravity, and depth variation.

When the tetrahedral fluid element becomes infinitesimal, it shrinks to a point, that is $dz \to 0$, the relation in Eq. 2.1 becomes $p_z = p_n$. Then, we have

$$p_x = p_y = p_z = p_n \tag{2.2}$$

Considering the tetrahedral fluid element again, p_n is the pressure on a surface with an arbitrary angle θ, it could be any orientation. Again, the tetrahedral fluid element is so small that it can be considered as a point so the derived expression, Eq. 2.2, indicates that the pressure at a point in a static fluid has an identical magnitude in all directions, that is, the pressure at a point in a static fluid, or in motion, is independent of orientation as long as there are no shearing stresses present. This is known as *Pascal's Law*, after Blaise Pascal (1623-1662), and applies to fluids at rest, or the fluid in motion in the absence of shear forces.

2.1.2 Pressure at a Point in Motion

For ideal fluid ($\mu = 0$) in motion, the shearing stress is also absent, thus the pressure at a point in an ideal fluid in motion is also independent of direction, that is $p = p_x = p_y = p_z$.

However, if there are strain rates in a moving viscous fluid ($\mu \neq 0$), there will be viscous stresses, both shear and normal in general. In that case the *dynamic pressure* is defined as the average of the three mutually perpendicular *normal stresses* σ_{ii} on the element as shown in Fig. 2-2

$$p = -\frac{1}{3}(\sigma_{xx} + \sigma_{yy} + \sigma_{zz}) \qquad (2.3)$$

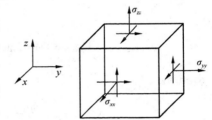

Fig. 2-2 Stresses on the surface of fluid element in motion

where the sign '−' presents the pressure acting along the inward normal direction, while σ_{xx}, σ_{yy}, σ_{zz} along the outward normal direction as shown in Fig. 2-2.

Equation 2.3 is subtle and rarely needed since the great majority of viscous flows have negligible viscous normal stresses.

2.2 Equilibrium of a Fluid Element

2.2.1 Euler Equilibrium Equation

The pressure in a fluid in which there are no shearing stresses in a gravitational field will vary spatially and vertically from point to point as indicated in Eq. 2.1, but how? To answer this question, consider a small rectangular fluid element removed from some arbitrary position within the mass of fluid of interest as illustrated in Fig. 2-3. Two types of forces acting on this element: *surface forces* due to the pressure, and a *body force* equal to the weight of the element. For simplicity the x-direction surface forces are not shown in the figure.

Let the pressure acting on the center point of element varies arbitrarily, $p = p(x, y, z, t)$. Thus, the pressures a short distance away from the center point can be approximated by using a Taylor series expansion of the pressure at the element center and neglecting higher order terms that will vanish as we let dx, dy, and dz approach zero. Thus, the net surface force in the y-direction

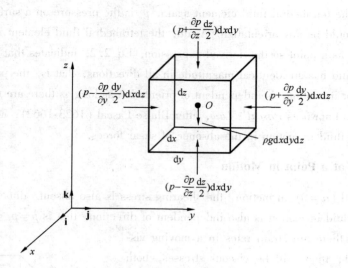

Fig. 2-3 Forces on a fluid element due to pressure variation

is given by

$$\Delta F_y = \left(p - \frac{\partial p}{\partial y}\frac{dy}{2}\right)dxdz - \left(p + \frac{\partial p}{\partial y}\frac{dy}{2}\right)dxdz = -\frac{\partial p}{\partial y}dxdydz$$

Similarly, the net surface forces in the x- and z-direction can be obtained as

$$\Delta F_x = -\frac{\partial p}{\partial x}dxdydz \qquad \Delta F_z = -\frac{\partial p}{\partial z}dxdydz$$

Therefore, the total net surface force vector on the element due to pressure is

$$\Delta \mathbf{F}_s = \left(-\frac{\partial p}{\partial x}\mathbf{i} - \frac{\partial p}{\partial y}\mathbf{j} - \frac{\partial p}{\partial z}\mathbf{k}\right)dxdydz \tag{2.4}$$

We recognize the term in parentheses as the negative vector gradient of p. Then Eq. 2.4 can be rewritten as

$$\Delta \mathbf{F}_s = -\nabla p \, dxdydz$$

It shows that the pressure gradient is a *surface* force which acts on the sides of the element.

There may also be a *body* force, due to gravitational potentials, acting on the entire mass of the element. The mass force acting on the fluid element can be given as:

$$\Delta \mathbf{F}_b = \mathbf{f}\rho dxdydz = (f_x\mathbf{i} + f_y\mathbf{j} + f_z\mathbf{k})\rho dxdydz$$

The total vector resultant of these surface forces and body forces must keep the element in equilibrium or cause it to move with acceleration $\mathbf{a}$. From Newton's second law, we have $\sum d\mathbf{F} = dm\mathbf{a}$, that is

$$\sum d\mathbf{F} = d\mathbf{F}_s + d\mathbf{F}_b = (-\nabla p + \mathbf{f}\rho)dxdydz = \rho dxdydz \, \mathbf{a}$$

or

$$\mathbf{f} - \frac{1}{\rho}\nabla p = \mathbf{a} \tag{2.5}$$

or in term of components

$$\left.\begin{aligned} f_x - \frac{1}{\rho}\frac{\partial p}{\partial x} &= a_x \\ f_y - \frac{1}{\rho}\frac{\partial p}{\partial y} &= a_y \\ f_z - \frac{1}{\rho}\frac{\partial p}{\partial z} &= a_z \end{aligned}\right\} \quad (2.6)$$

Equation 2.5 or 2.6 is the general equation of motion for a fluid in which there are no shearing stresses, is called the *Euler differential equation of motion*.

EULER EQUILIBRIUM EQUATION OF MOTION

If the fluid is at rest or at constant velocity, $\mathbf{a} = 0$. Equation 2.5 for the pressure distribution reduces to

$$\mathbf{f} - \frac{1}{\rho}\nabla p = 0 \quad (2.7)$$

or

$$\left.\begin{aligned} f_x - \frac{1}{\rho}\frac{\partial p}{\partial x} &= 0 \\ f_y - \frac{1}{\rho}\frac{\partial p}{\partial y} &= 0 \\ f_z - \frac{1}{\rho}\frac{\partial p}{\partial z} &= 0 \end{aligned}\right\} \quad (2.8)$$

Equation 2.7 or 2.8 is called the *Euler equilibrium equation of motion*, proposed by the Swiss mathematician Leonhard Euler (1707 – 1783) in 1775. It indicates that for a fluid in hydrostatic equilibrium, the total net surface force acting on the fluid element is balanced by the mass force of fluid element. It should be noted that it is not the pressure but the pressure *gradient* causing a net force which must be balanced by gravity or acceleration or some other effect in the fluid.

DIFFERENTIAL FORM OF EULER EQUILIBRIUM EQUATION

Times dx, dy and dz to the equations in the x-, y- and z- directions in Eq. 2.8 respectively and adds them up to give:

$$\frac{\partial p}{\partial x}dx + \frac{\partial p}{\partial y}dy + \frac{\partial p}{\partial z}dz = \rho(f_x dx + f_y dy + f_z dz)$$

The pressure $p = p(x, y, z)$ is the consistent function of the coordinates. According to the definition of total derivative, the left side of the above equation is the total derivative of pressure p. Thus, we have

$$dp = \rho(f_x dx + f_y dy + f_z dz) = \rho \mathbf{f} \cdot d\mathbf{s} \quad (2.9)$$

where d**s** is the differential length vector.

Equation 2.9 is the differential form of *Euler equilibrium equation of motion*. Consider the unit mass force on a fluid to be known, we can obtain the pressure distribution in the fluid by integrating Eq. 2.9.

2.2.2 Equipressure Surface

Equipressure surface is a surface in a fluid at rest, or in equilibrium, on which the pressure at any point has the identical magnitude, that is, $dp = 0$. Thus, Equation 2.9 reduces to

$$\mathbf{f} \cdot d\mathbf{s} = 0 \qquad (2.10)$$

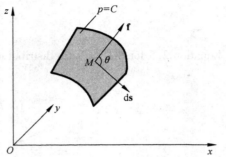

Fig. 2-4 Characteristic of equipressure surface

It indicates that the unit mass force is perpendicular to the equipressure surface because the dot product of the unit mass force vector **f** and the differential length vector d**s** on the equipressure surface is zero (see Fig. 2-4).

If the only mass force on the fluid is gravity, thus, $f_x = f_y = 0$, $f_z = -g$. Equation 2.10 becomes

$$\mathbf{g} \cdot d\mathbf{s} = 0 \qquad (2.11)$$

where $\mathbf{g} = -g\mathbf{k}$ because the customary coordinate system z is 'up.'

It states that a fluid in hydrostatic equilibrium will align its equipressure surface (constant-pressure surface) everywhere normal to the local-gravity vector. For a homogeneous, continuous and incompressible fluid at rest under gravity, the equipressure surfaces must be horizontal planes since the direction of gravity is downwards, i. e. the free surface and surface *MN* in Fig. 2-5. If there are two unmixed fluids as shown in last tank, the surface EF is not an equipressure surface because the continuous fluids are not the same: there are two kinds of fluids-water and mercury.

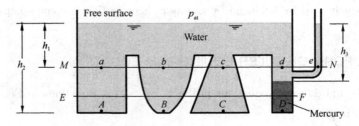

Fig. 2-5 Equipressure surfaces

If there are other mass forces other than gravity on a fluid, the equipressure surfaces would be non-horizontal surfaces normal to the resultant force vector of the mass forces. For example, when the fluid is in motion, the equipressure surface may be inclined surface or curved surface due to the extra unit mass force: accelerations (Further discussion is given in Section 2.4).

2.3 Hydrostatic Pressure Distribution in Gravitational Field

2.3.1 Hydrostatic Pressure Distribution

If the mass force of fluid at rest, or in equilibrium has only gravity, we have $f_x = f_y = 0$, $f_z = -g$, thus Euler equilibrium equation of motion (Eq. 2.8) gives the pressure distributions in the x-, y- and z- directions

$$\frac{\partial p}{\partial x} = 0 \qquad \frac{\partial p}{\partial y} = 0 \qquad \frac{\partial p}{\partial z} = -\rho g$$

This leads to the conclusion that for liquids or gases at rest, or in equilibrium, the pressure in the horizontal direction does not change, that is, is independent of x or y; and the pressure gradient in the vertical direction at any point in a fluid depends only on the specific weight of the fluid at that point. Hence $\partial p/\partial z$ can be replaced by the total derivative dp/dz, and the hydrostatic condition reduces to

$$\frac{dp}{dz} = -\rho g \tag{2.12}$$

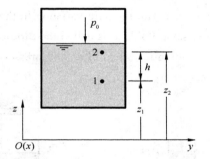

Fig. 2-6 Pressure variation in a fluid at rest with a free surface in a tank

For most engineering applications, the variation in g is negligible, so our main concern is with the possible variation in the fluid density. Liquids, however, are so nearly incompressible that we can neglect their density variation in hydrostatics. Thus, we can integrate Eq. 2.12 referred to a datum plane O-O in Fig. 2-6 to give the pressure at any point in a fluid at rest

$$p = -\rho g z + C$$

or

$$z + \frac{p}{\rho g} = C \tag{2.13}$$

Also we can integrate Eq. 2.12 from point 2 to point 1 to get the variation of pressure between points 2 and 1 in a fluid

$$\int_2^1 dp = \int_{z_2}^{z_1} -\rho g dz$$

to yield

$$p_1 - p_2 = -\rho g(z_1 - z_2) = \rho g h$$

or

$$p_1 = p_2 + \rho g h \tag{2.14}$$

or

$$z_1 + \frac{p_1}{\rho g} = z_2 + \frac{p_2}{\rho g} \tag{2.15}$$

where h is the difference of depths between points 1 and 2, $h = z_2 - z_1$ as shown in Fig. 2-6.

Equations 2.13 and 2.15 are the *hydrostatic pressure distribution* for all fluids regardless of their viscosity because there is no viscosity term in the equations.

From Eq. 2.13 to 2.15 the following statements about a hydrostatic condition can be concluded:

(1) Pressure in a continuously distributed uniform fluid at rest varies only with vertical distance and is independent of the shape of the container, as in Fig. 2-5. The pressures at all points on a given horizontal plane (equipressure surface) in the fluid have the same magnitude. For example, points a, b, c, d and e on the horizontal plane MN are at equal depth below the free surface and are interconnected by the identical fluid, water; therefore all points have the same pressure: $p_a = p_b = p_c = p_d = p_e = p_{at} + \rho_w g h_1$ (where p_{at} is the atmospheric pressure) based on Eq. 2.14. The same is true for points A, B, and C on the bottom, which all have the same higher pressure $p_A = p_B = p_C = p_{at} + \rho_w g h_2$ than at plane MN since the deeper depth of $h_2 > h_1$. However, point D, although at the same depth as points A, B, and C, has a larger pressure, $p_D = p_{at} + \rho_w g h_3 + \rho_{Hg} g (h_2 - h_3)$, because it is in a heavier fluid, mercury.

(2) The pressure increases with depth in the fluid. The pressure difference between two points in a fluid at rest can be obtained from Eq. 2.14. If the point 2 is located at the free surface, then $p_2 = p_0$ as shown in Fig. 2-6, Eq. 2.14 becomes

$$p_1 = p_0 + \rho g h \tag{2.16}$$

where p_0 is the pressure at the free surface, if the container is opened to the air, $p_0 = p_{at}$.

(3) The pressure variation in the fluid in balance can be transmitted equally to any points in the same fluid, it is called the *transmission of fluid pressure*. It is quite important for the operation of hydraulic jacks, lifts, and presses, and hydraulic controls on aircraft and other types of heavy machinery.

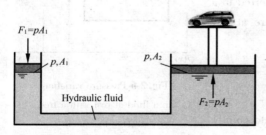

Fig. 2-7 Transmission of fluid pressure

Figure 2-7 shows two connected tanks which appear to connect to each other by a tube. Based on the definition of the equipressure surface, the pressures on the bottoms of both small and large pistons at same horizontal plane equal as the fluid is in balance

$$p_1 = p_2$$

thus we have

$$F_2 = \frac{A_2}{A_1} F_1$$

Since the piston area A_2 can be made much larger than A_1 and therefore a large mechanics advantage can be developed; that is, a small force applied at the smaller piston can be used to develop a large force at the larger piston. For example, using a hydraulic car jack with a piston area ratio of $A_2/A_1 = 10$, a person can lift a 1000kg car by applying a force of just 980N.

Meanwhile, when the force acting on the smaller piston increases ΔF_1, the pressure increment is $\Delta p = \Delta F_1 / A_1$, which is equally transmitted through tube to the larger piston, which can cause a force increment as

$$\Delta F_2 = \frac{A_2}{A_1}\Delta F_1$$

2.3.2 Measurement of Pressure

ABSOLUTE PRESSURE AND GAGE PRESSURE

In the engineering practice, the pressure are apt to be specified as the *absolute pressure*, p_{abs}, the actual or true pressure of a fluid relative to a *perfect vacuum* (zero pressure) (Fig. 2-8). From Eq. 2.16 gives the absolute pressure as

$$p_{abs} = p_{at} + \rho g h \tag{2.17}$$

As we live constantly under the pressure of the atmosphere, and everything else exists under this pressure. Therefore, it is convenient (and often done) to take atmospheric pressure as the datum, that is in engineering applications, *gage pressure*, p, the value *relative* to the local ambient atmosphere (Fig. 2-8), is often used

$$p = p_{abs} - p_{at}$$

Considering Eq. 2.17 we have

$$p = \rho g h \tag{2.18}$$

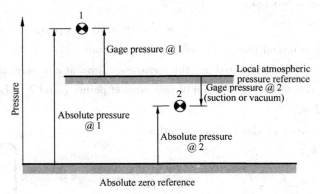

Fig. 2-8 Graphical representation of gage and absolute pressures

If the actual pressure is equal to atmospheric pressure, then the gage pressure equals zero. The measured pressure may be either higher or lower than the local atmosphere. Gage pressure is positive for absolute pressures greater than atmospheric and negative for those less than atmospheric (when the term *suction pressure* or *vacuum pressure* may be used) as shown in Fig. 2-8. When the gage pressure is negative, we say vacuum happens. The *vacuum pressure* is designated as p_v:

$$p_v = p_{at} - p_{abs} = -p \tag{2.19}$$

In this book, pressure will be assumed to be gage pressure, and local atmospheric pressure is *Engineering barometric pressure*, at (1at = 98kPa), unless specially designated absolute.

HEIGHT OF LIQUID COLUMN

As gravitational acceleration g is (approximately) constant, the gage pressure can be given by stating the vertical height of any fluid of density which is equal to this pressure.

$$h = \frac{p}{\rho g} \tag{2.20}$$

This vertical height h is known as *head* of fluid and the liquid rise (*height of liquid column*) in the tube due to the pressure p, which can also be used to interpret the magnitude of pressure in the engineering application. *Note*: If pressure is quoted in *head*, the density of the fluid *must* also be given.

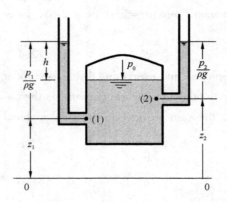

Fig. 2-9 Illustration of hydrostatic pressure distribution

Therefore, the terms in Eqs. 2.13 and 2.15 can be interpreted as (Fig. 2-9):

- the quantity z is the *elevation height* above the datum plane 0-0, called the *elevation head*, is related to the elevation potential energy per unit weight of fluid above datum plane;
- the quantity $p/\rho g$ is a length called the *pressure head* of the fluid, represents the height of a column related to the pressure potential energy per unit weight of fluid above atmospheric pressure, also called *piezometric height*.
- the quantity $z + p/\rho g$ is called the *piezometric head*, represents total potential energy per unit weight of fluid.

Thus, Eq. 2.13 can be interpreted as: the *piezometric head* at any point in a fluid at rest in a gravitational field are constant. The relation of pressures at points 1 and 2 in Eq. 2.15 can thus be illustrated as in Fig. 2-9.

The vacuum pressure can also be interpreted as the *height of vacuum* h_v:

$$h_v = \frac{p_{at} - p_{abs}}{\rho g} = \frac{p_v}{\rho g} \tag{2.21}$$

ATMOSPHERIC PRESSURE

The *atmospheric pressure* is usually measured with a mercury *barometer*, which in its simplest form consists of a glass tube closed at one end with the open end immersed in a container of mercury as shown in Fig. 2-10. The first mercury barometer was constructed in 1643-1644 by Evangelista Torricelli (1608-1647), an Italian physicist and mathematician.

Since atmospheric pressure forces a mercury column to rise a distance h into the tube and causes

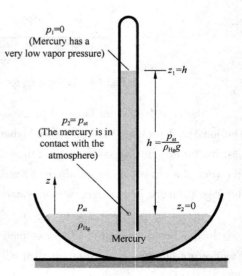

Fig. 2-10 A barometer measures local absolute atmospheric pressure

a near vacuum in the closed upper end, the upper mercury surface is at the vapor pressure. From Eq. 2.16 applies:

$$p_{at} = p_{vapor} + \rho_{Hg} g h$$

Because mercury has an extremely small vapor pressure at room temperatures (0.16 Pa at 20℃), it reduces to

$$p_{at} \approx \rho_{Hg} g h \qquad \text{or} \qquad h = \frac{p_{at}}{\rho_{Hg} g}$$

For standard atmospheric pressure, 1 atm = 101.35 kPa = 760 mmHg = 10.33 mH$_2$O; for engineering barometric pressure, 1 at = 98 kPa = 736 mmHg = 10 mH$_2$O.

EXAMPLE 2-1   Absolute and gage pressures

A newfound freshwater lake has a maximum depth of 60m. The mean atmospheric pressure in this area is 91kPa. Estimate the absolute and gage pressures in kPa at this maximum depth.

Solution:

Take the density of fresh water in the lake to be $\rho = 998 \text{kg/m}^3$. With $g = 9.81 \text{m/s}^2$, $p_{at} = 91 \text{kPa}$ and $h = 60$m, Eq. 2.17 predicts that the pressure at this depth will be

$$p_{abs} = p_{at} + \rho g h = 91 \text{kPa} + (998 \text{kg/m}^3)(9.81 \text{m/s}^2)(60 \text{m}) = 91 \text{kPa} + 587.4 \text{kPa} = 678.4 \text{kPa}$$

By omitting p_{at} we could state the gage pressure as $p = 587.4 \text{kPa}$.

EXAMPLE 2-2   Elevation, pressure and piezometric heads in a tank

Sketch the elevation, pressure and piezometric head of points A, B and C in a closed container as shown in Fig. E2-2a. Take the line 0-0 to be datum plane.

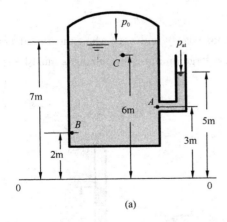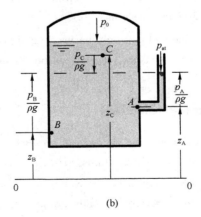

Fig. E2-2

Solution:

For point A, the elevation head $z_A = 3$m, the pressure head can be directly obtained from Fig. E2-2a, $\frac{p_A}{\rho g} = 5\text{m} - 3\text{m} = 2\text{m}$. Thus the piezometric head $z_A + \frac{p_A}{\rho g} = 5\text{m}$.

For point B, the elevation head $z_B = 2$m, the piezometric head $z_B + \frac{p_B}{\rho g} = z_A + \frac{p_A}{\rho g} = 5$m based on the hydrostatic pressure distribution $z + \frac{p}{\rho g} = C$, thus the pressure head $\frac{p_B}{\rho g} = 5\text{m} - 2\text{m} = 3\text{m}$.

For point C, the elevation head $z_C = 6$m, the piezometric head $z_C + \frac{p_C}{\rho g} = z_A + \frac{p_A}{\rho g} = 5$m based on the hydrostatic pressure distribution $z + \frac{p}{\rho g} = C$, thus the pressure head $\frac{p_C}{\rho g} = 5\text{m} - 6\text{m} = -1\text{m} < 0$, which indicates vacuum occuring at point C.

The heads at points A, B and C have been sketched in Fig. E2-2b.

2.3.3 Manometers

Manometry is a standard technique for measuring pressure using liquid columns in vertical or inclined tubes. The devices used in this manner are known as *manometers*. In this section, the barometer and the most commonly used manometers will be introduced.

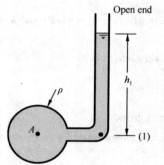

Fig. 2-11 Piezometer tube

PIEZOMETER TUBE

The simplest manometer is a tube opened at the top, and attached to the vessel containing liquid in which the pressure (higher than atmospheric) to be measured, as illustrated in Fig. 2-11. This simple device is known as a *piezometer tube*. As the tube is open to the atmosphere the pressure measured is relative to atmospheric so its reading, h_1, is gage pressure, which can be obtained directly from Eq. 2.18

$$p_A = \rho g h_1$$

This method can only be used for liquids (not for gases) and only when the liquid height is convenient to measure. It must not be too small or too large and pressure changes must be detectable.

U-TUBE MANOMETER

To overcome the shortage of piezometer tube on the measured fluids, *U-tube manometer*, as in Fig. 2-12, is widely used to measure the pressure of both liquids and gases. The U-shape tube is connected as in the figure and filled with a fluid called the *manometric fluid*, or *gage fluid*. The fluid whose pressure is being measured should have a mass density not equal to that of the gage fluid and the two fluids should not be able to mix readily, that is, they must be

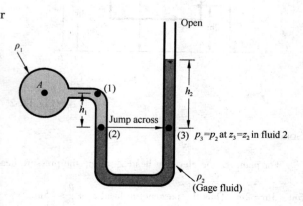

Fig. 2-12 U-tube manometer

immiscible.

As indicated in Eq. 2.15, any two points at the same elevation in a continuous mass of the same homogeneous static fluid will be at the same pressure. Thus, the pressures at point (2) and (3) shown in Fig. 2-12 are equal, $p_2 = p_3$. In other words, the horizontal plane crossing points (2) and (3) is an equipressure surface.

Thus, consider Eq. 2.14 in the left and right hand arms of the U-tube manometer to give

$$p_A + \rho_1 g h_1 = \rho_2 g h_2$$

Solving the pressure at point A in terms of the column height yields

$$p_A = \rho_2 g h_2 - \rho_1 g h_1$$

We also can find the pressure p_A by starting at one end of the system and work our way around to the other end and simply utilizing Eq. 2.16

$$p_A + \rho_1 g h_1 - \rho_2 g h_2 = 0$$

Thus again we have $p_A = \rho_2 g h_2 - \rho_1 g h_1$. In this way, we directly jump across from point (2) to (3) as shown in Fig. 2-12 because $p_3 = p_2$.

If the fluid being measured is a gas, the density will probably be very low in comparison to the density of the gage fluid, i.e., $\rho_2 \gg \rho_1$. In this case the term $\rho_1 g h_1$ can be neglected, and the gage pressure at point A is given by

$$p_A \approx \rho_2 g h_2$$

The U-tube manometer is also widely used to measure the *difference* in pressures between two containers or two points in a given system. The measured fluid in the system can be any gas, vapor or liquid provided it does not mix or react chemically with the gage liquid, which is usually mercury, water or light oil.

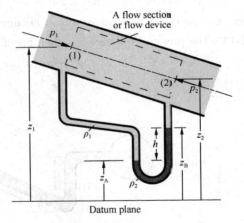

Fig. 2-13 Differential U-tube manometer

As in Fig. 2-13, start from point 1 and work around the way to point 2 to give

$$p_1 + \rho_1 g(z_1 - z_A) - \rho_2 g(z_B - z_A) - \rho_1 g(z_2 - z_B) = p_2$$

which gives the pressure difference between points 1 and 2

$$p_1 - p_2 = \rho_1 g(z_2 - z_1) + (\rho_2 - \rho_1) g h$$

If the pipe is horizontal, $z_1 = z_2$, the above equation will reduce to

$$p_1 - p_2 = (\rho_2 - \rho_1) g h$$

EXAMPLE 2-3 Measuring pressure with a multifluid manometer

In Fig. E2-3 the pressure in the air-pressurized water is measured by a multifluid manometer. The local atmospheric pressure is 85.6kPa. Determine the absolute air pressure in the tank if $h_1 = 0.1$m, $h_2 = 0.2$m, and $h_3 = 0.35$m. Take the densities of water, oil, and mercury to be 1000kg/m³, 850kg/m³, and 13600kg/m³, respectively.

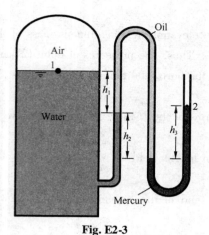

Fig. E2-3

Solution:
Since the variation of the air pressure with elevation is negligible due to its low density, the air pressure in the tank equals the pressure at the air-water interface. Thus, starting with the pressure at point 1 at the air-water interface, moving along the tube by adding or subtracting the $\rho g h$ terms until we reach point 2, and setting the result equal to p_{at} since the tube is open to the atmosphere gives

$$p_1 + \rho_w g h_1 + \rho_o g h_2 - \rho_{Hg} g h_3 = p_{at}$$

Solving p_1 and substituting gives

$$\begin{aligned} p_1 &= p_{at} - \rho_w g h_1 - \rho_o g h_2 + \rho_{Hg} g h_3 \\ &= 85.6 \text{kPa} - (9.81 \text{m/s}^2)[(1000 \text{kg/m}^3)(0.1 \text{m}) \\ &\quad + (850 \text{kg/m}^3)(0.2 \text{m}) - (13600 \text{kg/m}^3)(0.35 \text{m})] \\ &= 129.6 \text{kPa} \end{aligned}$$

DISCUSSION Note that jumping horizontally from one tube to the next and realizing that pressure remains the same in the same fluid simplifies the analysis considerably.

INCLINED-TUBE MANOMETER

If the pressure to be measured is very small, then tilting the arm, *the inclined-tube manometer*, can provide a convenient way of obtaining a larger reading of the manometer as shown in Fig. 2-14. The pressure difference is then given by

$$p_A - p_B = \rho_2 g l_2 \sin\theta + \rho_3 g h_3 - \rho_1 g h_1$$

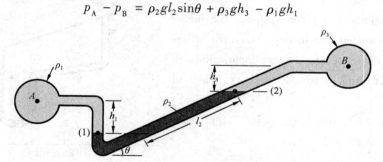

Fig. 2-14 Inclined-tube manometer

The sensitivity to pressure change can be increased further by a greater inclination of the manometer arm, alternatively the density of the manometric fluid may be changed.

Besides the widely used manometers, mechanical and electronic pressure measuring devices are also commonly used in the engineering practices, such as *Bourdon tubes* and *pressure transducers*.

2.4 Pressure Distribution in Rigid-Body Motion

In *rigid-body motion*, all particles are in combination of translation and rotation, and there is no relative motion between particles, which indicates no strains or strain rates between fluid particles. Thus Eq. 2. 5 applies for analyzing the pressure distribution in a fluid with rigid-body motion

$$\mathbf{f} - \frac{1}{\rho}\nabla p = \mathbf{a}$$

When the mass force of fluid in rigid-body motion is under only gravity, we have $f_x = f_y = 0$, $f_z = -g$, the above equation becomes

$$-\nabla p - \rho g \mathbf{k} = \rho \mathbf{a} \qquad (2.22)$$

or in the components:

$$-\frac{\partial p}{\partial x} = \rho a_x \quad -\frac{\partial p}{\partial y} = \rho a_y \quad -\frac{\partial p}{\partial z} = \rho g + \rho a_z \qquad (2.23)$$

2.4.1 Linear Rigid-Body Motion

For the general case of uniform rigid-body acceleration, an open container of liquid moves with a constant acceleration of **a**, as in Fig. 2-15. There is no movement in the x-direction, and thus $a_x = 0$. Eq. 2.23 reduces to

$$\frac{\partial p}{\partial x} = 0 \quad \frac{\partial p}{\partial y} = -\rho a_y \quad \frac{\partial p}{\partial z} = -\rho g - \rho a_z$$

It indicates the pressure is independent of the x-direction, which gives the total derivate of pressure as

$$dp = \frac{\partial p}{\partial y}dy + \frac{\partial p}{\partial z}dz = (-\rho a_y)dy + (-\rho g - \rho a_z)dz \qquad (2.24)$$

For the constant density of fluid, ρ = constant, integral of Eq. 2.24 gives the general form of pressure distribution in the rigid-body liquid

$$p = -\rho a_y y - \rho(g + a_z)z + C \qquad (2.25)$$

in which the integral constant C can be determined by considering the boundary condition at the free surface once the origin of coordinate system is settled.

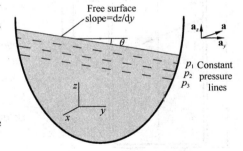

Fig. 2-15 Linear acceleration of a rigid-body liquid with a free surface

Along the line of constant pressure (equipressure surface), $dp = 0$, Eq. 2.24 becomes

$$\frac{dz}{dy} = -\frac{a_y}{g + a_z} \qquad (2.26)$$

It indicates that the constant pressure surface (*equipressure surface*) in the liquid in a tank with linear rigid-body motion is an *inclined plane* if $a_y \neq 0$. Along the free surface the pressure is atmospheric pressure and constant, so that the free surface will be inclined and is an equipressure surface when the fluid retains its volume unless it spills out. In addition, all lines of constant pressure will be parallel to the free surface as illustrated in Fig. 2-15 and are tilted at a downward angle θ given by

$$\theta = \tan^{-1}\frac{a_y}{g + a_z} \qquad (2.27)$$

Now consider the special case of $a_y = 0$ and $a_z \neq 0$, thus

$$\frac{\partial p}{\partial x} = \frac{\partial p}{\partial y} = 0$$

$$\frac{\partial p}{\partial z} = -\rho(g + a_z) \tag{2.28}$$

and

$$p = -\rho(g + a_z)z + C$$

It indicates the pressure distribution in the z-direction is non-hydrostatic, but still vary *linearly* with depth, but variation is the *combination of gravity and externally developed acceleration*. For example, a tank of water in an elevator moving upward will have slightly greater pressure at the bottom; If a tank of water falls freely, $a_z = -g$, thus Eq. 2.28 gives $\partial p/\partial z = 0$, which indicates the pressure gradient between any points is zero. If the container is closed with a pressure of p_0 on the liquid surface, the pressure on the all liquid particles is p_0; if the container is opened to the air, the pressure is zero.

EXAMPLE 2-4 Linear rigid-body motion of coffee in a mug

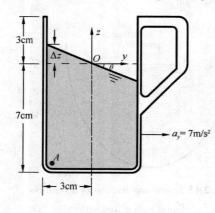

Fig. E2-4

A drag racer rests her coffee mug on a horizontal tray while she accelerates at $7 m/s^2$. The mug is 10cm deep and 6cm in diameter and contains coffee 7cm deep at rest. Assuming rigid body acceleration of the coffee and the symmetric of mug about its central axis, (a) determine whether coffee will spill out of the mug; (b) calculate the gage pressure in the corner at point A shown in Fig. E2-4 if the density of coffee is $1010 kg/m^3$.

Solution:

(a) The free surface tilts at the angle θ given by Eq. 2.27 regardless of the shape of the mug. With $a_z = 0$ and standard gravity,

$$\theta = \tan^{-1}\frac{a_y}{g} = \tan^{-1}\frac{7 m/s^2}{9.81 m/s^2} = 35.4°$$

Since the mug is symmetric about its central axis, the volume of coffee is conserved if the tilted surface intersects the original rest surface exactly at the centerline, as shown in Fig. E2-4. Thus the deflection at the left side of the mug is

$$\Delta z = R \tan\theta = \frac{6cm}{2}\tan 35.4° = 2.14 cm$$

This is less than the clearance available (10cm − 7cm = 3cm), so the coffee will not spill unless it was sloshed during the start-up of acceleration.

(b) Since the mug is symmetric about its central axis, take the center point O at the free surface as the original point of coordinate system. Thus, $y = z = 0$, $p = 0$, we have $C = 0$ in Eq. 2.25. The gage pressure at point A when in a linear acceleration can then be given by Eq. 2.25 with $y = -3cm$, $z = -7cm$ and $a_z = 0$

$$p = -\rho a_y y - \rho(g + a_z)z$$
$$= -(1010 kg/m^3)(7 m/s^2)(-0.03m) - (1010 kg/m^3)(9.81 m/s^2)(-0.07m)$$
$$= 905.7 Pa$$

DISCUSSION When at rest, the gage pressure at point A is given by
$$p = \rho g h = (1010 \text{kg/m}^3)(9.81 \text{m/s}^2)(0.07 \text{m}) = 695 \text{Pa}$$
which is 31 percent lower than the pressure when in a linear motion.

2.4.2 Rigid-Body Rotation

As a second special case, consider rotation of a liquid in a vertical cylindrical container about the axis of cylinder (z-axis) at a constant rotation velocity of ω without any translation, as sketched in Fig. 2-16. For the rigid-body rotation, there is no deformation, and thus no shear stress.

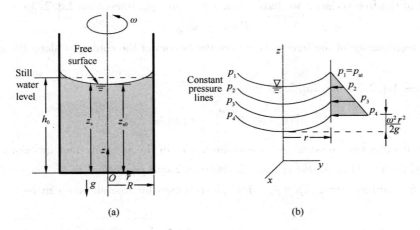

Fig. 2-16 (a) Rigid-body rotation of a liquid in a tank; (b) Parabolic constant-pressure surfaces

For convenience, the acceleration is written in the cylindrical coordinate system, of which the origin is located at the center of the bottom surface of cylinder:
$$a_r = -r\omega^2 \quad a_\theta = 0 \quad a_z = 0$$
Thus, the equation of motion for rotating liquids, Eq. 2.22, reduces to
$$\frac{\partial p}{\partial r} = \rho r \omega^2 \quad \frac{\partial p}{\partial \theta} = 0 \quad \frac{\partial p}{\partial z} = -\rho g \tag{2.29}$$
It indicates the pressure in the liquid in rigid-body rotation is independent of the angle θ and dependent only of r and z. Thus, the total derivative of the pressure is
$$dp = \frac{\partial p}{\partial r} dr + \frac{\partial p}{\partial z} dz = \rho r \omega^2 dr - \rho g dz \tag{2.30}$$
Along the equipressure surface, the pressure at any point in this surface has the same magnitude, $dp = 0$. So that from Eq. 2.30 we have
$$\frac{dz}{dr} = \frac{r\omega^2}{g} \tag{2.31}$$
Taking the center of the bottom surface of the rotating vertical cylinder as the origin ($r = 0$, $z = 0$), integrating Eq. 2.31 gives the equation of equipressure surfaces
$$z(r) = \frac{\omega^2 r^2}{2g} + C \tag{2.32}$$

It indicates that the equipressure surfaces in a rotating rigid-body liquid, including the free surface, are *paraboloid* surfaces, as illustrated in Fig. 2-16b.

To determine the pressure at any point in a rotating rigid-body liquid, integration of Eq. 2.30 yields

$$\int dp = \rho\omega^2 \int r\,dr - \rho g \int dz$$

or
$$p(r) = \frac{\rho\omega^2 r^2}{2} - \rho g z + C \tag{2.33}$$

The integral constant C in Eq. 2.33 can be obtained at specified point. For example, in the center point of the free surface, we have $r=0$, $z=z_{s0}$, $p=p_0$, then from Eq. 2.33

$$C = p_0 + \rho g z_{s0} \tag{2.34}$$

where z_{s0} is the distance of the free surface from the bottom of the container along the axis of rotation.

Substitute Eq. 2.34 into Eq. 2.33 to give

$$p(r) = p_0 + \frac{\rho\omega^2 r^2}{2} + \rho g(z_{s0} - z) \tag{2.35}$$

It indicates that the pressure varies *hydrostatically* in the vertical, and increases radially as shown in Fig. 2-16b. That is, *the pressure is linear in z and paraboloid in r*.

At the free surface, $z = z_s$, $p = p_0$, thus the equation for free surface can be obtained from Eq. 2.35

$$z_s(r) = z_{s0} + \frac{\omega^2 r^2}{2g} \tag{2.36}$$

where z_s is the distance of the free surface from the bottom of the container at radius r.

Substituting Eq. 2.36 into Eq. 2.35 yields

$$p = p_0 + \rho g(z_s - z_{s0}) + \rho g(z_{s0} - z) = p_0 + \rho g(z_s - z) = p_0 + \rho g h \tag{2.37}$$

where $h = z_s - z$.

Again, Eq. 2.37 confirms that the pressure in a rotating rigid-body liquid varies hydrostatically in the vertical.

Since mass is conserved and density is constant, the volume in rotation must be equal to the original volume of the liquid in the container, which is

$$\pi R^2 h_0 = \int_{r=0}^{R} 2\pi r\, z_s\, dr = \int_{r=0}^{R} 2\pi r\left(z_{s0} + \frac{\omega^2 r^2}{2g}\right)dr = \pi R^2\left(z_{s0} + \frac{\omega^2 R^2}{4g}\right)$$

Thus
$$z_{s0} = h_0 - \frac{\omega^2 R^2}{4g} \tag{2.38}$$

where h_0 is the original depth of the liquid at rest in the container with no rotation.

Substituting into Eq. 2.36 gives the general form of the equation of free surface of liquid in rigid-body rotation

$$z_s(r) = h_0 - \frac{\omega^2}{4g}(R^2 - 2r^2) \tag{2.39}$$

Here, as in the linear-acceleration case, it should be emphasized that the *paraboloid pressure distribution* (Eq. 2.35) sets up in *any* liquid under rigid-body rotation, regardless of the shape or size of the container. The container may even be closed and filled with liquid. It is only necessary that the liquid be continuously interconnected throughout the container.

EXAMPLE 2-5 Rising of a liquid during rotation

A cylindrical container with 20-cm-diameter and 60-cm-high shown in Fig. 2-16a is partially filled with 50-cm-high liquid with a density of $850 kg/m^3$. Now the cylinder is rotated at a constant speed. Determine the rotational speed at which the liquid will start spilling from the edges of the container.

Solution:

Assume the bottom surface of the container remains covered with liquid during rotation (no dry spots).

Just before the liquid starts spilling, the height of the liquid at the edge of the container equals the height of the container, that is $r = R = 0.1m$, $z_s = 0.6m$. From Eq. 2.39 with $h_0 = 0.5m$ we have the rotational speed

$$\omega = \sqrt{\frac{4g(h_0 - z_s)}{R^2 - 2R^2}} = \sqrt{\frac{4(9.81 m/s^2)[(0.5m) - (0.6m)]}{-(0.1m)^2}} = 19.8 rad/s = 189 rpm$$

Therefore, the rotational speed of this container should be limited to 189 rpm to avoid any spill of liquid as a result of the centrifugal effect.

DISCUSSION Note that the analysis is valid for any liquid since the result is independent of density or any other fluid property. We should also verify that our assumption of no dry spots is valid. The liquid height at the center is

$$z_s(0) = h_0 - \frac{\omega^2 R^2}{4g} = 0.4m$$

Since $z_s(0)$ is positive, the assumption is validated.

2.5 Hydrostatic Force on a Submerged Plane Surface

In engineering practice, the forces developing on the submerged surface, such as storage tanks, ships, dams, and other hydraulic structures, exerted by the fluid, is an important problem in the design of structures which interact with fluids. The magnitude, direction and the acting line of the resultant forces acting on these surfaces can be determined by the theory of fluid static discussed above. We consider the forces acting on the plane surface of submerged body in this section and on the curved surfaces in next section.

2.5.1 Force on a Submerged Plane with General Shape

A plate of arbitrary shape making an arbitrary angle θ with the horizontal free surface is completely submerged in a liquid at rest as shown in Fig. 2-17. A xOy coordinate system in the plane of the plate with the origin at the free surface is set. Its axis Ox is set as the interactive line of the free surface and the extensive surface of the plane into the paper, and the axis Oy is perpendicular to the axis Ox and down from the free surface in the plane of the plate.

From Eqs. 2.11 and 2.16 we know that the pressure on any submerged surface must be per-

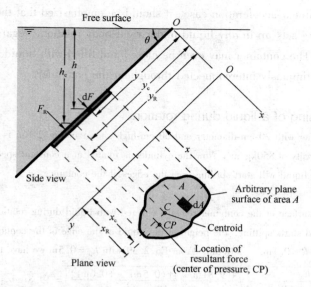

Fig. 2-17 Schematic of theoretical method of hydrostatic force on an inclined plane surface of arbitrary shape

pendicular to the surface since there are no shear stresses presented, and varies linearly with the depth if the fluid is incompressible. Thus, for an differential area dA of the plate with a depth of h, from Eq. 2.16 the gage pressure there is $p = \rho g h$, and then the force on this differential area is

$$dF = \rho g h dA = \rho g y \sin\theta dA \qquad (2.40)$$

where y is the distance of the differential area dA from the axis Ox.

Integrating Eq. 2.40 over the entire plate area A gives the total hydrostatic force on one side of the plate

$$F = \int_A dF = \rho g \sin\theta \int_A y dA \qquad (2.41)$$

The integral term, $\int_A y dA$, in Eq. 2.41 is the *first moment of area A* about axis Ox, which is related to the y-cooordinate of the centroid of the plate by $\int_A y dA = A y_c$. Substituting into Eq. 2.41 gives the resultant force acting on the plate

$$F = \rho g \sin\theta\, y_c A = \rho g h_c A = p_c A \qquad (2.42)$$

where y_c is the perpendicular distance of the centroid C of plate from the axis Ox and depends only on the shape of the area and is the same for all values of θ, even zero; h_c is the vertical distance of the centroid C of plate from the liquid surface, $h_c = y_c \sin\theta$ (Fig. 2-17); p_c is the pressure at the centroid of the plane surface.

Equation 2.42 presents that the magnitude of the resultant hydrostatic force on a submerged plate in a homogenous fluid at rest equals the product of the pressure at the plate centroid and the plate area, and is independent of the shape of plate and the angle θ the plate slants.

The point on the plate passed through by the acting line of the resultant force F is called the *center of pressure* (*CP*) of the plate. It usually does not act through the centroid but below it toward the high-pressure side. To find the distance to the CP, y_R, take moments of the elemental force pdA about the axis Ox and equate to the moment of the resultant F to give

$$Fy_R = \int_A y dF = \rho g \sin\theta \int_A y^2 dA$$

The integral term $\int_A y^2 dA$ defines the *second moment of area*, also called the *area moment of inertia*, of the plate area A about axis Ox, given the symbol I_x, so

$$Fy_R = \rho g \sin\theta\, I_x$$

Substituting Eq. 2.42 for F gives the results

$$y_R = \frac{I_x}{y_c A} \qquad (2.43)$$

The second moment of area is widely available for common shapes in engineering handbooks, but they are usually given about the axes passing through the centroid of the area by the *parallel axis theorem*, which in this case is expressed as

$$I_x = I_{xc} + y_c^2 A$$

where I_{xc} is the second moment of area with respect to an axis passing through its centroid and parallel to the Ox axis. Thus, Eq. 2.43 becomes

$$y_R = y_c + \frac{I_{xc}}{y_c A} \qquad (2.44)$$

For a given shape of area A, $I_{xc}/y_c A > 0$, from Eq. 2.44 gives $y_R > y_c$, which clearly indicates that the center of pressure, CP, is always below the centroid of the plate, C, and the resultant force does not pass through the centroid. For a submerged horizontal plate, $\theta = 0°$, $I_{xc} = 0$, thus $y_R = y_c$, the center of pressure CP lies at the centroid C. If we move the plate deeper, that is, with an increasing y_c, the center of pressure approaches to the centroid of the area because of constant I_{xc} and A.

The x coordinate, x_R for the resultant force can be determined exactly similarly to that of y_R by summing moments about the Oy axis. Thus

$$Fx_R = \rho g \sin\theta \int_A xy dA$$

and, therefore,

$$x_R = \frac{\int_A xy dA}{y_c A} = \frac{I_{xy}}{y_c A}$$

where I_{xy} is the *product of inertia* with respect to the x and y axes.

Again, using the parallel axis theorem, we have

$$x_R = x_c + \frac{I_{xyc}}{y_c A} \qquad (2.45)$$

where x_c is the distance of the centroid of the plate from the axis Ox, I_{xyc} is the product of inertia with respect to an orthogonal coordinate system passing through the centroid of the area and formed by a translation of the x-y coordinate system.

For the engineering practices, most of the submerged area usually has symmetrical axis passing through the centroid and parallel to the Oy axes. Thus, center of pressure must lie along the symmetrical axis $x = x_c$, since I_{xyc} is identically zero in this case. To determine the location of the center of pressure in this case, only y_R has to be determined.

Centroidal coordinates and moments of inertia for some common areas are given in Fig. 2-18.

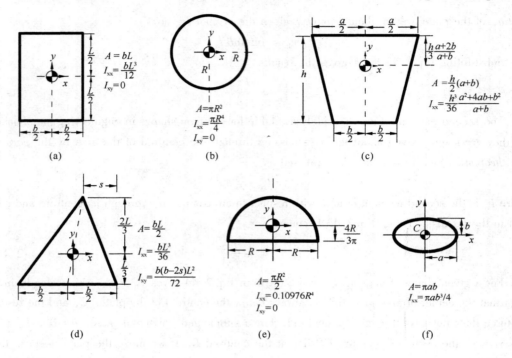

Fig. 2-18 Centroidal moments of inertia for various cross sections: (a) rectangle, (b) circle, (c) trapezoidal, (d) triangle, (e) semicircle and (f) ellipse

2.5.2 Graphical Approach

PRESSURE PRISM

Another approach to determine the resultant force and the center of pressure the force on a regular plane surface submerged in a liquid with a free surface is given by the graphical representation of gage pressure: the *pressure prism*.

Consider the pressure distribution along a vertical wall AB of a tank of width b, as in Fig. 2-19a, which contains a liquid having a specific weight ρg. Since the pressure must vary linearly with depth, the pressure equals zero at point A located at free surface and equals $\rho g h$ at the bottom point B. Since the pressure distribution applies across the vertical surface, we can draw the three-dimensional representation of the pressure distribution as shown in Fig. 2.19b. The base of this 'volume' in pressure-area space is the plane surface of interest, and its altitude at each point is the pressure p. Such a volume is called the *pressure prism*. Other examples are shown in Fig.

2-19c, d and e.

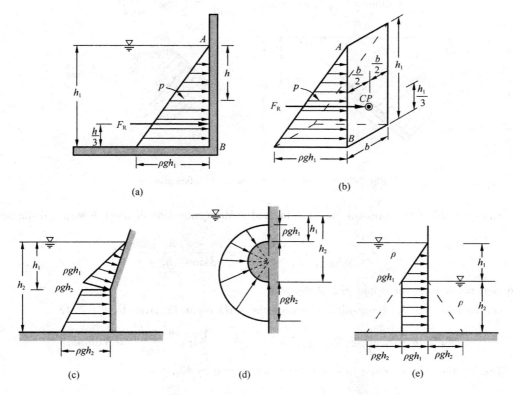

Fig. 2-19 Examples of pressure prisms

GRAPHICAL APPROACH

This virtual pressure prism has an interesting physical interpretation: the volume of the pressure prism equals the magnitude of the resultant hydrostatic force perpendicularly acting on the plate since

$$\volume = \int p\,dA = \int \rho g h\,dA$$

in which h is the vertical distance at any point to the free surface (see Fig. 2-19b), and the line of action of this force passes through the centroid of this homogeneous prism. The projection of the centroid on the plate is the *center of pressure, CP*. Therefore, with the concept of pressure prism, the determination of the resultant hydrostatic force on a plane surface reduces to finding the volume and the two coordinates of the centroid of this pressure prism.

To verify the above conclusion, consider an inclined surface with a width of b submerged in a liquid of density ρ to a depth of h_1 and h_2 as shown in Fig. 2-20. To the left of the surface can be seen a graphical representation of the (gage) pressure change with depth on the surface. Pressure increases from $p = \rho g h_1$ at the upper end of the surface linearly by $p = \rho g h$, to a maximum at the bottom end of $p = \rho g h_2$. The pressure prism is shown in Fig. 2-20b, of which the volume is

51

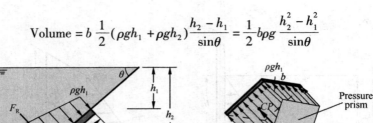

$$\text{Volume} = b \frac{1}{2}(\rho g h_1 + \rho g h_2)\frac{h_2 - h_1}{\sin\theta} = \frac{1}{2}b\rho g \frac{h_2^2 - h_1^2}{\sin\theta}$$

Fig. 2-20 Pressure prism for inclined rectangular area

The centroid of this pressure prism is located at the center line of width b with a distance of

$$l_c = \frac{h_2 - h_1}{3\sin\theta}\frac{\rho g h_1 + 2\rho g h_2}{\rho g h_1 + \rho g h_2} = \frac{h_2 - h_1}{3\sin\theta}\frac{h_1 + 2h_2}{h_1 + h_2}$$

away and below from the upper end of the plate.

On the other hand, the resultant force on the surface can be given by Eq. 2.42

$$F_R = p_c A = \frac{1}{2}\rho g(h_2 + h_1) \cdot b\frac{h_2 - h_1}{\sin\theta} = \frac{1}{2}b\rho g\frac{h_2^2 - h_1^2}{\sin\theta} = \text{Volume}$$

The location of the acting line of force can be given by Eq. 2.44

$$l'_c = y_R - \frac{h_1}{\sin\theta} = y_c + \frac{I_{xc}}{y_c A} - \frac{h_1}{\sin\theta}$$

$$= \frac{1}{2}\frac{h_1 + h_2}{\sin\theta} + \frac{\frac{1}{12}b\left(\frac{h_2 - h_1}{\sin\theta}\right)^3}{\frac{1}{2}\frac{h_1 + h_2}{\sin\theta}b\frac{h_2 - h_1}{\sin\theta}} - \frac{h_1}{\sin\theta} = \frac{h_2 - h_1}{3\sin\theta}\frac{h_1 + 2h_2}{h_1 + h_2} = l_c$$

The above shows that the results of the analytical method are identical to that of the pressure prism method. That is, the magnitude of the resultant force acting on the surface is equal to the volume of the pressure prism and the line of force passes through its centroid.

EXAMPLE 2-6 Hydrostatic force acting on an inclined rectangular plate

An inclined rectangular plate is submerged in water with an angle 30° with the horizontal free surface and to a depth of $h = 1$m at the upper end and $H = 3$m at the lower end as shown in Fig. E2-6a. The width of plate is $b = 5$m. Determine the hydrostatic force on the plate and the location of the pressure center.

Solution:

Take the density of water to be 1000kg/m³ and the gravitational acceleration to be 9.81m/s².

(a) Analytical method

From Fig. E2-6a we have the length of plate, the location of centroid and the area of plate

$$l = \frac{H - h}{\sin 30°} = \frac{3\text{m} - 1\text{m}}{\sin 30°} = 4\text{m}$$

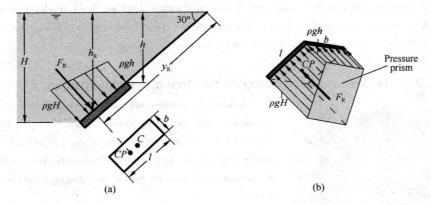

Fig. E2-6

$$h_c = \frac{H+h}{2} = \frac{3m+1m}{2} = 2m$$

$$y_c = \frac{h_c}{\sin 30°} = \frac{2m}{\sin 30°} = 4m$$

$$A = lb = (4m)(5m) = 20\ m^2$$

We then have the resultant force and the location of the pressure center from Eqs. 2.42 and 2.44

$$F = \rho g h_c A = (1000 kg/m^3)(9.81 m/s^2)(2m)(20m^2) = 392.4 kN$$

$$y_R = y_c + \frac{I_{xc}}{y_c A} = y_c + \frac{\frac{1}{12}bl^3}{y_c A} = 4m + \frac{\frac{1}{12}(5m)(4m)^3}{(4m)(20\ m^2)} = 4.333m$$

$$h_R = y_R \sin 30° = (4.333m)(\sin 30°) = 2.166m$$

Thus, the magnitude of the resultant hydrostatic force on the plate is 392.4kN and the location of the pressure center is located to a depth of 2.166m from the free surface.

(b) Graphical method

The pressure prism is drawn as in Fig. E2-6b, its volume is

$$F_R = \frac{1}{2}(\rho g h + \rho g H)\frac{H-h}{\sin\theta}b = \frac{1}{2}\rho g(h+H)lb$$

$$= \frac{1}{2}(1000 kg/m^3)(9.81 m/s^2)(1m+3m)(4m)(5m) = 392.4 kN$$

From Fig. 2-18c we have

$$y_R = \frac{h}{\sin 30°} + \frac{H-h}{3\sin 30°}\frac{\rho g h + 2\rho g H}{\rho g h + \rho g H} = \frac{h}{\sin 30°} + \frac{H-h}{3\sin 30°}\frac{h+2H}{h+H}$$

$$= \frac{1m}{\sin 30°} + \frac{3m-1m}{3\sin 30°}\frac{1m+2(3m)}{1m+3m} = 4.333m$$

or

$$h_R = y_R \sin 30° = 2.166m$$

DISCUSSION The use of pressure prisms for determining the force on submerged plane areas is convenient if the area is regular, i.e. trapezoidal and rectangular, so the volume and centroid can be easily determined. However, for other irrectangular shapes, integration would generally be needed to determine the volume and centroid. In these circumstances it is more convenient to use the equations 2.42 and 2.44 developed in the last section, in which the necessary integrations have been made and the results presented in a convenient and compact form that is applicable to submerged plane areas of any shape.

EXAMPLE 2-7 Hydrostatic force acting on the door of a submerged car

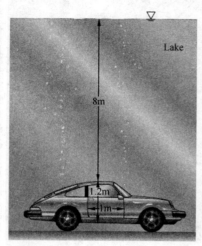

Fig. E2-7

A heavy car plunges into a lake during an accident and lands at the bottom of the lake on its wheels (Fig. E2-7). The door is 1.2m high and 1m wide, and the top edge of the door is 8m below the free surface of the water. Determine the hydrostatic force on the door and the location of the pressure center, and discuss if the driver can open the door.

Solution:

Assume: The bottom surface of the lake is horizontal; The passenger cabin is well-sealed so that no water leaks inside; The door can be approximated as a vertical rectangular plate; The pressure in the passenger cabin remains at atmospheric value since there is no water leaking in, and thus no compression of the air inside. Therefore, atmospheric pressure cancels out in the calculations since it acts on both sides of the door. Finally, the weight of the car is larger than the buoyant force acting on it.

The centroid (midpoint) of the door is

$$h_c = y_c = 8\text{m} + \frac{1.2\text{m}}{2} = 8.6\text{m}$$

Take the density of lake water to be 1000kg/m^3 throughout. Thus the pressure at the centroid (midpoint) of the door is determined to be

$$p_c = \rho g h_c = (1000\text{kg/m}^3)(9.81\text{m/s}^2)(8.6\text{m}) = 84.4\text{kPa}$$

Then the resultant hydrostatic force on the door becomes

$$F_R = p_c A = (84.4\text{kPa})(1.2\text{m})(1\text{m}) = 101.28\text{kN}$$

The pressure center is directly under the midpoint of the door, and its distance from the surface of the lake is determined from Eq. 2.44

$$y_R = y_c + \frac{I_{xc}}{y_c A} = y_c + \frac{\frac{1}{12}bl^3}{y_c bl} = 8.6\text{m} + \frac{\frac{1}{12}(1.2\text{m})^2}{8.6\text{m}} = 8.614\text{m}$$

DISCUSSION   A strong person can lift 100kg, whose weight is 981N or about 1kN. Also, the person can apply the force at a point farthest from the hinges (1m farther) for maximum effect and generate a moment of 1kN·m. The resultant hydrostatic force acts under the midpoint of the door, and thus a distance of 0.5m from the hinges. This generates a moment of $(101.28\text{kN})(0.5\text{m}) \approx 50.6\text{kN·m}$, which is about 50 times the moment the driver can possibly generate. Therefore, it is impossible for the driver to open the door of the car. The driver's best bet is to let some water in (by rolling the window down a little, for example) and to keep his or her head close to the ceiling. The driver should be able to open the door shortly before the car is filled with water since at that point the pressures on both sides of the door are nearly the same and opening the door in water is almost as easy as opening it in air.

2.6 Hydrostatic Force on a Curved Surface

2.6.1 Forces on the Curved Surface

For a submerged curved surface, the determination of the resultant hydrostatic force is more complicated since it typically requires the integration of the pressure forces that change direction along the curved surface. The concept of the pressure prism in this case is not much help either because of the complicated shapes involved. However, when the resultant force on a curved surface is separated into horizontal and vertical components, it can be computed very easily.

Consider a two-dimensional surface $\widehat{ab}$ with area of A sketched in Fig. 2-21a. The incremental pressure forces, being normal to the local area element, vary in direction along the surface and thus cannot be added numerically. Thus, we separate resultant force of these elemental pressure forces into two directions, F_H and F_V, as shown in Fig. 2-21a. Consider the free-body diagram of a fluid column abe contained in the vertical and horizontal projection planes passing through the two ends of the curved surface, we can see that the magnitude of F_H and F_H', F_V and F_V' are equal but act in opposite directions (*Newton's third law*) as shown in Fig. 2-21a and b, where F_H' and F_V' are the forces exerted by the surface $\widehat{ab}$ on the fluid column abe.

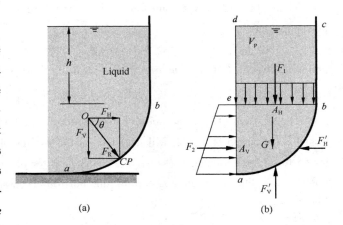

Fig. 2-21 Computation of hydrostatic force on a curved surface: (a) submerged curved surface; (b) the free-body diagram of fluid enclosed by the curved surface and the pressure force prism

The column of fluid in Fig. 2-21b is in static equilibrium. In the horizontal direction, the force F_H' due to the curved surface $\widehat{ab}$ is exactly equal to the force F_2 on the vertical projection ae of the curved surface, which can be computed by the plane surface formula, Eq. 2.42, based on a vertical projection of the curved surface:

$$F_H = F_H' = p_c A_V \qquad (2.46)$$

where p_c is the pressure on the centroid of the vertical projection plane of the curved surface, A_V is the area of the vertical projection plane of the curved surface.

Equation 2.46 presents that the horizontal component of force on a curved surface equals the hydrostatic force acting on the vertical projection of the curved surface.

In the vertical direction, the balance of vertical force on the free body of fluid gives

$$F_V = F_V' = F_1 + G = \rho g h A_H + \rho g V_{abe} = \rho g (V_{bcde} + V_{abe}) = \rho g V_P \qquad (2.47)$$

where V_p is the volume of the *pressure force prism*, a space volume enclosed by the surface of interest, the vertical surfaces originating from the edges of the curved surface to the free surface or the extension of free surface, and the free surface or the extension free surface. For example, the pressure force prism $abcdea$ in Fig. 2-21b is enclosed by the curved surface of interest $\overset{\frown}{ab}$, the vertical surfaces ad and bc and the free surface dc.

Equation 2.47 indicates that the vertical component of the resultant force on a curved surface is equal to the hydrostatic force acting on the horizontal projection of the curved surface, plus (or minus if acting in the opposite direction) the weight of the fluid block enclosed by the curved surface, or equals in magnitude and direction the weight of the entire fluid column of the pressure force prism above the curved surface.

Then, the overall resultant force can be found directly by combining the vertical and horizontal components

$$F_R = \sqrt{F_H^2 + F_V^2} \tag{2.48}$$

and acts through O at an angle of θ making to the horizontal

$$\theta = \tan^{-1}\left(\frac{F_V}{F_H}\right) \tag{2.49}$$

The position of O is the point of intersection of the horizontal line of action of F_H and the vertical line of action of F_V. It will not meet on the surface of the curved surface, and cannot be referred to as the center of pressure of F_R. The center of pressure of F_R would be the intersection point of the line with an angle θ to the horizontal passing through the point O as shown in Fig. 2-21a.

2.6.2 Pressure Force Prism

Curved surfaces can be concave or convex with the fluid above or below it, corresponcling to which the pressure force prisms are named *real* or *virtual pressure force prism* respectively.

The volume bounded by the curved surface of interest, the vertical surfaces originating from the edges of the curved surface and reaching the free surface, and the free surface as shown in Fig. 2-21b and Fig. 2-22a, is termed the *real pressure force prism*, because the vertical force acts

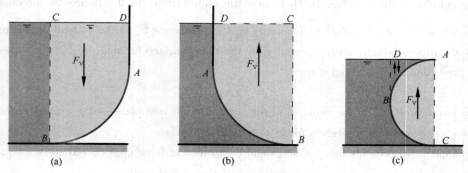

Fig. 2-22 (a) Real pressure force prism; (b) virtual pressure force prism and (c) superposition of real and virtual pressure force prisms

downward and the weight of fluid really acting on the curved surface. On the other hand, for the volume bounded by the curved surface of interest, the vertical surfaces and the extension surface of free surface as shown in Fig. 2-22b, is termed the *virtual pressure force prism*, because the vertical force acts upward and the weight of fluid actually does not directly act on the curved surface.

The real and virtual pressure force prism can be superposed and cancelled if they occupy the same space volume, because they have the same magnitude of vertical forces but act in opposite directions. For example, the vertical force on the curved surface ABC in Fig. 2-22c is the weight of fluid in semicircle, the force in the space volume ABD has been cancelled out.

2.6.3 Hydrostatic Force in Layered Fluids

Finally, hydrostatic forces acting on a plane or curved surface submerged in a multilayered fluid of different densities can be determined by considering different parts of surfaces in different fluids as different surfaces, finding the force on each part, and then adding them using vector addition. For an example, for a plane surface shown in Fig. 2-23, it can be expressed as

$$F_R = \Sigma F_{Ri} = \Sigma p_{ci} A_i \quad (2.50)$$

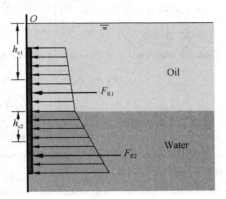

Fig. 2-23 The hydrostatic force on a surface submerged in multilayered fluids

where $p_{ci} = p_0 + \rho g h_{ci}$ is the pressure at the centroid of the portion of the surface in fluid i and A_i is the area of the plate in that fluid, p_0 is the pressure at the upper surface of fluid i; h_{ci} is the depth of centroid of the plate in the fluid i below the upper surface of fluid i.

The line of action of this equivalent force can be determined from the requirement that the moment of the equivalent force about any point is equal to the sum of the moments of the individual forces about the same point, such as point O.

EXAMPLE 2-8 A gravity-controlled cylindrical gate

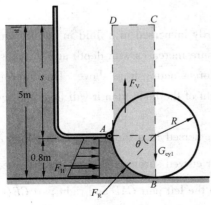

Fig. E2-8

A long solid cylinder of radius 0.8m hinged at point A is used as an automatic gate, as shown in Fig. E2-8. When the water level reaches 5m, the gate opens by turning about the hinge at point A. Determine: (a) the hydrostatic force acting on the cylinder and its line of action when the gate opens; (b) the weight of the cylinder per meter length of the cylinder.

Solution:

Assume friction at the hinge is negligible, and the atmospheric pressure acts on both sides of the gate, and thus it cancels out. Take the density of water to be 1000kg/m³ throughout.

(a) Consider the vertical projection of the circular surface AB of the cylinder contacting with water, we then have horizontal force component on the curved surface per m length equaling to the force on its vertical projection plane surface by Eq. 2.42

$$F_H = \rho g \left(s + \frac{R}{2}\right) A = (1000 \text{kg/m}^3)(9.81 \text{m/s}^2)\left(4.2\text{m} + \frac{0.8\text{m}}{2}\right)(0.8\text{m})(1\text{m})$$
$$= 36.10 \text{kN}$$

Plot the pressure force prism of the circular surface AB of the cylinder contacting with water, $ABCD$ as shown in Fig. E2-8, we then have the vertical force component on the curved surface per m length from Eq. 2.47:

$$F_V = \rho g V_{ABCD} = \rho g \left(\frac{\pi R^2}{4} + Rs\right) l$$
$$= (1000 \text{kg/m}^3)(9.81 \text{m/s}^2)\left[\frac{(3.14)(0.8\text{m})^2}{4} + (0.8\text{m})(4.2\text{m})\right](1\text{m})$$
$$= 37.89 \text{ kN}$$

Then the magnitude and direction of the resultant hydrostatic force acting on the cylindrical surface become

$$F_R = \sqrt{F_H^2 + F_V^2} = \sqrt{(36.10 \text{kN})^2 + (37.89 \text{kN})^2} = 52.33 \text{kN}$$

$$\theta = \tan^{-1}\left(\frac{F_V}{F_H}\right) = \tan^{-1}\left(\frac{37.89 \text{kN}}{36.10 \text{kN}}\right) = 46.4°$$

Therefore, the magnitude of the hydrostatic force acting on the cylinder is 52.33 kN per m length of the cylinder, and its line of action passes through the center of the cylinder making an angle 46.4° with the horizontal.

(b) When the water level is 5m high, the gate is about to open and thus the reaction force at the bottom of the cylinder is zero. Then the forces other than those at the hinge acting on the cylinder are its weight, acting through the center, and the hydrostatic force exerted by water. Taking a moment about point A at the location of the hinge and equating it to zero gives

$$F_R R \sin\theta - G_{cyl} R = 0$$

which yields

$$G_{cyl} = F_R \sin\theta = (52.33 \text{kN})(\sin 46.4°) = 37.9 \text{kN}$$

DISCUSSION The weight of the cylinder per m length is determined to be 37.9 kN. It can be shown that this corresponds to a mass of 3863 kg per m length and to a density of 1921 kg/m³ for the material of the cylinder.

2.7 Buoyancy, Flotation, and Stability

2.7.1 Buoyance

It is a common experience that any body wholly or partly immersed in a fluid in equilibrium subjects to a net upward vertical force results because pressure increases with depth and the pressure forces acting from below are larger than the pressure forces acting from above. This force is known as the *buoyant force* and if this force equals the weight of the body then it will float and the body is said to be *buoyant*.

Fig. 2-24 shows an arbitrary body $ABCD$ completely immersed in a fluid, in which points A and B separate the curved surface of the body into the upper curved surface $\widehat{ADB}$ and lower curved surface $\widehat{ACB}$, points C and D separate the body surface into the left part $\widehat{CAD}$ and right part $\widehat{CBD}$.

For the horizontal forces, F_1 and F_2, their magnitudes are the resultant hydrostatic forces on the vertical projection plane $C'D'$ on the left and $C''D''$ on the right sides of body, which are all equal and cancelled out since the depth of centroids and the areas of projection plane $C'D'$ and $C''D''$ are same and the force directions are in opposite, so the equilibrium equation of interest about the immersed body is in the z direction.

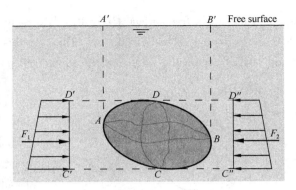

Fig. 2-24 The buoyant forces on an arbitrary immersed body

For the vertical forces, the upward thrust on the lower surface $\overparen{ACB}$ of the body corresponds to the weight of fluid, *virtual*, vertically above that surface, i. e., to the fluid volume $ACBB'A'A$ of the virtual pressure force prism. The downward thrust on surface $\overparen{ADB}$ equals the weight of fluid volume $ADBB'A'A$ of real pressure force prism. Thus, the net upward vertical force on the body is

$$F_B = \rho g \, V_{ACBB'A'A} - \rho g \, V_{ADBB'A'A} = \rho g \, V_{ACBD} = \rho g \, V_{\text{displaced volume}} (\uparrow) \qquad (2.51)$$

where $V_{\text{displaced volume}}$ is the fluid volume displaced by the immersed body or partly immersed body (floating body); F_B is the net resultant force upward acting on the immersed body and is called the *buoyant force*.

Equation 2.51 shows that a body immersed in a fluid experiences a vertical buoyant force equal to the weight of the fluid it displaces, and the line of action of the buoyant force upwardly passes through the centroid of the fluid volume displaced by the body, the *center of buoyancy*. This is known as *Archimedes' principle*, after the Greek mathematician Archimedes (287-212B. C.).

Since the fluid is in equilibrium, the buoyancy force acting on the object must therefore be equal and opposite to the weight of the fluid displaced by the body and must also pass through the *center of gravity* of that fluid. This point, corresponding to the *center of buoyancy*, does not, in general, correspond to the center of gravity of the body, because the position of the center of buoyancy depends on the shape of the volume involved.

Similar considerations apply to a partly immersed body-floating body. *Floating bodies* are a special case; only a portion of the body is submerged, with the remainder poking up out of the free surface. As illustrated in Fig. 2-25, the enclosed body by the curved surface

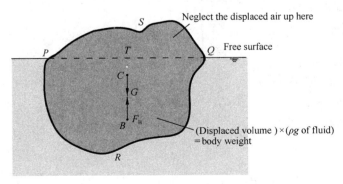

Fig. 2-25 Partially immersed body-Floating body

59

$\widehat{PRQ}$ and the origin free surface PTQ, $PRQTP$, is the displaced volume, while the enclosed body $PSQT$ is the volume of body in the air. Thus, in this case the buoyancy force corresponds to the weight of fluid within the volume of the submerged portion of the floating body, $PRQTP$.

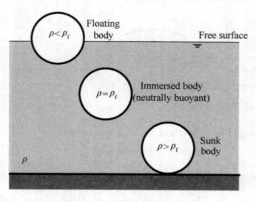

Fig. 2-26 Illustration of sunk, immersed and floating body

For a body in a fluid to be in equilibrium, the buoyancy force must equal the weight of the body. If the buoyancy force exceeds the weight of the body then the body rises (bubbles for example) to the free surface until the weight of the displaced fluid volume equals to the weight of the body, this body is called the *floating body* as in Fig. 2-26, where ρ_f is the density of the body. If the weight exceeds the buoyancy force then the body sinks (a stone in water for example) to the bed, this body is called the *sunk body*. If the weight equals the buoyancy force then the body can be floating at the any place in the fluid, this body is called the *immersed body*, or *submerged body*.

For summary, the buoyant force acting on a submerged, or partially submerged object has a magnitude equal to the weight of fluid displaced by the object and a direction directly opposite the direction of gravity giving rise to body forces.

EXAMPLE 2-9 Force on a cubical object

Consider a cubical object with sides of length h that is floating in water in such a way that $h/4$ of its vertical side is above the surface of the water, as indicated in Fig. E2-9. Find the density of this cube.

Solution:

From Archimedes' principle, Eq. 2.51, it follows directly that the buoyancy force must be

$$F_B = \rho_w g \frac{3}{4} h^3$$

and it must be pointing upward as indicated in the figure. The weight of cube is given by

$$G = \rho_{cube} g h^3$$

and it must be pointing downward as indicated in the figure.

Now in static equilibrium (the cube floating as indicated), the forces acting on the cube must sum to zero

$$F_B - G = \rho_w g \frac{3}{4} h^3 - \rho_{cube} g h^3 = 0$$

to get the density of the cube

$$\rho_{cube} = \frac{3}{4} \rho_w$$

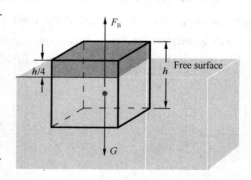

Fig. E2-9

EXAMPLE 2-10 Measuring specific gravity by a hydrometer

If you have a seawater aquarium, you have probably used a small cylindrical glass tube with some lead-weight at its bottom to measure the salinity of the water by simply watching how deep the tube sinks. Such a device that floats in a vertical position and is used to measure the specific gravity of a liquid is called a *hydrometer* (Fig. E2-10). The top part of the hydrometer extends above the liquid surface, and the divisions on it allow one to read the specific gravity directly. The hydrometer is calibrated such that in pure water it reads exactly 1.0 at the air-water interface. (a) Obtain a relation for the specific gravity of a liquid as a function of distance Δz from the mark corresponding to pure water; (b) When a hydrometer having an average stem diameter of 0.75 cm is placed in water, the stem protrudes 7.87 cm above the water surface. If the water is replaced with a liquid having a specific gravity of 1.10, how much of the stem would protrude above the liquid surface if the hydrometer totally weighs 0.19 N. Take the density of water to be 1000 kg/m³.

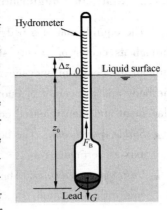

Fig. E2-10

average

Solution:

(a) Noting that the hydrometer is in static equilibrium, the buoyant force F_B exerted by the liquid must always be equal to the weight G of the hydrometer.

In pure water, let the vertical distance between the bottom of the hydrometer and the free surface of water be z_0. Setting $F_B = G$ in this case gives

$$G = F_{B,w} = \rho_w g \mathcal{V} = \rho_w g A z_0 \qquad (1)$$

where A is the average cross-sectional area of the tube, and ρ_w is the density of pure water.

In a fluid lighter than water ($\rho_f < \rho_w$), the hydrometer will sink deeper, and the liquid level will be a distance of Δz above z_0. Again setting $F_B = G$ gives

$$G = F_{B,f} = \rho_f g \mathcal{V} = \rho_f g A (z_0 + \Delta z) \qquad (2)$$

Setting Eqs. (1) and (2) here equal to each other since the weight of the hydrometer is constant and rearranging gives

$$\rho_w g A z_0 = \rho_f g A (z_0 + \Delta z)$$

or

$$S_G = \frac{\rho_f}{\rho_w} = \frac{z_0}{z_0 + \Delta z} \qquad (3)$$

which is the relation between the specific gravity of the fluid and Δz.

(b) The liquid is heavier than water because its specific gravity is 1.10, then Δz would be negative. From Eqs. (1) and (2) we have

$$\Delta z = (z_0 + \Delta z) - z_0 = \frac{G}{\rho_f g A} - \frac{G}{\rho_w g A}$$

$$= \frac{0.19\,\text{N}}{(9.81\,\text{m/s}^2)\frac{1}{4}(3.14)(0.0075\,\text{m})^2} \left[\frac{1}{(1.1)(1000\,\text{kg/m}^3)} - \frac{1}{1000\,\text{kg/m}^3} \right]$$

$$= -3.99\,\text{cm}$$

Thus, with the new liquid the stem will protrude $7.87\,\text{cm} + 3.99\,\text{cm} = 11.86\,\text{cm}$ above the liquid surface.

DISCUSSION The relation of Eq. (3) is valid for fluids both lighter or heavier than water, in which for a given hydrometer, Δz is negative for heavier fluids and positive for lighter fluids.

2.7.2 States of Equilibrium

The equilibrium of a body submerged in a liquid requires that the weight of the body acting through its center of gravity should be colinear with an equal hydrostatic upward force acting through the center of buoyancy. Depending upon the relative locations of the center of gravity C of the body and the center of buoyancy B, a floating or submerged body given a small angular displacement and then released could attain three different states of equilibrium:

- *Stable equilibrium*: The body returns to its original position by retaining the originally vertical axis as vertical.
- *Unstable equilibrium*: The body does not return to its original position but shifts to a new equilibrium position.
- *Neutral equilibrium*: The body neither returns to its original position nor increases its displacement further, but simply adopt its new position.

2.7.3 Stability of Immersed Bodies

Stability considerations are particularly important for submerged or floating bodies since the centers of buoyancy and gravity do not necessarily coincide. A small rotation can result in either a restoring or overturning couple. For example, for the completely submerged body shown in Fig. 2-27a, which has a center of gravity, C, below the center of buoyancy, B, a rotation from its equilibrium position will create a restoring couple formed by the weight, G, and the buoyant force, F_B, which causes the body to rotate back to its original position. Thus, for this configuration the body is *stable*. It is to be noted that as long as the center of gravity falls *below* the center of buoyancy, this will always be true; that is, the body is in a *stable equilibrium* position with respect to small rotations. However, as is illustrated in Fig. 2-27b, if the center of gravity of the completely submerged body is above the center of buoyancy, the resulting couple formed by the weight and the buoyant force will cause the body to overturn and move to a new equilibrium position. Thus, a completely submerged body with its center of gravity *above* its center of buoyancy is in an *unstable equilibrium* position. However, when the center of gravity C and center of buoyancy B coincides, the body will always assume the same position in which it is placed (Fig. 2-27c) and hence it is in *neutral equilibrium*. Therefore, it can be concluded that a submerged body will be in stable, un-

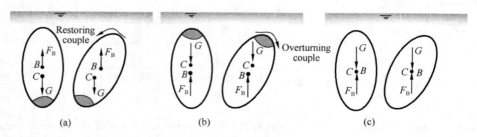

Fig. 2-27 An immersed buoyant body is (a) stable if C is directly below B, (b) unstable if C is directly above B, and (c) neutrally stable if C and B are coincident

stable or neutral equilibrium if its center of gravity is below, above or coincident with the center of buoyancy respectively.

2.7.4 Stability of Floating Bodies

For a floating body in fluid, if the body is bottom-heavy and thus the center of gravity C is below the centroid B of the body, is stable. However, when the body undergoes an angular displacement about a horizontal axis, the shape of the immersed volume changes and so the center of buoyancy moves relative to the body. As a result of above observation stable equilibrium may achieved, under certain condition, even when C is above B.

Figure 2-28 illustrates the states of equilibrium of a symmetric floating body. For Fig. 2-28a, the body is stable even when the center of gravity C is above the center of buoyancy B, action lines of gravity and buoyancy are collinear. For Fig. 2-28b, the body is tilted a small angle $\Delta\theta$, and a new waterline is established for the body to float at this angle. The new position B' of the center of buoyancy is established, from which a vertical line drawn upward intersects the line of symmetry (the original colinearity of gravity and buoyancy) at a point M, called the *metacenter*. The metacenter is independent of $\Delta\theta$ for small angles, and may be considered to be a fixed point for most hull shapes for small rolling angles up to about 20°. The *metacentric height*, designated as ρ, is the distance between the center of gravity C and the metacenter M. If point M is above C (Fig. 2-28b), that is, the metacentric height ρ is positive, a restoring moment is presented and the original position is *stable*. If M is below C (negative ρ) (Fig. 2-28c), the body is *unstable* and will overturn if disturbed. Stability increases with increasing positive ρ.

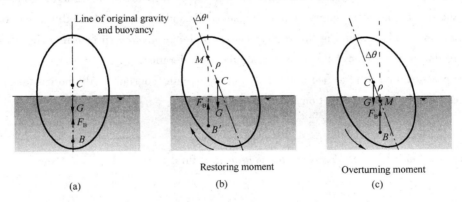

Fig. 2-28 The floating body in (a) original position is (b) stable if M is above C, or (c) unstable if M is below C when it is tilted a small angle

Hence the condition of stable equilibrium for a floating body can be expressed in terms of the metacentric height:

$\rho > 0$ Stable equilibrium, M is above C

$\rho = 0$ Neutral equilibrium, M coincides with C

$\rho < 0$ Unstable equilibrium, M is below C

For a floating body carrying liquid with a free surface undergoes an angular displacement, the liquid will also move to keep its free surface horizontal. Thus not only does the center of buoyancy B move, but also the center of gravity C of the floating body and its contents move in the same direction as the movement of B. Hence the stability of the body is reduced. For this reason, liquid which has to be carried in a ship is put into a number of separate compartments so as to minimize its movement within the ship.

Summary

Fluid statics discusses the fluid at rest or in motion with no relative motion between fluid particles, which leaves only the *pressure* normal to the fluid surface. The pressure is termed *absolute pressure* relative to absolute vacuum, or the *gage pressure* referred to the local atmospheric pressure. The latter is also called the *vacuum pressure* if it is negative. The pressure can be measured by the *manometers* in the height of liquid column, or by the mechanical or electronic pressure measuring devices. In engineering practices, most pressures are specified as gage pressure because the atmospheric pressure can usually be cancelled out.

The pressure at a point in a static fluid has the same magnitude in all directions. It varies from point to point in a fluid field by obeying the *Euler equilibrium equation of motion*. The constant pressure surface in the fluid, the *equipressure surface*, is proved to be perpendicular to the unit mass force

$$\mathbf{f} \cdot \mathbf{ds} = 0$$

It indicates that the *equipressure surface* in a liquid at rest under only gravity is a *horizontal plane* if the liquids the plane intersected are homogenous and interconnected. But it may be an *inclined plane* in a tank of liquid in linear-rigid motion and a *paraboloid* surface in a rigid-body rotating liquid due to the effect of the extra mass force: acceleration.

The pressure in a fluid at rest under only gravity remains constant in the horizontal direction, and varies linearly in the z-direction

$$\frac{dp}{dz} = -\rho g$$

which leads to the *hydrostatic pressure distribution* in a fluid at rest

$$z + \frac{p}{\rho g} = C$$

and the pressure at any point below the liquid surface

$$p_1 = p_0 + \rho g h$$

For the surfaces of structure in contact with fluids at rest, the resultant hydrostatic force on it is so important that its magnitude and the center of pressure should be determined in the design. For the submerged *plane* surface in a homogeneous fluid, the magnitude of the *resultant force* on it is equal to the product of the pressure at the centroid of the surface and the area of the surface

$$F = p_c A = \rho g h_c A$$

and the *center of pressure* is given by

$$y_R = y_c + \frac{I_{xc}}{y_c A}$$

If the plane is in a regular shape, the resultant force on it has a magnitude equating the volume of the *pressure prism*, and acts through the centroid of the pressure prism.

The resultant force on a *curved surface* is usually determined by separating it into horizontal and vertical components. The *horizontal component* equals the force on the vertical projection plane of the curved surface of interest, and the *vertical component* equals in magnitude and direction of the weight of the entire fluid column of the *pressure force prism* above the curved surface:

$$F_V = \rho g V_p$$

The *equilibrium* of a submerged or floating body in a liquid requires that the weight of the body acting through its center of gravity should be colinear with an equal upward buoyancy force acting through the center of buoyancy. With a small angular displacement given on a floating or submerged body and then released may attain *stable, unstable* or *neutral equilibrium*. A submerged body is said to be *stable* when its center of gravity is below than the center of buoyancy, while for a floating body the *metacenter* should be higher than the center of gravity to maintain its stable, even if the center of buoyancy is also higher than the center of gravity.

Word Problems

W2-1 Name the forces acting on (a) the ideal fluid and (b) the real fluid at rest. How about the ideal or real fluid in motion?

W2-2 What is the Euler equilibrium equation of motion? Explain its physical interpretation.

W2-3 Can Euler equilibrium equation of motion be used for the fluid with rigid-body motion? Explain.

W2-4 What is the equipressure surface? How can a surface be called an equipressure surface?

W2-5 Is the equipressure surface in a fluid with rigid-body motion a horizontal plane? Why? How can an equipressure surface be a horizontal plane?

W2-6 Define the hydrostatic pressure distribution in a fluid at rest.

W2-7 A tiny steel cube is suspended in water by a string. If the lengths of the sides of the cube are very small, how would you compare the magnitudes of the pressures on the top, bottom, and side surfaces of the cube?

W2-8 The two open tanks shown in Fig. W2-8 have the same bottom area A but different shapes. When the depth, h, of the liquid in the two tanks is the same, the pressure on the bottom of the two tanks should be equal in accord-

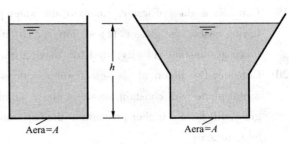

Fig. W2-8

ance with Eq. 2.18. However, the weight of the liquid in each of the tanks is different. How do you account for this apparent paradox?

W2-9 Define the elevation, pressure and piezometric heads. What are the specified physical meanings of these heads? How are they related?

W2-10 What is the minimum length of the piezometric tube if the piezometric head is 1m and the pressure head is 1.5m?

W2-11 Does the piezometric head at every point equal in a fluid at rest? How about in a fluid in motion?

W2-12 Define absolute pressure, gage pressure and vacuum. How are they related?

W2-13 The pressure measured by manometers is (a) absolute pressures or (b) gage pressures.

W2-14 Someone claims that the absolute pressure in a liquid of constant density doubles when the depth is doubled. Do you agree? Explain.

W2-15 Mercury is always used as the gage fluid in the manometers in the traditional experiments. Why?

W2-16 Two different fluids with $\rho_1 < \rho_2$ layered in a contained as in Fig. W2-16. Are the free surfaces in piezometer tubes sketched in the figure correct? Explain.

W2-17 In Fig. W2-17 the water in the container is at rest. The level difference, h_m, between the free surfaces in the U-tube manometer is (a) $\dfrac{p_A - p_B}{\rho_m}$, (b) $\dfrac{p_A - p_B}{\rho_m g - \rho g}$, or (c) 0.

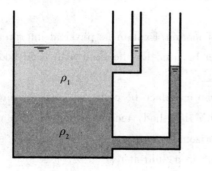

Fig. W2-16

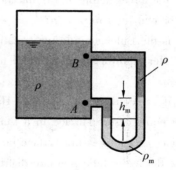

Fig. W2-17

W2-18 Under what conditions can a moving body of fluid be treated as a rigid body?

W2-19 Consider a glass of water. Compare the water pressures at the bottom surface for the following cases: the glass is (a) stationary, (b) moving up at constant velocity, (c) moving down at constant velocity, and (d) moving horizontally at constant velocity.

W2-20 Consider two identical glasses of water, one stationary and the other moving on a horizontal plane with constant acceleration. Assuming no splashing or spilling occurs, which glass will have a higher pressure at the (a) front, (b) midpoint, and (c) back of the bottom surface?

W2-21 Consider a vertical cylindrical container partially filled with water. Now the cylinder is rotated about its axis at a specified angular velocity, and rigid-body motion is established. Discuss how the pressure will be affected at the midpoint and at the edges of the bottom surface due to rotation.

W2-22 Define the resultant hydrostatic force acting on a submerged plane surface, and the center of pressure. How to compute?

W2-23 For a regular plane surface submerged in a liquid, the graphical method can be used to determine the magnitude and acting direction of the resultant hydrostatic force acting on it. How?

W2-24 Someone claims that she can determine the magnitude of the hydrostatic force acting on a plane surface submerged in water regardless of its shape and orientation if she knew the vertical distance of the centroid of the surface from the free surface and the area of the surface. Is this a valid claim? Explain.

W2-25 A submerged horizontal flat plate is suspended in water by a string attached at the centroid of its upper surface. Now the plate is rotated 45° about an axis that passes through its centroid. Discuss the change on the hydrostatic force acting on the top surface of this plate as a result of this rotation. Assume the plate remains submerged at all times.

W2-26 Consider three completely immersed planes with an identical area of surface and the identical immersed depth of the centroids as in Fig. W2-26. Do all the magnitude of resultant forces equal? Which center of pressure locates deepest in water?

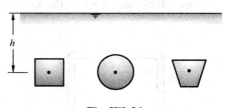

Fig. W2-26

W2-27 What is the pressure force prism? How to determine the vertical component of the resultant hydrostatic force acting on a submerged curved surface?

W2-28 What is buoyant force? What causes it? What is the magnitude of the buoyant force acting on a submerged body whose volume is $\forall$? What are the direction and the line of action of the buoyant force?

W2-29 Consider two identical spherical balls submerged in water at different depths. Will the buoyant forces acting on these two balls be the same or different? Explain. How about a aluminum ball and an iron ball submerged at the different depth?

W2-30 Consider a 3-kg copper cube and a 3-kg copper ball submerged in a liquid. Will the buoyant forces acting on these two bodies be the same or different? Explain.

W2-31 Discuss the stability of (a) a submerged and (b) a floating body whose center of gravity is above the center of buoyancy.

W2-32 An inverted test tube partially filled with air floats in a plastic water-filled soft drink bottle as shown in Fig.

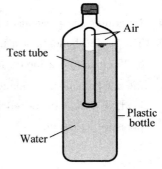

Fig. W2-32

W2-32. The amount of air in the tube has been adjusted so that it just floats. The bottle cap is securely fastened. A slight squeezing of the plastic bottle will cause the test tube to sink to the bottom of the bottle. Explain this phenomenon.

W2-33 What are the conditions of stable equilibrium of immersed body? How about floating body?

Problems

Pressure, Manometer, and Barometer

2-1 Determine the absolute and gage pressures of the point at a depth of 2m under the free surface. Assume the absolute pressure at the free surface is 1at.

2-2 Figure P2-2 shows a siphon to sucking oil ($\gamma = 8.5 \text{kN/m}^3$) from tank B to tank A. Assume at the very beginning the oil is at rest. Determine the pressure in the height of water colume at point C and point D.

2-3 In Fig. P2-3 the 20℃ water and gasoline surfaces are open to the atmosphere and at the same elevation. What is the height h of the third liquid in the right leg?

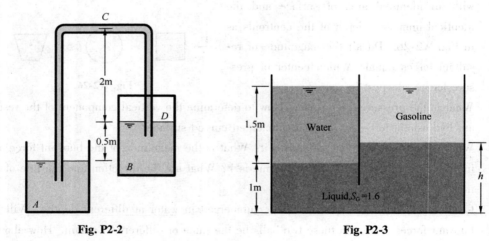

Fig. P2-2   Fig. P2-3

2-4 Determine the piezometer heights and piezometer heads at points A, B and C in Fig. P2-4.

2-5 The basic elements of a hydraulic press are shown in Fig. P2-5. The plunger has an area of 6.25 cm², and a force, F_1, can be applied to the plunger through a lever mechanism having a mechanical advantage of 8 to 1. If the large piston has an area of 937.5 cm², what load, F_2, can be raised by a force of 134N applied to the lever? Neglect the hydrostatic pressure variation.

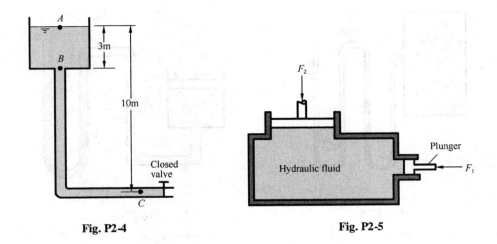

Fig. P2-4   Fig. P2-5

2-6 The U-tube in Fig. P2-6 has a 1-cm-diameter and contains mercury as shown. If 20 cm³ of water is poured into the right hand leg, what will the free-surface height in each leg be after the sloshing has died down?

2-7 Consider a 70-kg woman who has a total foot imprint area of 400 cm². She wishes to walk on the snow without sinking, but the snow cannot withstand pressures greater than 0.5kPa. Determine the minimum size of the snowshoes needed (imprint area per shoe) to enable her to walk on the snow without sinking when one foot carries her entire weight.

2-8 A simple experiment has long been used to demonstrate how negative pressure prevents water from being spilled out of an inverted glass. A glass that is fully filled by water and covered with a thin paper is inverted, as shown in Fig. P2-8. Determine the pressure at the bottom of the glass, and explain why water does not fall out.

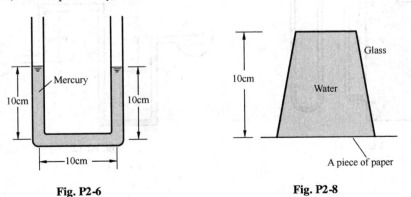

Fig. P2-6   Fig. P2-8

2-9 An open multi-U-tube manometer is used to measure the pressure at the free surface in a closed tank as in Fig. P2-9, determine the pressure p_0 in the tank. Units in m.

2-10 A piston having a cross-sectional area of 0.07 m² is located in a cylinder containing water as shown in Fig. P2-10. An open U-tube manometer is connected to the cylinder as shown, determine the applied force, F, acting on the piston? The weight of the piston is negligible.

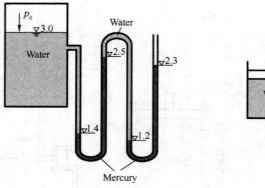

Fig. P2-9

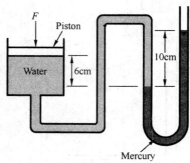

Fig. P2-10

2-11 The water in a tank is pressurized by air, and the pressure is measured by a multifluid manometer as shown in Fig. P2-11. Determine the gage pressure of air in the tank if $h_1 = 0.2$m, $h_2 = 0.3$m, and $h_3 = 0.46$m. Take the densities of water, oil, and mercury to be 1000kg/m^3, 850kg/m^3, and 13600kg/m^3, respectively.

2-12 Freshwater and seawater flowing in parallel horizontal pipelines are connected to each other by a double U-tube manometer, as shown in Fig. P2-12. Determine the pressure difference between the two pipelines. Take the density of seawater to be $\rho = 1035$kg/m^3. Can the air column be ignored in the analysis?

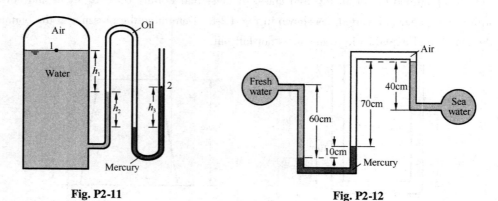

Fig. P2-11　　　　　　　　　　Fig. P2-12

2-13 For the inverted manometer of Fig. P2-13, all fluids are at 20℃. If $p_B - p_A = 97$kPa, what must the height H be in cm?

2-14 Water flows upward in a pipe slanted at 30°, as in Fig. P2-14. The mercury manometer reads $h = 12$cm. Both fluids are at 20℃. What is the pressure difference $p_1 - p_2$ in the pipe?

2-15 There are five points, points 1, 2, 3, 4 and 5, locating on a same horizontal plane as shown in Fig. P2-15. Compare the pressures at these points and explain. Assume $\rho_1 > \rho_2$.

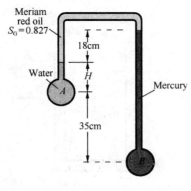

Fig. P2-13

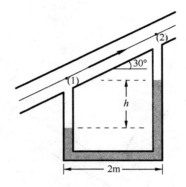

Fig. P2-14

2-16 In Fig. P2-16 air, oil and water are confined in a closed tank. The reading on the pressure gage A is $-1.47\text{N}/\text{cm}^2$. (a) Calculate the elevation different in the mercury manometer; (b) Plot the pressure distribution along the vertical tank wall qualitatively.

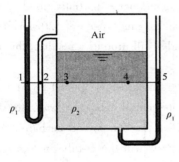

Fig. P2-15

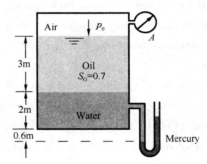

Fig. P2-16

Fluids in Rigid-Body Motion

2-17 A 60-cm-high, 40-cm-diameter cylindrical water tank is being transported on a level road at a constant acceleration of $4\text{m}/\text{s}^2$. Determine the allowable initial water height in the tank if no water is to spill out during acceleration.

2-18 The tank of liquid is moving with constant acceleration up a 30° inclined plane, as in Fig. P2-18. Assuming rigid-body motion, compute (a) the magnitude of the acceleration **a**, (b) whether the acceleration is up or down, and (c) the gage pressure at point A if the fluid is mercury at 20°C.

2-19 The open U-tube of Fig. P2-19 is partially filled with a liquid. When this device is accelerated with a horizontal acceleration a, a differential reading h develops between the manometer legs which are spaced a distance l apart. Determine the relationship between a, l, and h.

71

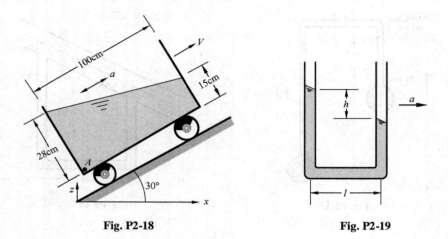

Fig. P2-18   **Fig. P2-19**

2-20 The U-tube of Fig. P2-20 is partially filled with water and rotates around the axis O-O. Determine the angular velocity that will cause the water to start to vaporize at the bottom of the tube (point A).

2-21 A 40-cm-diameter, 90-cm-high vertical cylindrical container is partially filled with 60-cm-high water. Now the cylinder is rotated at a constant angular speed of 120rpm. Determine how much the liquid level at the center of the cylinder will drop as a result of this rotational motion.

2-22 A 16-cm-diameter and 27-cm-high vertical cylinder container is opened to air and full of water. Compute the rigid-body rotation rate about its central axis, in r/min, (a) for which one-third of the water will spill out and (b) for which the bottom will be barely exposed.

2-23 For what uniform rotation rate in r/min about axis O-O will the U-tube in Fig. P2-23 take the configuration shown? The fluid is mercury at 20℃.

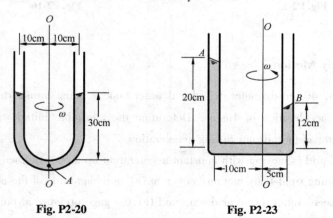

Fig. P2-20   **Fig. P2-23**

2-24 In Fig. P2-24a a 0.6-m-diameter closed cylinder 0.5-m high is filled with 0.4-m-high water at the lower part and 0.1-m-high oil ($S_G = 0.8$) in the upper part. The tank has a vent hole at the center of the cover and is rotated in rigid-body motion about its central z-axis. If the interface between water and oil at the central axis just touch the center point of the tank bottom (Fig. 2-24b), determine: (a) the rotation velocity ω; (b) the pressures at points a, b, c, and d in the height of water column; (c) the forces acting on the tank cover and the bottom by the liquids.

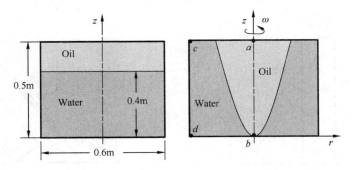

Fig. P2-24

Hydrostatic Forces on Plane surface

2-25 Sketch the pressure prism on surface AB in Fig. P2-25.

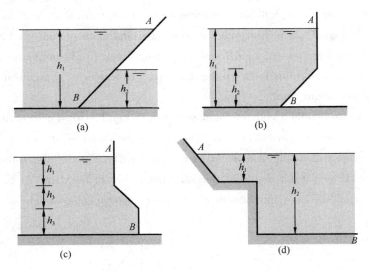

Fig. P2-25

2-26 A retaining wall against a mud slide is to be constructed by placing 0.8-m-high and 0.2-wide rectangular concrete blocks (the density is 2700kg/m³) side by side, as shown in Fig. P2-26. The friction coefficient between the ground and the concrete blocks is 0.3, and the density of the mud is about 1800kg/m³. There is concern that the concrete blocks may slide or tip over the lower left edge as the mud level rises. Determine the mud height at which (a) the blocks will overcome friction and start sliding and (b) the blocks will tip over.

2-27 A 1.0-m-wide and 2.0-m-long rectangular gate is located in a sloping surface of 45° in a tank as in Fig. P2-27. The centroid of gate is at a depth of 2.0m below the free surface.

The gate is hinged along its top edge and held in position by a vertical force F. Disregarding the friction at the hinge and the weight of the gate, determine the required magnitude of force F.

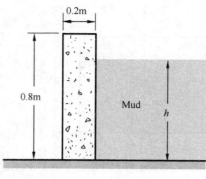

Fig. P2-26   Fig. P2-27

2-28 An open rectangular tank is 2m wide and 4m long. The tank contains water to a depth of 2m and oil ($S_G = 0.8$) on top of the water to a depth of 1m. Determine the magnitude and location of the resultant fluid force acting on one of the vertical ends of the tank.

2-29 Dam ABC in Fig. P2-29 is 30m wide into the paper and made of concrete ($S_G = 2.4$). Find the hydrostatic force on surface AB and its moment about C. Assuming no seepage of water under the dam, could this force tip the dam over? How does your argument change if there is seepage under the dam?

2-30 Water backs up behind a concrete dam as shown in Fig. P2-30. Leakage under the foundation gives a pressure distribution under the dam as indicated. If the water depth, h, is too great, the dam will topple over about its toe (point A). For the dimensions given, determine the maximum water depth for the following widths of the dam: $l = 6$, 12, and 18m. Base your analysis on a unit length of the dam. The specific weight of the concrete is 2403kg/m^3.

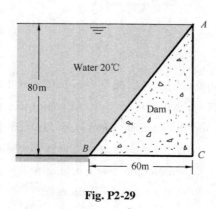 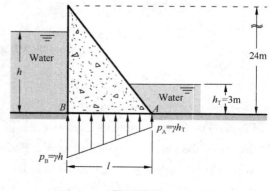

Fig. P2-29   Fig. P2-30

2-31 Gate ABC in Fig. P2-31 is 1m square and is hinged at B. It will open automatically when the

water level h becomes high enough. Determine the lowest height for which the gate will open. Neglect atmospheric pressure. Is this result independent of the liquid density?

2-32 A rectangular gate made of metal having a width of 1.0m into the paper and a height of 3.0m is supported by two I beams as shown in Fig. P2-32. The free surface of water in the tank levels with the top of the gate. If the forces acting on the I beams are required to be equal, that is $F_1 = F_2$, determine the location of the I beams, y_1 and y_2.

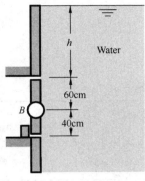
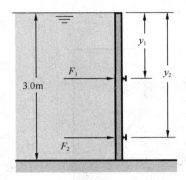

Fig. P2-31   Fig. P2-32

Hydrostatic Forces on Curved surface

2-33 Sketch the pressure force prism on surface ABC in Fig. P2-33.

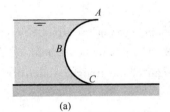 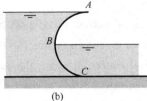 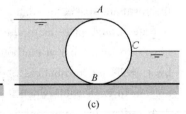

Fig. P2-33

2-34 Determine the resultant force on the arc sluice as shown in Fig. P2-34: (a) horizontal component and the location of its acting line; (b) Vertical component and the location of its acting line and (c) the resultant force and the angle made with the horizontal plane. The width of gate is 1m into paper.

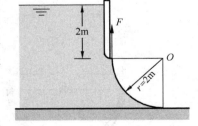

Fig. P2-34

2-35 Three gates of negligible weight are used to hold back water in a channel of width b as shown in Fig. P2-35. The force of the gate against the block for gate (b) is R. Determine (in terms of R) the force against the blocks for the other two gates.

2-36 The uniform body A in Fig. P2-36 has width b into the paper and is in static equilibrium when pivoted about hinge O. What is the specific gravity of this body if (a) $h = 0$ and (b) $h = R$?

75

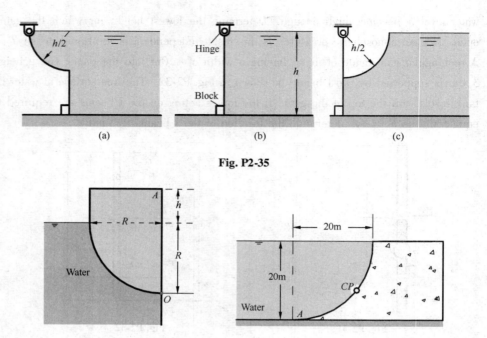

Fig. P2-35

Fig. P2-36

Fig. P2-37

2-37 The dam in Fig. P2-37 is a quarter circle 50m wide into the paper. Determine the horizontal and vertical components of the hydrostatic force against the dam and the point CP where the resultant strikes the dam.

2-38 A water trough of semicircular cross section of radius 0.5m consists of two symmetric parts hinged to each other at the bottom, as shown in Fig. P2-38. The two parts are held together by a cable and turnbuckle placed every 3m along the length of the trough. Calculate the tension in each cable when the trough is filled to the rim.

2-39 A 4-m-diameter water tank consists of two half cylinders, each weighing 4.5kN/m, bolted together as shown in Fig. P2-39. If the support of the end caps is neglected, determine the force induced in each bolt.

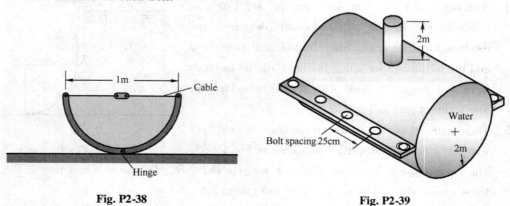

Fig. P2-38

Fig. P2-39

2-40 Gate *ABC* is a circular arc, sometimes called a *Tainter gate*, which can be raised and lowered by pivoting about point *O*, as in Fig. P2-40. For the position shown, determine (a) the hydrostatic force of the water on the gate and (b) its line of action. Does the force pass through point *O*?

2-41 The 4-m-diameter cylindrical gate in Fig. P2-41 is 10-m long into the paper and dams water as shown. The upstream and downstream water depths are 4m and 2m respectively. Determine the resultant hydrostatic force on the gate and its direction.

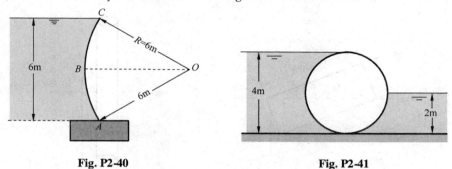

Fig. P2-40   Fig. P2-41

2-42 A 0.8-m-diameter cylindrical log in Fig. P2-42 is used to dam the oil ($S_G = 0.8$) floating on the top of water to the left. Assume the log is in the force balance, determine: (a) the force on the bank by per unit length of log; (b) the weight per unit length log; (c) the specify gravity of log.

2-43 The 0.6-m-diameter cylindrical gate in Fig. P2-43 is 2m wide into the paper and pivoted about hinge *O* by a 0.5-m-long cable. Neglect the friction at hinge *O*, compute the moment *M* of all forces about the hinge *O*.

2-44 An oil tank has a 0.6-m-diameter semi-spherical cover on the side wall, of which the center is immersed at a depth of 2.0m as in Fig. P2-44. Take the specify weight of oil to be 6867N/m^3, determine the horizontal and vertical forces acting on the semi-spherical cover.

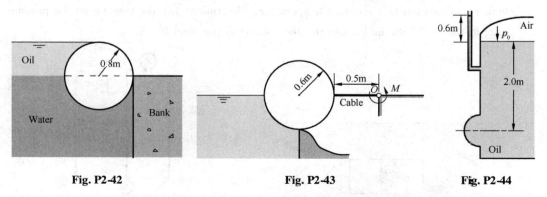

Fig. P2-42   Fig. P2-43   Fig. P2-44

Buoyancy

2-45 The homogeneous wooden block *A* in Fig. P2-45 is 0.7m by 0.7m by 1.3m and weighs 2.4kN. The concrete block *B* (specific weight $\gamma = 23.6 \text{kN/m}^3$) is suspended from *A* by means of the slender cable causing A to float in the position indicated. Determine the volume of B.

2-46 The density of a liquid is to be determined by an old 1-cm-diameter cylindrical hydrometer whose division marks are completely wiped out. The hydrometer is first dropped in water, and the water level is marked. The hydrometer is then dropped into the other liquid, and it is observed that the mark for water has risen 0.5cm above the liquid-air interface (Fig. P2-46). If the height of the water mark is 10cm, determine the density of the liquid.

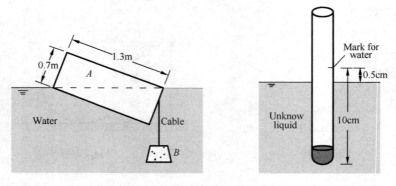

Fig. P2-45　　　　　Fig. P2-46

2-47 Consider a large cubic ice block floating in seawater. The specific gravities of ice and seawater are 0.92 and 1.025, respectively. If a 10-cm-high portion of the ice block extends above the surface of the water, determine the height of the ice block below the surface.

2-48 A uniform block of steel ($S_G = 7.85$) will "float" at a mercury-water interface as in Fig. P2-48. What is the ratio of the distances a and b for this condition?

2-49 The thin-walled, 1-m-diameter tank of Fig. P2-49 is closed at one end and has a mass of 90kg. The open end of the tank is lowered into the water and held in the position shown by a steel block having a density of 7840kg/m³. Assume that the air that is trapped in the tank is compressed at a constant temperature. Determine: (a) the reading on the pressure gage at the top of the tank, and (b) the volume of the steel block.

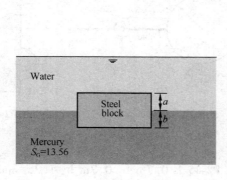
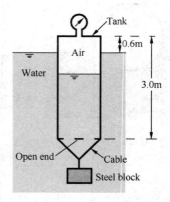

Fig. P2-48　　　　　Fig. P2-49

3 Fluid Kinematics

CHAPTER OPENING PHOTO: Typhoon Saomai 2006 was considered the most powerful typhoon on record to strike the eastern coast of the People's Republic of China, which brought heavy rain and wind to this areas and caused enormous damages. Typhoon is a forced vortex swirling counterclockwise in the Northern Hemisphere due to the Coriolis acceleration caused by the rotation of the Earth. The vortex (rotational flow) and its counter pair irrotational flow are discussed in the present chapter. (*Courtesy of NASA image by Jeff Schmaltz, MODIS Rapid Response Team, Goddard Space Flight Center*)

Fluid kinematics deals with the motion of fluids without necessarily considering the forces and moments that cause the motion. In this chapter, we start with a discussion of the descriptions of fluid flow—the Lagrangian and the Eulerian methods, the numerous ways of classification of fluid flow, such as viscous versus inviscid flow, one-, two- and three-dimensional flow, compressible versus incompressible flow, steady versus unsteady flow and uniform versus nonuniform flow, and the various ways to visualize flow fields—streamlines, streaklines and pathlines. It follows the fundamental kinematic properties of fluid motion and deformation, the concepts of rotationality and irrotationality of fluid flows, and the Navier-Stokes equations of motion for viscous fluid and the Euler's equation for ideal fluid. By integrating the Euler equation to a fluid element along a streamline and in an irrotational flow, we obtain the famous Bernoulli equation in a variety of applications. Finally, we discuss the concepts of stream function, velocity potential function and flow rate for the potential flow, present the method of superposition, the basic plane potential flows.

3.1 Description Approaches of the Flow Field

In the study of fluid motion there are two main approaches to describe what is happening: Lagrangian approach and Eulerian approach.

3.1.1 The Lagrangian Approach

The *Lagrangian description*, named after the Italian mathematician Joseph Louis Lagrange (1736-1813), involves watching the behavior and the trajectory of each individual fluid particle as it moves through space from some initial location and getting the whole flow field by integrating the all motions of the all particles. At each instant in time the fluid particle(s) being studied will have a different set of coordinates within some global coordinate systems, but each particle will be associated with a specific initial set of coordinates.

In Cartesian coordinate system as shown in Fig. 3-1, Lagrange adopted the original coordinate (a, b, c) as the marker of each individual particle and t as independent variables. Thus the motion of each fluid particle can be completely specified by the continuous function of the original coordinate and the temporal variable:

$$\left. \begin{array}{l} x = x(a, b, c, t) \\ y = y(a, b, c, t) \\ z = z(a, b, c, t) \end{array} \right\} \quad (3.1)$$

where a, b, c and t are called the *Lagrangian variables*.

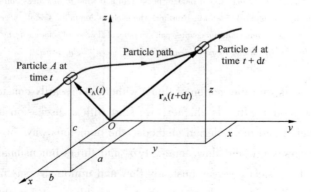

Fig. 3-1 The Lagrangian approach (particle location in terms of its position vector)

For the Lagrangian approach, the fluid parcel does not necessarily retain its size and shape during its motion, because the changes in shape due to interactions with neighboring fluid elements could be arisen. Eqs. 3.1 describe the exact spatial position (x, y, z) of any fluid particle at different times in terms of its *initial position* ($x_0 = a$, $y_0 = b$, $z_0 = c$) at the given initial time $t = t_0$. They are usually referred to as *parametric equations of the path* of fluid particles. The attention here is focused on the paths of different fluid particles as time goes on. After the equations describing the paths of fluid particles are determined, the instantaneous velocity components and acceleration components at any instant of time can be determined in the usual manner by taking derivatives with respect to time.

The velocity is

$$\left.\begin{aligned} u_x &= \frac{\partial x}{\partial t} = \frac{\partial x(a,\ b,\ c,\ t)}{\partial t} \\ u_y &= \frac{\partial y}{\partial t} = \frac{\partial y(a,\ b,\ c,\ t)}{\partial t} \\ u_z &= \frac{\partial z}{\partial t} = \frac{\partial z(a,\ b,\ c,\ t)}{\partial t} \end{aligned}\right\} \quad (3.2)$$

The acceleration is

$$\left.\begin{aligned} a_x &= \frac{\partial u_x}{\partial t} = \frac{\partial^2 x}{\partial t^2} \\ a_y &= \frac{\partial u_y}{\partial t} = \frac{\partial^2 y}{\partial t^2} \\ a_z &= \frac{\partial u_z}{\partial t} = \frac{\partial^2 z}{\partial t^2} \end{aligned}\right\} \quad (3.3)$$

in which u_x, u_y, u_z and a_x, a_y, a_z are respectively the x, y, z components of velocity and acceleration; ∂ is the *partial derivative operator*.

3.1.2 The Eulerian Approach

The alternative approach to describe the flow field is the *Eulerian description*, named after the Swiss mathematician Leonhard Euler (1707-1783). This corresponds to a coordinate system fixed in space, and in which fluid properties are studied as functions of time as the flow passes fixed spatial locations, namely we compute the pressure field $p(x_0,\ y_0,\ z_0,\ t)$ of the flow pattern, not the pressure changes $p(t)$ which a particle experiences as it moves through the field.

In the Eulerian method, the individual fluid particles are not identified. Instead, a finite space at a fixed position, through which fluids in and out, called a *flow domain* or *control volume*, is chosen, and the velocity of particles at this position as a function of time is sought. It is evident that in this case we need not be explicitly concerned with individual fluid particles or their trajectories. Instead, we define field variables, functions of space and time, within the control volume. For example, the pressure and density fields are scalar field variables, the velocity and acceleration fields as vector field variables

$$p = p(x,\ y,\ z,\ t)$$
$$\rho = \rho(x,\ y,\ z,\ t)$$
$$\mathbf{a} = \mathbf{a}(x,\ y,\ z,\ t)$$
$$\mathbf{u} = \mathbf{u}(x,\ y,\ z,\ t)$$

or in terms of components

$$\left.\begin{array}{l} u_x = u_x(x, y, z, t) \\ u_y = u_y(x, y, z, t) \\ u_z = u_z(x, y, z, t) \end{array}\right\} \qquad (3.4)$$

where x, y, z and t are independent variables and called the *Eulerian variables*.

The difference between Lagrangian and Eulerian descriptions is made clearer by imagining a person standing beside a river to measure its properties. In the Lagrangian approach, he throws in a probe that moves downstream with the water. In the Eulerian approach, he anchors the probe at a fixed location in the water.

The relationship between Eulerian and Lagrangian methods also can be shown. According to the Lagrangian method, we have a set of Eqs. 3.2 for each particle which can be combined with Eqs. 3.4 as follows

$$\left.\begin{array}{l} \dfrac{dx}{dt} = u_x(a, b, c, t) \\[4pt] \dfrac{dy}{dt} = u_y(a, b, c, t) \\[4pt] \dfrac{dz}{dt} = u_z(a, b, c, t) \end{array}\right\} \qquad (3.5)$$

where d is the *total derivative operator*.

The integrals of Eqs. 3.5 lead to three constants of integration, which can be considered as initial coordinates a, b, c of the fluid particle. Hence the solutions of Eqs. 3.5 give the equations of Lagrange (Eqs. 3.1).

Although the solution of Lagrangian equations yields the complete description of paths of fluid particles, the mathematical difficulty encountered in solving these equations makes the Lagrangian method impractical. In most fluid mechanics problems, rather the general state of motion expressed in terms of velocity components of flow and the change of velocity with respect to time at various points in the flow field are of greater practical significance. Therefore the Eulerian method is generally adopted in fluid mechanics.

3.1.3 The Acceleration Field

Foremost among the properties of a flow is the velocity field $\mathbf{u}(x, y, z, t)$. In general, velocity is a vector function of position and time and thus has three components u_x, u_y, and u_z, each a scalar field in itself:

$$\mathbf{u}(x, y, z, t) = \mathbf{i}\, u_x(x, y, z, t) + \mathbf{j}\, u_y(x, y, z, t) + \mathbf{k}\, u_z(x, y, z, t) \qquad (3.6)$$

In the Eulerian description, all field variables, such as velocity field in Eq. 3.6 are defined at any location (x, y, z) in the control volume and at any instant in time t. In the Eulerian description we don't really care what happens to individual fluid particles; rather we are concerned with the pressure, velocity, acceleration, etc., of whichever fluid particle happens to be at the location of interest at the time of interest.

To write Newton's second law for an infinitesimal fluid system, we need to calculate the acceleration vector field **a** of the flow. Thus we compute the total derivative of the velocity vector with respect to time

$$\mathbf{a} = \frac{d\mathbf{u}}{dt} = \frac{\partial \mathbf{u}}{\partial t} + \frac{\partial \mathbf{u}}{\partial x}\frac{dx}{dt} + \frac{\partial \mathbf{u}}{\partial y}\frac{dy}{dt} + \frac{\partial \mathbf{u}}{\partial z}\frac{dz}{dt}$$
$$= \frac{\partial \mathbf{u}}{\partial t} + \left(u_x \frac{\partial \mathbf{u}}{\partial x} + u_y \frac{\partial \mathbf{u}}{\partial y} + u_z \frac{\partial \mathbf{u}}{\partial z}\right) \quad (3.7)$$

in which dx/dt is the local velocity component u_x by definition, and $dy/dt = u_y$, and $dz/dt = u_z$.

Equation 3.7 can also be written in the scalar form

$$\left. \begin{array}{l} a_x = \dfrac{\partial u_x}{\partial t} + u_x \dfrac{\partial u_x}{\partial x} + u_y \dfrac{\partial u_x}{\partial y} + u_z \dfrac{\partial u_x}{\partial z} \\[6pt] a_y = \dfrac{\partial u_y}{\partial t} + u_x \dfrac{\partial u_y}{\partial x} + u_y \dfrac{\partial u_y}{\partial y} + u_z \dfrac{\partial u_y}{\partial z} \\[6pt] a_z = \dfrac{\partial u_z}{\partial t} + u_x \dfrac{\partial u_z}{\partial x} + u_y \dfrac{\partial u_z}{\partial y} + u_z \dfrac{\partial u_z}{\partial z} \end{array} \right\} \quad (3.8)$$

or in the compact form

$$\mathbf{a} = \frac{d\mathbf{u}}{dt} = \frac{\partial \mathbf{u}}{\partial t} + (\mathbf{u} \cdot \nabla)\mathbf{u} \quad (3.9)$$

where ∇ is the *gradient operator* or *del operator*, in Cartesian coordinates $\nabla = \mathbf{i}\frac{\partial}{\partial x} + \mathbf{j}\frac{\partial}{\partial y} + \mathbf{k}\frac{\partial}{\partial z}$.

The term $\partial \mathbf{u}/\partial t$ in Eq. 3.9 is called the *local acceleration*, which represents the change in velocity in time ('unsteady') and vanishes if the flow is steady, i.e., independent of time. The three terms in parentheses in Eq. 3.7, or $(\mathbf{u} \cdot \nabla)\mathbf{u}$ in Eq. 3.9 are called the *convective acceleration*, or the *advective acceleration*, which arises when the particle moves through regions of spatially varying velocity, as in a nozzle or diffuser. Convective effects may exist whether the flow is steady or unsteady. Flows which are nominally "steady" may have large accelerations due to the convective terms.

For example, Fig. 3-2 shows a flow from the tank to a serial pipe composed of a constant-diameter straight pipe and a converging-diverging pipe. If the water level in the tank is kept steady by pumping in water, the water depth over the pipe axis is a constant, the velocity of fluid particle in pipe will not change with time, thus the local acceleration in the whole combined pipe is zero. As for the convective acceleration, it is zero in the straight pipe because the velocity does not change from point to point due to the identical diameter, and greater than zero in the contraction pipe because of the increasing velocity and less than zero in the expansion pipe because of the decreasing velocity. Thus, the acceleration in the straight pipe $a_x = 0$, in the converging pipe $a_x = u_x \partial u_x/\partial x > 0$ and in the diverging pipe $a_x = u_x \partial u_x/\partial x < 0$.

If there is no water fed by the pump, the water level in the tank will decrease, so the water depth over the pipe axis decreases. Thus, the velocity of fluid particle changes with the time, which leads to a variation of the local acceleration in the whole combined pipe. However, at any instant the convective acceleration varies in the combined pipe just like the case of tank with a constant water level. Thus, the acceleration in the straight pipe is $a_x = \partial u_x / \partial t$, in the converging-diverging pipe is $a_x = \partial u_x / \partial t + u_x \partial u_x / \partial x$.

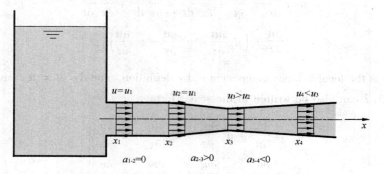

Fig. 3-2 The acceleration of steady flow in a combined pipe

The total time derivative, sometimes called the *substantial* or *material* derivative, concept may be applied to any variable, e. g., the density

$$\frac{\mathrm{d}\rho}{\mathrm{d}t} = \frac{\partial \rho}{\partial t} + u_x \frac{\partial \rho}{\partial x} + u_y \frac{\partial \rho}{\partial y} + u_z \frac{\partial \rho}{\partial z}$$

For the incompressible fluid, the material derivative of density is zero, i. e. $\mathrm{d}\rho/\mathrm{d}t = 0$.

3.2 Classification of Fluid Flows

There are many ways in which fluid flows can be classified. In this section we will present and discuss a number of these that will be of particular relevance. It is important to bear in mind throughout these discussions that the classifications we are considering are concerned with the flow, and not with the fluid itself. For example, when we distinguish between compressible and incompressible flows, we see that even though gases are generally very compressible substances, it is often very accurate to treat the flow of gases as incompressible.

3.2.1 Viscous and Inviscid Flows

As what discussed in Section 1.4, when two fluid layers move relative to each other, a friction force develops between them by the viscosity and the slower layer tries to slow down the faster layer. There is no fluid with zero viscosity, and thus all fluid flows involve viscous effects to some degree. Flows in which the frictional effects are significant are called *viscous flows*. However, in many flows of practical interest, there are regions (typically regions not close to solid surfaces) where viscous forces are negligibly small compared to inertial or pressure forces. In this situation it might be appropriate to treat the flow as *inviscid* by ignoring the effects of viscosity, which may greatly simplify the analysis without much loss in accuracy.

3.2.2 Flow Dimensionality

The dimensionality of a flow field corresponds to the number of spatial coordinates needed to describe all properties of the flow. Therefore, the flow can be termed *one-*, *two-*, or *three-dimensional* if the flow velocity varies in one, two, or three primary dimensions, respectively. However, at the same time we must recognize that all physical flows are really three dimensional (3D). Nevertheless, it is often convenient, and sometimes quite accurate, to view them as being of a lower dimensionality, e.g., being a two-dimensional one, even a one-dimensional one.

In *three-dimensional flow* it is assumed that all kinematic parameters, e. g., velocity, depend on three space variables. All three space components of kinematic parameters are important and of equal magnitude. For example, the flow past a wing is complex three-dimensional flow, and simplifying by eliminating any of the three velocities would lead to severe errors.

In *two-dimensional flow*, fluid flows mainly in two directions, and the third direction's flow is negligible, that is, fluid motion factors are functions of two space coordinates. In many situations one of the velocity components may be small relative to the other two, thus it is reasonable in this case to assume two-dimensional flow, e. g., the flow pasting a long circular cylinder(Fig. 3-3), or air plane wing.

Fig. 3-3 Two-dimensional flow past a circular cylinder at Re = 1.54.
(*Photographed by Sadatoshi Taneda and appears in the 'Album of Fluid Motion', by Milton Van Dyke.*)

In *one-dimensional flow*, also *total flow*, it is assumed that all properties are uniform over any plane perpendicular to the flow direction as indicated in Fig. 3-4a. One-dimensional flow is never reproduced exactly in practice. In practice a fluid flows along a circular pipe the velocity is zero at the wall due to *no-slip condition* and a maximum at the pipe center (Fig. 3-4b), while the velocity profile in an assumed one-dimensional flow is uniform (Fig. 3-4a). But both

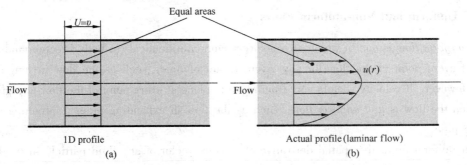

Fig. 3-4 The one-dimensional (a) uniform and (b) actual profile of velocity at the cross section of pipe flow

profiles have the equal areas because the identical flow rate in the pipe. Nevertheless, the assumption of one-dimensional flow simplifies analysis and often provides sufficiently accurate results.

3.2.3 Steady and Unsteady Flows

One of the most important, and often easiest to recognize, distinctions is associated with steady and unsteady flow. In the most general case all flow properties depend on time; for example the functional dependence of pressure at any point (x, y, z) at any instant might be denoted $p(x, y, z, t)$. Thus, if all kinematic variables of a flow (e. g., velocity, pressure and density) are independent of time, then the flow is *steady*; otherwise, it is *unsteady*. For example, the flow in the pipe in Fig. 3-2 is a steady flow if the water level in the tank is constant; otherwise, it is unsteady flow.

For the steady flow, the field equations are

$$\left.\begin{array}{l} \mathbf{u} = \mathbf{u}(x, y, z) \\ p = p(x, y, z) \\ \rho = \rho(x, y, z) \end{array}\right\} \quad (3.10)$$

or the temporal derivatives of the variables (designated as q), are zero

$$\frac{\partial q}{\partial t} = 0 \quad (3.11)$$

In general a steady flow process involves no change in properties, in mass flow rates or in heat transfer rates with respect to time. However, properties may vary from place to place within the control volume but are constant at any given place.

In fluid mechanics, *unsteady* is the most general term that applies to any flow that is not steady, but *transient* is typically used for developing flows. Clearly, a *transient flow* is time dependent, but the converse is not necessarily true. Transient behavior does not persist for 'long times.' In particular, a flow may exhibit a certain type of behavior, say oscillatory, for a few seconds, after which it might become steady.

3.2.4 Uniform and Non-uniform Flows

A *uniform flow* is one in which all velocity vectors are identical (in both direction and magnitude) at every point of the flow for any given instant of time, such as the flow in long straight pipe. However, if velocity vectors vary from place to place at every point of the flow at a given instant then the flow is *non-uniform flow*, such as the flow in expanding pipe, contracting pipe or bending pipe.

For uniform flow, the spatial derivative of velocity vector of any fluid particle in the field is zero, that is, the uniform flow is the flow with zero convective acceleration

$$(\mathbf{u} \cdot \nabla)\mathbf{u} = 0 \quad (3.12)$$

For non-uniform flow the convective acceleration is not zero

$$(\mathbf{u} \cdot \nabla)\mathbf{u} \neq 0 \tag{3.13}$$

For example, in Fig. 3-5a, the velocity at points a, b and c have the equal magnitude, $u_a = u_b = u_c$, and the identical direction at the same instant, so this flow is uniform, the velocity profile is similar to the one in Fig. 3-4a. However, the flows in the converging pipe and the bending pipes as in Fig. 3-5b is non-uniform. In fact, most actual flows are non-uniform.

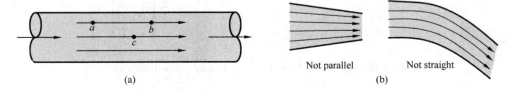

Fig. 3-5 Streamlines in (a) uniform flow and (b) non-uniform flow

GRADUALLY VARIED FLOW AND RAPIDLY VARIED FLOW

According to the variation of velocity with respect to the spatial distance, the non-uniform flow would be gradually varied flow or rapidly varied flow.

A *gradually varied flow* is a flow in which the velocity changes slowly with distance in the flow direction, that is, the convective acceleration of fluid particle is very small, $(\mathbf{u} \cdot \nabla)\mathbf{u} \approx 0$, or the streamlines are nearly parallel to each other. Vice versa, it is *rapidly varied flow*.

In gradually varied flow, the cross section is nearly a plane, the directions of velocities on the cross section are almost the same; the dynamic pressure distribution on the cross section is same as that of the *hydrostatic pressure distribution*:

$$z + \frac{p}{\gamma} = C \tag{3.14}$$

In rapidly varied flow, the dynamic pressure distribution on the cross section is not the same as that of the hydrostatic pressure distribution, the piezometric head is not a constant any more:

$$z + \frac{p}{\gamma} \neq C \tag{3.15}$$

Some examples of the gradually and rapidly varied flow are shown in Fig. 3-6. At section 1, the pizeometric heads at the top and bottom points are leveled as illustrated in the figure, which indicates this cross section is located in a gradually varied flow or an uniform flow. At section 5, the pizeometric head at the outside wall of the bend is larger than the one at the inner wall of the bend by Δh, which indicates a rapidly varied flow occurs in the bend. However, the section 2, the middle plane of the venturi pipe between section 1 and 3, is a gradually varied flow because of the nearly parallel streamlines at this section, even though the rapidly varied flows occur in the converging and diverging pipes.

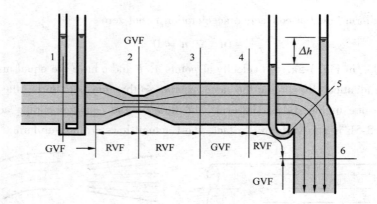

Fig. 3-6 Gradually varied flows (GVF) and rapidly varied flows (RVF)

3.2.5 Flow Types

Combining the above we can classify any flow into one of four types:

1. *Steady uniform flow.* Conditions do not change either with position in the stream or with time [$\partial q/\partial t = 0$, $(\mathbf{u} \cdot \nabla)\mathbf{u} = 0$], i.e., the flow of water in a constant-diameter straight pipe at constant velocity, as in Fig. 3-2, when the water level in the tank is constant.

2. *Steady non-uniform flow.* Conditions change from point to point in the stream but do not change with time [$\partial q/\partial t = 0$, $(\mathbf{u} \cdot \nabla)\mathbf{u} \neq 0$], i.e., the flow in the converging-diverging pipe in Fig. 3-2 when the water level in the tank is constant, or the flow in a tapering pipe with constant velocity at the inlet, while velocity will change as you move along the length of the pipe toward the exit.

3. *Unsteady uniform flow.* At a given instant in time the conditions at every point are the same, but will change with time [$\partial q/\partial t \neq 0$, $(\mathbf{u} \cdot \nabla)\mathbf{u} = 0$], i.e., the flow in the constant-diameter straight pipe at constant velocity in Fig. 3-2 when the water level in the tank varies with time, or the flow in a pipe of constant diameter connected to a pump pumping at a constant rate which is then switched off.

4. *Unsteady non-uniform flow.* Every condition of the flow may change from point to point [$(\mathbf{u} \cdot \nabla)\mathbf{u} \neq 0$] and with time at every point ($\partial q/\partial t \neq 0$). For example waves in a channel, the flow in the converging-diverging pipe in Fig. 3-2 when the water level in the tank varies with time.

EXAMPLE 3-1 Acceleration and flow patterns

The velocity field of a plane flow is given by $\mathbf{u} = t\mathbf{i} - y\mathbf{j} + z\mathbf{k}$ m/s. (a) Is this flow a steady or unsteady flow; (b) Is this flow an uniform or nonuniform flow? (c) Compute the acceleration at point (1, 2) when $t = 1$s.
Solution:

(a) The time derivative of the velocity vector for this flow is

$$\frac{\partial \mathbf{u}}{\partial t} = \mathbf{i} \neq 0$$

Thus this flow is unsteady flow.

(b) The convective derivative of the velocity vector for this flow is

$$(\mathbf{u} \cdot \nabla)\mathbf{u} = \left(u_x \frac{\partial u_x}{\partial x} + u_y \frac{\partial u_x}{\partial y} + u_z \frac{\partial u_x}{\partial z} \right)\mathbf{i} + \left(u_x \frac{\partial u_y}{\partial x} + u_y \frac{\partial u_y}{\partial y} + u_z \frac{\partial u_y}{\partial z} \right)\mathbf{j}$$

$$+ \left(u_x \frac{\partial u_z}{\partial x} + u_y \frac{\partial u_z}{\partial y} + u_z \frac{\partial u_z}{\partial z} \right)\mathbf{k}$$

$$= y\mathbf{j} + z\mathbf{k} \neq 0$$

Thus this flow is non-uniform flow.

(c) From Eq. 3.9 we have the acceleration vector is

$$\mathbf{a} = \frac{d\mathbf{u}}{dt} = \frac{\partial \mathbf{u}}{\partial t} + (\mathbf{u} \cdot \nabla)\mathbf{u} = \mathbf{i} + y\mathbf{j} + z\mathbf{k}$$

Thus the acceleration at point $(1, 1, 2)$ when $t = 1\,\text{s}$ is

$$a = \sqrt{a_x^2 + a_y^2 + a_z^2} = \sqrt{1 + 1 + 2^2} = 2.45\,\text{m/s}^2$$

3.3 Eulerian Concepts for Flow Description

3.3.1 Flow Visualization: Streamline, Pathline and Streakline

In the context of laboratory experiment flows, as well in numerical solutions of *computational fluid dynamics* (CFD) visualization is an absolute necessity. It is worth noting that even for theoretical analyses much can often be learned from a well-constructed sketch that emphasizes the key features of the flow physics in a situation of interest. The patterns of flow can be visualized in a dozen different ways, and you can view these sketches or photographs and learn a great deal qualitatively and often quantitatively about the flow. In this context, the visualization techniques we will describe are very standard and they are: streamlines, pathlines and streaklines.

1. A *streamline* is a line everywhere tangent to the local velocity vector at a given instant.
2. A *pathline* is the actual path traversed by a given fluid particle.
3. A *streakline* is the locus of fluid particles which have earlier sequentially passed through a prescribed point in the flow.

Note that a streamline is an instantaneous line, while the pathline and the streakline are generated by the passage of time.

STREAMLINE

It is often helpful to construct lines that indicate the direction and pattern of fluid flow. One such type of line (which is imaginary) is the streamline, which is a line along which all fluid particles have, at a given instant, velocity vectors that are tangential to the line (Fig. 3-7a). Streamlines provide a picture of the complete flow field at a given instant—an instantaneous picture of the directions taken by many particles as illustrated in Fig. 3-8.

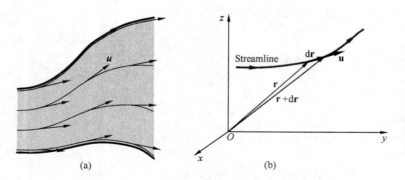
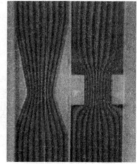

Fig. 3-7 Definition of streamline

Fig. 3-8 Photo of streamlines
(*Courtesy of Mao Genhai, Zhejiang University*)

Streamline is an Eulerian concept, of which the equation can be obtained from the definition:

$$d\mathbf{r} \times \mathbf{u} = 0 \tag{3.16}$$

or

$$\frac{dx}{u_x} = \frac{dy}{u_y} = \frac{dz}{u_z} \tag{3.17}$$

where $d\mathbf{r}$ is the infinitesimal length vector along the streamline, $d\mathbf{r} = dx\mathbf{i} + dy\mathbf{j} + dz\mathbf{k}$ as sketched in Fig. 3-7, $\mathbf{u}$ is the local velocity vector.

For a known velocity field in a two-dimensional flow as shown in Fig. 3-7a, the following differential equation for the streamline in a xy-plane can be obtained from Eq. 3.17

$$\left(\frac{dy}{dx}\right)_{\text{along a streamline}} = \frac{u_y}{u_x} \tag{3.18}$$

In some simple cases, Eq. 3.16 may be solvable analytically; in the general case, it must be solved numerically. Eq. 3.18 is also the slope of the curve in a xy-plane.

Based on the definition of the streamline, some physical attributes of streamlines should be considered:

(a) Streamlines display a snapshot of the entire flow field (or some selected portion of it) at a single instant in time with each streamline starting from a different selected point in the flow field. Thus, in a time-dependent flow a visualization based on streamlines will be constantly changing, possibly in a rather drastic manner, i. e., the outflow from a tank with a variable water level as illustrated in Fig. 3-9b.

(b) There is no component of velocity across a streamline. No fluid particle can cross a streamline and streamlines never cross. This follows from their definition, namely, they are in the direction of the velocity vector, and if they were to cross at any point in a flow field there would

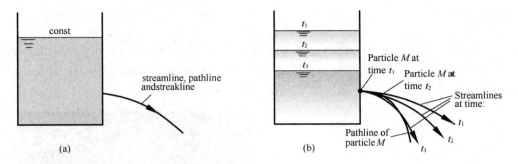

Fig. 3-9 Streamlines for (a) steady and (b) unsteady jet flow from a tank

have to be two velocity vectors at the same point. For the same reason, streamline is never a folding line. Actually, streamlines do cross at some special *singularities*, e. g. *stagnation point*, *sink* and *source* (Fig. 3-10).

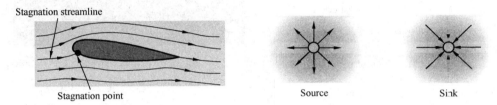

Fig. 3-10 Singularities: stagnation point, source and sink

(c) In flows bounded by solid walls or surfaces, the wall (or surface) can be defined to be a streamline. This is because flow cannot penetrate a solid boundary, implying that the only nonzero component(s) of the velocity vector very close to the surface is(are) the tangential one(s) —i. e. , those in the direction of the streamlines and there is no flow normal to it. For ideal fluid flow the fluid slips over the surface and follows it exactly. In any event, we can say that the surface of a solid body is a *streamline surface*. Sometimes, we pull out a bundle of streamlines from inside of a general flow for analysis as shown in Fig. 3-11. Such a bundle is called *streamtube*. By definition the fluid within a streamtube is confined there because it cannot cross the streamlines; thus the streamtube walls need not be solid but may be fluid surfaces.

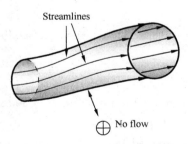

Fig. 3-11 Streamtube

Streamtube is very useful in analyzing flows. If one aligns a coordinate along the streamtube then the flow through it is *one-dimensional*.

For steady flow there are no changes with respect to time and hence the streamline pattern does not change. But the pattern does change from moment to moment when the flow is unsteady. For uniform flow streamlines must be straight and parallel (Fig. 3-5a). In the case of curved streamlines (see Fig. 3-5b) velocities at different points have different directions and the flow is nonuniform. However, for the velocities to be equal they must all have the same direction.

PATHLINE

A *pathline* is the trajectory line taken by a single fluid particle during a given time interval as shown in Fig. 3-12. It is equivalent to a time-exposure photograph which traces the movements of the particle marked, e. g., the trace of firework illustrated in Fig. 3-13.

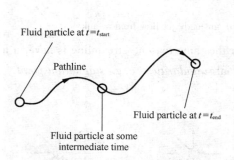

Fig. 3-12 Definition of pathline

Fig. 3-13 Pathline of firework
[Source: Mao at el. (2006)]

Pathline is a Lagrangian concept. For a differential length, from Eq. 3.2 it could be defined as

$$\left.\begin{array}{l} dx = u_x dt \\ dy = u_y dt \\ dz = u_z dt \end{array}\right\} \quad (3.19)$$

Thus, the equation for pathline is

$$\frac{dx}{u_x} = \frac{dy}{u_y} = \frac{dz}{u_z} = dt \quad (3.20)$$

where t is the variables, x, y and z are the dependent variables with respect to time t.

It is clear that if the velocity field is independent of time this is equivalent to the formulation for streamlines; but if it is time dependent the result will be different. That is, if the velocity field is steady, individual fluid particles will follow streamlines. Thus, for steady flow, pathlines are identical to streamlines (Fig. 3-9a). On the other hand, pathlines may cross. The speed and direction of particles which leave the fixed point may vary with time (see particle M in Fig. 3-9b), i. e., the flow is unsteady.

STREAKLINE

A *streakline* joins, at a given time, all particles that have passed through a given point (e. g., A in Fig. 3-14) over a given period of time. For example, a streakline could be a dye line or a smoke stream produced from a continuous supply of dye or smoke as illustrated in Fig. 3-15.

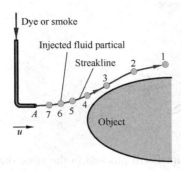

Fig. 3-14 Definition of streakline. Labeled tracer particles (1 through 7) were introduced sequentially.

Fig. 3-15 A model of a Cessna 182 with helium-filled bubbles showing streaklines of wingtip vortices tested in the Rensselaer Polytechnic Institute Subsonic Wind Tunnel(*Courtesy of Wikipedia*)

In comparing this with the case of pathlines we observe that the definition of a pathline involved only a single fluid element; but the current definition for a streakline concerns, evidently, a large number. Thus, in a sense, a streakline combines properties of streamlines and pathlines: it is made up of many pathlines (actually, fluid parcels each of which could produce a pathline), but all of these are observed simultaneously just as is a streamline traversing an entire flow field. In a time-dependent flow field each fluid element considered will have its trajectory determined not only by the velocity changes it encounters in traversing the flow field (as happened in the case of a pathline) but also by the changes in initial conditions at its point of origin. It is quite difficult to sketch such behavior except in very simple cases. But we shall note that the dye injection technique used in the Reynolds experiment in Chapter 5 results in a streakline.

Streaklines and pathlines are often compared by noting that the former corresponds to continuous injection of marker particles and instantaneous observation of them, whereas the pathline is formed by instantaneous injection and continuous observation.

FLOW VISUALIZATION

If the flow is steady, the path taken by a marked particle (a pathline) will be the same as the line formed by all other particles that previously passed through the point of injection (a streakline). For such cases these lines are tangent to the velocity field. Hence, pathlines, streamlines, and streaklines are the geometrically identical for steady flows. For unsteady flows none of these three types of lines need to be the same. Often one sees pictures of 'streamlines' made visible by the injection of smoke or dye into a flow as is shown in Fig. 3-8. Actually, such pictures show streaklines rather than streamlines. However, for steady flows the two are identical; only the nomenclature is incorrectly used. Streamlines are easily generated mathematically while pathlines and streaklines are obtained through experiments.

Consider the flow over an airfoil (Fig. 3-16). Assuming steady flow, the velocity u_1 at point 1 will always be the same in magnitude and direction, velocity u_2 at point 2 will always be the same and so on. But $u_1 \neq u_2 \neq \ldots \neq u_5$. The line drawn as a tangent to these directions forms a stream-

line S for steady non-uniform flow.

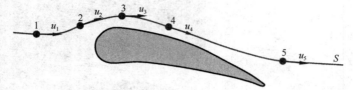

Fig. 3-16 Flow past an airfoil

Under steady conditions, every particle passing through point 1 will proceed in the same direction to point 2, and throug point 2 will always have the same velocity u_2, through point 3 will always have the same velocity u_3, and so on. Thus, starting at point 1, the route taken by every single particle (the pathline) must be the same as the route taken by a string of particles (the streakline) and these lines must coincide with the streamline.

During unsteady flow the streamline pattern (showing the instantaneous directions taken by 'all' particles) can change from moment to moment and will not be the same as a series of pathlines, each of which shows the historical route taken by a different particle over a given or different time period. Streaklines will also differ from streamlines and pathlines (Fig. 3-17).

For a two-dimensional flow in Fig. 3-17, the fluid particle M at time t_0 is located at the outlet section of the nozzle A as shown in Fig. 3-17a, the fluid out of nozzle A flows with a velocity of $\mathbf{u}_1$ from time t_0 to t_1 and of $\mathbf{u}_2$ from time t_1 to t_2. Thus, at time t_1, the fluid particle M arrives at point B, then the pathline of fluid particle M is AB. The streamline and streakline at time t_1 are also AB

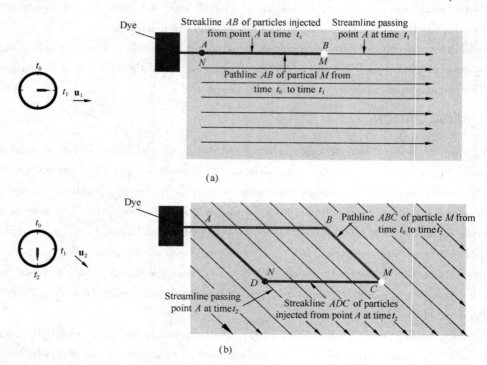

Fig. 3-17 Steamline, pathline and streakline in an unsteady flow(a) from time t_0 to t_1 and (b) from time t_1 to t_2.

because the flow is a steady flow with a velocity of $\mathbf{u}_1$ from time t_0 to t_1. At the end of time t_1, a fluid particle N is at the outlet of nozzle, and the flow suddenly changes its direction to $\mathbf{u}_2$. At time t_2, fluid particle M reaches point C while fluid particle N reaches point D, which indicates all fluid particles at line AB at time t_1 have reached to line DC at the end of time t_2. So, the flow is an unsteady flow from time t_0 to t_2, the pathline of fluid particle M is line ABC, the streakline and the streamline of all fluid particles out of the nozzle at time t_2 are line ADC and line AD respectively. It indicates that streamline, streakline and pathline do not coincide in an unsteady flow.

EXAMPLE 3-2 Streamlines for a steady plane flow

Given the steady two-dimensional velocity distribution

$$u_x = Kx, \qquad u_y = -Ky, \qquad u_z = 0 \tag{1}$$

where K is a positive constant. Compute and plot the streamlines of the flow, including directions, and give some possible interpretations of the pattern.

Solution:

Since time does not appear explicitly in Eq. 1, the motion is steady, so that streamlines, pathlines, and streaklines will coincide. Since $u_z = 0$ everywhere, the motion is two dimensional, in the xy plane. The streamlines can be computed by substituting the expressions for u_x and u_y into Eq. 3.17:

$$\frac{dx}{Kx} = \frac{dy}{-Ky}$$

or

$$\int \frac{dx}{x} = -\int \frac{dy}{y}$$

Integrate to give $\ln x = -\ln y + \ln C$ or

$$xy = C \tag{2}$$

This is the general expression for the streamlines, which are hyperbolas. The complete pattern is plotted in Fig. E3-2 by assigning various values to the constant C. The arrowheads can be determined only by returning to Eq. 1 to ascertain the velocity component directions, assuming K is positive. For example, in the upper right quadrant ($x > 0$, $y > 0$), u_x is positive and u_y is negative; hence the flow moves down and to the right, establishing the arrowheads as shown.

DISCUSSION Note that the streamline pattern is entirely independent of constant K. It could represent the impingement of two opposing streams, or the upper half could simulate the flow of a single downward stream against a flat wall. Taken in isolation, the upper right quadrant is similar to the flow in a 90° corner. It should also note the peculiarity that the two streamlines ($C = 0$) have opposite directions and intersect. This is possible only at a point where $u_x = u_y = u_z = 0$, which occurs at the origin in this case. Such a point of zero velocity is called a *stagnation point*.

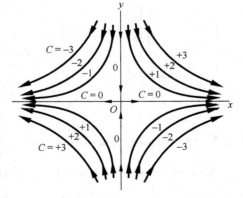

Fig. E3-2 Streamlines for the velocity distribution given by Eq. 1 for $K > 0$

EXAMPLE 3-3 Streamlines for an unsteady plane flow

Given the two-dimensional velocity distribution

$$u_x = x + t, \quad u_y = -y + t, \quad u_z = 0 \qquad (1)$$

Compute and plot: (a) the streamlines passing point $M(-1, -1)$ at time $t = 0$; (b) the pathline of fluid particle located at point M at time $t = 0$.

Solution:

Since time appears explicitly in Eq. 1, the motion is unsteady, so that streamlines and pathlines, will not coincide.

(a) The streamline can be computed by substituting the expressions of u_x and u_y into Eq. 3.17:

$$\frac{dx}{x + t} = \frac{dy}{-y + t}$$

Integrating it gives $\ln(x + t) = -\ln(y - t) + \ln C$ or

$$(x + t)(y - t) = C \qquad (2)$$

This is the general expression for the streamlines. At time $t = 0$, the streamline equation becomes

$$xy = C$$

Because the instant streamline at time $t = 0$ passing point $M(-1, -1)$, thus we have $C = 1$, and the instant streamline at time $t = 0$ passing point M would be

$$xy = 1$$

Thus, this streamline, AMB, is hyperbola as sketched in Fig. E3-3.

(b) The pathline can be computed by substituting the expressions for u_x and u_y into Eq. 3.20:

$$\frac{dx}{dt} = x + t, \quad \frac{dy}{dt} = -y + t$$

We solve these differential equations and obtain

$$\begin{cases} x = C_1 e^t - t - 1 \\ y = C_2 e^{-t} + t - 1 \end{cases} \qquad (3)$$

At $t = 0$, the fluid particle is located at point M, thus $x = -1$, $y = -1$. Substituting into Eq. 3 gives $C_1 = 0$, $C_2 = 0$. Thus the pathline equation of this fluid particle is

$$\begin{cases} x = -t - 1 \\ y = t - 1 \end{cases}$$

or

$$x + y + 2 = 0$$

The line MC sketched in Fig. E3-3 is the pathline ($t > 0$) of the particle located at point M at time $t = 0$, which does not coincide with streamline.

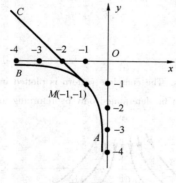

Fig. E3-3

3.3.2 Streamtube, Total Flow and Control Volume

A useful technique in fluid flow analysis is to consider only a part of the total fluid in isolation from the rest. This can be done by imagining a tubular surface formed by streamlines along which the fluid flows. This tubular surface is known as a *streamtube* as shown in Fig. 3-11. The walls of a streamtube are made of streamlines. As we have seen above, fluid cannot flow across a streamline, so fluid cannot cross a streamtube wall. When the streamtube is viewed as a solid wall of flow, the flow in it must be a steady flow, because the position and the shape of the streamtube will move with time if the flow is unsteady, whereas the solid wall never be.

The surface normal to the flow direction in a streamtube is termed the *cross section*. The cross section would be a plane if the flow is uniform, e. g. section 1-1 in Fig. 3-18 and would be a curved surface if the flow is non-uniform, e. g. section 2-2 in Fig. 3-18.

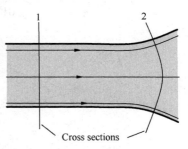

Fig. 3-18 Cross sections

The *element flow* is the fluid flow in a streamtube with an infinitesimal cross section. Its geometric features are same as the one for a streamtube, and the flow parameters at any point (such as velocity, pressure, elevation, etc.) on the cross section are same because it is assumed to be one dimensional with infinitesimal area of the cross section.

The *control volume* is a fixed volume in space (a geometric entity, independent of mass) through which the fluid may flow in or out as illustrated in Fig. 3-19. The surface area that completely encloses the control volume is *control surface*. And the cross-sections through which fluids flow into and out of the control volume are *control cross sections*.

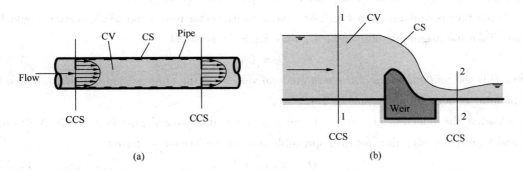

Fig. 3-19 Control volume (CV), control surface (CS) and control cross sections (CCS) in
(a) pipe flow and (b) open channel flow passing weir

The *total flow* is the flow in a control volume with a limited size cross section and a solid boundary, e. g. a flow in a pipe. It is an assemble of numerous element flows. The flow parameters vary crossing the cross section.

3.3.3 Flow Rate

VOLUME FLOW RATE

In fluid mechanics, more commonly we need to know the flow rate and the velocity of a flow.

The *volume flow rate*, more commonly known as *discharge*, designated as Q, is the volume of fluid past a cross section per unit time.

Suppose a flow passing an arbitrary surface as shown in Fig. 3-20a. Typically, **u** varies with position and pass through dA at an angle θ off the normal vector. Let **n** be defined as the unit vector normal to dA. Then the volume amount of fluid through the element surface dA in time dt is

$$d V = (\mathbf{u} \cdot \mathbf{n})\, dt\, dA = u\, dt\, dA \cos\theta$$

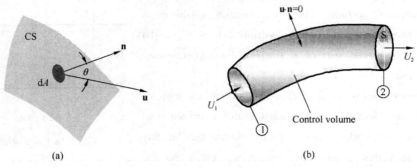

Fig. 3-20 (a) Volume flow rate of flow through an arbitrary surface; (b) one dimensional incompressible uniform flow in a pipe

Integrating dV/dt gives the total volume rate of flow Q through the control surface

$$Q = \int_{CS}(\mathbf{u} \cdot \mathbf{n})\, dA = \int_{CS} u_n\, dA \qquad (3.21)$$

where u_n is the velocity component of **u** normal to the control surface.

If the flow is uniform as in Fig. 3-4a, the velocity vector **u** over the whole section would be same. Take the magnitude of u_n to be U, thus Eq. 3.21 becomes

$$Q = UA \qquad (3.22)$$

where U is the *uniform velocity*, the magnitude of velocity at the cross section of uniform flow with an uniform velocity profile as in Fig. 3-4a.

Assume the velocities at sections 1 and 2 of a one dimensional pipe flow in Fig. 3-20b are U_1 and U_2 respectively, thus for incompressible flow at any instant we have

$$Q = U_1 A_1 = U_2 A_2 \qquad (3.23)$$

because of the identical flow rate passing through section 1 and section 2.

Equation 3.23 is the *continuity equation* for the one dimensional incompressible uniform flow.

MASS FLOW RATE

The *mass flow rate*, designated as $\dot{m}$, is the mass of fluid past a cross section per unit time. It can be measured by catching all fluid coming out of the section in a container over a fixed time period, and then measuring the weight of fluid in the container and dividing this by the time taken to collect this fluid, which may give the rate of accumulation of mass, the mass flow rate.

Volume flow can be multiplied by density to obtain the mass flow. If density varies over the surface, it must be part of the surface integral

$$\dot{m} = \int_{CS} \rho(\mathbf{u} \cdot \mathbf{n})\, dA = \int_{CS} \rho u_n\, dA \qquad (3.24)$$

If the flow is incompressible, the density is constant, it comes out of the integral and a directly proportional result:

$$\dot{m} = \rho Q = \rho \int_{CS} u_n \mathrm{d}A$$

If the flow is uniform, it becomes

$$\dot{m} = \rho U A \tag{3.25}$$

3.4 Kinematic Description of Fluid Element

3.4.1 Motion of Fluid Element

By definition, a fluid cannot sustain a shear stress no matter how small. Thus, in the area of fluid kinematics, it is necessary to relate the *deformation* of a fluid element to the *stresses* that act on it and describe the relationship between the magnitude of a shear stress and the accompanying fluid deformation for so-called '*Newtonian fluids.*'

The possible motions of a small element of fluid are illustrated in Fig. 3-21. They are *translation, rotation, shear strain,* and *linear strain (dilatation)*. Since the translation and rotation are both solid body type motions, in which no part of the element moves relative to any other part, these types of motion do not, by themselves, give rise to 'viscous' stresses. On the other hand, it is clear from the figure that shear strain and dilatation involve motions in which parts of the fluid element move relative to one another, resulting in a *deformation* of the fluid element. Therefore, these latter types of motion are expected to be responsible for the generation of viscous stress in a flowing fluid.

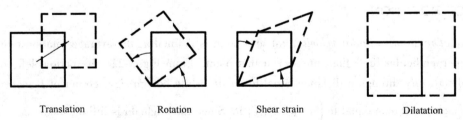

Fig. 3-21 Motions of flow element

The study of fluid dynamics is further complicated by the fact that all four types of motion or deformation usually occur simultaneously. Because fluid elements may be in constant motion, it is preferable in fluid dynamics to describe the motion and deformation of fluid elements in terms of *rates*. In particular, we discuss *velocity* (rate of translation), *angular velocity* (rate of rotation), *linear strain rate* (rate of linear strain), and *shear strain rate* (rate of shear strain).

By considering a differential fluid element, the rates of rotation, shear strain, and dilatation can be expressed in terms of the flow velocity field and derivatives of velocity, while the rate of translation is simply equal to the velocity.

Consider a fluid element in plane xOy illustrated in Fig. 3-22, take the velocity at the bottom left corner O to be u_x and u_y respectively. A short time dt later, the bottom left corner of the fluid element has been *translated* along the x axis by a distance that is equal to $u_x dt$ and along the y axis by $u_y dt$. Two fluid lines OA and OB of fluid element, initially perpendicular at time t, move and deform so that at $t + dt$ they have slightly different lengths $O'A'$ and $O'B'$ and are slightly off the perpendicular by angles $d\alpha$ and $d\beta$. Such deformation occurs kinematically because A, O, and B have slightly different velocities when the velocity field **u** has spatial gradients. All these differential changes in the velocities at points of A and B are noted in Fig. 3-22.

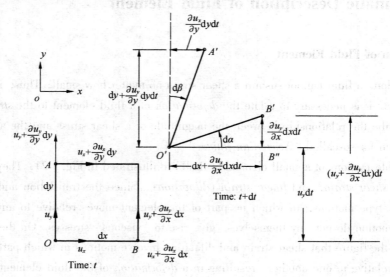

Fig. 3-22 Angular velocity and strain rate of two fluid lines deforming in the xOy plane

RATE OF ROTATION

Rate of rotation (angular velocity, ω) at a point is defined as the average rotation rate of two initially perpendicular lines that intersect at that point. From Fig. 3-22, the bottom left corner of O moved *vertically* through a distance equal to $u_y dt$, while bottom *right* corner of B moved vertically through a distance equal to $\left(u_y + \dfrac{\partial u_y}{\partial x}dx\right)dt$. Since the angle $d\alpha$ is infinitesimal, $d\alpha$ is directly related to velocity derivatives in the limit of small dt

$$d\alpha \approx \tan d\alpha = \frac{\left(u_y + \dfrac{\partial u_y}{\partial x}dx\right)dt - u_y dt}{dx} = \frac{\partial u_y}{\partial x}dt$$

Similarly, $d\beta$ can be obtained

$$d\beta \approx \tan d\beta = \frac{\left(u_x + \dfrac{\partial u_x}{\partial y}dy\right)dt - u_x dt}{dy} = \frac{\partial u_x}{\partial y}dt$$

Thus, according to the definition of the angular velocity ω_z about the z axis, the average rate of counterclockwise turning of the two lines, from Fig. 3-22 we have

$$\omega_z = \frac{1}{2}\left(\frac{d\alpha}{dt} - \frac{d\beta}{dt}\right) = \frac{1}{2}\left(\frac{\partial u_y}{\partial x} - \frac{\partial u_x}{\partial y}\right) \quad (3.26)$$

In exactly similar manner we determine the other two rates

$$\omega_x = \frac{1}{2}\left(\frac{\partial u_z}{\partial y} - \frac{\partial u_y}{\partial z}\right) \qquad \omega_y = \frac{1}{2}\left(\frac{\partial u_x}{\partial z} - \frac{\partial u_z}{\partial x}\right) \quad (3.27)$$

Thus, the angular velocity vector in Cartesian coordinate system is

$$\boldsymbol{\omega} = \mathbf{i}\,\omega_x + \mathbf{j}\,\omega_y + \mathbf{k}\,\omega_z = \frac{1}{2}\begin{vmatrix} \mathbf{i} & \mathbf{j} & \mathbf{k} \\ \frac{\partial}{\partial x} & \frac{\partial}{\partial y} & \frac{\partial}{\partial z} \\ u_x & u_y & u_z \end{vmatrix} = \frac{1}{2}(\nabla \times \mathbf{u}) \quad (3.28)$$

RATE OF SHEAR STRAIN

Shear strain rate, at a point is defined as half of the rate of decrease of the angle between two initially perpendicular lines that intersect at the point. From Fig. 3-22 we have

$$\varepsilon_{xy} = \frac{1}{2}\left(\frac{d\alpha}{dt} + \frac{d\beta}{dt}\right) = \frac{1}{2}\left(\frac{\partial u_y}{\partial x} + \frac{\partial u_x}{\partial y}\right)$$

The above equation can be easily extended to three dimensions. The shear strain rates in Cartesian coordinates thus are

$$\varepsilon_{xy} = \frac{1}{2}\left(\frac{\partial u_y}{\partial x} + \frac{\partial u_x}{\partial y}\right) \qquad \varepsilon_{yz} = \frac{1}{2}\left(\frac{\partial u_z}{\partial y} + \frac{\partial u_y}{\partial z}\right) \qquad \varepsilon_{zx} = \frac{1}{2}\left(\frac{\partial u_x}{\partial z} + \frac{\partial u_z}{\partial x}\right) \quad (3.29)$$

RATE OF LINEAR STRAIN

Linear strain rate is defined as the rate of increase in length per unit length. For the line OB in Fig. 3-22, we have

$$\varepsilon_{xx} = \frac{d}{dt}\left(\frac{O'B'' - OB}{OB}\right) = \frac{d}{dt}\left\{\frac{[dx + (u_x + \frac{\partial u_x}{\partial x}dx)dt - u_x dt] - dx}{dx}\right\} = \frac{\partial u_x}{\partial x}$$

Thus, we can have the linear strain rate in Cartesian coordinates as

$$\varepsilon_{xx} = \frac{\partial u_x}{\partial x} \qquad \varepsilon_{yy} = \frac{\partial u_y}{\partial y} \qquad \varepsilon_{zz} = \frac{\partial u_z}{\partial z} \quad (3.30)$$

3.4.2 Rotational and Irrotational Flows

Twice of the angular velocity of a fluid particle, Eq. 3.28, is called the *vorticity*, defined mathematically as the curl of the velocity vector

$$\boldsymbol{\zeta} = \text{curl}\,\mathbf{u} = \nabla \times \mathbf{u} = 2\boldsymbol{\omega} \quad (3.31)$$

Vorticity is a measure of rotation of a fluid particle. If the vorticity in a region of the flow is negligible or zero, fluid particles there are not rotating. Thus, the flow in that region is called *ir-*

rotational

$$\zeta = \text{curl}\,\mathbf{u} = 2\boldsymbol{\omega} = 0 \qquad (3.32\text{a})$$

or

$$\left.\begin{array}{ll} \omega_x = \dfrac{1}{2}\left(\dfrac{\partial u_z}{\partial y} - \dfrac{\partial u_y}{\partial z}\right) = 0 & \dfrac{\partial u_z}{\partial y} = \dfrac{\partial u_y}{\partial z} \\[6pt] \omega_y = \dfrac{1}{2}\left(\dfrac{\partial u_x}{\partial z} - \dfrac{\partial u_z}{\partial x}\right) = 0 & \dfrac{\partial u_x}{\partial z} = \dfrac{\partial u_z}{\partial x} \\[6pt] \omega_z = \dfrac{1}{2}\left(\dfrac{\partial u_y}{\partial x} - \dfrac{\partial u_x}{\partial y}\right) = 0 & \dfrac{\partial u_y}{\partial x} = \dfrac{\partial u_x}{\partial y} \end{array}\right\} \qquad (3.32\text{b})$$

If any one of ω_x, ω_y, ω_z cannot be neglected, the fluid particle that happens to occupy that point in space is rotating, this flow is called *rotational*. Such flows can be incompressible or compressible, steady or unsteady.

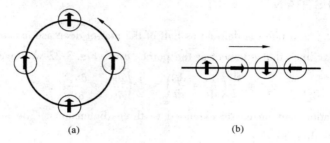

Fig. 3-23 (a) Irrotational flow and (b) rotational flow

For the *irrotational* flow (also called *potential flow*), $\nabla \times \mathbf{u} = 0$, the *angular velocity* of any infinitesimal fluid element is *zero* (i. e. is not rotating) as indicated by the above derivation of angular velocity, Eq. 3-32, and in Fig. 3-23a. Some examples of irrotational flows are uniform flow and ' potential vortex' flow. *Rotational flow* is also called *vortex flow*, in which fluid particles (elements) will not only translate (or deform) but also rotate about its instant axis as shown in Fig. 3-23b. The most common case of rotational and irrotational flows in engineering practice always occurs near a boundary layer as illustrated in Fig. 3-24: fluid particles within the viscous *boundary layer* near a solid wall are rotational(and thus have nonzero vorticity), while fluid particles outside

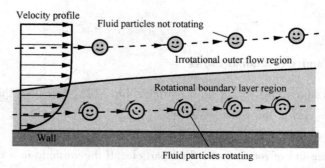

Fig. 3-24 Rotational flow and irrotational flow inside and outside a boundary layer

the boundary layer are irrotational (and their vorticity is zero)

Rotation does not bring about a relative motion between different parts of the fluid and therefore does not contribute to the generation of viscous stress.

3.5 Differential Equations of Motion

3.5.1 Differential Continuity Equation

Let us imagine an infinitesimally small fluid element (control volume) in the flow, with a differential volume, $d\mathcal{V}$. The fluid element is infinitesimal in the same sense as differential calculus; however, it is large enough to contain a huge number of molecules so that it can be viewed as a *continuum* medium. The fluid element with six faces (dx, dy, dz) is fixed in space with the fluid moving through it as shown in Fig. 3-25, and the fundamental physical principles are applied to just the fluid element itself.

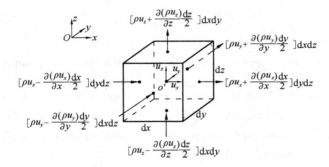

Fig. 3-25 Infinitesimal control volume

Take the coordinates Ox, Oy and Oz along the directions of the element edges, and the velocity components in the x-, y- and z-directions at the center point O' of the element to be u_x, u_y and u_z. Thus, the velocities in the centers of surfaces can be obtained by the Taylor series expansion, e.g.

Velocity on the left surface $= u_x - \dfrac{\partial u_x}{\partial x}\dfrac{dx}{2}$

Velocity on the right surface $= u_x + \dfrac{\partial u_x}{\partial x}\dfrac{dx}{2}$

Thus, the net mass flowing out of the element in the x-direction is

$$\Delta m_x = \left[\rho u_x + \frac{\partial(\rho u_x)}{\partial x}\frac{dx}{2}\right]dydzdt - \left[\rho u_x - \frac{\partial(\rho u_x)}{\partial x}\frac{dx}{2}\right]dydzdt = \frac{\partial(\rho u_x)}{\partial x}dxdydzdt$$

In a similar manner the net mass flowing out of the element in the y and z-direction can be obtained as

$$\Delta m_y = \frac{\partial(\rho u_y)}{\partial y}dxdydzdt$$

$$\Delta m_z = \frac{\partial(\rho u_z)}{\partial z} dx\,dy\,dz\,dt$$

Thus, the total net mass flowing out of the element is

$$\Delta m_x + \Delta m_y + \Delta m_z = \left[\frac{\partial(\rho u_x)}{\partial x} + \frac{\partial(\rho u_y)}{\partial y} + \frac{\partial(\rho u_z)}{\partial z}\right] dx\,dy\,dz\,dt$$

Based on the *continuum assumption* of the fluid, the *conservation law of mass* can be used in the analysis of flowing fluids. Thus, the total net mass out of the control volume in the unit time dt must equal to the decrease of the mass in the control volume due to the change of the fluid density

$$\left[\frac{\partial(\rho u_x)}{\partial x} + \frac{\partial(\rho u_y)}{\partial y} + \frac{\partial(\rho u_z)}{\partial z}\right] dx\,dy\,dz\,dt = -\frac{\partial \rho}{\partial t} dx\,dy\,dz\,dt$$

so

$$\frac{\partial \rho}{\partial t} + \frac{\partial(\rho u_x)}{\partial x} + \frac{\partial(\rho u_y)}{\partial y} + \frac{\partial(\rho u_z)}{\partial z} = 0 \tag{3.33}$$

or

$$\frac{\partial \rho}{\partial t} + \nabla \cdot (\rho \mathbf{u}) = 0 \tag{3.34}$$

where ∇ is the *vector-gradient operator*, $\nabla = \mathbf{i}\frac{\partial}{\partial x} + \mathbf{j}\frac{\partial}{\partial y} + \mathbf{k}\frac{\partial}{\partial z}$ in Cartesian coordinates.

Equations 3.33 and 3.34 are the general forms of the *differential continuity equations*. The flow may be either steady or unsteady, viscous or frictionless, compressible or incompressible. However, the equation does not allow for any source or sink singularities within the element.

In cylindrical coordinates Eq. 3.34 becomes

$$\frac{\partial \rho}{\partial t} + \frac{1}{r}\frac{\partial(\rho r u_r)}{\partial r} + \frac{1}{r}\frac{\partial(\rho u_\theta)}{\partial \theta} + \frac{\partial(\rho u_z)}{\partial z} = 0 \tag{3.35}$$

STEADY FLOW

For the steady flow, $\partial/\partial t = 0$ and all properties are functions of position only regardless of whether the flow is compressible or incompressible, thus Eq. 3.33 reduces to

$$\frac{\partial(\rho u_x)}{\partial x} + \frac{\partial(\rho u_y)}{\partial y} + \frac{\partial(\rho u_z)}{\partial z} = 0 \tag{3.36}$$

or

$$\nabla \cdot (\rho \mathbf{u}) = 0 \tag{3.37}$$

INCOMPRESSIBLE FLOW

For the incompressible flow, the density is constant, ρ = constant and $\partial \rho/\partial t \approx 0$ regardless of whether the flow is steady or unsteady, thus Eq. 3.33 yields

$$\frac{\partial u_x}{\partial x} + \frac{\partial u_y}{\partial y} + \frac{\partial u_z}{\partial z} = 0 \tag{3.38}$$

or

$$\nabla \cdot \mathbf{u} = 0 \tag{3.39}$$

In cylindrical form

$$\frac{1}{r}\frac{\partial(ru_r)}{\partial r} + \frac{1}{r}\frac{\partial u_\theta}{\partial \theta} + \frac{\partial u_z}{\partial z} = 0 \tag{3.40}$$

Equation 3.38 is the continuity equation established by Euler in 1755 and is the expression formula of the conversation law of mass. It is the fundamental differential equation controlling the fluid motion and must always be satisfied for a rational analysis of a flow pattern.

EXAMPLE 3-4 Finding a missing velocity component

Two velocity components of a steady, incompressible, three-dimensional flow field are known, namely, $u_x = ax^2 + by^2 + cz^2$ and $u_z = axz + byz^2$, where a, b, and c are constants. The y velocity component is missing. Generate an expression for u_y as a function of x, y, and z.

Solution:

Any physically possible velocity distribution for an incompressible fluid must satisfy conservation of mass as expressed by the continuity equation Eq. 3.38

$$\frac{\partial u_x}{\partial x} + \frac{\partial u_y}{\partial y} + \frac{\partial u_z}{\partial z} = 0$$

which gives

$$\frac{\partial u_y}{\partial y} = -\frac{\partial u_x}{\partial x} - \frac{\partial u_z}{\partial z} = -3ax - 2byz$$

Integrate it with respect to y to give

$$u_y = -3axy - by^2z + f(x, z)$$

where $f(x, z)$ is the integral arbitrary function of x and z.

DISCUSSION   The second velocity component cannot be explicitly determined since there are no derivatives of u_y with respect to x and z in the continuity equation, and the conservation of mass will still be satisfied with any form function $f(x, z)$. The specific form of this function will be governed by the flow field described by these velocity components—that is, some additional information is needed for completely determining u_y.

3.5.2 Differential Equations of Motion

Figure. 3-26 shows an infinitesimal cubic control volume in the flow field of motion. In general, two types of forces need to be considered: surface forces, which act on the surfaces of the differential element, and body forces, which are distributed throughout the element.

SURFACE FORCES

Surface forces are due to the stresses on the sides of the control surface and consist of both normal and tangential components. The normal components, σ_{xx}, σ_{yy}, and σ_{zz} (see Fig. 2-2), called *normal stresses*, are composed of pressure p (which always acts inwardly normal) and *viscous stresses* τ_{ii} (see Fig. 3-26a), namely

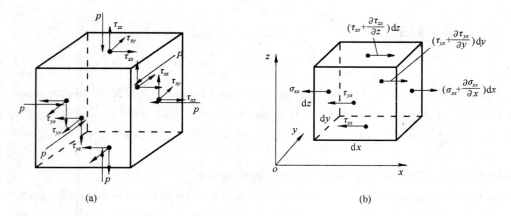

Fig. 3-26 (a) Notation for stresses and (b) surface forces in the x-direction acting on a fluid element

$$\left.\begin{array}{l}\sigma_{xx} = -p + \tau_{xx} = -p + 2\mu\dfrac{\partial u_x}{\partial x} \\[6pt] \sigma_{yy} = -p + \tau_{yy} = -p + 2\mu\dfrac{\partial u_y}{\partial y} \\[6pt] \sigma_{zz} = -p + \tau_{zz} = -p + 2\mu\dfrac{\partial u_z}{\partial z}\end{array}\right\} \quad (3.41)$$

where p is called the *mechanical pressure*, or *dynamic pressure*, at a point in the viscous flow. It is defined as the mean normal stress acting inwardly on a fluid element as in Section 2.1.2:

$$p = -\frac{1}{3}(\sigma_{xx} + \sigma_{yy} + \sigma_{zz}) \quad (3.42)$$

where the sign ' $-$ ' presents the pressure acting along the inward normal direction, while the normal stresses, σ_{xx}, σ_{yy}, σ_{zz}, along the outward normal direction as shown in Fig. 2-2 and Fig. 3-26b.

The tangential components, $\tau_{ij}(i \neq j)$, are called *shear stresses*. It is because pressure can act only normal to a surface, shear stresses are composed entirely of viscous stresses. For a Newtonian fluid, as discussed in Section. 1.4.1, the viscous stresses are proportional to the element strain rates and the coefficient of viscosity. For incompressible flow, the shear stress in three dimensional viscous flow can be determined by the *Newton's law of viscosity*

$$\left.\begin{array}{l}\tau_{xy} = \mu\left(\dfrac{\partial u_x}{\partial y} + \dfrac{\partial u_y}{\partial x}\right) = \tau_{yx} \\[6pt] \tau_{yz} = \mu\left(\dfrac{\partial u_y}{\partial z} + \dfrac{\partial u_z}{\partial y}\right) = \tau_{zy} \\[6pt] \tau_{zx} = \mu\left(\dfrac{\partial u_z}{\partial x} + \dfrac{\partial u_x}{\partial z}\right) = \tau_{xz}\end{array}\right\} \quad (3.43)$$

where μ is the dynamic viscosity.

Thus, the stresses on the sides of the control surface can be described as the *stress tensor*

$$\sigma_{ij} = \begin{pmatrix} \sigma_{xx} & \sigma_{xy} & \sigma_{xz} \\ \sigma_{yx} & \sigma_{yy} & \sigma_{yz} \\ \sigma_{zx} & \sigma_{zy} & \sigma_{zz} \end{pmatrix} = \begin{pmatrix} -p + \tau_{xx} & \tau_{xy} & \tau_{xz} \\ \tau_{yx} & -p + \tau_{yy} & \tau_{yz} \\ \tau_{zx} & \tau_{zy} & -p + \tau_{zz} \end{pmatrix}$$

in which σ_{ij} is the stress in j direction on a face normal to i axis.

NAVIER-STOKES EQUATIONS

Newton's second law, when applied to the infinitesimal cubic fluid element in Fig. 3-26, says that the net force on the fluid element equals its mass times the acceleration of the element. This is a vector relation,

$$\Sigma \mathbf{F} = m\mathbf{a} = \rho \frac{d\mathbf{u}}{dt} dxdydz \qquad (3.44)$$

and hence can be split into three scalar relations along the x-, y-, and z-axes. First we consider only the x-component of Newton's second law

$$\Sigma F_x = ma_x \qquad (3.45)$$

The left side of Eq. 3.45 is the resultant force of surface and body forces on the element in the x-direction. Body forces are due to external fields (gravity, magnetism, electric potential) which act upon the entire mass within the element. The only body force of interest in this book is gravity. The gravity force on the differential mass $\rho dxdydz$ within the control volume is

$$d\mathbf{F}_b = \rho \mathbf{f} dxdydz \qquad (3.46)$$

where the unit mass force vector $\mathbf{f}$ has components (f_x, f_y, f_z).

The surface forces are due to the stresses on the sides of the control surface. They are somewhat more complicated because, on each of the six sides of the parallelepiped, there is a normal component and two tangential components, together forming a three-dimensional vector (Fig. 3-26a). For example, on the top side (perpendicular to the z-axis), the stress vector τ_z has the normal component σ_{zz}, and the two tangential components τ_{zx} in the x-direction and τ_{zy} in the y-direction. Note that the stresses must be multiplied by the area on which they act to obtain the forces. Summing all the forces in the x-direction shown in Fig. 3-26b yields the net surface force acting in the x-direction:

$$dF_{sx} = \left(\frac{\partial \sigma_{xx}}{\partial x} + \frac{\partial \tau_{yx}}{\partial y} + \frac{\partial \tau_{zx}}{\partial z} \right) dxdydz \qquad (3.47a)$$

In a similar manner the net surface forces acting in the y- and z-directions can be obtained as

$$dF_{sy} = \left(\frac{\partial \tau_{xy}}{\partial x} + \frac{\partial \sigma_{yy}}{\partial y} + \frac{\partial \tau_{zy}}{\partial z} \right) dxdydz \qquad (3.47b)$$

$$dF_{sz} = \left(\frac{\partial \tau_{xz}}{\partial x} + \frac{\partial \tau_{yz}}{\partial y} + \frac{\partial \sigma_{zz}}{\partial z} \right) dxdydz \qquad (3.47c)$$

From Fig. 3-26b, Newton's second law in the x direction, Eq. 3.45, combined with Eqs. 3-46 and 3.47a, is applied

$$\rho f_x + \frac{\partial \sigma_{xx}}{\partial x} + \frac{\partial \tau_{yx}}{\partial y} + \frac{\partial \tau_{zx}}{\partial z} = \rho \frac{du_x}{dt} \qquad (3.48)$$

Substituting Eqs. 3.41 and Eqs. 3.43 and differential continuity equation of Eq. 3.38 into the above equation gives the differential momentum equation for a Newtonian fluid with constant density and viscosity in the x-direction

$$f_x - \frac{1}{\rho}\frac{\partial p}{\partial x} + \nu \nabla^2 u_x = \frac{du_x}{dt} = \frac{\partial u_x}{\partial t} + u_x\frac{\partial u_x}{\partial x} + u_y\frac{\partial u_x}{\partial y} + u_z\frac{\partial u_x}{\partial z} \quad (3.49\text{a})$$

In a similar manner we have the equations in the y and z direction

$$f_y - \frac{1}{\rho}\frac{\partial p}{\partial y} + \nu \nabla^2 u_y = \frac{du_y}{dt} = \frac{\partial u_y}{\partial t} + u_x\frac{\partial u_y}{\partial x} + u_y\frac{\partial u_y}{\partial y} + u_z\frac{\partial u_y}{\partial z} \quad (3.49\text{b})$$

$$f_z - \frac{1}{\rho}\frac{\partial p}{\partial z} + \nu \nabla^2 u_z = \frac{du_z}{dt} = \frac{\partial u_z}{\partial t} + u_x\frac{\partial u_z}{\partial x} + u_y\frac{\partial u_z}{\partial y} + u_z\frac{\partial u_z}{\partial z} \quad (3.49\text{c})$$

In vector form:

$$\mathbf{f} - \frac{1}{\rho}\nabla p + \nu \nabla^2 \mathbf{u} = \frac{\partial \mathbf{u}}{\partial t} + (\mathbf{u} \cdot \nabla)\mathbf{u} \quad (3.50)$$

where ∇^2 is *Laplacian operator*, $\nabla^2 = \frac{\partial^2}{\partial x^2} + \frac{\partial^2}{\partial y^2} + \frac{\partial^2}{\partial z^2}$.

These are the *Navier-Stokes equations*, named after the French engineer Louis Marie Henri Navier (1785—1836) and the English mathematician Sir George Gabriel Stokes (1819-1903), who both developed the viscous terms, although independently of each other. The *Navier-Stokes equations* are second-order nonlinear partial differential equations and are quite formidable, but surprisingly many solutions have been found to a variety of interesting viscous-flow problems. Eqs. 3.49 have four unknowns: p, u_x, u_y and u_z. Therefore, they should be combined with the incompressible continuity relation (Eq. 3.38) to form four equations in these four unknowns.

3.5.3 Euler's Equation for Inviscid Flow

The simplest expression of Eq. 3.50 is the one for inviscid flow (frictionless flow), $\nu = 0$, thus Eq. 3.50 reduces to

$$\mathbf{f} - \frac{1}{\rho}\nabla p = \frac{\partial \mathbf{u}}{\partial t} + (\mathbf{u} \cdot \nabla)\mathbf{u} \quad (3.51)$$

This is *Euler's equation for inviscid flow*. It is a relatively simple expression that allows only for the effect of elevation, density and acceleration on fluid pressure. The following assumptions are made:

- steady flow, i.e., properties at a given point are constant;
- frictionless (inviscid) flow-no shear stress;
- zero shaft work.

Although Eqs. 3.51 are considerably simpler than the general equations of motion, Eqs. 3.49, they are still not amenable to a general analytical solution that would allow us to determine the pressure and velocity at all points within an inviscid flow field. However, under some circumstances we can use them to obtain useful information about inviscid flow fields. For example, as shown in the following section we can integrate Eq. 3.51 to obtain a relationship (the Bernoulli equation) between elevation, pressure, and velocity along a streamline or in an irrotational flow field.

3.6 The Bernoulli Equation

The *Bernoulli equation* is an approximate relation between pressure, velocity, and elevation, and is valid in regions of steady, incompressible flow where net frictional forces are negligible. Despite its simplicity, it has been proven to be a very powerful tool in fluid mechanics. In this section, we derive the Bernoulli equation by direct application of Euler's equations to a fluid particle moving along a streamline and in an inviscid flow.

3.6.1 Bernoulli Equation along a Streamline

Consider the motion of a fluid particle in a flow field in steady flow described in detail in Fig. 3-27. In regions of flow where net frictional forces are negligible, the significant forces acting in the s-direction are the pressure (acting on both sides) and the component of the weight of the particle in the s-direction. That is, the only body force we consider is gravity. Take the conventional choice of z-coordinate being directed upwards to give

$$\mathbf{f} = -g\mathbf{k} \qquad (3.52)$$

When the flow is a steady, $\partial \mathbf{u}/\partial t = 0$. Also, it will be convenient to use the vector identity for the second term in the right side of Eq. 3.51

$$(\mathbf{u} \cdot \nabla)\mathbf{u} \equiv \frac{1}{2}\nabla(\mathbf{u} \cdot \mathbf{u}) - \mathbf{u} \times (\nabla \times \mathbf{u})$$

$$(3.53)$$

Thus, the Euler's equation, Eq. 3.51, can be rewritten and rearranged as

$$g\mathbf{k} + \frac{1}{\rho}\nabla p + \frac{1}{2}\nabla(u^2) = \mathbf{u} \times (\nabla \times \mathbf{u})$$

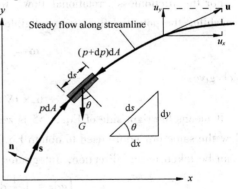

Fig. 3-27 Steady flow along a streamline

(3.54)

Dot the entire equation into an arbitrary infinitesimal length vector d**s** to give

$$g\mathbf{k} \cdot d\mathbf{s} + \frac{1}{\rho}\nabla p \cdot d\mathbf{s} + \frac{1}{2}\nabla(u^2) \cdot d\mathbf{s} = \mathbf{u} \times (\nabla \times \mathbf{u}) \cdot d\mathbf{s} \qquad (3.55)$$

in which

(a) $g\mathbf{k} \cdot d\mathbf{s} = gdz$.

(b) $\nabla p \cdot d\mathbf{s} = \dfrac{\partial p}{\partial x}dx + \dfrac{\partial p}{\partial y}dy + \dfrac{\partial p}{\partial z}dz = dp$ when the flow is steady.

(c) Since we are integrating Euler equation along a streamline, the velocity **u** is parallel to d**s**, $\mathbf{u} \times (\nabla \times \mathbf{u})$ is perpendicular to **u**, and so to d**s**. Thus $\mathbf{u} \times (\nabla \times \mathbf{u}) \cdot d\mathbf{s} = 0$.

Therefore a simplified expression of Eq. 3.55 along a streamline can be obtained

$$g\,dz + \frac{1}{\rho}dp + \frac{1}{2}d(u^2) = 0 \qquad (3.56)$$

which, for constant acceleration of gravity, can be integrated to give

$$\int g\,dz + \int \frac{1}{\rho}dp + \int \frac{1}{2}d(u^2) = C \qquad (3.57)$$

where C is a constant of integral along a streamline.

Equation 3.57 is called the *generalized Bernoulli equation*. It applies for inviscid, steady compressible or incompressible flow along a streamline, with zero shaft work.

However, a far more useful expression is obtained if it assumed that density is constant (i.e., the fluid is incompressible):

$$gz + \frac{p}{\rho} + \frac{u^2}{2} = C \text{ (along a streamline)} \qquad (3.58)$$

This is the *Bernouilli equation* for inviscid steady incompressible flow along a streamline with zero shaft work. It was stated in words by the Swiss mathematician Daniel Bernoulli (1700-1782) in a text written in 1738. A complete derivation of the equation was given in 1755 by Leonhard Euler.

3.6.2 Bernoulli Equation for Potential Flow

For the frictionless irrotational flow, we also can have Eq. 3.55 from Euler's equation. Meanwhile, the potential flow has negligible or zero vorticity, that is

$$\boldsymbol{\omega} = \frac{1}{2}(\nabla \times \mathbf{u}) = 0$$

which gives

$$\mathbf{u} \times (\nabla \times \mathbf{u}) \cdot d\mathbf{s} = 0$$

It means the right side of Eq. 3.55 is zero regardless of the direction of $d\mathbf{s}$. Then we can now follow the same procedure used to obtain Eq. 3.56, where the differential changes dp, $d(u^2)$, and dz can be taken in any direction. Integration of Eq. 3.56

$$\int g\,dz + \int \frac{1}{\rho}dp + \int \frac{1}{2}d(u^2) = C$$

again yields

$$gz + \frac{p}{\rho} + \frac{u^2}{2} = C \text{(in a potential flow)} \qquad (3.59)$$

which for irrotational flow the constant C is same throughout the flow field.

Equation 3.59 is exactly the same form as Eq. 3.58 but is not limited to application along a streamline. However, Eq. 3.59 is restricted to inviscid, steady, incompressible irrotational flow.

3.6.3 Other Forms of Bernoulli Equation

IN TERMS OF HEAD

Each term in the Bernouilli equation, Eq. 3.58 and Eq. 3.59, has the units of energy/mass and the equation is a special statement of the law of conservation of energy. If each term in the Bernouilli equation is divided by g, then each term has the units of length (elevation) and is called a *head*. The head form of the Bernouilli equation is:

$$z + \frac{p}{\rho g} + \frac{u^2}{2g} = C \qquad (3.60)$$

or be written between any two points 1 and 2 along the streamline or any two points in the potential flow

$$z_1 + \frac{p_1}{\rho g} + \frac{u_1^2}{2g} = z_2 + \frac{p_2}{\rho g} + \frac{u_2^2}{2g} \qquad (3.61)$$

The terms in the Bernouilli equations, Eqs. 3.60 and 3.61, are referred to as

1. z is the *elevation head* and represents the elevation potential energy per unit weight fluid;

2. $p/\rho g$ is the *pressure head*, or the height of a fluid column that produces the static pressure p and represents the pressure potential energy per unit weight fluid;

3. $u^2/2g$ is the *velocity head*, or the elevation needed for the fluid to reach the velocity u during frictionless free fall, and represents kinematic energy per unit weight fluid;

4. $z + p/\rho g$ is the *piezometric head* and represents the total potential energy of the fluid particle per unit weight;

5. $z + \frac{p}{\rho g} + \frac{u^2}{2g}$ is the *total head* and represents the total energy of the fluid particle per unit weight.

IN TERMS OF PRESSURE

Time the Bernoulli equation, Eq. 3.58 or 3.59 by ρ, we can obtained the Bernoulli equation in terms of pressure

$$\rho g z + p + \frac{\rho u^2}{2} = C \qquad (3.62)$$

Each term in this equation has *pressure* units, and thus each term represents some kind of pressure:

1. $\rho g z$ is the *hydrostatic pressure*, which is not pressure in a real sense since its value depends on the reference level selected; it accounts for the elevation effects, i.e., of fluid weight on pressure.

2. p is the *static pressure* (it does not incorporate any dynamic effects); it represents the actual thermodynamic pressure of the fluid, and can be measured by the piezometer tube as in Fig. 3-28.

3. $\rho u^2/2$ is the *dynamic pressure*; it represents the pressure rise when the fluid in motion is brought to a stop isentropically as sketched in Fig. 3-28.

4. $p + \frac{\rho u^2}{2}$ is the *stagnation pressure*; it is the sum of the static and dynamic pressures, and

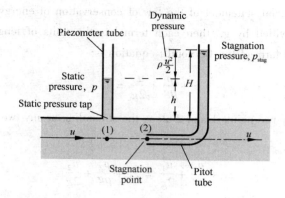

Fig. 3-28 The static, dynamic, and stagnation pressures

can be measured by the Pitot tube as in Fig. 3-28.

5. $\rho g z + p + \dfrac{\rho u^2}{2}$ is the *total pressure*; it is the sum of the static, dynamic, and hydrostatic pressures.

Therefore, the Bernoulli equation states that the total pressure along a streamline is constant (Eq. 3.58), or the total pressure at any point in the whole potential flow field is the same (Eq. 3.59).

3.6.4 Restrictions of the Bernoulli Equation

The Bernouilli equation Eq. 3.58 or 3.59 indicates the conservation of mechanics energy per unit weight along a streamline or in a steady incompressible inviscid flow. It is very famous and very widely used, but one should be wary of its restrictions:

1. *Steady flow* for $\partial \mathbf{u}/\partial t = 0$: a common assumption applicable to many flows.
2. *Incompressible flow* for $\rho =$ constant: acceptable if the flow Mach number is less than 0.3.
3. *Frictionless flow* for $\mu = 0$: very restrictive, solid walls introduce friction effects. Thus, one must confine it to regions of the flow which are nearly frictionless.
4. *Flow along a single streamline* for Eq. 3.58: the constant may vary from streamline to streamline unless the flow is also irrotational; or *flow in a potential flow* for Eq. 3.59.
5. *No shaft work* (pumps or turbines) between point 1 and point 2 along a streamline, or between section 1 and section 2 in the potential flow.
6. *No heat transfer*, either added or removed, between point 1 and point 2 along a streamline, or between section 1 and section 2 in the potential flow.

In an inviscid region of flow, the Bernoulli equation holds along streamlines, and the Bernoulli constant may change from streamline to streamline. In an irrotational region of flow, the Bernoulli constant is the same everywhere, so the Bernoulli equation holds everywhere in the irrotational region of flow, even across streamlines. Thus, the irrotational approximation is more restrictive than the inviscid approximation.

3.6.5 Application of Bernoulli Equation

PITOT-STATIC PROBE

Pitot-static probes are used to give velocity information at a point within a flow field by using the combination of a static pressure tap and a Pitot tube as illustrated in Fig. 3-28. They are widely used in experimental fluids and also have industrial applications. A *static pressure tap* is simply a small hole drilled into a wall such that the plane of the hole is parallel to the flow direction. It measures the static pressure. A *Pitot tube* is a small tube with its open end aligned into the flow so as to sense the full impact pressure of the flowing fluid. It measures the stagnation pressure. When vertical piezometer tubes are attached to the pressure tap and the Pitot tube, the liquid rises in the piezometer tube to a column height (head) that is proportional to the pressure being measured as illustrated in Fig. 3-28.

In Fig. 3-28, some of the liquid flowing along the horizontal pipe rises into the piezometer and some into the pitot tube to level. While fluid in the pipe flows past the piezometer opening, fluid flowing directly towards the pitot tube opening is brought suddenly to rest at point 2. Thus, when the equilibrium is reached the fluid in the tubes, including that at its tip, (2), will be stationary. This point with $u_2 = 0$ is *stagnation point*. The streamline that terminates at the stagnation point is *stagnation streamline* as shown in Fig. 3-10.

Apply Bernouilli's equation Eq. 3.61 to the streamline between point (1) and point (2)

$$z_1 + \frac{p_1}{\rho g} + \frac{u_1^2}{2g} = z_2 + \frac{p_2}{\rho g} + \frac{u_2^2}{2g}$$

Use $z_1 = z_2$, $u_1 = u$, $u_2 = 0$, we have

$$p_2 = p_1 + \frac{1}{2}\rho u^2 \tag{3.63}$$

in which p_2 is called the *stagnation pressure*, p_1 is the *static pressure* and $\frac{1}{2}\rho u^2$ is the *dynamic pressure*.

It indicates that the pressure at the stagnation point is greater than the static pressure by an amount of the dynamic pressure, $\frac{1}{2}\rho u^2$.

Take the readings of piezometric tube and Pitot tube in Fig. 3-28 to be h and H respectively, we have $p_1 = \rho g h$, $p_2 = \rho g H$. Rearrange Eq. 3.63 to give

$$u = \sqrt{2\frac{p_2 - p_1}{\rho}} = \sqrt{2g(H - h)} \tag{3.64}$$

which is known as the *Pitot formula*.

Equation 3.64 implies that the fluid speed can be calculated if the values of the static and stagnation pressures in a fluid are known. Those two tubes (piezometer and Pitot) can be combined in a single instrument called a *pitot-static probe* [Henri. de Pitot (1695-1771)]. This con-

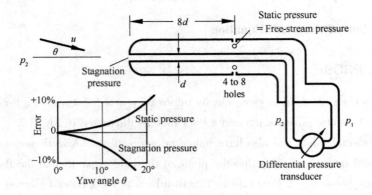

Fig. 3-29 Pitot-static probe (reproduced from 'Fluid Mechanics' by Frank M·CM. White, 1998)

sists of two concentric tubes, as shown in Fig. 3-29.

The inner tube is subject to the stagnation pressure while the radial holes in the outer tube are static pressure tapings. Thus the annular space between the two tubes is subject to a pressure that, by careful design of the probe, is very close to the static pressure in the upstream region. The rounded ellipsoidal front of the probe prevents separation of the flow and helps to ensure that the flow over the static pressure holes is fairly uniform. These holes must be located at a sufficient distance from the front, around which the fluid accelerates and the pressure decreases. The instrument must also be aligned accurately with the flow direction (see inset graph on Fig. 3-29). This can be a limitation if the predominant flow direction is not known in advance. Pitot-static probes can be used for fluids and gases where density can be assumed constant.

Besides, for the measurement of the static pressure, it requires a smooth hole with no burrs or imperfections. As indicated in Fig. 3-30, such imperfections can cause the measured pressure to be greater or less than the actual static pressure.

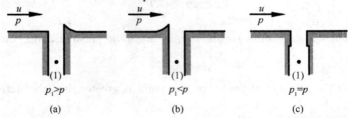

Fig. 3-30 Incorrect [(a) and (b)] and correct [(c)] design of static pressure taps

VENTURI METER

The *venturi meter* [G. B. Venturi (1746-1822)] is a well-known volume flowmeter, widely used in industry. The venturi meter can easily be installed in a pipeline and creates very little overall pressure loss. Essentially the venturi meter is a convergent-divergent pipe. The divergent section restores the fluid to almost its original state. Fig. 3-31 illustrates a venturi meter.

Assume the flow is horizontal ($z_1 = z_2$), steady, inviscid, and incompressible between points

(1) and (2), and the velocity profiles are uniform with a magnitude of U_1 and U_2 at section 1 and 2, the Bernoulli equation Eq. 3.61 then becomes

$$\frac{p_1}{\rho g} + \frac{U_1^2}{2g} = \frac{p_2}{\rho g} + \frac{U_2^2}{2g}$$

The continuity equation, Eq. 3.23, for the one dimensional incompressible uniform flow is

$$Q = A_1 U_1 = A_2 U_2$$

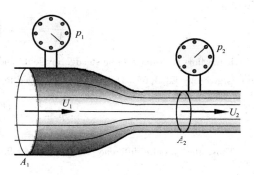

Fig. 3-31 Venturi meter

Combination of these two equations results in the following theoretical flowrate

$$Q = A_2 \sqrt{\frac{2(p_1 - p_2)}{\rho[1 - (A_2/A_1)^2]}} = A_2 \sqrt{\frac{2(p_1 - p_2)}{\rho(1 - \beta^4)}} \qquad (3.65)$$

where A_2 is the throat area, $A_2 = \frac{1}{4}\pi d_2^2$; d_2 is the diameter of the throat; β is the throat-to-pipe diameter ratio, $\beta = d_2/d_1$; d_1 is the diameter before the converging.

Thus, for a given flow geometry (A_1, A_2) the flowrate can be determined if the pressure difference, $p_1 - p_2$, is measured.

Actually, the measured volume flowrate will be less than the theoretical value because of the energy loss in the viscous flow. These differences are quite consistent and may be as small as 1 to 2% or as large as 40% depending on the geometry used. It is usual to account for this difference by multiplying the ideal volume flowrate in Eq. 3.65 by a discharge coefficient, C_d

$$Q = C_d A_2 \sqrt{\frac{2(p_1 - p_2)}{\rho[1 - (A_2/A_1)^2]}} = C_d A_2 \sqrt{\frac{2(p_1 - p_2)}{\rho[1 - \beta^4]}} \qquad (3.66)$$

where C_d is called the *Venturi discharge coefficient*.

EXAMPLE 3-5 Venturi meter

Kerosene ($S_G = 0.85$) flows through a Venturi meter as shown in Fig. E3-5 with flowrates between 0.005 and 0.050 m³/s. Determine the range in pressure difference, $p_1 - p_2$, needed to measure these flowrates.

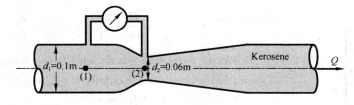

Fig. E3-5

Solution:

From the given parameters, we have the density of the flowing fluid

$$\rho = S_G \rho_{H_2O} = 0.85 \times 1000 \text{kg/m}^3 = 850 \text{kg/m}^3$$

and the area ratio

$$\frac{A_2}{A_1} = \frac{d_2^2}{d_1^2} = \left(\frac{0.06\text{m}}{0.1\text{m}}\right)^2 = 0.36$$

If the flow is assumed to be steady, inviscid, and incompressible, the relationship between flowrate and pressure is given by Eq. 3.65. This can be rearranged to give

$$p_1 - p_2 = \frac{Q^2\rho[1-(A_2/A_1)^2]}{2A_2^2}$$

Thus, the pressure difference for the smallest flowrate is

$$p_1 - p_2 = (0.005\text{ m}^3/\text{s})^2(850\text{kg/m}^3)\frac{1-0.36^2}{2\left[\left(\frac{\pi}{4}\right)(0.06\text{m})^2\right]^2} = 1.16\text{kPa}$$

and for the largest flowrate is

$$p_1 - p_2 = (0.050\text{ m}^3/\text{s})^2(850\text{kg/m}^3)\frac{1-0.36^2}{2\left[\left(\frac{\pi}{4}\right)(0.06\text{m})^2\right]^2} = 116\text{kPa}$$

Thus

$$1.16\text{kPa} \leq p_1 - p_2 \leq 116\text{kPa}$$

These values represent the pressure differences for inviscid, steady, incompressible conditions.

DISCUSSION It is seen from Eq. 3.65 that the flowrate varies as the square root of the pressure difference. Hence, as indicated by the numerical results, a 10-fold increase in flowrate requires a 100-fold increase in pressure difference. This nonlinear relationship can cause difficulties when measuring flowrates over a wide range of values. Such measurements would require pressure transducers with a wide range of operation. An alternative is to use two flow meters in parallel-one for the larger and one for the smaller flowrate ranges.

EXAMPLE 3-6 Jet

A stream of water of diameter $d = 0.1$m flows steadily from a tank of diameter D as shown in Fig. E3-6. Determine the flowrate, Q, of jet out of the tank if the water depth remains constant, $h = 2.0$m.

Solution:

For steady, inviscid, incompressible flow, the Bernoulli equation, Eq. 3.61, applies between points (1) and (2) along a streamline is

$$z_1 + \frac{p_1}{\rho g} + \frac{u_1^2}{2g} = z_2 + \frac{p_2}{\rho g} + \frac{u_2^2}{2g} \quad (1)$$

With the assumptions that the water level remains constant (h = constant), and a larger diameter of tank, $D \gg d$, thus $u_1 \approx 0$, $p_1 = p_2 = 0$, $z_1 = h$, and $z_2 = 0$, Eq. 1 becomes

$$u_2 = \sqrt{2gh} = \sqrt{2(9.81\text{m/s}^2)(2.0\text{m})} = 6.26\text{m/s}$$

The flowrate, Q, of the jet out of the tank is

$$Q = A_2 u_2 = \left(\frac{\pi}{4}\right)(0.1\text{m})^2(6.26\text{m/s}) = 0.049\text{ m}^3/\text{s}$$

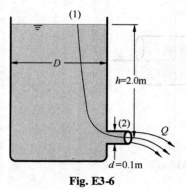

Fig. E3-6

3.7 Steady Plane Potential Flows

3.7.1 The Stream Function

The most common application of differential continuity equation is to the two dimensional incompressible flow, therefore Eq. 3.38 reduces to

$$\frac{\partial u_x}{\partial x} + \frac{\partial u_y}{\partial y} = 0 \tag{3.67}$$

Define a function $\psi(x, y)$ satisfying this continuity equation

$$\frac{\partial}{\partial x}\left(\frac{\partial \psi}{\partial y}\right) + \frac{\partial}{\partial y}\left(-\frac{\partial \psi}{\partial x}\right) = 0$$

we have

$$u_x = \frac{\partial \psi}{\partial y} \qquad u_y = -\frac{\partial \psi}{\partial x} \tag{3.68}$$

or

$$\mathbf{u} = -\mathbf{i}\frac{\partial \psi}{\partial y} - \mathbf{j}\frac{\partial \psi}{\partial x} \tag{3.69}$$

where ψ is the *stream function* in an *incompressible, two-dimensional flow* and must be defined such that

$$d\psi = -u_y dx + u_x dy = \frac{\partial \psi}{\partial x}dx + \frac{\partial \psi}{\partial y}dy \tag{3.70}$$

The stream function and the corresponding *velocity potential function* (Eq. 3.81, see next section) were invented by the Italian mathematician Joseph Louis Lagrange (1736-1813) and published in his treatise on fluid mechanics in 1781.

For the inviscid *irrotational* flow in the xy plane, where $\omega_z \equiv 0$, or $\frac{\partial u_x}{\partial y} - \frac{\partial u_y}{\partial x} = 0$, substitute Eqs. 3.68 into it to give

$$\nabla^2 \psi = \frac{\partial^2 \psi}{\partial x^2} + \frac{\partial^2 \psi}{\partial y^2} = 0 \tag{3.71}$$

where $\nabla^2(\cdot) = \nabla \cdot \nabla(\cdot)$ is the *Laplacian operator*.

This is the second-order *Laplace equation*, for which many solutions and analytical techniques are known.

Suppose an incompressible plane flow in polar coordinates, the important coordinates are r and θ, with $u_z = 0$, and the density is constant. Then the form of the incompressible polar coordinate stream function is

$$d\psi = u_r r d\theta - u_\theta dr \tag{3.72}$$

$$u_r = \frac{1}{r}\frac{\partial \psi}{\partial \theta} \qquad u_\theta = -\frac{\partial \psi}{\partial r} \tag{3.73}$$

GEOMETRIC INTERPRETATION OF ψ

One of the geometric interpretations is the *lines of constant ψ are streamlines of the flow*. This is easily proven by considering a streamline in the xy-plane, as sketched in Fig. 3-32. From Eq. 3.17 that along such a streamline in two-dimensional flow

Thus along a streamline we have
$$\frac{dx}{u_x} = \frac{dy}{u_y}$$

$$u_x dy - u_y dx = 0 \tag{3.74}$$

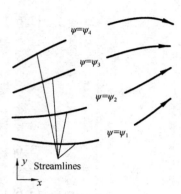

Fig. 3-32 Curves of constant stream function represent streamlines of the flow

Again with a constant ψ the definition equation of the stream function, Eq. 3.70 becomes

$$d\psi = -u_y dx + u_x dy \tag{3.75}$$

By comparing Eq. 3.74 with Eq. 3.75 we see that $d\psi = 0$ along a streamline; thus the statement that ψ is a constant along a streamline is true, or

$$\psi = \text{const along a streamline} \tag{3.76}$$

Having found a given solution $\psi(x, y)$, we can plot lines of constant ψ to give the streamlines of the flow.

Another physical interpretations of ψ is that the difference in the value of ψ from one streamline to another is equal to the volume flow rate per unit width between the two streamlines. Considering a control volume with unit depth into paper bounded by the two streamlines and by cross-sectional slice A and cross-sectional slice B (Fig. 3-33). The infinitesimal element dA with a length ds along slice B as in Fig. 3-33a is zoomed in Fig. 3-33b, accompanied with its unit normal vector $\mathbf{n}$ and velocity vector $\mathbf{u}$ through it. Thus, the volume flow rate dq through the element dA of control surfaceis

$$dq = (\mathbf{u} \cdot \mathbf{n}) dA = (\mathbf{i} u_x + \mathbf{j} u_y) \cdot \left(\mathbf{i} \frac{dy}{ds} - \mathbf{j} \frac{dx}{ds} \right) (ds)(1)$$

$$= u_x dy - u_y dx = \frac{\partial \psi}{\partial x} dx + \frac{\partial \psi}{\partial y} dy = d\psi \tag{3.77}$$

Thus the change in ψ across the element is numerically equal to the volume flow rate through the element. The total volume flow rate through cross-sectional slice B can be found by integrating Eq. 3.77 from streamline 1 to streamline 2

$$q = \int_B (\mathbf{u} \cdot \mathbf{n}) dA = \int_{\psi_1}^{\psi_2} d\psi = \psi_2 - \psi_1 \tag{3.78}$$

Equation 3.78 presents a constant flowrate between two streamlines, which indicates that when the streamlines are far apart (e. g. slice B of Fig. 3-33a), the magnitude of velocity (the fluid speed) in that vicinity is small relative to the speed in locations where the streamlines are close together (e. g. slice A of Fig. 3-33a). This is easily explained by conservation of mass. As the streamlines converge, the cross-sectional area between them decreases, and the velocity must in-

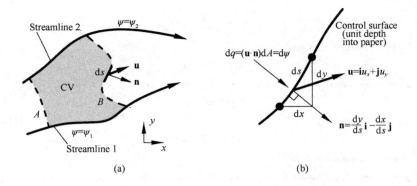

Fig. 3-33 Geometric interpretation of stream function: (a) control volume bounded by streamlines ψ_1 and ψ_2 and slices A and B in the xy-plane; (b) magnified view of the region around element dA

crease to maintain the flow rate between the streamlines.

Further, the direction of the flow can be ascertained by noting whether ψ increases or decreases. As sketched in Fig. 3-34, the flow is to the right if ψ_2 is greater than ψ_1; otherwise the flow is to the left. It is because the directions of **u** and **n** must be on the same side of infinitesimal length ds if $\psi_2 > \psi_1$ according to Eq. 3.78 and Fig. 3-33a; otherwise they are on appositive sides. Namely, if you are moving in the direction of the flow, the stream function increases to your left, which might be called the '*left-side convention*' (Fig. 3-34c).

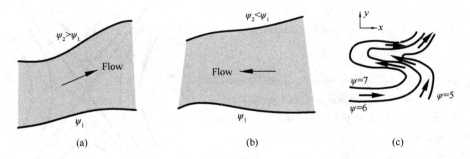

Fig. 3-34 Sign convention: (a) flow to the right if ψ_2 is greater; (b) flow to the left if ψ_1 is greater; (c) illustration of the 'left-side convention.'

EXAMPLE 3-7 Stream function

The velocity components in a steady, incompressible, two-dimensional flow field are
$$u_x = 2y \qquad u_y = 4x$$
Determine the corresponding stream function and show on a sketch several streamlines. Indicate the direction of flow along the streamlines.

Solution:

From the definition of the stream function, Eq. 3.68

$$u_x = \frac{\partial \psi}{\partial y} = 2y \tag{1}$$

119

$$u_y = -\frac{\partial \psi}{\partial x} = 4x \qquad (2)$$

Equation 1 can be integrated to give

$$\psi = y^2 + f(x) \qquad (3)$$

where $f(x)$ is an arbitrary function of x. Substituting Eq. 3 into Eq. 2 yields

$$f'(x) = -4x$$

Integrate to give

$$f(x) = -2x^2 + C \qquad (4)$$

where C is the arbitrary integral constant.

Substituting Eq. 4 into Eq. 3 yields the expression of stream function

$$\psi = -2x^2 + y^2 + C$$

Since the velocities are related to the derivatives of the stream function, an arbitrary constant can always be added to the function, and the value of the constant is actually of no consequence. Usually, for simplicity, we set $C = 0$ so that for this particular example the simplest form for the stream function is

$$\psi = -2x^2 + y^2 \qquad (5)$$

Streamlines can now be determined by setting ψ = constant and plotting the resulting curve. With the above expression for ψ (with $C = 0$) the value of ψ at the origin is zero so that the equation of the streamline passing through the origin (the $\psi = 0$ streamline) is

$$-2x^2 + y^2 = 0$$

or

$$y = \pm \sqrt{2} x$$

Other streamlines can be obtained by setting ψ be various constants. It follows from Eq. 5 that the equations of these streamlines (for $\psi \neq 0$) can be expressed in the form

$$\frac{y^2}{\psi} - \frac{x^2}{\psi/2} = 1$$

which we recognize as the equation of a hyperbola. Thus, the streamlines are a family of hyperbolas with the $\psi = 0$ streamlines as asymptotes. Several of the streamlines are plotted in Fig. E3-6. Since the velocities can be calculated at any point, the direction of flow along a given streamline can be easily deduced. For example, $u_y = -\partial \psi / \partial x = 4x$ so that $u_y > 0$ if $x > 0$ and $u_y < 0$ if $x < 0$. The direction of flow is indicated in the figure.

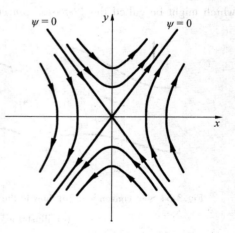

Fig. E3-7

3.7.2 Velocity Potential Function

For the steady incompressible inviscid *irrotational* flow, where $\omega \equiv 0$, or Eqs. 3.32, gives rise to a scalar function ϕ similar to the stream function ψ, of which the gradient can be expressed as the velocity components

$$u_x = \frac{\partial \phi}{\partial x} \qquad u_y = \frac{\partial \phi}{\partial y} \qquad u_z = \frac{\partial \phi}{\partial z} \qquad (3.79)$$

or in vector form

$$\mathbf{u} = \nabla\phi \tag{3.80}$$

where $\phi = \phi(x, y, z, t)$ is called the *velocity potential function* and defined as

$$d\phi = \frac{\partial \phi}{\partial x}dx + \frac{\partial \phi}{\partial y}dy + \frac{\partial \phi}{\partial z}dz \tag{3.81}$$

where ϕ has units of m²/s.

The velocity potential is a consequence of the irrotationality of the flow field, whereas the stream function is a consequence of conservation of mass. It is to be noted that the *velocity potential* ϕ, unlike the stream function, is *fully three-dimensional* and not limited to two coordinates. It reduces a velocity problem with three unknowns u_x, u_y and u_z to a single unknown potential ϕ.

For an incompressible fluid, we have the differential continuity equation from the conservation of mass, Eq. 3.39

$$\nabla \cdot \mathbf{u} = 0 \tag{3.82}$$

and therefore for incompressible, irrotational flow (with $\mathbf{u} = \nabla\phi$) it follows that

$$\nabla^2 \phi = 0 \tag{3.83}$$

In Cartesian coordinates

$$\frac{\partial^2 \phi}{\partial x^2} + \frac{\partial^2 \phi}{\partial y^2} + \frac{\partial^2 \phi}{\partial z^2} = 0 \tag{3.84}$$

This differential equation arises in many different areas of engineering and physics and is called *Laplace's equation*. Thus, inviscid, incompressible, irrotational flow fields are governed by Laplace's equation. This type of flow is commonly called a *potential flow*. Lines of constant ϕ are called the *equipotential lines* of the flow.

For some problems it will be convenient to use cylindrical coordinate, r, θ and z. In this coordinate system, the potential velocity is

$$d\phi = u_r dr + u_\theta r d\theta + u_z dz \tag{3.85}$$

$$u_r = \frac{\partial \phi}{\partial r} \qquad u_\theta = \frac{1}{r}\frac{\partial \phi}{\partial \theta} \qquad u_z = \frac{\partial \phi}{\partial z} \tag{3.86}$$

Also, Laplace's equation in cylindrical coordinates is

$$\frac{1}{r}\frac{\partial}{\partial r}\left(r\frac{\partial \phi}{\partial r}\right) + \frac{1}{r^2}\frac{\partial^2 \phi}{\partial \theta^2} + \frac{\partial^2 \phi}{\partial z^2} = 0 \tag{3.87}$$

EXAMPLE 3-8 Velocity potential function

Given the velocity distribution of steady two-dimensional incompressible flow

$$u_x = \frac{Cx}{x^2 + y^2}, \qquad u_y = \frac{Cy}{x^2 + y^2}$$

where C is a constant. Determine the stream function and the velocity potential function, and plot the streamlines and equipotential lines.

Solution:

The stream function and the velocity potential function can be obtained by integrate over the velocity field:

$$\psi = \int -u_y dx + u_x dy = C\int \frac{-y\,dx + x\,dy}{x^2 + y^2} = C\int d\left[\frac{y/x}{1 + (y/x)^2}\right] = C\tan^{-1}\left(\frac{y}{x}\right) = C\theta$$

$$\phi = \int u_x dx + u_y dy = C\int \frac{xdx + ydy}{x^2 + y^2} = C\int d[\ln\sqrt{x^2 + y^2}] = C\ln\sqrt{x^2 + y^2} = C\ln r$$

where $\theta = \tan^{-1}(y/x)$, $r = \sqrt{x^2 + y^2}$.

The streamlines and the equipotential lines with $C>0$ then are plotted in Fig. E3-8. It shows that the streamlines of the flow are radial spokes passing through the origin (lines of constant θ) while the equipotential lines are concentric circles (lines of constant r) because $u_r = \frac{\partial \phi}{\partial r} = \frac{C}{r}$ and $u_\theta = \frac{1}{r}\frac{\partial \phi}{\partial \theta} = 0$. The streamlines and equipotential lines are mutually orthogonal everywhere except at the origin, a singularity point. Such a flow is called the point *sink flow* ($C<0$) or the point *source flow* ($C>0$).

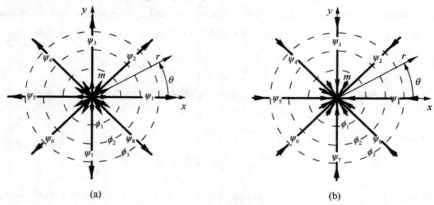

(a)   (b)

Fig. E3-8 (a) Point source flow ($C>0$) and (b) point sink flow ($C<0$)

DISCUSSION   The integral constant C can be obtained from the flow rate, the *flow strength*. Suppose the flow rate per unit length flowing into or out of the origin is m. Looking at the xy plane, we would see a cylindrical radial inward or outward flow as sketched in Fig. E3-8. Thus, the flowrate is

$$m = \int_0^{2\pi} u_r r d\theta = \int_0^{2\pi} \frac{\partial \phi}{\partial r} r d\theta = \int_0^{2\pi} \frac{C}{r} r d\theta = 2\pi C$$

which gives the integral constant C

$$C = \frac{m}{2\pi}$$

If m is positive, $C>0$, the flow is radially outward, and the flow is considered to be a *source* flow. If m is negative, $C<0$, the flow is toward the origin, and the flow is considered to be a *sink* flow. The flow rate, m, is the *strength* of the source or sink.

EXAMPLE 3-9 Stream function and velocity potential function

A two dimensional flow has a stream function $\psi = ax^2 - ay^2$, in which a is a constant with $a \neq 0$. Is this flow irrotational? If yes, determine the velocity potential function.

Solution:

For the two dimensional flow, the velocity filed can be determined from the stream function by Eq. 3.68

$$u_x = \frac{\partial \psi}{\partial y} = -2ay \qquad (1)$$

$$u_y = -\frac{\partial \psi}{\partial x} = -2ax \qquad (2)$$

Thus, the angular velocity for the two dimensional flow is

$$\omega_z = \frac{1}{2}\left(\frac{\partial u_y}{\partial x} - \frac{\partial u_x}{\partial y}\right) = \frac{1}{2}(-2a + 2a) = 0$$

It indicates this flow is an irrotational flow and has a velocity potential function.

In the x-direction, we have $u_x = \dfrac{\partial \phi}{\partial x} = -2ay$. Integrate it to give the velocity potential function

$$\phi = -\int 2ay\, dx = -2ayx + f(y) \tag{3}$$

where $f(y)$ is an arbitrary function of y.

Partial derivative of Eq. 3 with respect to y to give u_y, and combining Eq. 1 gives

$$u_y = \frac{\partial \phi}{\partial y} = -2ax + f'(y) = -2ax$$

Solving it gives $f'(y) = 0$. Integrating it yields $f(x) = C$, where C is the arbitrary integral constant. Substituting $f(x) = C$ back into Eq. 3 gives the required velocity potential function

$$\phi = -2ayx + C$$

Thus, a summary of the equations and solution procedure relevant to irrotational regions of flow is as follows. In a region of irrotational flow, the velocity field is obtained first by solution of the Laplace equation for velocity potential function ϕ (Eq. 3.83), followed by application of Eq. 3.79 to obtain the velocity field. To solve the Laplace equation, we must provide boundary conditions for ϕ everywhere along the boundary of the flow field of interest. Once the velocity field is known, we use the Bernoulli equation (Eq. 3.59) to obtain the pressure field, where the Bernoulli constant C is obtained from a boundary condition on pressure p somewhere in the flow.

The following example illustrates a situation in which the flow field consists of two separate regions—an inviscid, rotational region and an inviscid, irrotational region.

EXAMPLE 3-10   A two-region model of a tornado

A horizontal slice through a tornado (Fig. E3-10) is modeled by two distinct regions. The *inner* or *core region* ($0 < r < R$) is modeled by solid body rotation—a rotational but inviscid region of flow as discussed in Section 2.4. The *outer region* ($r > R$) is modeled as an irrotational region of flow. The flow is two dimensional in the $r\theta$-plane, and the components of the velocity field $\mathbf{u} = (u_r, u_\theta)$ are given by

$$u_r = 0 \qquad u_\theta = \begin{cases} \omega r & 0 < r < R \\ \dfrac{\omega R^2}{r} & r > R \end{cases} \tag{1}$$

where ω is the magnitude of the angular velocity in the inner region, R is the radius of boundary between inner and outer region. The ambient pressure (far away from the tornado) is equal to p_∞. Calculate the pressure field in a horizontal slice (as in Fig. E3-10a) of the tornado for $0 < r < \infty$. What is the pressure at $r = 0$? Plot the pressure and velocity fields.

Solution:

Assume:

1. The flow is steady and incompressible.
2. Although R increases and ω decreases with increasing elevation z, R and ω are assumed to be constants

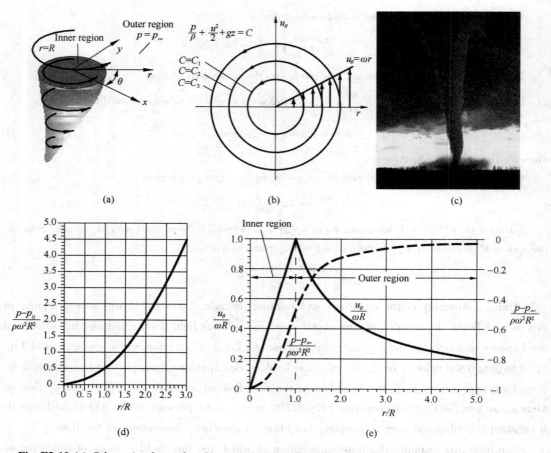

Fig. E3-10 (a) Schematic of tornado; (b) streamlines of solid body rotation ($r < R$); (c) Category F5 tornado viewed from the southeast as it approached Elie, Manitoba on June 22nd, 2007 (*Courtesy of Justin Hobson, Justin1569, at en. wikipedia*); (d) Nondimensional pressure as a function of nondimensional radial location for a fluid in solid body rotation with negligible gravity; (e) Nondimensional tangential velocity distribution (solid curve) and nondimensional pressure distribution (dash curve) along a horizontal radial slice through a tornado. The inner and outer regions of flow are marked when considering a particular horizontal slice.

3. The flow in the horizontal slice is two-dimensional in the $r\theta$-plane (no dependence on z and no z-component of velocity).

4. The effects of gravity are negligible within a particular horizontal slice (an additional hydrostatic pressure field exists in the z-direction, of course, but this does not affect the dynamics of the flow, as discussed in Section 2.4).

In the inner region ($r < R$), the pressure field can be found as that for solid body rotation with a negligible gravity as shown in Fig. E3.10b. Apply the Bernoulli equation along a streamline, Eq. 3.58, directly and neglect the gravity term to give

$$p = C\rho - \frac{\rho u_\theta^2}{2} = C\rho - \frac{\rho \omega^2 r^2}{2} \tag{2}$$

where C is the Bernoulli constant that varies with radius as illustrated in Fig. E3-10b.

For the rigid body rotation with a negligible gravity, the pressure at a point can be obtained from Eq. 2.35

$$p = p'_{at} + \frac{\rho \omega^2 r^2}{2} \tag{3}$$

where p'_{at} is the constant pressure at the top surface of tornado.

Partial derivative of Eq. 3 with respect to r gives

$$\frac{\partial p}{\partial r} = \rho \omega^2 r \tag{4}$$

Since the flow is steady and incompressible, and there is no variation of any flow variable in the θ-direction and gravity is neglected in the z-direction, p is not a function of t, θ or z. Thus p is a function of r only, and we replace the partial derivative in Eq. 4 with a total derivative. Integration yields

$$p = \frac{\rho \omega^2 r^2}{2} + C_1 \tag{5}$$

where C_1 is a constant of integration.

Assume the pressure at the origin ($r = 0$) to be p_0, thus, constant C_1 is found by applying the known pressure boundary condition at the origin. It turns out that $C_1 = p_0$ and Eq. 5 to be

$$p = \frac{\rho \omega^2 r^2}{2} + p_0 \tag{6}$$

It indicates that the pressure distribution is parabolic with respect to r as illustrated in Fig. E3-10d.

Finally, we equate Eqs. 2 and 6 to solve for the Bernoulli constant C

$$C = \frac{p_0}{\rho} + \omega^2 r^2$$

Substituting the Bernoulli constant C back into Eq. 2, we have the pressure field in the inner region ($r < R$)

$$p = p_0 + \frac{\rho \omega^2 r^2}{2} \tag{7}$$

Since the outer region is a region of irrotational flow, the Bernoulli equation is appropriate and the Bernoulli constant C is the same everywhere from $r = R$ outward to $r \to \infty$. The Bernoulli constant C is found by applying the boundary condition far from the tornado, namely, as $r \to \infty$, $u_\theta \to 0$ and $p \to p_\infty$ (Fig. E3-10a). The Bernoulli equation, Eq. 3.59, with neglected gravity yields

$$C = \frac{p_\infty}{\rho} \tag{8}$$

The pressure field anywhere in the outer region ($r > R$) with neglected gravity is obtained by substituting the value of constant C from Eq. 8 into the Bernoulli equation (Eq. 3.59)

$$p = C\rho - \frac{\rho u_\theta^2}{2} = p_\infty - \frac{\rho \omega^2 R^4}{2r^2} \tag{9}$$

At $r = R$, the interface between the inner and outer regions, the pressure must be continuous (no sudden jumps in p). Equating Eqs. 7 and 9 at this interface yields

$$p_{r=R} = p_0 + \frac{\rho \omega^2 R^2}{2} = p_\infty - \frac{\rho \omega^2 R^4}{2R^2}$$

from which the pressure p_0 at $r = 0$ is found

$$p_0 = p_\infty - \rho \omega^2 R^2 \tag{10}$$

Equation 10 provides the value of pressure in the middle of the tornado—the eye of the storm. This is the lowest pressure in the flow field.

Substitution of Eq. 10 into Eq. 7 enables us to rewrite Eq. 7, the pressure field in inner region ($r < R$), in terms of the given far-field ambient pressure p_∞

$$p = p_\infty - \rho\omega^2\left(R^2 - \frac{r^2}{2}\right) \tag{11}$$

Instead of plotting p as a function of r in this horizontal slice, we plot a nondimensional pressure distribution instead, so that the plot is valid for any horizontal slice. In terms of nondimensional variables

Inner region ($r < R$) $\quad \dfrac{u_\theta}{\omega R} = \dfrac{r}{R} \quad \dfrac{p - p_\infty}{\rho\omega^2 R^2} = \dfrac{1}{2}\left(\dfrac{r}{R}\right)^2 - 1$

Outer region ($r > R$) $\quad \dfrac{u_\theta}{\omega R} = \dfrac{R}{r} \quad \dfrac{p - p_\infty}{\rho\omega^2 R^2} = -\dfrac{1}{2}\left(\dfrac{R}{r}\right)^2$

Fig. E3-10e shows both nondimensional tangential velocity and nondimensional pressure as functions of nondimensional radial location.

DISCUSSION   In the outer region, pressure increases as speed decreases—a direct result of the Bernoulli equation, which applies with the *same* Bernoulli constant everywhere in the outer region. In the inner region, p increases parabolically with r even though speed also increases; this is because the Bernoulli constant changes from streamline to streamline (as also pointed out in Fig. E3-10b). The pressure is lowest in the center of the tornado and rises to atmospheric pressure in the far field (Fig. E3-10e). Finally, the flow in the inner region is *rotational* but *inviscid*, since viscosity plays no role in that region of the flow. The flow in the outer region is *irrotational* and *inviscid*. Note, however, that viscosity still acts on fluid particles in the outer region (Viscosity causes the fluid particles to shear and distort, even though the net viscous force on any fluid particle in the outer region is zero.).

3.7.3 Flow Net

If a flow is irrotational and incompressible in the xy plane, ϕ and ψ both exist and are related. We have previously shown that lines of constant ψ are streamlines; that is

$$\left.\frac{dy}{dx}\right|_{\text{along }\psi = \text{constant}} = \frac{u_y}{u_x} \tag{3.88}$$

The definition of the velocity potential ϕ in xy plane gives

$$d\phi = \frac{\partial\phi}{\partial x}dx + \frac{\partial\phi}{\partial y}dy = u_x dx + u_y dy$$

along a line of constant ϕ we have $d\phi = 0$ so that

$$\left.\frac{dy}{dx}\right|_{\text{along }\phi = \text{constant}} = -\frac{u_x}{u_y} \tag{3.89}$$

Thus,

$$\left.\frac{dy}{dx}\right|_{\text{along }\psi = \text{constant}} \cdot \left.\frac{dy}{dx}\right|_{\text{along }\phi = \text{constant}} = -1 \tag{3.90}$$

It indicates that the streamlines and equipotential lines are everywhere mutually perpendicular at all points where they intersect as shown in Fig. 3-5a except at a stagnation point because both u_x and u_y are zero. Thus, for any potential flow field a 'flow net' can be drawn that consists of a family of streamlines and equipotential lines. The *flow net* is useful in visualizing flow patterns and can be used to obtain graphical solutions by sketching in streamlines and equipotential lines and adjusting lines until the lines are approximately *orthogonal* at all point where they intersect.

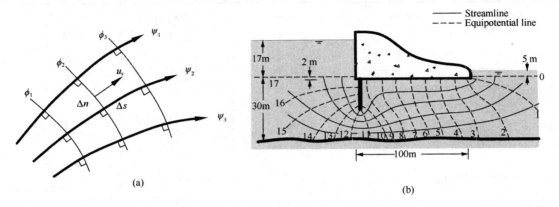

Fig. 3-35 Streamlines and equipotential lines are orthogonal: (a) typical inviscid-flow pattern; (b) flow net of seepage flow under a dam

The first recorded use of flow nets for engineering analysis was that of Philipi Forchheimer who used flow nets to map and calculate the seepage of water in a tunnel beneath the English Channel close to the end of the nineteenth century. Flow nets today can be used in many applications and ways. They can be used to calculate seepage, hydraulic gradient, factor of safety, resistance to quicksand failure and can be used as a quick and easy engineering analysis tool. An example of a flow net is displayed in Fig. 3-35b. This figure illustrates how streamlines and associated equipotential lines are combined together to form a flow net that can be used in seepage analysis. We then can compute the seepage velocity from Eq. 3.78

$$u_s = \frac{q}{\Delta A} = \frac{q}{(\Delta n)(1)} \approx \frac{\Delta \psi}{\Delta n} = \frac{\psi_1 - \psi_2}{\Delta n}$$

or from the definition of velocity potential function, Eq. 3.79

$$u_s = \frac{\partial \phi}{\partial s} \approx \frac{\Delta \phi}{\Delta s} = \frac{\phi_2 - \phi_3}{\Delta s}$$

where Δn, Δs are as shown in Fig. 3-35a.

If we take the increments of values of stream function and velocity potential function in plotting flow net to be same, namely $\Delta \psi = \Delta \phi$, that is we have

$$\Delta n = \Delta s$$

It indicates that the quadrilateral grid of flow net should be near to a square when the increments of ψ and ϕ in a flow net are same.

3.7.4 Some Basic, Plane Potential Flows

Several basic velocity potentials and stream functions can be used to describe simple plane potential flows. Table 3-1 provides a summary of those pertinent equations for the basic, plane potential flows.

127

Table 3-1 Some basic, plane potential flows

Description of Flow Field	Velocity Potential	Stream Function	Velocity Components
Uniform flow at angle α with the x axis (see Fig. 3-36)	$\phi = U(x\cos\alpha + y\sin\alpha)$	$\psi = U(y\cos\alpha - x\sin\alpha)$	$u_x = U\cos\alpha$ $u_y = U\sin\alpha$
Source or sink (see Fig. E3-8) $m > 0$ source $m < 0$ sink	$\phi = \dfrac{m}{2\pi}\ln r$	$\psi = \dfrac{m}{2\pi}\theta$	$u_r = \dfrac{m}{2\pi r}$ $u_\theta = 0$
Free vortex (see Fig. 3-37) * $\Gamma > 0$ counterclockwise motion $\Gamma < 0$ clockwise motion	$\phi = \dfrac{\Gamma}{2\pi}\theta$	$\psi = -\dfrac{\Gamma}{2\pi}\ln r$	$u_r = 0$ $u_\theta = \dfrac{\Gamma}{2\pi r}$
Doublet (see Fig. 3-38) **	$\phi = \dfrac{K\cos\theta}{r}$	$\psi = -\dfrac{K\sin\theta}{r}$	$u_r = -\dfrac{K\cos\theta}{r^2}$ $u_\theta = -\dfrac{K\sin\theta}{r^2}$

Note: *: Γ is called the *circulation* or the *vortex strength*; **: K is a constant equal to ma/π, called the *strength* of the doublet; U is the free stream flow velocity.

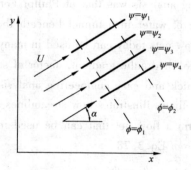

Fig. 3-36 Uniform flow in an arbitrary direction α

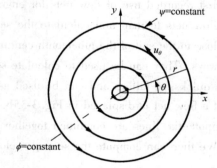

Fig. 3-37 The streamline pattern for a vortex

A source or sink illustrated in Fig. E3-8 represents a purely radial flow. As indicated in Example 3-8, if m is positive, the flow is radially outward, and the flow is considered to be a *source* flow. If m is negative, the flow is toward the origin, and the flow is considered to be a *sink* flow. The flowrate, m, is the *strength* of the source or sink. We note that at the origin where $r = 0$ the velocity becomes infinite, which is of course physically impossible. Thus, sources and sinks do not really exist in real flow fields, and the line representing the source or sink is a mathematical *singularity* in the flow field. However, some real flows can be approximated at points away from the origin by using sources or sinks. Also, the velocity potential representing this hypothetical flow can be combined with other basic velocity potentials to describe approximately some real flow fields. For example, the combination of a uniform flow and a source flow can be used to represent the flow around a half-body discussed in next section.

A *vortex* in Fig. 3-37 represents a flow in which the streamlines are concentric circles. Following the standard convention in mathematics, positive Γ represents a counterclockwise vortex,

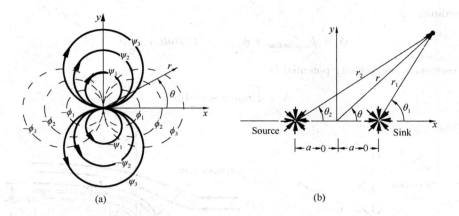

Fig. 3-38 Doublet. (a) Streamlines (solid) and equipotential lines (dashed) for a doublet of strength K located at the origin in the xy-plane and aligned with the x-axis; (b) The combination of a source and sink of equal strength located along the x-axis

while negative Γ represents a clockwise vortex. Vortex motion can be either rotational or irrotational. The rotational vortex is commonly called a *forced vortex*, whereas the irrotational vortex is usually called a *free vortex*. The swirling motion of the water as it drains from a bathtub is similar to that of a free vortex, whereas the motion of a liquid contained in a tank that is rotated about its axis with angular velocity ω corresponds to a forced vortex.

The so-called *doublet* illustrated in Fig. 3-38a is formed by letting the source and sink illustrated in Fig. 3-38b approach one another ($a \rightarrow 0$) while increasing the strength m ($m \rightarrow \infty$) so that the product ma/π remains constant. Just as sources and sinks are not physically realistic entities, neither are doublets. However, the doublet when combined with other basic potential flows provides a useful representation of some flow fields of practical interest. For example, the combination of a uniform flow and a doublet can be used to represent the flow around a circular cylinder.

3.7.5 Superposition of Basic, Plane Potential Flows

For incompressible plane potential flow, both the potential function and the stream function satisfy Laplace's equations in two dimensions. That is, potential flows are governed by linear partial differential equations, the solutions can be combined in superposition, e. g. if $\phi_1(x, y, z)$ and $\phi_2(x, y, z)$ are two solutions to Laplace's equation, then $\phi_3 = \phi_1 + \phi_2$ is also a solution. Meanwhile, any streamline in an inviscid flow acts as solid boundary, such that there is no flow through the boundary or streamline. Thus, some of the basic velocity potentials or stream functions can be combined to yield a streamline that represents a particular body shape of interest, that combination can be used to describe in detail the flow around that body. This method of solving some interesting flow problems, commonly called the *method of superposition*. In the following a superposition of a uniform flow and a source is discussed.

If we superimpose a x-directed uniform stream against an isolated source as illustrated in Fig. 3-39, a *half-body* shape appears. If the source is at the origin, the combined stream function in polar coordinates is

$$\psi = \psi_{\text{uniform flow}} + \psi_{\text{source}} = Ur\sin\theta + \frac{m}{2\pi}\theta \quad (3.91)$$

and the corresponding velocity potential is

$$\phi = Ur\cos\theta + \frac{m}{2\pi}\ln r \quad (3.92)$$

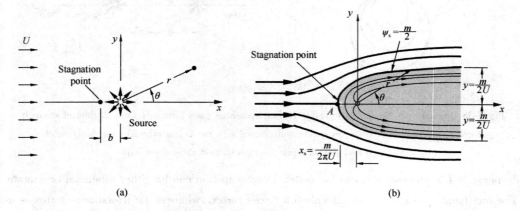

Fig. 3-39 The flow around a half-body: (a) superposition of a source and a uniform flow; (b) replacement of streamline $\psi_s = m/2$ with a solid boundary to form half-body

The velocity components are found by partial differentiate:

$$\left.\begin{array}{l} u_r = \dfrac{\partial \phi}{\partial r} = U\cos\theta + \dfrac{m}{2\pi r} \\[2mm] u_\theta = \dfrac{1}{r}\dfrac{\partial \phi}{\partial \theta} = -U\sin\theta \end{array}\right\} \quad (3.93)$$

To find the coordinates of the stagnation point, setting $u_\theta = 0$ gives $\theta = 0$ or $\theta = \pi$; setting $u_r = 0$ gives $r = -\dfrac{m}{2\pi U\cos\theta}$. If $\theta = 0$ we have $r < 0$, which is impossible. Therefore, $\theta = \pi$, $r = \dfrac{m}{2\pi U}$ is the only choice for the stagnation point. Thus, the single stagnation point locates at point A with a coordinate of

$$\theta = \pi, \quad r = \frac{m}{2\pi U} \quad (3.94)$$

Substitution into the stream function equation, Eq. 3.91, gives $\psi_s = \dfrac{m}{2}$ for the streamline passing through the stagnation point. Thus, from Eq. 3.91 the stagnation streamline equation is

$$Ur\sin\theta + \frac{m}{2\pi}\theta = \frac{m}{2} \quad (3.95)$$

or

$$y = r\sin\theta = \frac{m}{2U}\left(1 - \frac{\theta}{\pi}\right) \quad (3.96)$$

We can set θ in Eq 3.95 equal to various constants:

$$\theta = \frac{\pi}{2}, \frac{3\pi}{2} \qquad y = \pm \frac{m}{4U}$$

$$\theta = \pi \text{ (Eq. 3.94)} \qquad x_s = -r = -\frac{m}{2\pi U}$$

$$\theta \to 0, 2\pi \qquad r \to \infty, \text{ the streamline approaches to } y \to \pm \frac{m}{2U}$$

A plot of streamlines is shown in Fig. 3.39b. If we replace the streamline of $\psi_s = \frac{m}{2}$ with a solid boundary, as indicated in the figure, then a curved, roughly elliptical, *half-body* shape, opened at the downstream end, appears, which separates the source flow inside from the stream flow outside. This body shape is named after the Scottish engineer W. J. M. Rankine (1820-1872) and called *Rankine half-body*, which could be considered as the frontal half of the bridge pier and gate pier. Other streamlines in the flow field can be obtained by setting ψ = constant in Eq. 3.91 and plotting the resulting equation. A number of these streamlines are shown in Fig. 3-39b. Although the streamlines inside the body are shown, they are actually of no interest in this case, since we are concerned with the flow field outside the body. It should be noted that the singularity in the flow field (the source) occurs inside the body, and there are no singularities in the flow field of interest (outside the body).

Summary

This chapter considered several fundamental concepts of fluid kinematics and basic differential equations.

In fluid mechanics, both the Eulerian and Lagrangian approaches can be applied to describe the flow. *Lagrangian approach* follows each particle while *Eulerian approach* only concerned with the fluid properties in a finite space at a fixed position, the *control volume*. Since the description of Eulerian approach is of greater practical significant, it is generally adopted in engineering practice.

Based on the Eulerian approach, the Eulerian acceleration filed can be described as

$$\mathbf{a} = \frac{d\mathbf{u}}{dt} = \frac{\partial \mathbf{u}}{\partial t} + (\mathbf{u} \cdot \nabla) \mathbf{u}$$

where $\partial \mathbf{u}/\partial t$ is the *local acceleration* varying with time, $(\mathbf{u} \cdot \nabla) \mathbf{u}$ is the *convective acceleration*, or the *advective acceleration*, due to the spatial variations of the fluid properties.

All physical flows are *three-dimensional flows*. Some can be visualized qualitatively by the *streamlines*, *pathlines* and *streaklines*. Some flows can be simplified to be *two-dimensional flows*, i.e. the flow past a long circular cylinder or airfoil, or *one-dimensional flows* (also called *total flow*), i.e. the flow in a long constant-diameter straight pipe, of which the acceleration is $a_s = \partial u_s/\partial t + u_s \partial u_s/\partial x$.

For *steady flow*, the kinematic variables do not vary with respect to time, $\partial q/\partial t = 0$, and streamlines, streaklines, and pathlines are geometrically coincided. For *unsteady flow*, $\partial q/\partial t \neq 0$, streamlines, streaklines, and pathlines may change its shape or position from time to time.

For *uniform flow*, the velocity at any point in a the flow has the identical both magnitude and direction, in which streamlines parallel to each other, $(\mathbf{u} \cdot \nabla)\mathbf{u} = 0$. If any one of the magnitude or direction of velocity changes from point to point, it is the *nonuniform flow*, $(\mathbf{u} \cdot \nabla)\mathbf{u} \neq 0$, i. e. the flow in a bending pipe or converging-diverging pipe.

Element flow in a *streamtube*, and the *total flow* in *a control volume* are the commonly used flow in the analysis of fluid mechanics, while the streamline can be considered as the solid boundary because no flow can cross a streamline. The volume of fluid passing through the *control sections* of the control volume is usually accounted by the *volume flow rate* or *mass flow rate*.

Applying the *conservation law of mass* on the incompressible flow of a *Newtonian fluid* with constant properties gives the differential form of *continuity equation*

$$\frac{\partial u_x}{\partial x} + \frac{\partial u_y}{\partial y} + \frac{\partial u_z}{\partial z} = 0$$

Its common application in the steady, incompressible, two-dimensional flow in Cartesian coordinates leads to the *stream function*

$$d\psi = \frac{\partial \psi}{\partial x}dx + \frac{\partial \psi}{\partial y}dy$$

with

$$u_x = \frac{\partial \psi}{\partial y} \qquad u_y = -\frac{\partial \psi}{\partial x}$$

The *geometrical interpretations* of the stream function is that the line of constant ψ is a streamline and $d\psi = dq$.

Navier-Stokes equations is derived by the application of *Newton's second law* on the fluid element in an incompressible, viscous flow. For the inviscid flow, it is reduced to much simpler equations: *Euler's equation of motion*. Integrating it along a streamline or in a potential flow gives the famous *Bernoulli equation* in terms of head

$$z + \frac{p}{\rho g} + \frac{u^2}{2g} = C$$

in which z is the *elevation head*, $\frac{p}{\rho g}$ the *pressure head*, $\frac{u^2}{2g}$ the *velocity head*, C the integral constant along a streamline or in a potential flow.

Bernoulli equation can also be interpreted in terms of pressure, of which the terms are termed as *hydrostatic pressure*, *static pressure* and *dynamic pressure* respectively. It is a very powerful tool, meanwhile most abused tool in the fluid mechanics.

The deformation of a fluid element, *translation*, *rotation*, *shear strain* and *dilatation* can be described by the *velocity*, *angular velocity* (rate of rotation), *linear strain rate*, and *shear strain* rate. For the flow with zero *vorticity*, $\zeta = 0$, or zero *angular velocity*, $\omega = 0$, or

$$\zeta = \text{curl}\mathbf{u} = 2\omega$$

or
$$\frac{\partial u_y}{\partial x} = \frac{\partial u_x}{\partial y} \quad \frac{\partial u_z}{\partial y} = \frac{\partial u_y}{\partial z} \quad \frac{\partial u_x}{\partial z} = \frac{\partial u_z}{\partial x}$$

is called the *irrotational flow*, in which the fluid particle are not rotating about its own instant axis. For example, the flow outside of the boundary layer can be considered as the irrotational flow. Otherwise, it is a *rotational flow*.

For the irrotational flow, the *velocity potential function* can be defined

$$d\phi = \frac{\partial \phi}{\partial x}dx + \frac{\partial \phi}{\partial y}dy + \frac{\partial \phi}{\partial z}dz$$

with
$$u_x = \frac{\partial \phi}{\partial x} \quad u_y = \frac{\partial \phi}{\partial y} \quad u_z = \frac{\partial \phi}{\partial z}.$$

The lines of constant velocity potential functions are *equipotential lines*. In an inviscid, irrotational and incompressible flow in the xOy plane, the streamlines and equipotential lines are everywhere *mutually perpendicular* at all points where they intersect to configure a *flow net*. Both stream function and velocity potential satisfy the *Laplace's equation*, which indicates that the some complicated potential flows can be *superposed* by some basic velocity potential flows, uniform flow, source or sink, vortex, and doublet, e. g. the flow past the half-body, the circular.

Word Problems

W3-1 Define the Lagrangian and Eulerian approaches. How do they differ from each other? Which one is used to measure the flow velocity in hydrologic station?

W3-2 A tiny neutrally buoyant electronic pressure probe is released into the inlet pipe of a water pump and transmits 2000 pressure readings per second as it passes through the pump. Is this a Lagrangian or an Eulerian measurement? Explain.

W3-3 A Pitot-static probe can often be seen protruding from the underside of an airplane. As the airplane flies, the probe measures relative wind speed. Is this a Lagrangian or an Eulerian measurement? Explain.

W3-4 What are uniform flow and nonuniform flow? Give some examples. What about their Eulerian acceleration?

W3-5 What are steady flow and unsteady flow? Give some examples. What about their Eulerian acceleration?

W3-6 What are streamlines, pathlines and streaklines? Do they coincide in steady flow or unsteady flow?

W3-7 How does the streamlines in steady flow differ from the streamlines in unsteady flow?

W3-8 Define the streamtube and control volume. How are they related to each other?

W3-9 Define the element flow and total flow. How are they related to each other?

W3-10 Define mass and volume flow rates. How are they related to each other?

W3-11 How does the angular velocity of fluid element differ from the angular velocity of rigid rotation?

W3-12 Define the angular velocity and the vorticity.

W3-13 Define the rotational and irrotational flow. How to classify?

W3-14 Does the inviscid flow must be irrotational flow? Does viscous flow must be rotational flow? Explain.

W3-15 Write down the differential continuity equation in different forms. What is the physical meaning of the differential continuity equation for incompressible flow?

W3-16 In what way is the Euler equation an approximation of the Navier-Stokes equation? Where in a flow field is the Euler equation an appropriate approximation?

W3-17 Define static, dynamic, and hydrostatic pressure. Under what conditions is their sum constant for a flow stream?

W3-18 What flow property determines whether a region of flow is rotational or irrotational? Discuss.

W3-19 Express the Bernoulli equation in three different ways using (a) energies, (b) pressures, and (c) heads.

W3-20 What are the three major assumptions used in the derivation of the Bernoulli equation?

W3-21 Define the stagnation point and the stagnation pressure. How to measure the stagnation pressure?

W3-22 In a certain application, a siphon must go over a high wall. Can water or oil with a specific gravity of 0.8 go over a higher wall? Why?

W3-23 A student siphons water over a 8.5-m-high wall at sea level. She then climbs to the summit of Mount Shasta (elevation 4390m, $p_{at} = 58.5 \text{kPa}$) and attempts the same experiment. Comment on her prospects for success.

W3-24 A glass manometer with oil as the working fluid is connected to an air duct as shown in Fig. W3-24. Will the oil in the manometer move as in Fig. W3-24a or b? Explain. What would your response be if the flow direction is reversed?

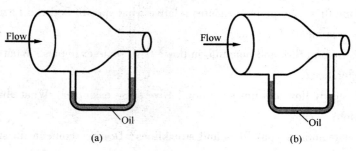

Fig. W3-24

W3-25 Write the Bernoulli equation, and discuss how it differs between an inviscid, rotational region of flow and a viscous, irrotational region of flow. Which case is more restrictive (in regards to the Bernoulli equation)?

W3-26 Define stream function. What are the geometric interpretations of stream function?

W3-27 What restrictions or conditions are imposed on stream function ψ so that it exactly satisfies the two-dimensional incompressible continuity equation by definition? Why are these restrictions necessary?

W3-28 Consider two-dimensional flow in the xy-plane. What is the significance of the difference in value of stream function ψ from one streamline to another?

W3-29 The solid boundary of the steady flow is an equipotational line, is it true? Explain.

W3-30 Define the velocity potential function. The variation of velocity potential function ϕ with respect to x or y must be linear if $\nabla^2\phi = 0$, is it true? Explain.

W3-31 What restriction conditions are imposed on the definitions of stream function ψ and velocity potential function ϕ? How about the restriction conditions for ψ and ϕ satisfying the Laplace equation?

W3-32 What is flow net? What are the flow properties of flow net? How can it be used in engineering practice?

W3-33 Streamlines in a steady, two-dimensional, incompressible flow field are sketched in Fig. W3-33. The flow in the region shown is also approximated as irrotational. Sketch what a few equipotential curves might look like in this flow field. Explain how you arrive at the curves you sketch.

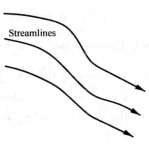

Fig. W3-33

Problems

Lagrangian and Eulerian Descriptions

3-1 The velocity field of a flow is given by

$$\mathbf{u} = 20\frac{y}{\sqrt{x^2+y^2}}\mathbf{i} - 20\frac{x}{\sqrt{x^2+y^2}}\mathbf{j} \quad \text{m/s}$$

Determine the fluid speed at points along the x axis and the y axis. What is the angle between the velocity vector and the x axis at points $(x, y) = (5, 0)$, $(5, 5)$, and $(0, 5)$?

3-2 The velocity in a certain flow field is given by the equation

$$\mathbf{u} = x\mathbf{i} + x^2 z\mathbf{j} + yz\mathbf{k}$$

Determine the expressions for the three rectangular components of acceleration.

3-3 The velocities of uniform flow in a pipe shown in Fig. P3-3 is given by $v_1 = 0.5t$ m/s and $v_2 = 1.0t$ m/s, where t is in seconds. Determine the local acceleration at points (1) and (2). Is the average convective acceleration between these two points negative, zero, or positive?

Explain.

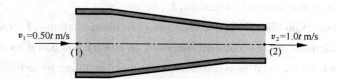

Fig. P3-3

3-4 Consider steady flow of air through the diffuser portion of a wind tunnel. Along the centerline of the diffuser ($x = 0$ to $x = L$), the air speed decreases from $u_{entrance}$ to u_{exit}. Measurements reveal that the centerline air speed decreases parabolically through the diffuser as $u(x) = a + bx^2$. Determine an equation for centerline speed $u(x)$ from $x = 0$ to $x = L$ based on the given parameters.

3-5 For the velocity field of Prob. 3-4, calculate the fluid acceleration along the diffuser centerline as a function of x and the given parameters. For $L = 2.0$ m, $u_{entrance} = 30$ m/s, and $u_{exit} = 5.0$ m/s, calculate the acceleration at $x = 0$ and $x = 1.0$ m.

Flow Patterns and Flow Visualization

3-6 The velocity field of a plane flow is given by $\mathbf{u} = (4y - 6x)t\mathbf{i} + (6y - 9x)t\mathbf{j}$ (m/s). (a) Is this flow a steady or unsteady flow; (b) Is this flow an uniform or nonuniform flow? (c) Compute the acceleration at point (2, 4) when $t = 1$s; (d) Determine the streamline equation when $t = 1$s.

3-7 A velocity filed is given by

$$u_x = xy^2 \qquad u_y = -\frac{1}{3}y^3 \qquad u_z = xy$$

(a) Calculate the acceleration at point (1, 2, 3); (b) Is this flow a one-, two-, or three-dimensional flow? (c) Is this flow a steady or an unsteady flow? (d) Is this flow an uniform or nonuniform flow?

3-8 A velocity field for a plane flow is given by

$$u_x = \frac{-Cy}{x^2 + y^2} \qquad u_y = \frac{Cx}{x^2 + y^2}$$

where C is the constant. Determine the streamline equation and plot the streamlines.

3-9 A velocity field of plane flow is given by $\mathbf{u} = x\mathbf{i} + x(x - 1)(y + 1)\mathbf{j}$. Plot the streamline that passes through $x = 0$ and $y = 0$. Compare this streamline with the streakline through the origin.

3-10 The velocity field for a incompressible plane flow is given by $\mathbf{u} = x^2t\mathbf{i} - 2xyt\mathbf{j}$. Determine the streamline equation and the pathline equation passing point $(-2, 1)$ at time $t = 1$s.

3-11 A two-dimensional unsteady velocity field is given by $\mathbf{u} = x(1 + 2t)\mathbf{i} + y\mathbf{j}$. (a) Find the time-varying streamlines which pass through reference point (x_0, y_0) and sketch some. (b) Find the equation of pathline passing through reference point (x_0, y_0) at $t = 0$ and sketch this pathline.

3-12 Water flows from a rotating lawn sprinkler as shown in Fig. P3-12. The end of the sprinkler arm moves at a speed of ωR, where $\omega = 10$ rad/s is the angular velocity of the sprinkler arm and $R = 0.15$m is its radius. The water exits the nozzle at a speed of $u = 3$m/s relative to the rotating arm. Gravity and the interaction between the air and the water are negligible. (a) Show that the pathlines for this flow are straight radial lines. *Hint:* Consider the direction of flow (relative to the stationary ground) as the water leaves the sprinkler arm. (b) Show that at any given instant the stream of water that came from the sprinkler forms an arc given by $r = R + (u_a/\omega)\theta$, where the angle θ is as indicated in the figure and u_a is the water speed relative to the ground. Plot this curve for the data given.

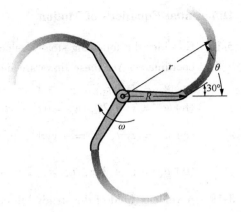

Fig. P3-12

3-13 An uniform air flow at 1 atm and at 20℃ is flowing through a 0.5-m-diameter pipe at a velocity of 30m/s. Determine the volumetric, mass and weight flow rates of this air flow.

Deformation of Fluid Element

3-14 The velocity filed of an incompressible, two-dimensional flow is given by $\mathbf{u} = [y^2 - x(1+x)]\mathbf{i} + [y(2x+1)]\mathbf{j}$. Show that the flow is irrotational and satisfies conservation law of mass.

3-15 Consider the following two steady velocity fields in Cartesian coordinates:
(a) $u_x = -ay$, $u_y = ax$, $u_z = 0$
(b) $u_x = -\dfrac{cy}{x^2+y^2}$, $u_y = \dfrac{cx}{x^2+y^2}$, $u_z = 0$

where a, c are the constants. Are they irrotational or rotational flows? Do the flows have the angular deformation (shear strain rate)?

3-16 The velocity filed for a steady incompressible rotational flow in Cartesian coordinates is given by $u_x = 2y + 3z$, $u_y = 2z + 3x$, $u_z = 2x + 3y$. Determine the angular velocity and the rate law of shear strain.

3-17 An incompressible viscous fluid is placed between two large parallel plates as shown in Fig. P3-17. The bottom plate is fixed and the upper plate moves with a constant velocity, U. For these conditions the velocity distribution between the plates is linear and can be expressed as $u = Uy/b$. (a) Does this flow satisfy the conservation law of mass? (b) Determine the rotation

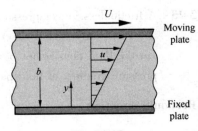

Fig. P3-17

vector, the vorticity, and the rate of angular deformation.

Differential Equations of Motion

3-18 Consider the following steady velocity fields in Cartesian coordinates. Are these flows are incompressible flow?
(a) $u_x = 2x^2 + y^2$, $u_y = x^3 - x(y^2 - 2y)$
(b) $u_x = xt + 2y$, $u_y = xt^2 - yt$
(c) $u_x = xyzt$, $u_y = -xyzt^2$, $u_z = \frac{1}{2}(xt^2 - yt)z^2$
(d) $u_x = y^2 + 2xz$, $u_y = -2yz + x^2yz$, $u_z = \frac{1}{2}x^2z^2 + x^3y^4$

3-19 A velocity field of the steady, three-dimensional flow in Cartesian coordinate system is given by
$$\mathbf{u} = (axy^2 - b)\mathbf{i} + cy^3\mathbf{j} + dxy\mathbf{k}$$
where a, b, c, and d are constants. Under what conditions is this flow field incompressible?

3-20 The velocity component in the x-direction of a steady, two-dimensional, incompressible flow is $u_x = ax + b$, where a and b are constants. Velocity component u_y is unknown. Generate an expression for u_y as a function of x and y.

3-21 A frictionless, incompressible steady flow field is given by
$$\mathbf{u} = 2zy\mathbf{i} - y^2\mathbf{j}$$
Let the density be constant and neglect gravity. Find an expression for the pressure gradient in the x-direction.

3-22 If z is 'up', what are the conditions on constants a and b for which the velocity field $u_x = ay$, $u_y = bx$, $u_z = 0$ is an exact solution to the continuity and Navier-Stokes equations for incompressible flow?

3-23 The velocity field of a steady, two-dimensional incompressible flow of a Newtonian fluid is given by $u_x = 2xy$, $u_y = y^2 - x^2$, $u_z = 0$. (a) Does this flow satisfy conservation law of mass? (b) Find the pressure field, $p(x, y)$ if the pressure at the point $(x = 0, y = 0)$ is equal to p_{at}.

3-24 Consider a steady, incompressible flow of an ideal fluid in which the pressure field is known as $p = 4x^3 - 2y^2 - yz^2 + 5z$ (N/m²). Let the fluid density be $\rho = 1000$ kg/m³ and $g = 9.8$ m/s². Determine the acceleration of fluid particle located at point $(3, 1, -5)$ (m).

3-25 A frictionless, incompressible steady-flow field is given by
$$\mathbf{u} = (3x^2 - 2zy)\mathbf{i} + (y^2 - 6xy + 3yz^2)\mathbf{j} - (z^3 + xy^2)\mathbf{k} \text{ (m/s)}$$
Let the fluid density be $\rho = 1000$ kg/m³ and $g = 9.8$ m/s². Find an expression for the pressure gradient of fluid particle at point of $x = 2$m, $y = 3$m and $z = 1$m.

Bernoulli equation

3-26 Water flows through the pipe contraction shown in Fig. P3-26. For the given 0.2m difference in manometer level, determine the flowrates in both cases as a function of the diameter of the small pipe, D.

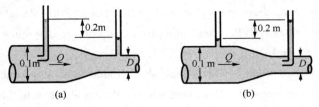

Fig. P3-26

3-27 A Pitot-static probe connected to a water manometer shown in Fig. P3-27 is used to measure the velocity of air. If the deflection (the vertical distance between the fluid levels in the two arms) is 7.3 cm, determine the air velocity. Take the density of air to be 1.25 kg/m^3.

3-28 The specific gravity of the manometer fluid shown in Fig. P3-28 is 1.07. Determine the volume flowrate, Q, if the flow is inviscid and incompressible and the flowing fluid is (a) water, (b) gasoline, or (c) air at standard conditions.

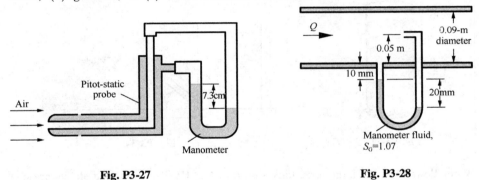

Fig. P3-27   **Fig. P3-28**

3-29 Determine the flowrate through the Venturi meter shown in Fig. P3-29 if ideal conditions exist.

3-30 A necked-down section in a *venturi* develops a low throat pressure which can aspirate fluid upward from a reservoir, as in Fig. P3-30. Using Bernoulli's equation to derive an expression for the velocity U_1 which is just sufficient to bring reservoir fluid into the throat. Assume the velocity profile at the throat section is uniform.

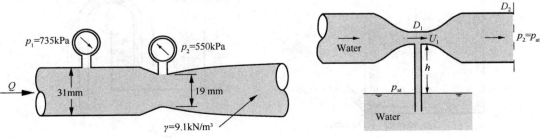

Fig. P3-29   **Fig. P3-30**

3-31 A pressurized tank of water has a 10-cm-diameter orifice at the bottom, where water discharges to the atmosphere(Fig. P3-31). The water level in the tank is 3 m above the outlet. The air pressure above the water level in the tank is 300 kPa (absolute) while the atmospheric pressure is 100kPa. Neglecting frictional effects, determine the initial discharge of water from the tank.

3-32 The water in a 10-m-diameter, 2-m-high aboveground swimming pool is to be emptied by unplugging a 3-cm-diameter, 25-m-long horizontal pipe attached to the bottom of the pool. Determine the maximum discharge of water through the pipe. Also, explain why the actual flow rate will be less.

3-33 Reconsider Prob. 3-32. Determine how long it will take to empty the swimming pool completely.

3-34 Water flows from a large tank of depth H, through a pipe of length L, and strikes the ground as shown in Fig. P3-34. Viscous effects are negligible. Determine the distance h as a function of θ.

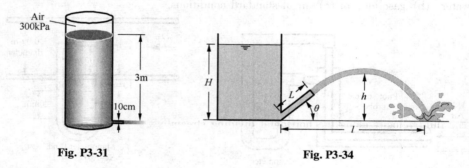

Fig. P3-31   Fig. P3-34

3-35 Water flows from the large open tank shown in Fig. P3-35. If viscous effects are neglected, determine the heights, h_1, h_2, and h_3, to which the three streams rise.

3-36 Once it has been started by sufficient suction, the *siphon* in Fig. P3-36 will run continuously as long as reservoir fluid is available. Using Bernoulli's equation, show (a) that the exit velocity U_2 depends only upon gravity and the distance H and (b) that the lowest (vacuum) pressure occurs at point 3 and depends on the distance $L + H$.

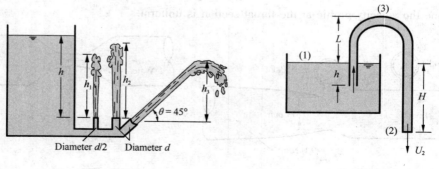

Fig. P3-35   Fig. P3-36

3-37 Two jets flow out from two holes on the tank walls shown in Fig. P3-37. Prove $h_1 y_1 = h_2 y_2$ at the intersected point of two jets.

3-38 A handheld bicycle pump can be used as an atomizer to generate a fine mist of paint or pesticide by forcing air at a high velocity through a small hole and placing a short tube between the liquid reservoir and the high-speed air jet whose low pressure drives the liquid up through the tube. In such an atomizer, the hole diameter is 0.3cm, the vertical distance between the liquid level in the tube and the hole is 10cm, and the bore (diameter) and the stroke of the air pump are 5cm and 20cm, respectively. If the atmospheric conditions are 20°C and 95kPa, determine the minimum speed that the piston must be moved in the cylinder during pumping to initiate the atomizing effect. The liquid reservoir is open to the atmosphere.

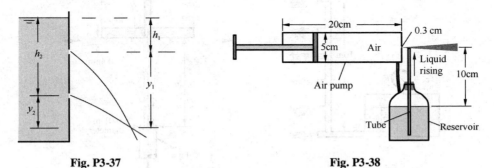

Fig. P3-37   Fig. P3-38

3-39 An incompressible fluid with density ρ flows steadily past the object shown in Fig. P3-39. The fluid velocity along the horizontal dividing streamline ($-\infty \leq x \leq -a$) is found to be $u = U(1 + a/x)$, where a is the radius of curvature of the front of the object and U is the free upstream flow velocity prior to the object. (a) Determine the pressure gradient along this streamline. (b) If the upstream pressure is p_0, integrate the pressure gradient to obtain the pressure $p(x)$ for $-\infty \leq x \leq -a$. (c) Show from the result of part (b) that the pressure at the stagnation point ($x = -a$) is $p_0 + \rho U^2/2$, as expected from the Bernoulli equation.

3-40 In Fig. P3-40 an inviscid incompressible fluid flows steadily along the stagnation streamline, of which the far upstream velocity is U. Upon leaving the stagnation point, point (1), the fluid speed along the surface of the object is assumed to be given by $u = 2U\sin\theta$, where θ is the angle indicated. At what angular position, θ_2, should a hole be drilled to give a pressure difference of $p_1 - p_2 = \rho U^2/2$? Gravity is negligible.

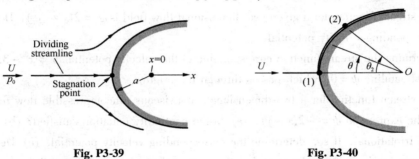

Fig. P3-39   Fig. P3-40

3-41 Water flows from a pressurized tank through six equally spaced outlets on the vertical spray tower shown in Fig. P3-41. The diameter of the lowest outlet is 13mm. Determine the diameters of the other five outlets if the flowrate is to be the same from each of the outlets when the pressure in the tank is $p_1 = 200$ kPa.

3-42 Water flows from the pipe shown in Fig. P3-42 as a free jet and strikes a circular flat plate. The flow geometry shown is axisymmetrical. Determine the flowrate and the manometer reading, H.

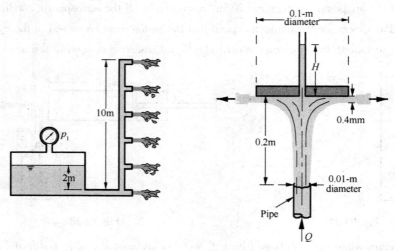

Fig. P3-41   Fig. P3-42

Plan Potential Flow

3-43 Consider the following velocity fields for steady flows. Are the flows incompressible? Are the flow irrotational?

(a) $u_x = 4y, \quad u_y = -3x$

(b) $u_x = 4xy, \quad u_y = 0$

(c) $u_r = \dfrac{C}{r}, \quad u_\theta = 0$

(d) $u_r = 0, \quad u_\theta = \dfrac{C}{r}$

3-44 The stream function for a given two-dimensional flow field is $\psi = 2(x^2 - y^2)$. Determine the corresponding velocity potential.

3-45 Determine the stream function corresponding to the velocity potential $\phi = x^3 - 3xy^2$. Sketch the streamline $\psi = 0$, which passes through the origin.

3-46 The stream function for a two-dimensional, nonviscous, incompressible flow field is given by the expression $\psi = -2(x - y)$. (a) Is the continuity equation satisfied? (b) Is the flow field irrotational? If so, determine the corresponding velocity potential. (c) Determine the pressure gradient in the horizontal x direction at the point $x = 2$m, $y = 2$m.

3-47 A steady, two-dimensional, incompressible flow field in the xy-plane has a stream function given by $\psi = ax^2 - by^2 + cx + dxy$ where a, b, c, and d are constants. (a) Obtain expressions for velocity components u_x and u_y. (b) Verify that the flow field satisfies the incompressible continuity equation.

3-48 The stream function for a two-dimensional, nonviscous, incompressible potential flow field is given by the expression $\psi = 5xy - 4x + 3y + 10$. (a) Determine the velocity components u_x and u_y; (b) Find the velocity potential function; (c) Determine the velocity and pressure at the point $x = 1\,\text{m}$, $y = 2\,\text{m}$ if the fluid density is $\rho = 850\,\text{kg/m}^3$ and the pressure at the stagnation point is $p_0 = 10^5\,\text{N/m}^2$.

3-49 Consider fully developed, two-dimensional channel flow between two infinite parallel plates separated by distance h, with both the top plate and bottom plate stationary, and a forced pressure gradient dp/dx (dp/dx is constant and negative.) driving the flow as illustrated in Fig. P3-49. The flow is steady, incompressible, and two-dimensional in the xy-plane. The velocity components are given by $u_x = \dfrac{1}{2\mu}\dfrac{dp}{dx}(y^2 - hy)$ and $u_y = 0$, where μ is the fluid's viscosity. Generate an expression for stream function ψ along the vertical dashed line in Fig. P3-49. For convenience, let $\psi = 0$ along the bottom wall of the channel. What is the value of ψ along the top wall?

3-50 As a follow-up to Prob. 3-49, calculate the volume flow rate per unit width into the page of Fig. P3-49 from first principles (integration of the velocity field). Compare your result to that obtained directly from the stream function. Discuss.

3-51 A 90°-bend pipe with a rectangular section is placed horizontally as illustrated in Fig. P3-51. The flow in the straight pipe prior to the bend is uniform. The velocity and pressure at section 0-0 are u_0 and p_0 respectively. The bend is symmetric about section A-A, and has the inner and outer radii of r_1 and r_2 respectively. Assume the streamlines of flow in the bend are concentric arcs about center M and meet the conditions of potential flow with $u_\theta r$ = constant. Determine the distributions of velocity u and pressure p at section A-A.

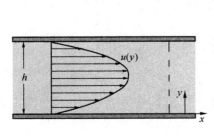

Fig. 3-49

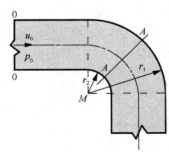

Fig. P3-51

3-52 In Fig. P3-52 a sampling probe aligned with the flow and used to sample the quality of the moving air stream is sketched. For simplicity consider a two dimensional case in which the sampling probe height is $h = 4.5$mm and its width (into the page) is 52mm. The values of the stream function corresponding to the lower and upper dividing streamlines are $\psi_l = 0.105\,\text{m}^2/\text{s}$ and $\psi_u = 0.150\,\text{m}^2/\text{s}$, respectively. Calculate the volume flow rate through the probe (in units of m^3/s) and the speed of the air sucked through the probe. Assume the velocity profile in the probe is uniform.

3-53 One end of a pond has a shoreline that resembles a half-body as shown in Fig. P3-53. A vertical porous pipe is located near the end of the pond so that water can be pumped out. When water is pumped at the rate of $0.08\,\text{m}^3/\text{s}$ through a 3-m-long pipe, what will be the velocity at point A? *Hint:* Consider the flow *inside* a half-body.

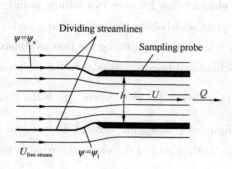
Fig. P3-52

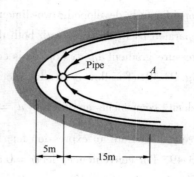
Fig. P3-53

3-54 A steady, incompressible plane potential flow is superposed by a point sink and a point source. The point source is located at point $x = -1$m, $y = 0$ with a source strength $20\,\text{m}^2/\text{s}$. The point sink is located at point $x = 2$m, $y = 0$ with a source strength $40\,\text{m}^2/\text{s}$. Let the fluid density be $1.8\,\text{kg/m}^3$. If the pressure at the origin point ($x = 0$, $y = 0$) is zero, determine the velocities and pressures at point $x = 0$, $y = 1$m and point $x = 1$m, $y = 1$m.

4 Fundamental Equations for Steady Total Flow

CHAPTER OPENING PHOTO: Each man is suspended above sea by two water jets, of which water is supplied by water ski through a long hose, and then is split and redirected 180 degree into two streams to producing the thrust holding the weight of the man. The force, due to the change of the direction of a flow, can be obtained by the control volume momentum principle discussed in the present chapter. (ⓒ*JetLev*) ❶

Many practical problems in fluid mechanics require specific analysis of the behavior of the *total flow* of a finite flow region (*control volume*) in the flow direction, such as pipe flow and open channel flow, by neglecting the spatial variations of physical variables on the cross sections. That is, the flow is simplified as a *one-dimensional flow* with uniform or *averaged* flow variable distributions over the whole section. Thus, in this chapter we will deal with three fundamental governing equations most commonly used in fluid mechanics, the *continuity*, *energy*, and *linear momentum equations*, in integral form (the conservation form) in the finite control volume fixed in space. They are the mathematical statements of three fundamental physical principles upon which all of fluid dynamics is based: conservation law of mass, the first laws of thermodynamics (law of energy conservation) and Newton's second law of motion. The purpose of this chapter is to derive and discuss these equations.

4.1 Average of Sectional Parameters

For the purpose of analyzing flow, the *control volume*, V, an imagined closed reasonably large, fixed finite space drawn within a finite region of the flow, and *control surface*(CS), the closed surface which bounds the control volume, as in Fig. 3-19 are employed in the total flow. In the con-

---

❶ http://thecoolgadgets.com/jetlev-r200-33-feet-long-hose-water-jetpack/

trol volume the mass of fluid may vary with time or may be constant and the flow may be uniform or non-uniform and laminar or turbulent.

For simplicity we assume that flow variables, i.e., velocity, pressure and et ac., are uniformly distributed or averaged over control cross-sectional areas where fluid enters or leaves the control volume. This uniform flow or averaged flow is called *one-dimensional flow*. The effects of nonuniformities will be discussed by the correction factors in energy and momentum equations. Thus, the control volume is a one-dimensional concept, e.g. the simplification of the pipe flow.

The *mean velocity*, or *average velocity*, v, over the control cross section is defined as the average value of the component of velocity normal to the section area involved

$$v = \frac{\int_{CS} \rho(\mathbf{u} \cdot \mathbf{n}) \, dA}{\rho A} = \frac{\dot{m}}{\rho A} \tag{4.1}$$

If the flow is incompressible, the density is constant, Eq. 4.1 reduces to

$$v = \frac{\int_{CS} (\mathbf{u} \cdot \mathbf{n}) \, dA}{A} = \frac{\int_{CS} u_n \, dA}{A} = \frac{Q}{A} \tag{4.2}$$

or

$$Q = Av \tag{4.3}$$

where u_n is the normal component of velocity normal to the section area involved, $u_n = U$ in one-dimensional uniform flow.

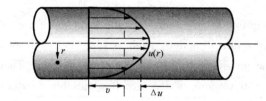

Fig. 4-1 Geographical interpretation of averaged velocity

Equation 4.2 indicates that the cylindrical volume with A-bottom, v-height is equal to the volume enclosed by the curved surface of velocity distribution and the cross section, $\int_{CS} u_n \, dA$, as shown in Fig. 4-1 and Fig. 3-4. It also indicates that for the incompressible fluid, mean velocity is inverse proportional to the cross-sectional area. That is, high velocity occurs at the region with dense streamlines; while low velocity occurs at the region with sparse streamlines as illustrated in Fig. 3-8.

For flow in a duct, the normal velocity is usually nonuniform, as shown in Fig. 4-1. For this case the simple substitution of the average velocity to the point velocity in the control volume analysis would be somewhat in error and should be corrected. Therefore, the *kinematic correction factor*, α, and the *momentum-flux correction factor*, β, are introduced in the section of energy equation and momentum equation accounting for the variation of u across the duct section.

EXAMPLE 4-1 Average velocity

The velocity profile of an incompressible flow in the circular pipe as in Fig. 4-1 is

$$u = u_{max} \left(1 - \frac{r}{R}\right)^{1/7}$$

where u_{max} is the maximum velocity, r is the distance of point away from pipe axis, R is the radius of pipe. Determine: (a) the volume flow rate and the average velocity v, and (b) the distance from the pipe wall to the point where the velocity equals to the average velocity.

Solution:

(a) For a r-radius and dr-width element cross section, the element cross-sectional area is $dA = 2\pi r dr$. Thus the volume flow rate is

$$Q = \int u dA = \int_0^R u_{max} \left(1 - \frac{r}{R}\right)^{1/7} 2\pi r dr = \frac{49}{60}\pi R^2 u_{max}$$

and the average velocity is

$$v = \frac{Q}{A} = \frac{49}{60} u_{max}$$

(b) Take the distance from the pipe wall to the point where the velocity equals to the average velocity be y, thus we have $y = R - r$. At this point the velocity equals to the average velocity, thus we have

$$u_{max} \left(\frac{y}{R}\right)^{1/7} = \frac{49}{60} u_{max}$$

Solving for y gives

$$y = 0.242R$$

It indicates that the average velocity occurs at the point $0.242R$ away from the pipe wall, not the midway ($0.5R$) from the pipe wall.

4.2 Continuity Equation

In Section 3.5.1 we have derived the partial differential equation representing conservation of mass in a fluid flow, the so-called *differential continuity equation*. In this section, we will then recover an integral form, often called the *integral form*, that can be applied to engineering calculations in an approximate, but very useful, way.

A steady total flow is flowing through a finite control volume enclosed by two control cross sections, 1-1 and 2-2, and the lateral boundary as in Fig. 4-2. The control volume has one inlet and one outlet with a volume of V.

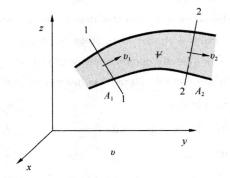

Fig. 4-2 Steady total flow in a finite control volume

Integrating the differential continuity equation, Eq. 3.36, all over the control volume and employing the Gauss law gives

$$\iiint_V \left[\frac{\partial(\rho u_x)}{\partial x} + \frac{\partial(\rho u_y)}{\partial y} + \frac{\partial(\rho u_z)}{\partial z}\right] dV$$

$$= \iint_A \rho u_n dA = 0 \tag{4.4}$$

where A is the area of control surface of the control volume, u_n is the velocity component normal to the control surface.

In Fig. 4-2, there is no flow rate through the lateral surface of the control volume because the lateral surface is composed of streamlines, or is the solid boundaries with $u_n = 0$. Thus, the fluid only flow into and out of the control volume through the control cross sections, 1-1 and 2-2, it follows Eq. 4.4 as

$$\int_{A_1} \rho_1 u_1 \, dA = \int_{A_2} \rho_2 u_2 \, dA$$

or
$$\dot{m}_1 = \dot{m}_2 \tag{4.5}$$

where u is the velocity normal to the control cross section (from now on the subscription 'n' of the normal velocity, u_n, to the control cross section will be omitted); $\dot{m}$ is the mass flow rate, $\dot{m} = \int \rho u \, dA = \rho v A$, where v is the area-averaged velocity over the control cross section.

Equation 4.5 is the *continuity equation* for the steady compressible or incompressible, viscous or inviscid total flow (one dimensional flow) with only one inlet and one outlet.

For the steady and incompressible flow, ρ = constant, Eq. 4.5 reduces to

$$Q_1 = Q_2 \tag{4.6}$$

or
$$v_1 A_1 = v_2 A_2 \tag{4.7}$$

where v_1, v_2 are the area-averaged velocities at the inlet and outlet control cross sections respectively, $v = Q/A$.

Equation 4.7 indicates that for the incompressible fluid, mean velocity is inverse proportional to the cross-sectional area, that is, high velocity occurs at the region with dense streamlines, while low velocity occurs at the region with sparse streamlines as shown in Fig. 3-8.

If the control volume has more than one inlet or outlet, Eqs. 4.5 and 4.6 become

Steady flow $\qquad \Sigma \dot{m}_{in} = \Sigma \dot{m}_{out}$ (4.8)

Steady incompressible flow $\qquad \Sigma Q_{in} = \Sigma Q_{out}$ (4.9)

or $\qquad \Sigma Q_i = 0$ (4.10)

where i is the No. of the branch pipes. In Eq. 4.10 the value of Q_i can be assumed to be positive if the fluid flows out of the junction, or be negative when it flows in.

For the steady and incompressible flow in a junction as illustrated in Fig. 4-3, the continuity equation would be $Q_2 + Q_3 - Q_1 = 0$, or $v_2 A_2 + v_3 A_3 - v_1 A_1 = 0$.

If the flow in Fig. 4-3 is unsteady, Eqs. 4.5 to 4.10 are also available at any instant because the flow can be considered as steady at any instant.

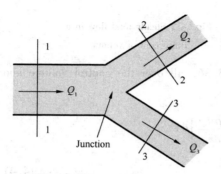

Fig. 4-3 Flows near the junction

EXAMPLE 4-2 Nonuniform velocity profile

Incompressible, laminar water flow developing in a straight pipe has a radius of R as indicated in Fig. E4-2. At section (1), the velocity profile is uniform; the velocity is equal to a constant value U and is parallel to the pipe axis everywhere. At section (2), the velocity profile is axisymmetric and parabolic, $u_2 = u_{max}\left[1 - \left(\frac{r}{R}\right)^2\right]$, with zero velocity at the pipe wall and a maximum value of u_{max} at the centerline. How are U and u_{max} related? How are the average velocity at section (2), v_2, and u_{max} related?

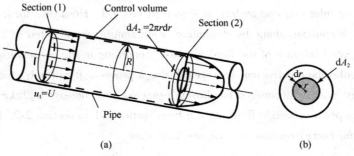

Fig. E4-2

Solution:

An appropriate control volume is sketched (dashed lines) in Fig. E4-2a. Since the flow is incompressible, we have $\rho_1 = \rho_2 = \rho$. Thus, the application of Eq. 4.6 to the contents of this control volume yields

$$UA_1 = \int_{A_2} u_2 \, dA_2 \tag{1}$$

Since the element cross-sectional area, dA_2, is equal to $2\pi r dr$ (see shaded area element in Fig. E4-2b), with the application of the parabolic velocity relationship for flow through section (2), Eq. 1 becomes

$$U\pi R^2 = \int_0^R u_{max}\left[1 - \left(\frac{r}{R}\right)^2\right] 2\pi r \, dr \tag{2}$$

Integrating of Eq. 2 gives

$$U\pi R^2 = 2\pi u_{max}\left(\frac{r^2}{2} - \frac{r^4}{4R^2}\right)\Bigg|_0^R = \frac{1}{2}\pi u_{max} R^2$$

or
$$u_{max} = 2U$$

Since this flow is incompressible, we conclude from Eq. 4.2 that U is the average velocity at all sections of the control volume. Thus, the average velocity at section (2), v_2, is one-half the maximum velocity, u_{max}, there, or

$$v_2 = U = \frac{1}{2}u_{max}$$

DISCUSSION   The average velocity at section (2) indicates that in engineering practice it may induce a somewhat error if the velocity is assumed to be one-dimensional, namely uniform, especially when the flow is laminar. Thus, the corrected factors are usually introduced to fix this error as discussed in Sections 4.3.2 and 4.4.2.

4.3 Energy Equation

4.3.1 Differential Energy Equation for Viscous Tube Flow

As discussed in Section 3.6.1, for a one dimensional steady uniform incompressible inviscid flow, the Bernoulli equation along a streamline is

$$z + \frac{p}{\rho g} + \frac{u^2}{2g} = \text{constant along a streamline} \quad (3.58)$$

or

$$z_1 + \frac{p_1}{\rho g} + \frac{u_1^2}{2g} = z_2 + \frac{p_2}{\rho g} + \frac{u_2^2}{2g} \quad (3.61)$$

Equation 3.58 is a widely used form of the Bernoulli equation for a steady incompressible frictionless streamtube flow. It is clearly related to the steady-flow energy equation for a streamtube (flow with one inlet and one outlet) as shown in Fig. 4-3. However, for a viscous flow, the total energy is not a constant along the streamline or streamtube as indicated in Eq. 3.58. The energy will be dissipated because of the fluid viscous. According to the conservation law of energy, the energy can neither be created nor destroyed during a process; it only change forms. Thus, every bit of energy must be accounted for during a process and conserved. Take h_w' to be the mechanic energy loss per unit weight fluid moving from section 1-1 to section 2-2, from Eq. 3.61 we have the *differential energy equation* for viscous tube flow

$$z_1 + \frac{p_1}{\rho g} + \frac{u_1^2}{2g} = z_2 + \frac{p_2}{\rho g} + \frac{u_2^2}{2g} + h_w' \quad (4.11)$$

where h_w' is the energy loss in terms of head for streamtube flow, called the *head loss*.

This relation is much more general than the Bernoulli equation, because it allows for friction and viscous work (another frictional effect).

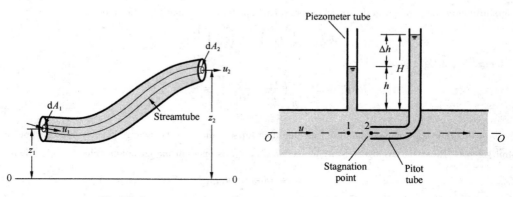

Fig. E4-3   Fig. 4-4 Streamtube flow in a total flow

EXAMPLE 4-3 Pitot-static probe

For the given level difference in the piezometer and pitot tubes, Δh, as illustrated in Fig. E4-4, determine the actual velocity, u, in the pipe as a function of Δh if the fluid viscous is not negligible.

Solution:

Along the streamtube line O-O apply the differential energy equation for viscous tube flow, Eq. 4.11

$$z_1 + \frac{p_1}{\rho g} + \frac{u_1^2}{2g} = z_2 + \frac{p_2}{\rho g} + \frac{u_2^2}{2g} + h_w' \quad (1)$$

where h_w' is related to the velocity as $h_w' = \zeta \frac{u^2}{2g}$ (see Section 6.1), in which ζ is called the *coefficient of minor loss*, its value is greater than zero and close to zero, and can be determined experimentally.

With the given parameters: $z_1 = z_2$, $u_1 = u$, $u_2 = 0$, and $\dfrac{p_2}{\rho g} - \dfrac{p_1}{\rho g} = \Delta h$, rearranging Eq. 1 gives

$$u = \dfrac{1}{\sqrt{1-\zeta}}\sqrt{2g\left(\dfrac{p_2}{\rho g} - \dfrac{p_1}{\rho g}\right)} = \dfrac{1}{\sqrt{1-\zeta}}\sqrt{2g\Delta h}$$

Let $c = \dfrac{1}{\sqrt{1-\zeta}}$, we have

$$u = c\sqrt{2g\Delta h} \tag{2}$$

where c is a coefficient depending on the energy loss and the configuration of Pitot probe. Its value is close to 1.0 and also can be determined by experiments.

DISCUSSION   As indicated in Section 3.6.5, if the flow is frictionless, the velocity at the stagnation point of the Pitot tube illustrated can be expressed as Eq. 3.64

$$u = \sqrt{2g(H-h)} = \sqrt{2g\Delta h} \tag{3}$$

It is larger than the actual velocity in Eq. 2 because the fluid viscous in a real flow causes the energy loss and slows down the flow.

4.3.2 General Form of Energy Equation

Consider a steady incompressible total flow with only one inlet of area A_1 and one outlet of area A_2 as indicated in Fig. 4-4. The flow rate in the arbitrary streamtube is $dQ = u_1 dA_1 = u_2 dA_2$.

To get the integral form of the conservation law of energy for a steady incompressible flow in a control volume, multiplying each term of the differential energy equation for the steady incompressible streamtube flow, Eq. 4.11, by the weight flowrate, $\rho g dQ$, yields the relation of energies through the streamtube per unit time

$$\left(z_1 + \dfrac{p_1}{\rho g} + \dfrac{u_1^2}{2g}\right)\rho g dQ = \left(z_2 + \dfrac{p_2}{\rho g} + \dfrac{u_2^2}{2g}\right)\rho g dQ + h'_w \rho g dQ \tag{4.12}$$

Integrating it over the control cross sections of total flow gives

$$\int_{A_1}\left(z_1 + \dfrac{p_1}{\rho g} + \dfrac{u_1^2}{2g}\right)\rho g u_1 dA_1 = \int_{A_2}\left(z_2 + \dfrac{p_2}{\rho g} + \dfrac{u_2^2}{2g}\right)\rho g u_2 dA_2 + \int_Q h'_w \rho g dQ \tag{4.13}$$

Each integral term in the above equation can be integrated easily based on some conditions as follows.

(1) Integral of potential energy

The control cross sections in the total flow should be chosen as sections in an uniform flow or gradually varied flow, thus we have $z + \dfrac{p}{\rho g} = C$ as indicated in Section 3.2.4. For the incompressible flow it follows

$$\int_A \left(z + \dfrac{p}{\rho g}\right)\rho g u dA = \left(z + \dfrac{p}{\rho g}\right)\rho g \int_A u dA = \left(z + \dfrac{p}{\rho g}\right)\rho g Q \tag{4.14}$$

(2) Integral of kinematic energy

$$\int_A \frac{u^2}{2g} \rho g u \, dA = \int_A \frac{u^3}{2g} \rho g \, dA \tag{4.15}$$

Often the flow entering or leaving a control volume is not strictly one-dimensional. In particular, the velocity may vary over the cross section, as in Fig. 4.1. In this case the kinetic energy term in Eq. 4.15 is difficult to be integrated. Thus, the average velocity introduced in Section 4.1 with a dimensionless correction factor α is used to replace the point velocity in the term of kinematic energy for the convenience of integration:

$$\int_A \frac{u^3}{2g} \rho g \, dA = \frac{\alpha v^2}{2g} \rho g Q \tag{4.16}$$

where α is called the *kinematic correction factor* and defined as

$$\alpha = \frac{\int_A \frac{u^2}{2g} \rho g u \, dA}{\int_A \frac{v^2}{2g} \rho g v \, dA} = \frac{\int_A u^3 \, dA}{v^3 A} = \frac{1}{A} \int_A \left(\frac{u}{v}\right)^3 dA \tag{4.17}$$

The value of the kinetic-energy correction factor depends on the velocity distribution. It has a value of about $\alpha = 2.0$ for fully developed laminar pipe flow (definition see Section 6.2.1) and $\alpha = 1.04 \sim 1.11$ for turbulent pipe flow (definition see Section 6.2.1). In practical problems, the kinematic energy correction factors are often ignored, i.e., α is set equal to the unity ($\alpha \approx 1.0$). It is because most flows encountered in practice are turbulent, for which the correction factor is near unity, and the kinetic energy terms are often small relative to the other terms in the energy equation, and multiplying them by a factor less than 2.0 does not make much difference. Besides, when the velocity and thus the kinetic energy are high, the flow turns turbulent. However, you should keep in mind that you may encounter some situations for which these factors are significant, especially when the flow is laminar. Therefore, we recommend that you always include the kinetic energy correction factor when analyzing fluid flow problems.

(3) Integral of head loss

Integral $\int_Q h'_w \rho g \, dQ$ is the mechanic energy loss of fluid per unit time flowing from control cross section 1-1 to 2-2 of the total flow. Take h_w to be the total *head loss* for the total flow to represent the averaged mechanic energy loss per unit weight fluid flowing from section 1-1 to 2-2, thus

$$\int_Q h'_w \rho g \, dQ = h_w \rho g Q \tag{4.18}$$

Substitution of Eqs. 4.14, 4.16 and 4.18 into Eq. 4.13 yields

$$\left(z_1 + \frac{p_1}{\rho g} + \frac{\alpha_1 v_1^2}{2g}\right)\rho g Q_1 = \left(z_2 + \frac{p_2}{\rho g} + \frac{\alpha_2 v_2^2}{2g}\right)\rho g Q_2 + h_w \rho g Q \tag{4.19}$$

For the steady incompressible total flow with one inlet and one outlet, the continuity equation is $Q_1 = Q_2 = Q$, then Eq. 4.19 reduces to

$$z_1 + \frac{p_1}{\rho g} + \frac{\alpha_1 v_1^2}{2g} = z_2 + \frac{p_2}{\rho g} + \frac{\alpha_2 v_2^2}{2g} + h_w \tag{4.20}$$

Equation 4.20 is the *general form of energy equation* for steady incompressible viscous total flow in terms of head, and represents the relation between pressure, velocity, and elevation in a frictional flow.

RESTRICTION CONDITIONS

The energy equation (Eq. 4.20) is usually to be used to determine the pressure in a flow field. However, before we reach there, it is important to understand the restrictions on its applicability and observe the limitations on its use, as explained here:

(1) Steady flow. The first limitation on the energy equation is that it is applicable to steady flow, which is introduced when the Bernoulli equation is derived.

(2) Incompressible fluid. The second assumption used in the derivation of the energy equation is that ρ = constant and thus the flow is incompressible. Thus, the density in the integral term can be lifted to the outside of integral symbol, and the volume flow rate in Eq. 4.19 can be cancelled out by using the continuity equation for the steady incompressible flow, Eq. 4.6.

(3) The only mass force is gravity. This limitation is introduced when the Bernoulli equation is derived in Section 3.6.1.

(4) Two control cross-sections must be located in the uniform flow or gradually varied flow field, but there may be rapidly varied flows between them. It makes sure the integral of potential energy is possible.

(5) Only one inlet and outlet. The energy equation with multi-inlets or outlets will be given in Section 4.3.4.

(6) Identical flow rate between two control cross sections. The energy equation in a flow section with varied flow rates cannot be expressed in the form of head, and only can be expressed in the form of weight fluid as discussed in Section 4.3.4.

(7) No shaft work and heat transfer between two control cross sections in the system. It is because we derive the energy equation only in a flow section that does not involve a pump, turbine, fan, or any other machine or impeller carrying out energy interactions with the fluid particles, also the heat transfer. We will consider the energy equation in the flow section involving any of these devices in Section 4.3.4.

SYMBOL INTERPRETATIONS

The physical and geometric meanings of the terms in the energy equations, Eq. 4.20, are same as that in the Bernoulli equation, Eq. 3.61. However, Eq. 4.20 emphases the *average meanings* of the variables. The physical and geometric meanings of terms in Eq. 4.20 can be interpreted as follows:

z : the *elevation head*, represents the elevation potential energy per unit weight fluid of the computational point at the control cross section;

$p/\rho g$: the *pressure head*, or the *piezometric height*, represents the pressure potential energy per unit weight fluid of the computational point at the control cross section;

$\dfrac{\alpha v^2}{2g}$: the *velocity head*, represents the *averaged* kinematic energy per unit weight fluid of the computational point at the control cross section;

$z + p/\rho g$: the *piezometric head*, represents the total potential energy per unit weight fluid of the computational point at the control cross section;

$z + \dfrac{p}{\rho g} + \dfrac{\alpha v^2}{2g} = H$: the *total head*, represents the total averaged mechanical energy per unit weight fluid of the computational point at the control cross section.

h_w : the *total head loss*, represents the averaged mechanical energy per unit weight fluid between two control cross sections; In Chapter 6 we shall develop methods of correlating h_w losses with flow parameters in pipes, valves, fittings, channels and other internal-flow devices.

PROCEDURE FOR PROBLEM SOLVING

When Eq. 4.20 is used to solve the pratical problems, the following steps should be followed:

1. Selecting the datum plane. The datum plane can be selected at random, but it must help to simplify the computation. So we may select the surface through the centroid of the cross-section ($z = 0$), i.e. in a pipe flow, or the free liquid surface ($p = 0$), i.e. for open channel flow, as the datum plane.

2. Selecting the control cross sections. These sections should be the cross-sections in uniform or gradually varied flows with variables already known as many as possible.

3. Selecting the computational point. For the pipe flow, it is usually located at the pipe axis; for the open channel flow, it is usually on the free surface. For the same equation, the same pressure datum should be introduced.

4. Writing the energy equation and solving the problem. Note: the continuity equation is always needed in the problem solving.

EXAMPLE 4-4 Frictional loss in a long pipe

Gasoline at 20℃ is pumped through a smooth 12-cm-diameter, 10-km long pipe at a flow rate of 75 m³/h. The inlet is fed by a pump at an absolute pressure of 24at. The exit is at standard atmospheric pressure and is 150m higher. Estimate the *frictional head loss* h_f(a head loss only due to the fluid viscous, detailed see Sections 6.1 and 6.6), and compare it to the velocity head of the flow $v^2/2g$. (These numbers are quite realistic for liquid flow through long pipelines.)

Solution:

For gasoline at 20℃, from Table A.3 in Appendix A, $\rho = 680 \text{kg/m}^3$. Take $\alpha_1 = \alpha_2 \approx 1.0$. The head loss in this problem is $h_w = h_f$.

Take the horizontal plane passing the axis of pipe inlet to be the datum plane and the computational point on the pipe axis, hence Eq. 4.20 applies:

$$z_1 + \frac{p_1}{\rho g} + \frac{\alpha_1 v_1^2}{2g} = z_2 + \frac{p_2}{\rho g} + \frac{\alpha_2 v_2^2}{2g} + h_f \tag{1}$$

The pipe is of uniform cross section, and thus the average velocity and velocity head everywhere are

$$v_1 = v_2 = \frac{Q}{A} = \frac{(75/3600) \text{ m}^3/\text{s}}{\frac{\pi}{4}(0.12\text{m})^2} = 1.84\text{m/s}$$

$$\frac{\alpha_1 v_1^2}{2g} = \frac{\alpha_2 v_2^2}{2g} = \frac{(1.0)(1.84\text{m/s})^2}{2(9.81\text{m/s}^2)} = 0.173\text{m}$$

Leaving the frictional loss and substituting all the given parameters, $z_1 = 0$, $z_2 = 150$m, and the above calculated parameters into Eq. 1 gives

$$0 + (24\text{at})\frac{98000\text{N/(m}^2\cdot\text{at)}}{(680\text{kg/m}^3)(9.81\text{m/s}^2)} + 0.173\text{m}$$

$$= 150\text{m} + (1\text{at})\frac{98000\text{N/(m}^2\cdot\text{at)}}{(680\text{kg/m}^3)(9.81\text{m/s}^2)} + 0.173\text{m} + h_f$$

$$h_f = 352.6\text{m} - 14.7\text{m} - 150\text{m} = 187.9\text{m}$$

So This friction head is larger than the elevation change $\Delta z(150\text{m})$, and the pump must drive the flow against both changes, hence the higher inlet pressure. The ratio of friction to velocity head is

$$\frac{h_f}{v^2/2g} = \frac{187.9\text{m}}{0.173\text{m}} = 1086$$

DISCUSSION This high ratio is typical of long pipelines (Note that we did not make direct use of the 10,000m pipe length, whose effect is hidden within h_f.) In Chap. 6 we can state this problem in a more direct fashion: Given the flow rate, fluid, and pipe size, what inlet pressure is needed? Our correlations for h_f will lead to the estimate $p_{\text{inlet}} = 24$at, as stated above.

EXAMPLE 4-5 Flow over a spillway

An overflow spillway was constructed crossing a river with a cross section, as in Fig. E4-5. The inverted elevation of the downstream river bed is 105.0m. When the upstream water level in the reservoir is 120.0m, the water depth at the dam toe (a gradually varied section C) is $h_c = 1.2$m. Assume the head loss of flow pasting the spillway is $0.1\frac{v_c^2}{2g}$, determine the average velocity at section C, v_c.

Solution:

Take the density of water to be $\rho = 1000\text{kg/m}^3$ and $\alpha_1 = \alpha_2 = 1.0$.

Take the downstream river bed to be the datum plane. The upstream control cross section 1-1 is selected at the section far away from the rapidly varied flow region around the inlet of the spillway, where $v_1 \approx 0$ since its large section area at section 1-1. The downstream control section 2-2 is chosen at the section of dam toe C, a nearly gradually varied section, where $v_2 = v_c$. Both the computational points are taken at the free surface, thus, $p_1 = p_2 = 0$ in gage pressure.

Equation 4.20 applies

$$z_1 + \frac{p_1}{\rho g} + \frac{\alpha_1 v_1^2}{2g} = z_2 + \frac{p_2}{\rho g} + \frac{\alpha_2 v_2^2}{2g} + h_w$$

with the given parameters yields

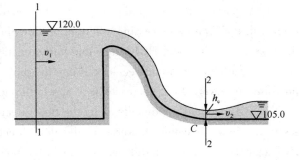

Fig. E4-5

$$(120.0\text{m} - 105.0\text{m}) + 0 + 0 = 1.2\text{m} + 0 + \frac{v_c^2}{2g} + 0.1\frac{v_c^2}{2g}$$

Solving it gives the velocity at the dam toe

$$v_c = \sqrt{\frac{2(9.81\text{m/s}^2)(15.0\text{m} - 1.2\text{m})}{1.0 + 0.1}} = 15.69\text{m/s}$$

4.3.3 Energy and Hydraulic Grade Lines

Grade lines are the graphical interpretations of energy equation, and a useful visual interpretation to sketch a flow.

The *energy grade line* (EGL) is defined as the connected curve of the liquid levels in the Pitot tubes (the liquid height of total head, $z + \frac{p}{\rho g} + \frac{\alpha v^2}{2g}$) at several locations along the flow direction as illustrated in Fig. 4-5. The *hydraulic grade line* (HGL) is defined as the connected curve of the liquid levels in the piezometer tubes (the liquid height of piezometer head, $z + \frac{p}{\rho g}$) at several locations in the flow direction. The difference between the heights of EGL and HGL is equal to the dynamic head, $\frac{\alpha v^2}{2g}$.

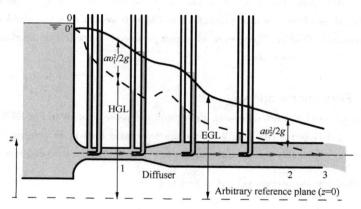

Fig. 4-5 The HGL and the EGL of a steady flow with a free jet to the air

From Fig. 4-5 we have the following statements.

1. At point 0 at the liquid surface just before the section of pipe inlet, EGL and HGL are even with the liquid surface since there is no flow there, and at point 0′ just after the section of the pipe inlet, EGL and HGL drops due to the energy loss around the inlet. After that HGL decreases rapidly as the liquid accelerates into the pipe, which induces a rapid decrease of the static pressure; however, EGL decreases very slowly through the well-rounded pipe inlet.

2. EGL declines continually along the flow direction due to friction and other irreversible losses in the flow. It cannot increase in the flow direction unless energy is supplied to the fluid. HGL can rise (i. e. in the diffuser) or fall in the flow direction, but can never exceed EGL.

3. HGL rises in the diffuser section as the velocity decreases, and the static pressure recovers

somewhat; the total pressure does *not* recover, however, and EGL decreases through the diffuser.

4. The difference between EGL and HGL is $\alpha v_1^2/2g$ at point 1, and $\alpha v_2^2/2g$ at point 2. Since $v_1 > v_2$, the difference between the two grade lines is larger at point 1 than at point 2.

5. The downward slopes of both grade lines are larger for the smaller diameter section of pipe since the frictional head loss (proportional to the velocity) is greater.

6. HGL decays to the liquid surface at the outlet since the pressure there is atmospheric. However, EGL is still higher than HGL by the amount $\alpha v_2^2/2g$ since $v_2 = v_3$ at the outlet.

FEATURES OF EGL AND HGL

Thus, we have the following features about the HGL and EGL:

- For *stationary bodies* such as reservoirs or lakes, the EGL and HGL coincide with the free surface of liquid. The elevation of free surface z in such cases represents both the EGL and the HGL since the velocity is zero and the static (gage) pressure is zero.

- The EGL is always a distance $\alpha v^2/2g$ above the HGL. These two curves approach each other as the velocity decreases (i. e. pipe diameter increases), and diverge as the velocity increases (i. e. pipe diameter decreases). The height of the HGL decreases as the velocity increases, and vice versa.

- For *open-channel flow*, the HGL coincides with the free surface of the liquid, and the EGL is a distance $\alpha v^2/2g$ above the free surface.

- At the *pipe exit*, the pressure head is zero (atmospheric pressure) and thus the HGL coincides with the axis at pipe exit (location 3 in Fig. 4-5).

- The *mechanical energy loss* due to frictional effects (conversion to thermal energy) causes the EGL and HGL to slope downward in the direction of flow. The slope, discussed follows, is a measure of the head loss in the pipe. A local change of shape, such as the entrance in Fig. 4-5, that generates significant frictional effects causes a sudden drop in both EGL and HGL at that location.

- A *steep jump* occurs in EGL and HGL whenever mechanical energy is added to the fluid (by a pump, for example) (Fig. 4-6). Likewise, a *steep drop* occurs in EGL and HGL whenever mechanical energy is removed from the fluid (by a turbine, for example). The related energy equation will be discussed in Section 4.3.4.

- The (gage) pressure of a fluid is zero at locations where the HGL *intersects* the fluid (Fig. 4-7). The pressure in a flow section that lies above the HGL is negative, and the pressure in a section that lies below the HGL is positive. Therefore, an accurate drawing of a piping system and the HGL can be used to determine the regions where the pressure in the pipe is negative (below the atmospheric pressure).

The last remark enables us to avoid situations in which the pressure drops below the vapor pressure of the liquid (which may cause *cavitation*). Proper consideration is necessary in the placement of a liquid pump to ensure that the suction side pressure does not fall too low, especially at elevated temperatures where vapor pressure is higher than it is at low temperatures.

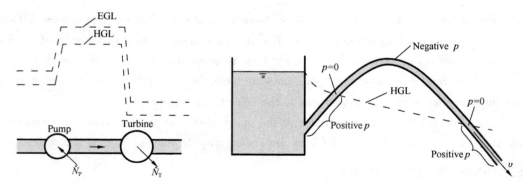

Fig. 4-6 Steep jump and a steep drop due to the pump and the turbine

Fig. 4-7 The HGL can be used to determine the vacuum region in the pipe system by judging whether the flow section lies above the HGL

For summary, in general flow conditions, the EGL will drop slowly due to friction losses and will drop sharply due to a substantial loss (a valve or obstruction) or due to work extraction (to a turbine). The EGL can rise only if there is work addition (as from a pump or propeller). The HGL generally follows the behavior of the EGL with respect to losses or work transfer, and it rises and/or falls if the velocity decreases and/or increases.

FRICTIONAL AND HYDRAULIC SLOPES

As discussed above, the EGL for viscous flow always drops slowly in the flow direction. Its slope is called the *frictional slope*, and defined as the energy loss per unit length in the flow direction

$$J = \frac{dh_w}{ds} = -\frac{dH}{ds} = \frac{d\left(z + \frac{p}{\rho g} + \frac{\alpha v^2}{2g}\right)}{ds} \qquad (4.21)$$

The hydraulic grade line (HGL) would rise, drop or be horizontal depending on the change of sectional shape of pipe. Its slope is called the *hydraulic slope*, and defined as the piezometric head drop per unit length in the flow direction

$$J_P = -\frac{d\left(z + \frac{p}{\rho g}\right)}{ds} \qquad (4.22)$$

The *frictional slope* for viscous flow is always greater than zero, while *hydraulic slope* can be greater, less than or equal to zero.

HGL AND EGL FOR INVISCID FLOW

For inviscid flow, the energy grade line is a horizontal line (Fig. 4-8), $J = 0$, due to the frictionless of flow, $h_w = 0$. In this case, the Bernoulli equation along a streamline, Eq. 3.58, can be applied, and the sum of pressure, velocity, and elevation heads along a streamline in a steady incompressible flow is constant when both the compressibility and frictional effects of flow are negligible.

The hydraulic grade line is also a horizontal line, $J_P = 0$ only when the flow velocity is constant (Fig. 4-8a). But if the flow velocity varies along the flow, it will not be the case for HGL any more (Fig. 4-8b).

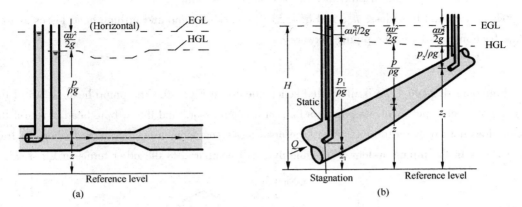

Fig. 4-8 HGL and EGL for inviscid flow with (a) constant velocity in each separated flow sections and (b) varied velocity along the flow direction

4.3.4 Other Forms of Energy Equation

ENERGY EQUATION WITH SHAFT WORK

If there are pumps, turbines or other shaft machines between the two control cross sections of total flow as in Fig. 4-6, the general form of the energy equation, Eq. 4.20 for the steady incompressible flow energy equation should be written as

$$z_1 + \frac{p_1}{\rho g} + \frac{\alpha_1 v_1^2}{2g} \pm h_s = z_2 + \frac{p_2}{\rho g} + \frac{\alpha_2 v_2^2}{2g} + h_w \qquad (4.23)$$

where h_s is the head of the shaft machine, e.g., *pump head* for pump, $h_s = h_P$, and *turbine head* for turbine, $h_s = h_T$. Pump adds mechanical power to the fluid, $+h_s$; whereas turbines extract mechanical power from the fluid, $-h_s$.

Equation 4.23 is the form of energy equation for steady viscous flow in terms of head with the shaft work in system.

Pump head (input) and turbine head (output) are defined by dividing their power by the weight flow rate:

$$h_T = \frac{\dot{N}_T}{\eta_T \rho g Q} = \frac{\dot{N}_T + \dot{E}_{LT}}{\rho g Q} \qquad (4.24)$$

$$h_P = \frac{\eta_P \dot{N}_P}{\rho g Q} = \frac{\dot{N}_P - \dot{E}_{LP}}{\rho g Q} \qquad (4.25)$$

where $\dot{N}_T$ is the *shaft power* output through the turbine's shaft; $\dot{N}_P$ is the shaft power input through the pump's shaft; η_T is the *turbine efficiency*; η_P is the *pump efficiency*; $\dot{E}_{LT}$ is the power of mechanical energy loss of fluid in turbine unit, $\dot{E}_{LT} = \rho g Q h_{wT}$; $\dot{E}_{LP}$ is the power of mechanical energy loss of fluid in pump unit, $\dot{E}_{LP} = \rho g Q h_{wP}$; h_{wT} is the energy loss in term of head in the turbine unit; h_{wP} is the energy loss of fluid in term of head in the pump unit.

Thus, the total head loss, h_w, in Eq. 4.23, consists of pump and turbine head losses as well as the frictional losses in the piping network h_{wD}:

$$h_w = h_{wD} + h_{wP} - h_{wT} \tag{4.26}$$

Equations 4.24 to 4.26 is illustrated schematically in Fig. 4-9. The pump head is zero if the piping system does not involve a pump, a fan, or a compressor, and the turbine head is zero if the system does not involve a turbine. Also, the head loss h_w can sometimes be ignored when the frictional losses in the piping system are negligibly small compared to the other terms in Eq. 4.23.

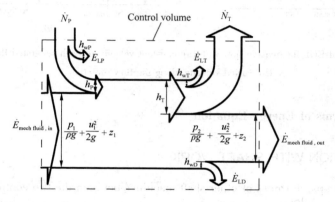

Fig. 4-9 Mechanical energy flow chart for a fluid flow system that involves a pump and a turbine. Vertical dimensions show each energy term expressed as *head* corresponding to each term of Eq. 4.23. (redrawn of Figure 5-51 from book of Çengel and Cimbala, 2006)

EXAMPLE 4-6 Water pumping

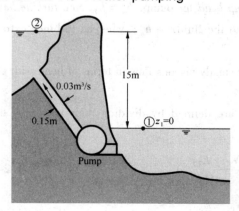

Fig. E4-6

Water is pumped from a lower reservoir to a higher reservoir through a 0.15-m-diameter pipe (Fig. E4-6). The free surface of the upper reservoir is 15m higher than the surface of the lower reservoir. The flow rate of water is measured to be 0.03 m³/s, and the irreversible head loss of the pumping system (excluding the pump unit) is $10v^2/2g$ (where v is the sectional average velocity in the pipe). If the pump efficiency is $\eta_P = 0.76$, determined the shaft work of the pump during this process.

Solution:

Assume the flow is steady and incompressible. Take the density of water to be $\rho = 1000 \text{kg/m}^3$, and $\alpha_1 = \alpha_2 \approx 1.0$.

Take the free surface of the lower reservoir to be the reference datum plane. The computational points (1) and (2) are located at the free surfaces of the upper and lower reservoirs respectively. Thus, we have $z_1 = 0$, $z_2 =$

15m, $p_1 = p_2 = 0$ due to points 1 and 2 are open to the atmosphere and $v_1 \approx v_2 \approx 0$ since the large free surfaces in the upper and lower reservoirs. Thus, the energy equation for steady and incompressible flow, Eq. 4.23, applies

$$z_1 + \frac{p_1}{\rho g} + \frac{\alpha_1 v_1^2}{2g} + h_P = z_2 + \frac{p_2}{\rho g} + \frac{\alpha_2 v_2^2}{2g} + h_w$$

With the given parameters the above equation becomes

$$0 + 0 + 0 + h_P = 15\text{m} + 0 + 0 + 10\frac{v^2}{2g}$$

with the velocity in the pipe

$$v = \frac{Q}{A} = \frac{0.03\,\text{m}^3/\text{s}}{\frac{\pi}{4}(0.15\text{m})^2} = 1.70\,\text{m/s}$$

yields

$$h_P = 15\text{m} + 10\frac{(1.70\,\text{m/s})^2}{2(9.81\,\text{m/s}^2)} = 16.47\text{m}$$

Thus, the shaft work of pump needed to pump water from lower reservoir to higher reservoir is

$$\dot{N}_P = \frac{\rho g Q\, h_P}{\eta_P} = \frac{(1000\,\text{kg/m}^3)(9.81\,\text{m/s}^2)(0.03\,\text{m}^3/\text{s})(16.47\text{m})}{0.76} = 6.38\,\text{kW}$$

DISCUSSION The 6.38kW of power is used to overcome both the elevation rise and the friction in the piping system. Note that a less power of pump ($h_P = 15$m, $\dot{N}_P = 5.81$kW) is needed if there is no mechanical energy loss in the pump system.

EXAMPLE 4-7 Hydroelectric power generation from a dam

In a hydroelectric power plant, 100 m³/s of water flows from an elevation of 120m to a turbine, where electric power is generated (Fig. E4-7). The total irreversible head loss in the piping system from point 1 to point 2 (excluding the turbine unit) is determined to be 35m. If the overall efficiency of the turbine-generator is 80 percent, estimate the electric power output.

Solution:

Assume the flow is steady and incompressible. We take the density of water to be $\rho = 1000\,\text{kg/m}^3$, and $\alpha_1 = \alpha_2 \approx 1.0$.

Take the free surface of the lower reservoir to be the datum plane. The computational points (1) and (2) are located at the free surfaces of the upper and lower reservoirs respectively. Thus, we have $z_2 = 0$, $z_1 = 120$m. Also, both points 1 and 2 are open to the atmosphere ($p_1 = p_2 = 0$) and the flow velocities are negligible at both points ($v_1 = v_2 = 0$). Then the energy equation for steady incompressible flow, Eq. 4.23, applies

$$z_1 + \frac{p_1}{\rho g} + \frac{\alpha_1 v_1^2}{2g} - h_T = z_2 + \frac{p_2}{\rho g} + \frac{\alpha_2 v_2^2}{2g} + h_w$$

With the given parameters the above equation becomes

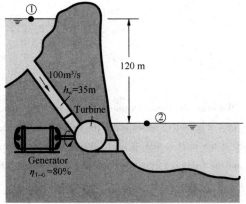

Fig. E4-7

$$120\text{m} + 0 + 0 - h_\text{T} = 0 + 0 + 0 + 35\text{m}$$

Thus, the turbine head is

$$h_\text{T} = 120\text{m} - 35\text{m} = 85\text{m}$$

The electric power output by the turbine-generator is

$$\dot{N}_\text{T} = \eta_\text{T-G} \rho g Q h_\text{T} = (0.80)(1000\text{kg/m}^3)(9.81\text{m/s}^2)(100 \text{ m}^3/\text{s})(85\text{m}) = 66.7\text{MW}$$

DISCUSSION   For a perfect turbine-generator ($\eta_\text{T-G} = 1.0$), this resource would generate 66.7MW/0.8 = 83.4MW of electricity. The power generation would increase by almost 0.834 MW for each percentage point improvement in the efficiency of the turbine-generator unit.

EXAMPLE 4-8   Fan selection for air cooling of a computer

A fan is to be selected to cool a computer case whose dimensions are 12cm × 40cm × 40cm (Fig. E4-8). Half of the volume in the case is expected to be filled with components and the other half to be air space. A 6-cm-diameter hole is available at the back of the case for the installation of the fan that is to replace the air in the void spaces of the case once every second. Small low-power fan-motor-combined units are available in the market and their efficiency is estimated to be 30 percent. Determine (a) the wattage of the fan-motor unit to be purchased and (b) the pressure difference across the fan. Take the air density to be 1.2kg/m³.

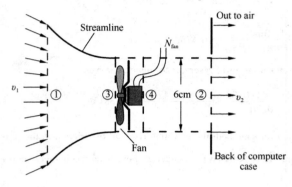

Fig. E4-8

Solution:

Assume the flow is steady and incompressible, the losses other than those due to the inefficiency of the fan-motor unit are negligible ($h_\text{w} = 0$). Take $\alpha_1 = \alpha_2 \approx 1.0$.

(a) Noting that half of the volume of the case is occupied by the components, the air volume in the computer case is

$$\forall = (\text{Void fraction})(\text{Total case volume}) = (0.5)(12\text{cm} \times 40\text{cm} \times 40\text{cm}) = 9600 \text{ cm}^3$$

Therefore, the volume flow rate through the case and the average air velocity through the outlet to the air are

$$Q = \frac{\forall}{\Delta t} = \frac{9600 \text{ cm}^3}{1\text{s}} = 9.6 \times 10^{-3} \text{ m}^3/\text{s}$$

$$v_2 = \frac{Q}{A} = \frac{9.6 \times 10^{-3} \text{ m}^3/\text{s}}{\frac{\pi}{4}(0.06\text{m})^2} = 3.40\text{m/s}$$

We draw the control volume around the fan such that both the inlet and the exit are at the atmospheric pressure ($p_1 = p_2 = 0$), as shown in the figure, and the inlet section 1 is large and far from the fan so that the flow velocity at the inlet section is negligible ($v_1 = 0$). Noting that $z_1 = z_2$ and the mechanical losses due to fan inefficiency and the frictional effects in flow are disregarded, the energy equation, Eq. 4.23

$$z_1 + \frac{p_1}{\rho g} + \frac{\alpha_1 v_1^2}{2g} + h_\text{fan} = z_2 + \frac{p_2}{\rho g} + \frac{\alpha_2 v_2^2}{2g}$$

yields

$$h_{\text{fan}} = \frac{\alpha_2 v_2^2}{2g} = \frac{(1.0)(3.40\text{m/s})^2}{(2)(9.81\text{m/s}^2)} = 0.589\text{m}$$

Then the required electrical power input to the fan is determined to be

$$\dot{N}_{\text{fan}} = \frac{\rho g Q h_{\text{fan}}}{\eta_{\text{fan}}} = \frac{(1.2\text{kg/m}^3)(9.81\text{m/s}^2)(9.6 \times 10^{-3}\text{ m}^3/\text{s})(0.589\text{m})}{0.3} = 0.22\text{W}$$

Therefore, a fan-motor rated at 0.22 W is adequate for this job.

(b) To determine the pressure difference across the fan unit, we take points 3 and 4 to be on the two sides of the fan on a horizontal line. This time again $z_3 = z_4$ and $v_3 = v_4$ since the fan is a narrow cross section. Disregarding mechanical losses, the energy equation reduces to

$$\frac{p_3}{\rho g} + h_{\text{fan}} = \frac{p_4}{\rho g}$$

Then we have

$$p_4 - p_3 = \rho g\, h_{\text{fan}} = (1.2\text{kg/m}^3)(9.81\text{m/s}^2)(0.589\text{m}) = 6.9\text{Pa}$$

Therefore, the pressure rise across the fan is 6.9 Pa.

DISCUSSION The efficiency of the fan-motor unit is given to be 30 percent, which means 30 percent of the electric power $\dot{N}_{\text{fan}}$ consumed by the unit will be converted to useful mechanical energy while the rest (70 percent) will be 'lost' and converted to thermal energy. Also, a much larger fan is required in an actual system to overcome frictional losses inside the computer case.

ENERGY EQUATION FOR DIVERGING FLOW

T-junction and Y-junction are common components in the pipe system of gas or water supply. When fluid flows through those junctions, diverging or converging flow occurs as in Fig. 4-10 and Fig. 4-11. The energy equation in terms of head, Eq. 4.23, for steady incompressible total flow with only one inlet and one outlet may not be applicable since the multi-inlets or multi-outlets in those cases. The detailed discussion is as follows.

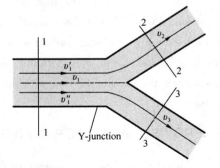

Fig. 4-10 Diverging flow in Y-junction

Consider the control volume enclosed by the inlet section 1-1, outlet sections 2-2 and 3-3, and the pipe wall as in Fig. 4-10. Fluid flowing into the control volume through section 1-1 is separated into two half fluid bunches smoothly by the wedged shape of the Y-junction wall: one flows out from section 2-2 and another from section 3-3. In the diverging flow, the energy mixture prior to the stagnation point of diverging junction between those two half fluid bunches are negligible because the small turbulence caused by the wedged joint of the branch walls. Thus, the energy equation is considered to be available between upper half of section 1-1 and section 2-2

$$z_1' + \frac{p_1'}{\rho g} + \frac{\alpha_1' v_1'^2}{2g} = z_2 + \frac{p_2}{\rho g} + \frac{\alpha_2 v_2^2}{2g} + h_{w1-2}$$

and the lower half of section 1-1 and section 3-3

$$z''_1 + \frac{p''_1}{\rho g} + \frac{\alpha''_1 v''^2_1}{2g} = z_2 + \frac{p_2}{\rho g} + \frac{\alpha_2 v_2^2}{2g} + h_{w1-3}$$

where superscripts ′ and ″ represent the properties of fluids in the upper and the lower part of section 1-1 respectively; h_{w1-2} is the head loss between upper half of section 1-1 and section 2-2; h_{w1-3} is the head loss between lower half of section 1-1 and section 3-3.

Assume the section 1-1 is taken to be the gradually varied section, the pizometric head at any point on the section 1-1 has the same magnitude:

$$z'_1 + \frac{p'_1}{\rho g} = z''_1 + \frac{p''_1}{\rho g} = z_1 + \frac{p_1}{\rho g}$$

For the total flow, the properties on a section are sectional averaged, then we have $\frac{\alpha'_1 v'^2_1}{2g} \approx \frac{\alpha_1 v_1^2}{2g}$ and $\frac{\alpha''_1 v''^2_1}{2g} \approx \frac{\alpha_1 v_1^2}{2g}$. Thus we can say

$$z_1 + \frac{p_1}{\rho g} + \frac{\alpha_1 v_1^2}{2g} = z_2 + \frac{p_2}{\rho g} + \frac{\alpha_2 v_2^2}{2g} + h_{w1-2} \quad (4.27)$$

and

$$z_1 + \frac{p_1}{\rho g} + \frac{\alpha_1 v_1^2}{2g} = z_3 + \frac{p_3}{\rho g} + \frac{\alpha_3 v_3^2}{2g} + h_{w1-3} \quad (4.28)$$

would be almost true.

It indicates that for the total flow with the division in the engineering practice, the energy equation, Eq. 4.27 or 4.28 is available if the control cross sections are located at the gradually varied section with a near uniform or averaged velocity distribution and the head loss between the sections are taken into consideration. It should be noted Eq. 4.27 and 4.28 are preferred to be used along the streamtube flow with the sectional-averaged properties instead of the properties at a point.

ENERGY EQUATION FOR CONVERGING FLOW

Consider the fluids are flowing in their own pipes and then converged through a Y-junction as in Fig. 4-11. The energies in each branch, $z_i + \frac{p_i}{\rho g} + \frac{\alpha_i v_i^2}{2g}$, do not equal. Besides the energy loss due to the turbulence caused by the emerging, this unequal indicates the energy exchanges between those two branch flows after they are emerged after the Y-junction. Thus, the energy in terms of head (per unit weight), Eq. 4.23, is not available any more. However, the conservation law of energy for the total flow is still available, that is, the energy for the fluid passing through the computational cross sections per unit time should be conserved. Then we have

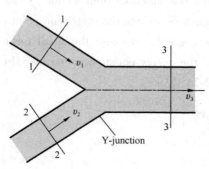

Fig. 4-11 Converging flow in Y-junction

$$\rho g Q_1\left(z_1 + \frac{p_1}{\rho g} + \frac{\alpha_1 v_1^2}{2g}\right) + \rho g Q_2\left(z_2 + \frac{p_2}{\rho g} + \frac{\alpha_2 v_2^2}{2g}\right) = \rho g Q_3\left(z_3 + \frac{p_3}{\rho g} + \frac{\alpha_3 v_3^2}{2g}\right)$$
$$+ \rho g Q_1 h_{wl-2} + \rho g Q_2 h_{wl-3} \qquad (4.29)$$

Equation 4.29 is the form of energy equation in terms of the mechanical energy.

In general, the energy exchange in the municipal pipe system is neglected. In this case, Eq. 4.23 is still available for the converging flow.

EXAMPLE 4-9 Injector

An *injector* as illustrated in the region of dash line rectangular in Fig. E4-9a, is a type of pump that uses the Venturi effect of a converging-diverging nozzle to convert the pressure energy of a motive fluid to velocity energy which creates a low pressure zone that draws in and entrains a suction fluid. After passing through the throat of the injector, the mixed fluid expands and the velocity is reduced which results in recompressing the mixed fluids by converting velocity energy back into pressure energy. The motive fluid may be a liquid, steam or any other gas. The entrained suction fluid may be a gas, a liquid, a slurry, or a dust-laden gas stream.

Assume the injector is used to drawn in water as shown in Fig. E4-9a. Take the flowrates of motive fluid and entrained suction water to be $Q_1 = 0.004$ m³/s and $Q_2 = 0.0005$ m³/s respectively, and the diameter of suction pipe and injector outlet to be $D_2 = 0.05$m and $D_3 = 0.08$m respectively. If the axis of injector is $h = 5$m above the free surface of lower tank, determine (a) the acting head of the upper tank, H, needed above the axis of injector, (b) the vacuum at the inlet of injector (section 5-5) and (c) the diameter of the nozzle, d. Assume the head loss is negligible.

Solution:

Assume the flow is steady and incompressible, and the outlet section is fully filled by water. We take the density of water to be $\rho = 1000$kg/m³ and $\alpha \approx 1.0$.

Take the axis of injector to be the datum plane. The computational points on sections 1-1 to 5-5 are selected to be located on the free surfaces or at the axis.

(a) Determine H

Because the injector has two inlets and one outlets, the energy equation in terms of the mechanical energy, Eq. 4.29, should be applied here. With the negligible head loss, it reduces to

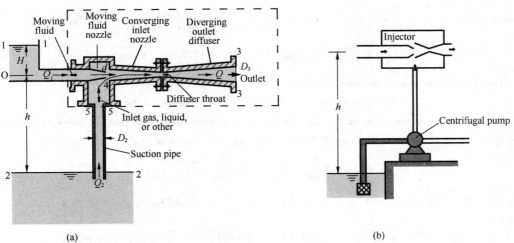

Fig. E4-9

$$\rho_1 g Q_1 \left(z_1 + \frac{p_1}{\rho g} + \frac{\alpha_1 v_1^2}{2g}\right) + \rho_2 g Q_2 \left(z_2 + \frac{p_2}{\rho g} + \frac{\alpha_2 v_2^2}{2g}\right) = \rho_3 g Q_3 \left(z_3 + \frac{p_3}{\rho g} + \frac{\alpha_3 v_3^2}{2g}\right)$$

where $\rho_1 = \rho_2 = \rho_3 = \rho$, $Q_3 = Q_1 + Q_2 = 0.0045 \text{ m}^3/\text{s}$, $z_1 = H$, $z_2 = -h = -5\text{m}$, $z_3 = 0$, $p_1 = p_2 = p_3 = 0$, $v_1 = v_2 = 0$, and

$$v_3 = \frac{Q_3}{\frac{\pi D_3^2}{4}} = \frac{0.0045 \text{ m}^3/\text{s}}{\frac{\pi}{4}(0.08\text{m})^2} = 0.896 \text{m/s}$$

Thus, we have

$$\rho g Q_1 H - \rho g Q_2 h = \rho g Q_3 \frac{\alpha_3 v_3^2}{2g}$$

which yields the acting head of the upper tank needed above the axis of injector

$$H = \frac{Q_3 \frac{\alpha_3 v_3^2}{2g} + Q_2 h}{Q_1}$$

$$= \frac{(0.0045 \text{ m}^3/\text{s}) \frac{(1.0)(0.896 \text{m/s})^2}{(2)(9.81 \text{m/s}^2)} + (0.0005 \text{ m}^3/\text{s})(5\text{m})}{0.004 \text{m}^3/\text{s}}$$

$$= 0.671 \text{m}$$

DISCUSSION Assume the head loss to be negligible, and the energy equation in term of the head, Eq. 4.23, is available between sections 1-1 and 3-3

$$z_1 + \frac{p_1}{\rho g} + \frac{\alpha_1 v_1^2}{2g} = z_3 + \frac{p_3}{\rho g} + \frac{\alpha_3 v_3^2}{2g}$$

which yields $H = \frac{\alpha_3 v_3^2}{2g} = 0.041\text{m}$. It is obviously wrong because the flow rate and energy at section 1-1 differ from the one at section 3-3 once the fluid is entrained into the injector through the suction pipe.

(b) Determine the vacuum at the inlet of injector

With the negligible head loss, the energy equation in term of the head, Eq. 4.23, between sections 2-2 and 5-5 reduces to

$$z_2 + \frac{p_2}{\rho g} + \frac{\alpha_2 v_2^2}{2g} = z_5 + \frac{p_5}{\rho g} + \frac{\alpha_5 v_5^2}{2g}$$

where $z_2 = -h = -5\text{m}$, $z_5 \approx 0$, $p_2 = 0$, $v_2 = 0$, $v_5 = \frac{Q_2}{\pi D_2^2/4} = \frac{0.0005 \text{ m}^3/\text{s}}{\pi(0.05\text{m})^2/4} = 0.255 \text{m/s}$. Thus,

$$\frac{p_5}{\rho g} = -h - \frac{\alpha_5 v_5^2}{2g} = -5\text{m} - \frac{(1.0)(0.255 \text{m/s})^2}{(2)(9.81 \text{m/s}^2)} = -5.0033\text{m}$$

Therefore, the vacuum at the inlet of injector is

$$p_{v5} = -p_5 = -(1000 \text{kg/m}^3)(9.81 \text{m/s}^2)(-5.0033\text{m}) = 49.08 \text{kPa}$$

(c) Assume the vacuum at the motive fluid nozzle is same as the vacuum at the inlet of injector, we can determine the diameter of the motive fluid nozzle by the energy equation in term of head:

$$z_1 + \frac{p_1}{\rho g} + \frac{\alpha_1 v_1^2}{2g} = z_4 + \frac{p_4}{\rho g} + \frac{\alpha_4 v_4^2}{2g}$$

where $z_1 = H = 0.671\text{m}$, $z_4 = 0$, $p_1 = 0$, $v_1 = 0$, $\frac{p_4}{\rho g} = \frac{p_5}{\rho g} = -5.0033\text{m}$, $\alpha_4 \approx 1.0$. Thus

$$v_4 = \sqrt{2g\left(H - \frac{p_4}{\rho g}\right)} = \sqrt{(2)(9.81 \text{m/s}^2)[0.671\text{m} - (-5.003\text{m})]} = 10.55 \text{m/s}$$

Then the diameter of the nozzle is

$$d = \sqrt{\frac{4Q_1}{\pi v_4}} = \sqrt{\frac{(4)(0.004 \text{m}^3/\text{s})}{\pi(10.55 \text{m/s})}} = 0.022 \text{m}$$

DISCUSSION In engineering practice, the remove of the air in the pump can be implemented by the injector. Fig. E4-8b illustrates the schematic system of this application. Once the injector works, the air in the shell of the centrifugal pump is drawn into the injector and the room of air is replaced by water entrained from the reservoir.

ENERGY EQUATION FOR INCOMPRESSIBLE GAS FLOW

The most common applications of energy equation are to gas flows. Clearly, gases are not incompressible fluids. Meanwhile gas flows are generally *adiabatic*. That is, there is no (appreciable) flow of thermal energy to or from the fluid during the flow. In this case, the energy equation for the steady incompressible flow can be used if the flow speed of the gas is much less than the speed of sound, such that the variation in density of the gas (due to this effect) along each streamline can be ignored. In the engineering practice, adiabatic flow at $Ma < 0.3$ (see Section 12.1) can be considered to be slow enough.

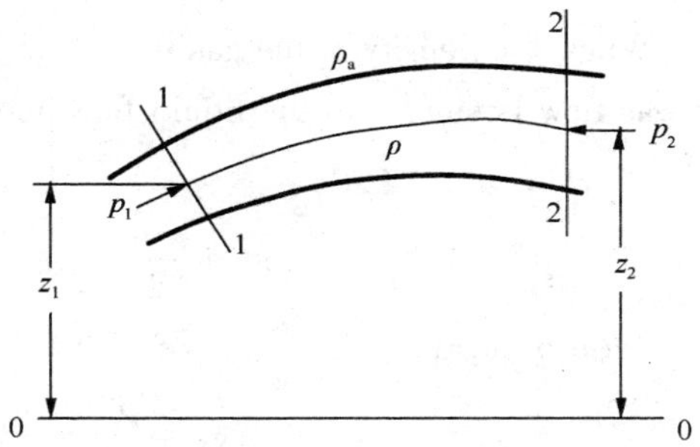

In the case of gas flow, the density of gas flow is in the same order of the gas outside of the gas flow. Thus, the elevation difference should be taken into consideration when the gage pressure is used in the energy equation.

Fig. 4-12 Steady incompressible gas flow

In Fig. 4-12 is a steady incompressible gas flow in a pipe. Take the density of the gas flow to be ρ, the density of air outside of the pipe be ρ_a, the absolute pressure at the computational points on sections 1-1 and 2-2 be $p_{1\text{abs}}$ and $p_{2\text{abs}}$ respectively. Thus, the energy equation in term of head for steady incompressible flow between sections 1-1 and 2-2 appears

$$z_1 + \frac{p_{1\text{abs}}}{\rho g} + \frac{\alpha_1 v_1^2}{2g} = z_2 + \frac{p_{2\text{abs}}}{\rho g} + \frac{\alpha_2 v_2^2}{2g} + h_w$$

Assuming $\alpha_1 = \alpha_2 = 1.0$, and rewriting the above equation in terms of pressure we have

$$\rho g z_1 + p_{1\text{abs}} + \frac{\rho v_1^2}{2} = \rho g z_2 + p_{2\text{abs}} + \frac{\rho v_2^2}{2} + p_w \qquad (4.30)$$

where p_w is the *loss of pressure*, $p_w = \rho g h_w$.

Express the absolute pressures at sections 1-1 and 2-2 by the gage pressure as

$$p_{1\text{abs}} = p_1 + p_{\text{at}}$$
$$p_{2\text{abs}} = p_2 + p_{\text{at}} - \rho_a g(z_2 - z_1)$$

where p_{at} is the atmospheric pressure at the elevation of z_1; $p_{\text{at}} - \rho_a g(z_2 - z_1)$ is the atmospheric pressure at the elevation of z_2.

Substitution of the above absolute pressures into Eq. 4.30 yields

$$p_1 + \frac{\rho v_1^2}{2} + (\rho_a - \rho)g(z_2 - z_1) = p_2 + \frac{\rho v_2^2}{2} + p_w \qquad (4.31)$$

where p_1 and p_2 are the *static pressures*; $\frac{\rho v_1^2}{2}$ and $\frac{\rho v_2^2}{2}$ are the *kinematic pressures*; $(\rho_a - \rho)g$ is the *effective buoyancy* per unit volume gas; $(\rho_a - \rho)g(z_2 - z_1)$ is the elevation energy per unit volume gas at section 2-2 referred to section 1-1, called the *elevation pressure*.

Equation 4.31 is the energy equation in term of pressure for steady incompressible gas flow.

When the density of gas flow is same as the air density outside the flow, $\rho = \rho_a$, or the elevations of the computational points are same, $z_1 = z_2$, the third term of the left side in Eq. 4.31 is zero, namely the elevation pressure is zero. Thus, Eq. 4.31 reduces to

$$p_1 + \frac{\rho v_1^2}{2} = p_2 + \frac{\rho v_2^2}{2} + p_w \qquad (4.32)$$

When the density of the gas flow is far large than the air density outside of the flow, $\rho \gg \rho_a$, the gas flow is similar to the liquid flow and the air density ρ_a can be neglected. Thus, Eq. 4.31 reduces to

$$p_1 + \frac{\rho v_1^2}{2} - \rho g(z_2 - z_1) = p_2 + \frac{\rho v_2^2}{2} + p_w \qquad (4.33)$$

or in term of head

$$z_1 + \frac{p_1}{\rho g} + \frac{v_1^2}{2g} = z_2 + \frac{p_2}{\rho g} + \frac{v_2^2}{2g} + h_w \qquad (4.34)$$

Equation 4.34 is the energy equation in term of head for the steady incompressible gas flow with much larger density than the air density, which is same as the expression of energy equation for the liquid flow, Eq. 4.23.

EXAMPLE 4-10 Boiler

The schematic of a boiler with a natural smoke exhaust ventilation system is illustrated in Fig. E4-10. It evacuates the smoke from the outlet section 1-1 of boiler to a 1.0-m-diameter vertical chimney then to the air. Take the smoke density to be $\rho = 0.7 \text{kg/m}^3$, the air density $\rho_a = 1.2 \text{kg/m}^3$, the smoke flowrate in the stovepipe $Q = 10 \text{m}^3/\text{s}$, and the pressure loss in the chimney $p_w = 0.035 \frac{H}{d} \frac{\rho v^2}{2}$, in which d is the chimney diameter, and v is the velocity in the chimney. If the area at section 1-1 is twice of the area of chimney section, the vacuum at section 1-1 is 10mmH$_2$O, determine the required height of chimney, H.

Solution:

Assume the flow is steady and incompressible. We take the water density to be $\rho_w = 1000 \text{kg/m}^3$ and $\alpha \approx 1.0$.

Take plane 0-0 to be the datum plane as illustrated in Fig. E4-10.

Thus, the energy equation in term of pressure for steady incompressible gas flow is

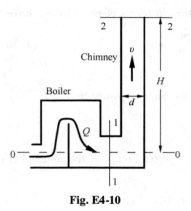

Fig. E4-10

$$p_1 + \frac{\rho v_1^2}{2} + (\rho_a - \rho)g(z_2 - z_1) = p_2 + \frac{\rho v_2^2}{2} + p_w \tag{1}$$

where $z_2 - z_1 = H$, $p_2 = 0$ since the smoke is removed to the air, and

$$p_1 = -\rho_w g h_{v1} = -(1000 \text{kg/m}^3)(9.81 \text{m/s}^2)(0.01 \text{m}) = -98.1 \text{Pa}$$

$$v_2 = v = \frac{Q}{A} = \frac{Q}{\pi d^2/4} = \frac{10 \text{ m}^3/\text{s}}{\pi (1.0 \text{m})^2/4} = 12.74 \text{m/s}$$

The continuity equation between sections 1-1 and 2-2, $v_1(2A) = v_2 A$, gives the velocity at section 1-1

$$v_1 = \frac{v_2}{2} = \frac{12.74 \text{m/s}}{2} = 6.37 \text{m/s}$$

With the given parameters and $p_w = 0.035 \frac{H}{d} \frac{\rho v^2}{2}$, Eq. 1 becomes

$$-98.1 \text{Pa} + \frac{(0.7 \text{kg/m}^3)(6.37 \text{m/s})^2}{2} + (1.2 \text{kg/m}^3 - 0.7 \text{kg/m}^3)(9.81 \text{m/s}^2)H$$

$$= 0 + \frac{(0.7 \text{kg/m}^3)(12.74 \text{m/s})^2}{2} + 0.035 \frac{H}{1.0 \text{m}} \frac{(0.7 \text{kg/m}^3)(12.74 \text{m/s})^2}{2}$$

Solving to give the required height of chimney

$$H = 48.24 \text{m}$$

DISCUSSION This example indicates that it is the elevation pressure, $(\rho_a - \rho)g(z_2 - z_1)$, providing the energy for smoke moving upwards through the chimney. The smoke and the air outside the chimney should have the temperature difference to retain the effective buoyancy, and a certain height of chimney $(z_2 - z_1)$ is required to keep the smoke flow upwards in the chimney.

4.4 Linear Momentum Equation

4.4.1 Form for Uniform Flow

It is often important to determine the force produced on a solid body by fluid flowing steadily over it. For example, there is a force exerted on a solid surface by a jet of fluid impinging on it. There are also the aerodynamic/hydrodynamic forces (lift and drag) on an aircraft wing or hydrofoil, the force on a pie bend caused by the fluid flowing within it, the thrust on a propeller and so on. All these forces are dynamic forces and they are associated with a change in the momentum of the fluid.

The magnitude of such a force is determined essentially by *Newton's second law of motion*. In its most general form, Newton's second law states that: The net force acting on a certain 'batch' of fluid in any fixed direction is equal to the total rate of increase of momentum of that fluid in that direction.

$$\Sigma \mathbf{F} = \frac{d\mathbf{K}}{dt} = \frac{d(\Sigma m \mathbf{u})}{dt} \tag{4.35}$$

Since force and momentum are both vector quantities, it is essential to specify both magnitudes and directions. Where we are concerned with a collection of 'batch' of fluid, such as a control volume, Newton's law may be applied (for a given direction) to each 'batch' of fluid individually. When the resulting equations are added, the total force in the given fixed direction corresponds to the net force acting in that direction at the boundaries of the control volume. Only these external, boundary forces are involved because any internal forces between separate 'batch' of fluid occur in pairs of action and reaction and therefore cancel in the total.

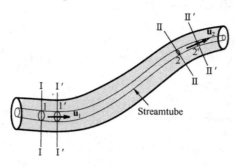

Fig. 4-13 Space change of control volume in a streamtube within time interval dt

To derive a relation by which force may be related to the fluid within a control volume, we begin by applying Newton's second law to a small element in a streamtube in a steady total flow with only one inlet and one outlet as illustrated in Fig. 4-13. After a short time interval dt the fluid that formerly occupied the space between sections I-I to II-II has moved forward to occupy the space between the sections I'-I' and II'-II'. In general, its momentum changes during this short time interval.

Take the averaged area from section I-I to I'-I' be dA_1 and from section II-II to II'-II' be dA_2, and the averaged normal velocity to be $\mathbf{u}_1$ and $\mathbf{u}_2$. Thus, the net increase of momentum of tube flow in streamtube during the time interval dt is:

$$d\mathbf{K} = \mathbf{K}_{1'-2'} - \mathbf{K}_{1'-2} = (\mathbf{K}_{1'-2} + \mathbf{K}_{2-2'})_{t+dt} - (\mathbf{K}_{1-1'} + \mathbf{K}_{1'-2})_t \qquad (4.36)$$

Because the flow is steady, the shape and the position of the control volume, and also the streamtube within it remains stationary with respect to the fixed coordinate axes and the variables in it remain unchanged. Thus, the momentum of tube flow in flow section from I'-I' to II-II keep unchanged during the time interval, that is $(\mathbf{K}_{1'-2})_{t+dt} = (\mathbf{K}_{1'-2})_t$. Eq. 4.36 reduces to

$$d\mathbf{K} = (\mathbf{K}_{2-2'})_{t+dt} - (\mathbf{K}_{1-1'})_t = m_2 \mathbf{u}_2 - m_1 \mathbf{u}_1 = \rho_2 u_2 dA_2 dt \mathbf{u}_2 - \rho_1 u_1 dA_1 dt \mathbf{u}_1 \qquad (4.37)$$

Equation 4.37 states that during the time interval dt the increase of momentum of fluid considered is equal to the momentum leaving the streamtube in that time minus the momentum of that entering.

The net increase of the momentum of total flow during the time interval dt can be obtained by adding up all the net momentums of all streamtubes in this total flow. Thus integrating Eq. 4.37 yields

$$\Sigma\, d\mathbf{K} = \int_{A_2} \rho_2 u_2 dA_2\, dt\, \mathbf{u}_2 - \int_{A_1} \rho_1 u_1 dA_1\, dt \mathbf{u}_1 \qquad (4.38)$$

Assume the control cross sections of the total flow are selected to be at the uniform or gradually varied flow sections, the normal velocity vectors at any point on those sections thus have nearly equal magnitude and identical direction. From Eq. 4.38 we then have the time rate of the momentum in the control volume

$$\frac{\Sigma\, d\mathbf{K}}{dt} = \int_{A_2} \rho_2 u_2 dA_2 \mathbf{u}_2 - \int_{A_1} \rho_1 u_1 dA_1\, \mathbf{u}_1 = \dot{m}_2 \mathbf{u}_2 - \dot{m}_1 \mathbf{u}_1 \qquad (4.39)$$

where $\dot{m}$ is the *mass flow rate* through the section per unit time, $\dot{m} = \rho u A$.

Combining Eqs. 4.35 and 4.39, we have the *linear momentum equation* for the steady incompressible or compressible uniform flow with only one inlet and one outlet

$$\Sigma \mathbf{F} = \dot{m}_2 \mathbf{u}_2 - \dot{m}_1 \mathbf{u}_1 \qquad (4.40)$$

Equation 4.40 states that the net force vector on a fixed control volume equals the vector sum of outlet momentum fluxes subtracted by the vector sum of inlet fluxes.

Since the total flow is steady, the time rate of mass flow entering the control volume through the inlet should equal to the time rate of mass flow leaving from the outlet, that is $\dot{m} = \rho_1 u_1 A_1 = \rho_2 u_2 A_2$. Eq. 4.40 becomes

$$\Sigma \mathbf{F} = \dot{m}(\mathbf{u}_2 - \mathbf{u}_1) \qquad (4.41)$$

For the incompressible flow, $\rho_1 = \rho_2 = \rho$, $\dot{m} = \rho Q$, we have

$$\Sigma \mathbf{F} = \rho Q(\mathbf{u}_2 - \mathbf{u}_1) \qquad (4.42)$$

This is the *linear momentum equation* for steady incompressible uniform flow.

In Eq. 4.41 or 4.42, the force vector includes all the external force vectors acting on the fluid in the control volume

$$\Sigma \mathbf{F} = \Sigma \mathbf{F}_b + \Sigma \mathbf{F}_s + \mathbf{R} \qquad (4.43)$$

where $\mathbf{F}_b$ is the mass forces on fluid in the control volume, i.e. the gravity $\mathbf{G}$; $\mathbf{F}_s$ is the surface forces on the control volume surface, i.e. the pressure on the control cross sections; $\mathbf{R}$ is the resultant external force on the fluid by the surrounding boundaries of control volume.

4.4.2 General Form

However, the velocity profile on a section of flow is usually nonuniform. For this case the simple momentum-flux calculation $\int_A \rho u dA \mathbf{u} = \dot{m}\mathbf{u}$ in Eq. 4.39 is somewhat in error. It should be corrected to $\beta \dot{m} \mathbf{v}$, where $\mathbf{v}$ is the vector of sectional averaged velocity normal to the section, β is the dimensionless *momentum correction factor*, first shown by the French scientist Joseph Boussinesq(1842-1929).

Momentum correction factor β is dimensionless and defined as the ratio of the real momentum computed by the point velocity to the assumed momentum computed by the average velocity

$$\beta = \frac{\int_A \rho u dA u}{\dot{m} v} = \frac{\int_A \rho u u dA}{\rho v A v} = \frac{1}{A} \int_A \left(\frac{u}{v}\right)^2 dA \qquad (4.44)$$

It turns out that the value of β depends on the velocity profile in the flow section, and is always greater than or equal to unity for any velocity profile you can imagine: $\beta = 4/3$ for laminar flow and $\beta = 1.02 \sim 1.05$ for turbulent flow. This means that the values of β for turbulent flow are so close to unity that the non-uniformity of velocity is normally neglected, while the laminar correction may sometimes be important. In the first approximate estimation, β can be taken to be unity, namely $\beta \approx 1.0$.

Hence, for the steady incompressible or compressible non-uniform flow Eq. 4.41 would be given as

$$\Sigma \mathbf{F} = \dot{m}(\beta_2 \mathbf{v}_2 - \beta_1 \mathbf{v}_1) \qquad (4.45)$$

For the steady incompressible non-uniform flow, Eq. 4.45 becomes

$$\Sigma \mathbf{F} = \rho Q(\beta_2 \mathbf{v}_2 - \beta_1 \mathbf{v}_1) \qquad (4.46)$$

or in the components of the coordinate system

$$\left.\begin{array}{l} \Sigma F_x = \rho Q(\beta_2 v_{2x} - \beta_1 v_{1x}) \\ \Sigma F_y = \rho Q(\beta_2 v_{2y} - \beta_1 v_{1y}) \\ \Sigma F_z = \rho Q(\beta_2 v_{2z} - \beta_1 v_{1z}) \end{array}\right\} \qquad (4.47)$$

Equation 4.47 is the *general form of the linear momentum equation* for the steady incompressible total flow with one inlet and one outlet. It can be applied to both the viscous and inviscid flow, uniform ($\beta = 1.0$, $v = U$) and nonuniform flow. It states that the resultant force on a control volume equals to the net time rate of the momentum out of the control volume.

If the steady incompressible flow has more than one inlet or outlet, Eq. 4.46 should be rewritten as

$$\Sigma \mathbf{F} = \Sigma(\rho Q \beta \mathbf{v})_{out} - \Sigma(\rho Q \beta \mathbf{v})_{in} \qquad (4.48)$$

The restricted conditions for Eq. 4.46 or 4.48 are: steady incompressible viscous or inviscid flow.

When the linear momentum equation is used to solve the practical problem, it should be noted that:

- the control cross sections must be located at gradually varied flow or uniform flow, even if the rapidly varied flow may exist between those sections;
- the only mass force is gravity;
- the external forces have an algebraic sign, either positive or negative;
- only external forces acting on the control volume are considered.

It also should be noted that the solution of a typical problem involving momentum equation almost always leads to a consideration of the continuity equation and energy equation as the equal partners in the analysis. The only exception is when the complete velocity distribution or the pressure at the section is already known from a previous or given analysis, but that means that the continuity relation or energy relation have already been used to obtain the given information. The point is that the continuity relation and energy relation are always the important elements in a force analysis of flow.

4.4.3 Applications of Momentum Equation

To apply the linear momentum equation determining the force on a fluid or a body, first of all is to select an appropriate control volume and a coordinate system leading to a simplest momentum equation, then analyze the forces acting on the fluid, and finally apply the momentum equation, always together with the continuity equation and the energy equation, to solve the problem.

EXAMPLE 4-11   Force by a jet

When a steady water jet strikes at a solid surface a stream of fluid is formed which moves over the surface until the boundaries are reached, and the fluid then leaves the surface tangentially as in Fig. E4-11. Assume water flows in the horizontal plane. The jet velocity and flow rate out of the nozzle are v_1 and Q_1 respectively. Determine: (a) the force acting on the solid surface by the jet, and (b) the outflow flowrate Q_2 and Q_3. Assume the friction water flowing through the plate is negligible.

Solution:

A suitable control volume, enclosed by the sections 1-1 to 3-3, the air-water interfaces and the solid surface, is illustrated by the dash dot line in the figure. The coordinate system is taken as shown in the

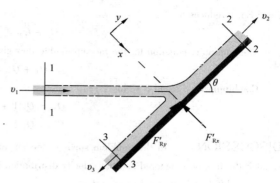

Fig. E4-11

figure, of which the x-direction perpendicular to the plane, and the y-direction parallel to the plane. Since the pressure around the fluid flow is atmospheric, the gage pressures on both where the fluid enters the control volume and where it leaves are zero.

Take the flow to be steady and incompressible, the resultant external forces on the control volume as shown.

(a) The momentum equation for steady incompressible total flow with multiple outlets is given by Eq. 4.48

$$\Sigma \mathbf{F} = \Sigma(\rho Q \beta \mathbf{v})_{\text{out}} - (\rho Q \beta \mathbf{v})_{\text{in}}$$

Since the x-direction is perpendicular to the plane, thus, the fluid, after striking the surface and moving along it, have no component of velocity and therefore no momentum in the x-direction, that is $v_{2x} = v_{3x} = 0$. The external force on the fluid can be provided only by the solid surface since the effect of gravity is neglected due to the horizontal flow and the pressure around the fluid flow is atmospheric. Therefore, the force acting on the fluid in the x-direction is:

$$-F'_{Rx} = 0 - \beta_1 \rho Q_1 v_{1x} = -\beta_1 \rho Q_1 v_1 \sin\theta$$
$$F'_{Rx} = \beta_1 \rho Q_1 v_1 \sin\theta \ (\leftarrow)$$

Since the flow over the surface with a negligible friction, no shear force is possible exerted by the surface on the fluid. Based on the balance forces, the external force acting on the fluid in the y-direction is zero:

$$F'_{Ry} = 0$$

Therefore, the resultant force exerted by the fluid on the solid surface equals in magnitude and is opposite to the force in the x-direction, F'_{Rx}, and is thus:

$$F_R = \beta_1 \rho Q_1 v_1 \sin\theta \ (\rightarrow)$$

DISCUSSION The plate absorbs the full brunt of the momentum of the water jet since the x-direction momentum at the outlet of the control volume is zero. If the control volume were drawn instead along the interface between the water and the plate, there would be additional (unknown) pressure forces in the analysis.

(b) Take the momentum-flux correction factor to be unity, namely $\beta = 1.0$. With $F'_{Ry} = 0$ in the y-direction as discussed above the momentum equation, Eq. 4.48, applies

$$\rho Q_2 v_2 - \rho Q_3 v_3 - \rho Q_1 v_1 \cos\theta = 0 \qquad (1)$$

The energy equation with negligible loss along the streamline in a steady compressible flow between sections 1-1 to 2-2 gives

$$z_1 + \frac{p_1}{\rho g} + \frac{v_1^2}{2g} = z_2 + \frac{p_2}{\rho g} + \frac{v_2^2}{2g}$$

where $z_1 = z_2$ since the plane is horizontal, and $p_1 = p_2 = 0$ since the pressure around the fluid flow is atmospheric. Thus, we have $v_1 = v_2$. With the same manner, we can have $v_1 = v_3$. Thus, for the steady incompressible

flow, Eq. 1 reduces to
$$Q_2 - Q_3 = Q_1 \cos\theta \tag{2}$$
The continuity equation for the incompressible flow gives
$$Q_2 + Q_3 = Q_1 \tag{3}$$
Combining Eqs. 2 and 3 yields
$$Q_2 = Q_1(1 + \cos\theta)/2$$
$$Q_3 = Q_1(1 - \cos\theta)/2$$

DISCUSSION   In this problem solving, the magnitudes of outflow velocities at sections 2-2 and 3-3 are equal if the friction is ignored and the velocity distribution in section is taken to be uniform. Actually, it would be somewhat different in the real viscous flow.

EXAMPLE 4-12 Thrust force on an elbow

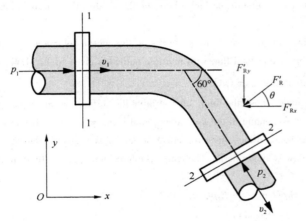

Fig. E4-12

In Fig. E4-12 a pipe elbow is used to deflect water flow at a rate of $1.57 \text{ m}^3/\text{s}$ in a horizontal pipe downward 60° while accelerating it. The diameter of the elbow is 1.0m at the inlet and 0.75m at the outlet. The gage pressure at the inlet of elbow is 5.05×10^4 Pa. Assume the head loss in the elbow is negligible, determine the thrust force on the elbow exerted by the deflected water.

Solution:

Assume the flow is a steady incompressible flow.

Since the axis of the bend is in the horizontal plane, elevation changes along the pipe axis are thus negligible, the weights of the pipe and the fluid in it act in a direction perpendicular to this horizontal plane and have no effects on the momentums of water in the elbow.

Take the fluid body in the elbow as the control volume and designate the inlet by 1 and the outlet by 2. We also take the x-and y-coordinates and the forces on the fluid by the elbow wall as shown. Take the momentum correction factor to be $\beta = 1.0$ and the water density $\rho = 1000 \text{kg/m}^3$.

The momentum equation for steady incompressible total flow, Eq. 4.46, is given as
$$\Sigma \mathbf{F} = \rho Q(\beta_2 \mathbf{v}_2 - \beta_1 \mathbf{v}_1)$$

Thus, the components of momentum equation in the x-direction and y-direction are

x-direction: $\quad p_1 A_1 - p_2 A_2 \cos 60° - F'_{Rx} = \rho Q v_2 \cos 60° - \rho Q v_1$

y-direction: $\quad p_2 A_2 \sin 60° - F'_{Ry} = -\rho Q v_2 \sin 60° - 0$

Thus
$$F'_{Rx} = p_1 A_1 - p_2 A_2 \cos 60° - \rho Q v_2 \cos 60° + \rho Q v_1 \tag{1}$$
$$F'_{Ry} = p_2 A_2 \sin 60° + \rho Q v_2 \sin 60° \tag{2}$$

where the inlet and outlet velocities of water are
$$v_1 = \frac{Q}{\pi D_1^2/4} = \frac{1.57 \text{ m}^3/\text{s}}{\pi(1.0\text{m})^2/4} = 2.0 \text{m/s}$$

$$v_2 = \frac{Q}{\pi D_2^2/4} = \frac{1.57 \text{ m}^3/\text{s}}{\pi(0.75\text{m})^2/4} = 3.56\text{m/s}$$

The energy equation between sections 1-1 to 2-2 with a negligible loss

$$z_1 + \frac{p_1}{\rho g} + \frac{v_1^2}{2g} = z_2 + \frac{p_2}{\rho g} + \frac{v_2^2}{2g}$$

gives the pressure at the outlet of elbow with $z_1 = z_2$ and $p_1 = 5.05 \times 10^4 \text{Pa}$:

$$p_2 = p_1 + \frac{\rho}{2}(v_1^2 - v_2^2)$$

$$= 5.05 \times 10^4 \text{Pa} + \frac{1000\text{kg/m}^3}{2}[(2.0\text{m/s})^2 - (3.56\text{m/s})^2]$$

$$= 4.616 \times 10^4 \text{Pa}$$

Substituting all the given and computed values of parameters into Eqs. 1 and 2 gives

$F'_{Rx} = (5.05 \times 10^4 \text{Pa})[\pi(1.0\text{m})^2/4] - (4.616 \times 10^4 \text{Pa})[\pi(0.75\text{m})^2/4]\cos60°$
   $- (1000\text{kg/m}^3)(1.57 \text{ m}^3/\text{s})[(3.56\text{m/s})\cos60° - 2.0\text{m/s}]$
   $= 29.8\text{kN} (\leftarrow)$

$F'_{Ry} = (4.616 \times 10^4 \text{Pa})[\pi(0.75\text{m})^2/4]\sin60° + (1000\text{kg/m}^3)(1.57\text{m}^3/\text{s})(3.56\text{m/s})\sin60°$
   $= 22.5\text{kN} (\downarrow)$

Thus, the thrust forces on the elbow by water equals in magnitude but with opposite directions

$$F_{Rx} = F'_{Rx} = 29.8\text{kN} (\rightarrow)$$
$$F_{Ry} = F'_{Ry} = 22.5\text{kN} (\uparrow)$$

The resultant thrust force is

$$F_R = \sqrt{F_{Rx} + F_{Ry}} = \sqrt{(29.8\text{kN})^2 + (22.5\text{kN})^2} = 37.3\text{kN}$$

which acts along the line with an angle to the horizontal

$$\theta = \tan^{-1}\frac{F_{Ry}}{F_{Rx}} = \tan^{-1}\frac{22.5\text{kN}}{29.8\text{kN}} = 37.1°$$

DISCUSSION   The nonzero pressure distribution along the inside walls of the elbow appears as the resultant force on the elbow in our analysis. The actual value of p_2 will be lower than the calculated one, $4.616 \times 10^4 \text{Pa}$, because of the frictional and other irreversible losses in the elbow.

EXAMPLE 4-13   Force on a dam

The cross section of a dam in a horizontal channel is illustrated in Fig. E4-13. The water depth upstream of the dam is $H_1 = 1.5\text{m}$, and the downstream is $H_2 = 0.6\text{m}$. Assume the head loss of water flowing over the dam is negligible, determine the horizontal force exerting on the dam per meter width into paper.

Solution:

Assume the flow is a steady incompressible flow.

We take the flow between sections 1-1 to 2-2 as the control volume shown in the figure. We also take the x- and z-coordinates and the forces on the fluid by the dam as shown. Take the momentum correction factor to be $\beta = 1.0$ and the water density $\rho = 1000\text{kg/m}^3$.

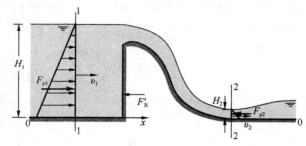

Fig. E4-13

For the steady incompressible total flow, the momentum equation in the x-direction, Eq. 4.47, applies

$$p_{1c}A_1 - p_{2c}A_2 - F'_R = \rho Q(\beta_2 v_2 - \beta_1 v_1)$$

Thus

$$F'_R = p_{1c}A_1 - p_{2c}A_2 - \rho Q(\beta_2 v_2 - \beta_1 v_1) \qquad (1)$$

where p_{1c} and p_{2c} are the pressures at the centroids of sections 1-1 and 2-2 respectively, $p_{1c} = \rho g H_1/2$, $p_{2c} = \rho g H_2/2$.

The energy equation for a steady incompressible flow with a negligible loss and referred to the channel bed 0-0 gives

$$H_1 + \frac{p_1}{\rho g} + \frac{v_1^2}{2g} = H_2 + \frac{p_2}{\rho g} + \frac{v_2^2}{2g}$$

with $p_1 = p_2 = 0$ at the free surface it follows

$$1.5\text{m} + \frac{v_1^2}{2(9.81\text{m/s}^2)} = 0.6\text{m} + \frac{v_2^2}{2(9.81\text{m/s}^2)} \qquad (2)$$

The continuity equation for a steady incompressible flow gives

$$H_1 b_1 v_1 = H_2 b_2 v_2$$

with $b_1 = b_2 = b = 1.0\text{m}$ yields

$$(1.5\text{m}) v_1 = (0.6\text{m}) v_2 \qquad (3)$$

Combining Eqs. 2 and 3 gives $v_1 = 1.83\text{m/s}$ and $v_2 = 4.58\text{m/s}$. Substituting into Eq. 1 yields

$$F'_R = \frac{1}{2}\rho g H_1^2 b_1 - \frac{1}{2}\rho g H_2^2 b_2 - \rho H_1 b_1 v_1 (\beta_2 v_2 - \beta_1 v_1)$$

$$= \frac{1}{2}(1000\text{kg/m}^3)(9.81\text{m/s}^2)(1.0\text{m})[(1.5\text{m})^2 - (0.6\text{m})^2]$$

$$- (1000\text{kg/m}^3)(1.5\text{m})(1.0\text{m})(1.83\text{m/s})[(1)(4.58\text{m/s}) - (1)(1.83\text{m/s})]$$

$$= 1.72\text{kN}(\leftarrow)$$

Thus, the thrust force on the dam by the water is equal in magnitude and opposite in direction to this force and is thus:

$$F_R = F'_R = 1.72\text{kN} \; (\rightarrow)$$

Summary

In this chapter the flow of a fluid is analyzed by using the principles of *conservation of mass, momentum,* and *energy* applied to *control volume,* an *Eulerian description* of a finite volume in a fixed space described by the sectional *averaged properties* with its *control cross sections* be selected in the uniform flow or gradually varied flow. In this case, the nonuniformity of the fluid velocity on a section of real flow is corrected by the *kinematic correction factor*

$$\alpha = \frac{\int_A u^3 dA}{v^3 A} = \frac{1}{A}\int_A \left(\frac{u}{v}\right)^3 dA$$

and the *momentum-flux correction factor*

$$\beta = \frac{\int_A \rho u dA u}{\dot{m} v} = \frac{1}{A}\int_A \left(\frac{u}{v}\right)^2 dA$$

The kinematic and the momentum-flux correction factors have relative larger value in fully developed laminar flow ($\alpha = 2.0$, $\beta = 4/3$) due to the larger nonuniformity of the velocity distribution, and near to the unity in the fully developed turbulent flow ($\alpha = 1.04 \sim 1.11$, $\beta = 1.02 \sim 1.05$). Thus, the correction factor in the engineering problems (usually to be the turbulent flow) is normally ignored.

The *continuity equation*, a statement of the fact that mass is conserved, is obtained in a form that can be applied to any steady or unsteady, incompressible or compressible, viscous or inviscid total flow

$$\Sigma \dot{m}_{in} = \Sigma \dot{m}_{out}$$

Its simplified form

$$\Sigma(vA)_{in} = \Sigma(vA)_{out}$$

is usually used to solve steady incompressible problems dealing with the volume flowrates.

The steady-state *energy equation*, obtained from the differential energy equation for streamtube flow originating from *Bernoulli equation*, consists of the Bernoulli equation with extra terms that account for the energy losses due to the fluid viscous and the change of geometrically shape, as well as terms accounting for the existence of pumps or turbines in the flow. For the steady incompressible non-uniform total flow with shaft work and only one inlet and one outlet, the *energy equation* is

$$z_1 + \frac{p_1}{\rho g} + \frac{\alpha_1 v_1^2}{2g} \pm h_s = z_2 + \frac{p_2}{\rho g} + \frac{\alpha_2 v_2^2}{2g} + h_w$$

The applications of this equation are subjected to the restriction conditions discussed in detail in Section 4.3.

Both the *energy gradient line* (EGL) and *hydraulic gradient line*(HGL) are the graphical interpretations of energy equation, and a useful visual interpretation to sketch a flow. In general flow conditions, the EGL drops continuously while the HGL may rise or fall referred to the variation of velocity. The location where the HGL *intersects* the fluid indicates a zero gage pressure of the fluid. Thus, an accurate drawing of a piping system and the HGL can be used to determine the regions where the gage pressure in the pipe is negative.

The *linear momentum equation*, a form of *Newton's second law* of motion applied to flow through a control volume, is obtained and used to solve flow problems dealing with surface forces and body forces acting on the fluid and its surroundings. For the steady incompressible non-uniform total flow with only one inlet and one outlet, the linear momentum equation is given by

$$\Sigma \mathbf{F} = \rho Q(\beta_2 \mathbf{v}_2 - \beta_1 \mathbf{v}_1)$$

In this relation, the net flux of momentum through the control surface is directly related to the net external force exerted on the contents of the control volume.

The continuity, energy, and momentum equations are three of the most fundamental relations in fluid mechanics, and they are used extensively in the chapters that follow. All of them can be used *together* to determine the forces and torques acting on fluid systems, or only continuity and energy equations to determine the *flowrate* or *pressure* in the flow system.

Word Problems

W4-1 Define the control volume. Is it a one-dimensional concept? Explain.

W4-2 For the nonuniformities of the variables over the cross-section of control volume, the averaged concept has been introduced in the control volume analysis. But how is this nonuniformity considered in the control volume analysis?

W4-3 Define the gradually varied flow and rapidly varied flow. How about their difference in the dynamic pressure distribution over the cross-sectional area?

W4-4 The piezometer height at any point in a cross section of gradually varied flow is same. Is it true? Why?

W4-5 What is the physical meaning of the continuity equation for total flow: $v_1 A_1 = v_2 A_2$?

W4-6 Does the amount of mass entering a control volume have to be equal to the amount of mass leaving during an unsteady flow process?

W4-7 When is the flow through a control volume steady?

W4-8 Consider a device with one inlet and one outlet. If the volume flow rates at the inlet and at the outlet are the same, is the flow through this device necessarily steady? Why?

W4-9 How does the energy equations for a steady incompressible total flow differ from the differential energy equation for a steady incompressible tube flow?

W4-10 Two papers are placed close to each other with a small gap. If the air in this gap is suddenly blown, do the two papers move closer, apart or stay stationary? Explain.

W4-11 Water flows from point A to point B in a tilted pipe with constant diameter. Determine the tilting direction of pipe if the relation of pressures are as follows: (a) $p_A > p_B$; (b) $p_A = p_B$; (c) $p_A < p_B$. Neglect the frictional loss in the pipe.

W4-12 What are the restricted conditions of the energy equation for steady incompressible total flow? How to select the computational cross section, the reference level (datum plane), the computational point and the working pressure?

W4-13 Define the energy gradient line and the hydraulic gradient line. How do they differ from each other? Under what conditions do both lines coincide with the free surface of a liquid?

W4-14 How is the location of the hydraulic grade line determined for open-channel flow? How is it determined the outlet of a pipe discharging to the atmosphere?

W4-15 For a horizontal pressure pipe, how can the hydraulic gradient line be rising, falling or horizontal? Give examples. Neglect the energy loss in the pipe.

W4-16 Why can the energy loss always be neglected in momentum equation for the steady incompressible total flow?

W4-17 What does it mean if the computed force from the momentum equation is negative? Is the magnitude of force under determination related to the size of the control volume? How to select the control volume?

W4-18 What is the kinematic correction factor? Is it significant?

W4-19 The water level in a tank is 20m above the ground. A hose is connected to the bottom of the tank, and the nozzle at the end of the hose is pointed straight up. The water stream from the nozzle is observed to rise 25m above the ground. Explain what may cause the water from the hose to rise above the tank level.

W4-20 Is momentum a vector? If so, in what direction does it point?

W4-21 Is fluid weight a body force or surface force? How about pressure?

W4-22 How can we minimize the number of surface forces exposed during analysis?

W4-23 What is the importance of the momentum-flux correction factor in the momentum analysis of slow systems? For which type of flow is it significant and must it be considered in analysis: laminar flow, turbulent flow, or jet flow?

W4-24 In the application of the momentum equation, explain why we can usually disregard the atmospheric pressure and work with gage pressures only.

W4-25 Two firefighters are fighting a fire with identical water hoses and nozzles, except that one is holding the hose straight so that the water leaves the nozzle in the same direction it comes, while the other holds it backward so that the water makes a U-turn before being discharged. Which firefighter will experience a greater reaction force?

W4-26 A constant-velocity horizontal water jet from a stationary nozzle of constant exit cross section impinges normally on a stationary vertical flat plate. A certain force F is required to hold the plate against the water stream. If the water velocity is doubled, will the necessary holding force also be doubled? Explain.

W4-27 A constant-velocity horizontal water jet from a stationary nozzle of constant exit cross section impinges normally on a vertical flat plate that is held in a nearly frictionless track, as shown in Fig. W4-27. As the water jet hits the plate, the track begins to move due to the water force. Will the acceleration of the plate remain constant or change? Explain.

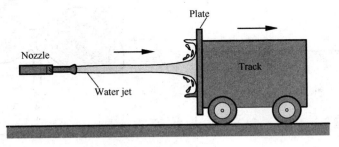

Fig. W4-27

Problems

Averaged Properties and Correction Factors

4-1 The velocity profile of fully developed laminar flow in a circular pipe is

$$u = u_{max}\left[1 - \left(\frac{r}{R}\right)^2\right]$$

where u_{max} is the maximum velocity at the pipe axis, r is the distance of point away from pipe axis, R is the radius of pipe. Determine: (a) the average velocity v; (b) the kinematic correction factor α; (c) the momentum-flux correction factor β.

4-2 The velocity profile of fully developed turbulent flow in a circular pipe is

$$u = u_{max}\left(1 - \frac{r}{R}\right)^{1/9}$$

where u_{max} is the maximum velocity at the pipe axis, r is the distance of point away from pipe axis, R is the radius of pipe. Determine: (a) the average velocity v; (b) the kinematic correction factor α; (c) the momentum-flux correction factor β.

Continuity Equation

4-3 A contraction pipe element for oil transportation has a 200-mm-diameter at the entrance and 60-mm-diameter at the exit. Let the oil density be $860 kg/m^3$. If the velocity at the entrance is $2m/s$, determine the velocity at the exit of this contraction pipe and the mass flow rate through the pipe.

4-4 In Fig. P4-4 a water jet pump involves a jet cross-sectional area of $0.01\ m^2$, and a jet velocity of $30m/s$. The jet is surrounded by entrained water. The total cross-sectional area associated with the jet and entrained streams is $0.075\ m^2$. These two fluid streams leave the pump thoroughly mixed with an average velocity of $6m/s$ through a cross-sectional area of $0.075 m^2$. Determine the pumping rate (i.e., the entrained fluid flowrate) involved.

4-5 Oil having a specific gravity of 0.9 is pumped as illustrated in Fig. P4-5 with a water jet pump. The water volume flowrate is $2\ m^3/s$. The water and oil mixture has an average specific gravity of 0.95. Calculate the flow rate at which the pump moves oil.

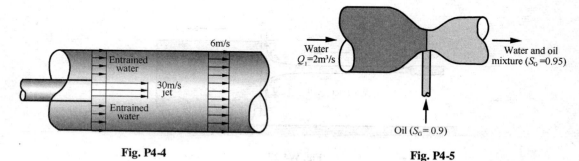

Fig. P4-4   Fig. P4-5

4-6 The open tank in Fig. P4-6 contains water at $20°C$ and is being filled through section 1. Assume incompressible flow. First derive an analytic expression for the water-level change dh/dt in terms of arbitrary volume flows (Q_1, Q_2, Q_3) and tank diameter d. Then, if the water level h is constant, determine the exit velocity v_2 for the given data $v_1 = 3m/s$ and $Q_3 = 0.01\ m^3/s$.

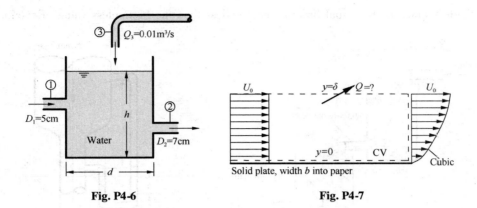

Fig. P4-6   Fig. P4-7

4-7 An incompressible fluid flows past an impermeable flat plate, as in Fig. P4-7, with a uniform inlet profile $u = U_0$ and a cubic polynomial exit profile

$$u \approx U_0\left(\frac{3\eta - \eta^3}{2}\right)$$

where $\eta = y/\delta$. Compute the volume flow rate Q across the top surface of the control volume.

4-8 The converging-diverging nozzle is used to expand and accelerate dry air to supersonic speeds at the exit. The diameter at the converging-diverging throat (the entrance of diverging) is $d_1 = 1$ cm, where $v_1 = 517$ m/s, $p_1 = 284$ kPa, $T_1 = 665$ K. The diameter at the exit of diverging is $d_2 = 2.5$ cm. For steady compressible flow of an ideal gas, estimate (a) the mass flow rate and (b) the velocity v_2 at the exit of diverging.

4-9 Water at 20°C flows steadily through the piping junction in Fig. P4-9 at a flow rate of 1.262 L/s at section 1. The average velocity at section 2 is 2.5 m/s. A portion of the flow is diverted through the showerhead, which contains 100 holes of 1-mm diameter. Assuming uniform shower flow, estimate the exit velocity from the showerhead jets.

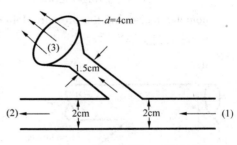

Fig. P4-9

4-10 A 1-m³ rigid tank initially contains air whose density is 1.18 kg/m³. The tank is connected to a high-pressure supply pipe through a valve. When the valve is opened, air is allowed to enter the tank until the density in the tank rises to 7.20 kg/m³. Determine the mass of air that has entered the tank.

Energy Equation

4-11 Water flows steadily through the large tanks shown in Fig. P4-11. Determine the water depth, h_A. Neglect viscous effects.

4-12 Oil ($S_G = 0.9$) flows downward through a vertical pipe contraction as shown in Fig. P4-12. If the mercury manometer reading is $h = 100$ mm, determine the volume flowrate for frictionless flow. Is the actual flowrate more or less than the frictionless value? Explain.

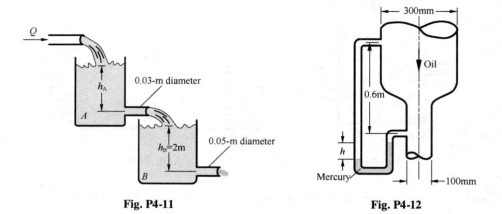

Fig. P4-11   Fig. P4-12

4-13 Water flowing in a pipe with a velocity of 25m/s and a static pressure of 940kPa is split into two branches as indicated in Fig. P4-13. If the measured static pressure at sections (2) and (3) and the velocity measured at section (3) are as shown in Fig. P4-13, determine (a) the velocity at section (2), and (b) the amount of available power lost in this horizontal Y-connection.

4-14 Air flows through a serial pipe at a rate of 200L/s. The serial pipe consists of two sections of diameters 20cm and 10cm with a smooth reducing section that connects them shown in Fig. P4-14. Neglecting frictional effects, determine the differential height in the water manometer measuring the pressure dufference between two pipe sections. Take the air density to be 1.2kg/m^3.

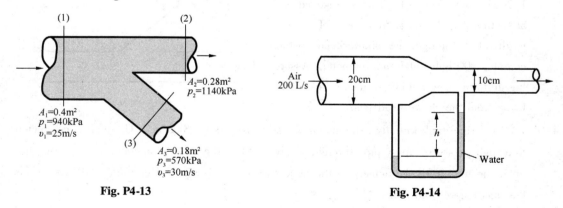

Fig. P4-13   Fig. P4-14

4-15 In Fig. P4-15 the gage pressure on the water surface in a closed tank is $p_0 = 4.9 \times 10^4 \text{N/m}^2$. A 0.05-m-diameter 4-m-long pipe is connected to the bottom of the tank with an angle of 30° with the horizontal plane. Let the entrance center of the pipe be 5m under the water surface in tank, the total head loss in the pipe be 2.3m. Determine the initiate flowrate in the pipe.

4-16 In Fig. P4-16 shows a simplified water supply system by the pressurized air in a closed tank. The water level difference between two tanks is 15m. Assume the irreversible head loss in the whole piping system is $20v^2/2g$ (where v is the average velocity in the pipe), and the required flowrate in the pipe is $0.012 \text{ m}^3/\text{s}$, determine the required air pressure in the closed tank, p_0.

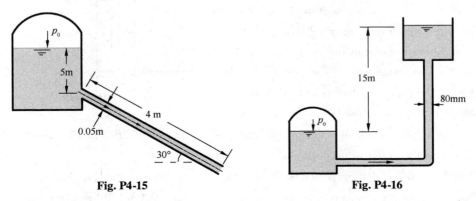

Fig. P4-15   Fig. P4-16

4-17 A large space is ventilated by drawing atmospheric air through a centrifugal ventilator fan located near the exit of the 200-mm-diameter pipe as in Fig. P4-17. A glass tube with two open ends is used to measure the air flow rate by connecting the pipe at one end and inserting another end into a water tank. When the air flow in pipe is steady, the raised water column height in the glass tube is 150mm. Let the air density be 1.29kg/m^3, determine the air flow rate drawn by the fan.

4-18 In Fig P4-18 water is pumped from a lower reservoir by a 75 percent efficient 13.3kW pump to a upper reservoir 20m higher tham the free surface of lower reservoir. The irreversible head loss in the piping system is $8v^2/2g$ (where v is the average velocity in the pipe). (a) Determine the flow rate in the pipe and the pump head h_P; (b) Sketch the energy grade line and hydraulic grade line of the whole system.

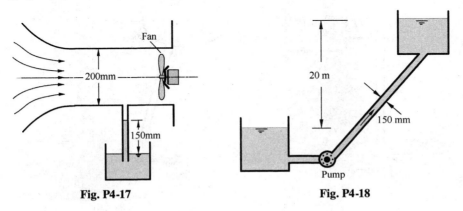

Fig. P4-17   Fig. P4-18

4-19 When the pump in Fig. P4-19 draws $220 \text{ m}^3/\text{h}$ of water at 20℃ from the reservoir, the irreversible head loss in the piping system is 5m. The flow discharges through a nozzle to the atmosphere. Determine (a) the pump head and (b) the pump power delivered to the water.

4-20 Underground water is to be pumped by a 70 percent efficient 3kW submerged pump to a pool whose free surface is 30m above the underground water level(Fig. P4-20). The diameter of the pipe is 7cm on the intake side and 5cm on the discharge side. The irreversible head loss of the piping system is 5m. Determine (a) the maximum flow rate of water and (b) the pressure difference across the pump. Assume the elevation difference between the pump inlet and the outlet to be negligible.

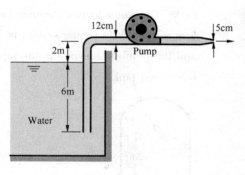

Fig. P4-19

4-21 The demand for electric power is usually much higher during the day than it is at night. Suppose a utility company is selling electric power for $0.03/kWh at night and is willing to pay $0.08/kWh for power produced during the day. To take advantage of this opportunity, an entrepreneur is considering building a large reservoir 40m above the lake level as in Fig. P4-21, pumping water from the lake to the reservoir at night using cheap power, and letting the water flow from the reservoir back to the lake during the day, producing power as the pump-motor operates as a turbine-generator during reverse flow. Preliminary analysis shows that a water flow rate of 2 m^3/s can be used in either direction, and the irreversible head loss of the piping system is 4m. The combined pump-motor and turbine-generator efficiencies are expected to be 75 percent each. Assuming the system operates for 10hr each in the pump and turbine modes during a typical day, determine (a) the pump power and the turbine power and (b) the potential revenue this pump-turbine system can generate per year.

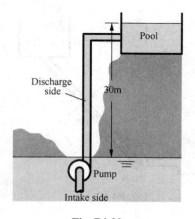

Fig. P4-20

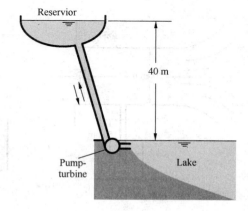

Fig. P4-21

4-22 The turbine shown in Fig. P4-22 develops 2500kW electricity when the water flowrate is 20 m³/s. The head loss across the turbine from (1) to (2) is negligible, but the irreversible head loss for the entire flow is 2.5m. (a) Determine the pressure difference, $p_1 - p_2$, across the turbine; (b) Determine the elevation h.

4-23 In Fig. P4-23 both the entrance and the exit of the ventilating tunnel system are opened to the atmospheric air. The diameters of tunnel prior to and after the fan are identical. The the velocity at the exit of the ventilating system is

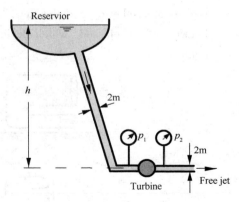

Fig. P4-22

40m/s. Let the air density be 1.29kg/m³, determine the pressures at the sections just prior to and after the fan, p_1 and p_2. Disregarding the head losses in the system.

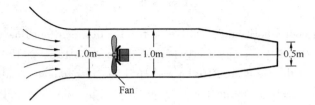

Fig. P4-23

Momentum Equation

4-24 In Fig. P4-24 the nozzle with a 50-mm-diameter exit is connected to a 150-mm-diameter pipe by the flange bolts. When the flow in the piping system is steady, the irreversible head loss in the constant-diameter pipe is $h_{w1,2} = 5v^2/2g$, where v is average velocity in the pipe, and in the nozzle is $h_{w2,3} = 0.05v_3^2/2g$, where v_3 is the average velocity at the nozzle exit. Determine (a) the jet flow rate; (b) the pressure at section 2-2; (c) the resultant forces on the flange bolts.

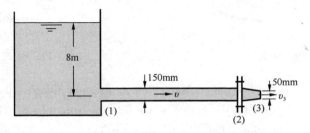

Fig. P4-24

4-25 The four devices shown in Fig. P4-25 rest on frictionless wheels, are restricted to move in the horizontal only and are initially held stationary. The pressure at the inlets and outlets of each is atmospheric, and the flow is incompressible. The contents of each device is not known. When released, which devices will move to the right and which to the left? Explain.

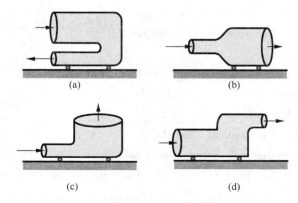

Fig. P4-25

4-26 Water flows through a horizontal, 180° pipe bend with a constant flow area of 9000 mm² as in Fig. P4-26. The flow velocity everywhere in the bend is 15m/s. The pressures at the entrance and exit of the bend are 210kPa and 165kPa, respectively. Calculate the horizontal (x and y) components of the anchoring force needed to hold the bend in place.

4-27 Determine the magnitude and direction of the anchoring force needed to hold the horizontal elbow and nozzle combination shown in Fig. P4-27 in place. Water exits to the atmosphere at section (2). Take the atmospheric pressure be 100kPa.

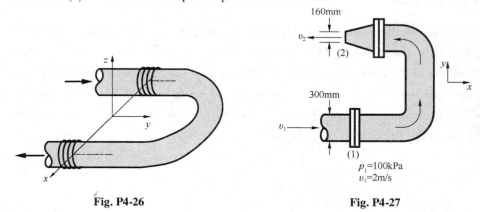

Fig. P4-26   Fig. P4-27

4-28 A reducing elbow is used to deflect water flow at a rate of 30kg/s upward by an angle 45° from the horizontal pipe and discharges water into the atmosphere. The sectional areas of the elbow and the elevation difference are shown in Fig. P4-28. The total mass of the elbow and the water in it is 50kg. Determine the anchoring force needed to hold the elbow in place. Take the momentum-flux correction factor to be 1.03.

4-29 In Fig. P4-29 water is flowing through a pipe with a reducing section from a diameter of 8 cm to 5cm. All fluids are at 20℃. The velocity of water at section 1 is 5m/s, the pressure at section 2 is 101 kPa. If the manometer reading is 58 cm, estimate the total force resisted by the flange bolts.

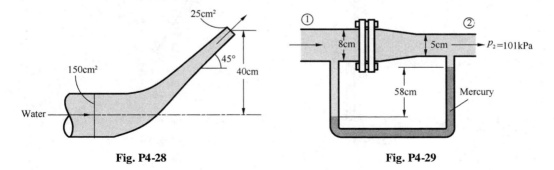

Fig. P4-28   Fig. P4-29

4-30 A 10-mm-diameter jet of water is deflected by a homogeneous rectangular block (15mm by 200mm by 100mm) that weighs 6N as in Fig. P4-30. Determine the minimum volume flowrate needed to tip the block.

4-31 Water accelerated by a nozzle to 15m/s strikes the vertical back surface of a cart moving horizontally at a constant velocity of 5m/s in the flow direction. The mass flow rate of water out of nozzle is 25kg/s. After the strike, the water stream splatters off in all directions in the plane of the back surface. (a) Determine the force that needs to be applied on the brakes of the cart to prevent it from accelerating. (b) If this force were used to generate power instead of wasting it on the brakes, determine the maximum amount of power that can be generated.

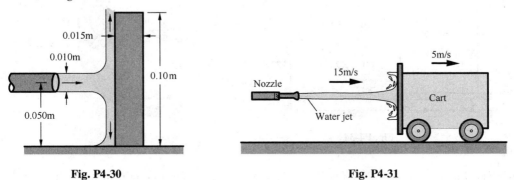

Fig. P4-30   Fig. P4-31

4-32 The 6-cm-diameter circular water jet at 20℃ in Fig. P4-32 strikes a plate containing a circular hole of 4-cm diameter. Part of the jet passes through the hole, and part is deflected. Determine the horizontal force required to hold the plate.

4-33 A vertical, circular cross-sectional jet of air strikes a conical deflector as indicated in Fig. P4-33. A vertical anchoring force of 0.1N is required to hold the deflector in place. Determine the mass (in kg) of the deflector. Assume the magnitude of velocity of the air remains constant.

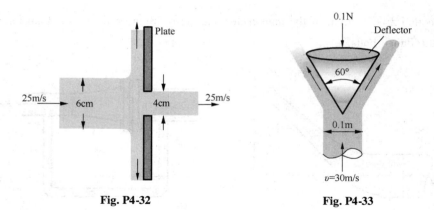

Fig. P4-32   Fig. P4-33

4-34 The 4-m-diameter water tank in Fig. P4-34 stands on a frictionless cart and feeds a jet of diameter 4 cm and velocity 8 m/s, which is deflected 60° by a vane. Compute the tension in the supporting cable.

4-35 A Pelton wheel vane directs a horizontal, 25-mm-diameter circular water jet symmetrically in a horizontal plane as indicated in Fig. P4-35. The flowrate of water jet is 3.34×10^{-3} m^3/s. Determine the x-direction component of anchoring force required to hold the vane stationary. The fluid speed magnitude remains constant along the vane surface. Disregarding gravity and friction.

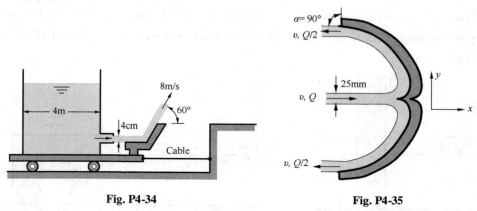

Fig. P4-34   Fig. P4-35

4-36 The water jet in Fig. P4-36 strikes at a speed of v and flowrate of Q to three different fixed plates and is symmetrically deflected in both sides at an angle of α. Disregarding gravity and friction. (a) Compare the force magnitudes required to hold the plates fixed; (b) Determine the value of angle α to make the holding force maximum; and (c) determine the ratio of maximum force to the force in Fig. 4-36b?

4-37 The water jet with a per unit width into the paper and a thickness of 50mm strikes an inclined fixed plate at a velocity of 18m/s and an angle of 30°, as in Fig. P4-37, and breaks into two jets along the plate surface. The length of plate in y-direction is 1.2m. Disregarding the frictions between the air, water and plate. (a) Determine the flow rates at the exit sections 1 and 2, Q_1 and Q_2; (b) Determine the thrust force on the plate by the jet; (c) If the plate moves at a velocity of 8m/s in the direction of jet, determine the thrust force on the plate by the jet, again.

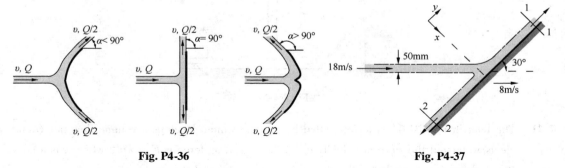

Fig. P4-36   Fig. P4-37

4-38 Consider a jet from a nozzle connected to a 180° bend pipe by four flange bolts at location A as in Fig. P4-38. The bolts were arranged around the flange with a uniform distance from the clock locations 1:30, 4:30, 7:30 to 10:30. The vertical distance between the centers of the upper and lower bolts is 175mm. The diameters of the bend pipe and the nozzle exit are 100mm and 25mm, respectively. The reading of pressure gage M prior to the flange is 196.5kPa. Assume the total weight of bend pipe, nozzle and water inside is 150N, determine the forces on the upper flange bolts and lower flange bolts.

4-39 Water is flowing into and discharging from a pipe U section as shown in Fig. P4-39. At flange (1), the total absolute pressure is 200kPa, and 30kg/s water flows into the pipe. At flange (2), the total absolute pressure is 150kPa. At location (3), 8kg/s of water discharges to the atmosphere, which is at 100kPa. Determine the total x- and z-forces at the two flanges connecting the pipe. Discuss the significance of gravity force for this problem. Take the momentum-flux correction factor to be 1.03.

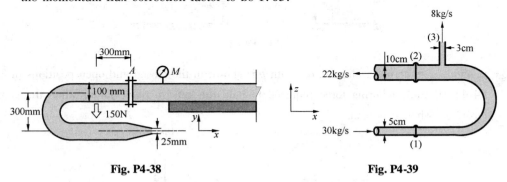

Fig. P4-38   Fig. P4-39

4-40 The sluice gate in Fig. P4-40 can control and measure flow in open channels. At sections 1 and 2, the flow is uniform and the pressure is hydrostatic. The channel width is b into the paper. Neglecting bottom friction, derive an expression for the force F required to hold the gate. For what condition h_2/h_1 is the force largest? For very low velocity $v_1^2 \ll gh_1$, for what value of h_2/h_1 will the force be one-half of the maximum?

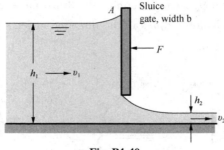

Fig. P4-40

4-41 The boat in Fig. P4-41 is jet-propelled by a pump which develops a volume flow rate Q and ejects water out the stern at velocity v_j. If the boat drag force is $F = kv^2$, where k is a constant, develop a formula for the steady forward speed v of the boat.

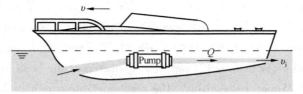

Fig. P4-41

4-42 The cross section of a board-crested weir in a 2-m-wide (into the paper) rectangular channel with a horizontal bed is shown in Fig. P4-42. Water flows over the weir at a flowrate of 5.2 m³/s. Determine the horizontal thrust force exerting on the weir.

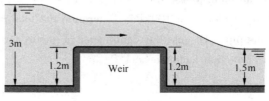

Fig. P4-42

4-43 A sluice gate across a channel of width b is shown in the closed and open positions in Fig. P4-43. Is the anchoring force required to hold the gate in place larger when the gate is closed or when it is open?

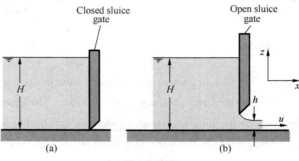

Fig. P4-43

5 Similitude, Dimensional Analysis, and Modeling

CHAPTER OPENING PHOTO: When finished, Baihetan hydropower station will be the second largest hydroelectric power plant after the Three Gorges Dam, and its dam be the third largest dam in China. Due to the complex design and the expensive cost, the comprehensive model of the Baihetan station has been constructed, tested and analyzed by using the principles of dimensional analysis and modeling from the present chapter. (ⓒ 杨潘, 王浩宇. 中国水利网站[1])

Many important engineering problems are too complex, both geometrically and physically, to be solved completely by theoretical or mathematical methods. They must be tested by experiment, and interpreted by experimental data in *dimensionless* form. Experiments which might result in tables of output, or even multiple volumes of tables, might be reduced to a single set of curves, or even a single curve when suitably nondimensionalized. The technique for doing this is *dimensional analysis*.

In this chapter, we begin with the concepts of dimensions and units and the principle of dimensional homogeneity, which indicates that equation must be dimensionally homogeneous if a theoretical equation does exist among the variables affecting a physical process. Then we present two step-by-step methods of dimensional analysis, *Rayleigh's method* and *Buckingham pi theorem*, to group the dimensional variables into a smaller number of dimensionless groups of variables. We also discuss the concept of similarity between a model and a prototype, and some commonly used

---

[1] http://www.chinawater.com.cn/ztgz/xwzt/2012sltj/5/201201/t20120120_211790.html

dimensionless groups. Finally, we apply this technique to several practical problems to illustrate both its utilities and its limitations.

5.1 Introduction

5.1.1 Dimensions and Units

In fluid mechanics, any flow situation can be described by certain physical variables, e. g. length, velocity, area, volume, acceleration etc. All of those variables can be categorized by the *dimension* and measured in the term of *unit*. A *dimension* indicates the property and the category of the physical variables, it is qualitative. A *unit* is the standard for measuring the magnitudes of the physical variables, it is quantitative. Thus length is a dimension associated with such variables as distance, displacement, width, and height, while meters are the numerical unit for expressing length. Therefore, dimensions are properties which can be measured. Units are the standard elements we use to quantify these dimensions. For summary, a *dimension* is the measure by which a physical variable is expressed qualitatively, that is, to identify the nature, or type, of the physical variable (such as length, time, stress, and velocity); a *unit* is the magnitude assigned to the dimension, which provides a numerical measure of the physical variable.

5.1.2 Basic and Derived Dimensions

Some basic quantities, such as mass, length, time, and temperature, are selected as *primary quantities*, and used to provide a qualitative description of any other *secondary quantities*. In fluid mechanics, the dimensions of those four primary quantities, mass, length, time, and temperature, are also referred to as *basic dimensions*, denoted as M, L, T and Θ, from which all other dimensions can be derived. All other dimensions derived as combinations of these four primary dimensions are *secondary dimensions*, or *derived dimensions*. For example, the dimension of velocity $[u]$ is LT^{-1}, where the braces around the variable 'u' mean 'the dimension' of velocity u.

For a wide variety of problems involving fluid mechanics, the four basic dimensions are usually taken to be mass M, length L, time T, and temperature Θ, or a $MLT\Theta$ system for short. Alternatively, $FLT\Theta$ system could be used, with force replacing mass. The common dimensions in fluid mechanics in both systems of $MLT\Theta$ and $FLT\Theta$ are given in Table 5-1.

The dimensions for all physical variables can be described as a combination of the primary dimensions: M, L and T

$$[q] = M^\alpha L^\beta T^\gamma \tag{5.1}$$

where q represents the physical variable; α, β, γ are the exponents of the primary dimensions of M, L and T, respectively.

Equation 5.1 indicates that the properties of the physical variable q are determined by the exponents α, β, γ:

Geometrical variables (i. e., area, volume) $\qquad \alpha = 0, \beta \neq 0, \gamma = 0$

Table 5-1 Dimensions of fluid mechanics variables

Quantity	Symbol	Dimension MLTΘ	Dimension FLTΘ	Quantity	Symbol	Dimension MLTΘ	Dimension FLTΘ
Length*	L	L	L	Angular velocity	ω	T^{-1}	T^{-1}
Area	A	L^2	L^2	Viscosity	μ	$ML^{-1}T^{-1}$	FTL^{-2}
Volume	V	L^3	L^3	Kinematic viscosity	ν	L^2T^{-1}	L^2T^{-1}
Time*	t	T	T	Surface tension	σ	MT^{-2}	FL^{-1}
Velocity	U, u, v	LT^{-1}	LT^{-1}	Force	F	MLT^{-2}	F
Acceleration	a	LT^{-2}	LT^{-2}	Moment, torque	M	ML^2T^{-2}	FL
Speed of sound	c	LT^{-1}	LT^{-1}	Power	$\dot{N}$	ML^2T^{-3}	FLT^{-1}
Volume flowrate	Q	L^3T^{-1}	L^3T^{-1}	Energy	E	ML^2T^{-2}	FL
Mass*	m	M	$FL^{-1}T^2$	Density	ρ	ML^{-3}	FT^2L^{-4}
Mass flowrate	$\dot{m}$	MT^{-1}	$FL^{-1}T$	Temperature*	T	Θ	Θ
Pressure, stress	p, σ	$ML^{-1}T^{-2}$	FL^{-2}	Specific heat	c_v	$L^2T^{-2}\Theta^{-1}$	$L^2T^{-2}\Theta^{-1}$
Angle	θ	None	None	Specific weight	γ	$ML^{-2}T^{-2}$	FL^{-3}

Note: * indicates those primary quantities are referred to as basic dimensions.

Kinematic variables (i.e., velocity, acceleration) $\alpha = 0,\ \beta \neq 0,\ \gamma \neq 0$

Dynamics variables (i.e., force, pressure) $\alpha \neq 0,\ \beta \neq 0,\ \gamma \neq 0$

If $\alpha = \beta = \gamma = 0$, or $[q] = 1$, this variable should be a *pure number*, or *dimensionless group*, *dimensionless produce*, which arise from mathematical manipulations and have no dimensions and never did. There are some physical variables that are naturally dimensionless by virtue of their definition as ratios of dimensional quantities, such as specific gravity (ratio of density to standard water density), or a dimensionless product of a combination of few variables, i.e.

$$\mathrm{Re} = \frac{vD}{\nu} \sim \frac{(LT^{-1})L}{L^2T^{-1}} = 1$$

All angles are dimensionless (ratio of arc length to radius), however, they should be taken in radians.

The pure number is independent of the scaling effects between two flows in different scales or between model and prototype. It indicates that the same quantities with $[q] = 1$ in two flows should be equal if those two flows are similitude.

5.1.3 Dimensional Homogeneity

The *principle of dimensional homogeneity*, a fundamental for dimensional analysis, can be stated as follows: If an equation truly expresses a proper relationship between variables in a physical process, it will be *dimensionally homogeneous*; i.e., each of its additive terms will have the same dimensions.

All the equations which are derived from the theory of mechanics are of this form. For example, the Bernoulli's equation for incompressible flow

$$z + \frac{p}{\rho g} + \frac{u^2}{2g} = \text{Constant}$$

Each term, including the constant, has dimensions of length, or L. Thus, the equation is dimensionally homogeneous and gives proper results for any consistent set of units.

5.2 Dimensional Analysis

Basically, *dimensional analysis* is a method for obtaining the dependent variables expressed by the combination of other variables or reducing a number of dimensional variables into a smaller number of dimensionless groups. The former method is the *Rayleigh's method*, which is directly based on the fundamental principle of dimensional homogeneity of physical variables involved in a problem; the latter is the *Buckingham pi theorem*, proposed in 1914 by Edgar Buckingham (1867-1940), which reduces the number and complexity of experimental variables which affect a given physical phenomenon, by using a sort of compacting technique. If a phenomenon depends upon n dimensional variables, *Buckingham pi theorem* will reduce the problem to only k dimensionless variables. Then, express the dependent π *term* (or *pi term*, the dimensionless group), which includes the dependent variable, as the function of other dimensionless π term.

For example, the pressure loss, Δp, from friction in a long, round, straight, smooth pipe depends on all these variables: the length l and diameter of the pipe D, the flow rate Q, and the density ρ and viscosity μ of the liquid. That is, $\Delta p = f(U, D, \rho, \mu, l)$. If any one of these variables is changed, the pressure drop also changes. The empirical method of obtaining an equation relating these factors to pressure drop requires that the effect of each separate variable be determined in turn by systematically varying that variable (i. e. U) while keep all others (i. e. D, ρ, μ, l) constant. The procedure is laborious, and is difficult to organize or correlate the results so obtained into a useful relationship for calculations. Thus, the motive behind dimensional analysis is that any dimensionally homogeneous equation can be written in an entirely equivalent nondimensional form which is more compact.

5.2.1 Rayleigh's method

Rayleigh's method is a method directly based on the fundamental principle of dimensional homogeneity, for which the number of physical variables involved in a problem should not exceed 5.

Consider a physical process concerned with physical variables $x_1, x_2, \ldots, x_n$, then we can express the nondimensional parameters of the dependent variable x_i by combining other variables as follows.

Step 1: Identify the dependent variable and express it as a product of all the independent variables raised to the unknown exponents

$$x_i = K x_1^a x_2^b \ldots x_{n-1}^t \tag{5.2}$$

where K is a dimensionless constant, and $a, b, \ldots, t$ are exponents, whose values are not yet known.

Step 2: Rewrite the equations in terms of dimensions

$$[x_i] = [x_1]^a [x_2]^b \ldots [x_{n-1}]^t \tag{5.3}$$

Step 3: Equating the indices of $n - 1$ fundamental dimensions of the variables involved, and $n - 1$ independent equations are obtained.

Step 4: Solve these $n - 1$ equations to obtain the exponents of $a, b, \ldots, t$, then express the dependent variable x_i by the combination of other variables.

EXAMPLE 5-1 Force on an immersed body

Suppose the force F on a particular body immersed in a stream of fluid depends only on the body length l, stream velocity U, fluid density ρ, and fluid viscosity μ, that is,

$$F = f(l, U, \rho, \mu)$$

where f is some function. Determine the expression of the force by the other variables.

Solution:

Step 1: With the Rayleigh's method, we assume that

$$F = K l^a \rho^b \mu^c U^d$$

Step 2: Note that the dimensions of the left side, force, must equal those on the right side. Here, we use only the three independent dimensions for the variables on the right side: M, L, and T. Rewriting the above equation in term of primary dimensions yields

$$MLT^{-2} = L^a (ML^{-3})^b (ML^{-1}T^{-1})^c (LT^{-1})^d$$

Step 3: Based on the principle of dimensional homogeneity, the exponents of M, L, and T in the right must be same as the one in the left. Thus equating the exponents to each other in terms of their respective fundamental dimensions gives

$$M: \quad 1 = b + c$$
$$L: \quad 1 = a - 3b - c + d$$
$$T: \quad -2 = -c - d$$

Step 4: It is seen that there are three equations, but four unknown variables. This means that a complete solution cannot be obtained. Thus, we choose to solve a, b, and d in terms of c. These choices are based on experience. Therefore, we have

$$a = 2 - c, \quad b = 1 - c, \quad d = 2 - c$$

Substitution of the exponents back into the original equation yields

$$F = K l^{2-c} \rho^{1-c} \mu^c U^{2-c} = K \rho l^2 U^2 \left(\frac{\rho U l}{\mu}\right)^{-c}$$

or

$$\frac{F}{\rho l^2 U^2} = g\left(\frac{\mu}{\rho U l}\right) \tag{1}$$

where $g(\ )$ is the function different mathematically from the original function $f(\ )$, but contains all the same information.

DISCUSSION Equation 1 indicates that the dimensionless *force coefficient* $F/(\rho l^2 U^2)$ is a function only of the dimensionless *Reynolds number*, $\rho U l/\mu$, which will be discussed in Section 5.4.2. And the function $g(\ )$ could be determined by the experiments.

EXAMPLE 5-2   Pressure drop in a long straight circular pipe

The pressure drop per unit length that develops along a pipe as the result of friction cannot be explained analytically without the use of experimental data. Determine the expression of the pressure drop per unit length on the variables: pipe diameter D, flow velocity U, fluid density ρ, and fluid viscosity μ.

Solution:

Step 1: Write the dependent variable, the pressure drop per unit length, Δp_l as

$$\Delta p_l = K U^a D^b \rho^c \mu^d$$

Step 2: Writing the dimensions of each variable in the above equation in *MLT* system yields

$$ML^{-2}T^{-2} = (LT^{-1})^a (L)^b (ML^{-3})^c (ML^{-1}T^{-1})^d$$

Step 3: Based on the principle of dimensional homogeneity, equating the exponents of M, L, and T on both sides gives

$$M: \quad 1 = c + d$$
$$L: \quad -2 = a + b - 3c - d$$
$$T: \quad -2 = -a - d$$

Step 4: There are three equations and four unknowns. Solving these equations in terms of the unknown d, we have

$$a = 2 - d, \quad b = -d - 1, \quad c = 1 - d$$

Thus, substitution of the exponents back into the original equation yields

$$\Delta p_l = K U^{2-d} D^{-d-1} \rho^{1-d} \mu^d = K \left(\frac{\rho U^2}{D}\right)\left(\frac{\rho U D}{\mu}\right)^{-d}$$

or

$$\frac{D \Delta p_l}{\rho U^2} = K \left(\frac{\rho U D}{\mu}\right)^{-d}$$

DISCUSSION   From the above examples we know, the numerical values of K and d can never be known from dimensional analysis. They have to be found out from experiments. It indicates that not all exponents of variables can be exactly determined if the number of variables is greater than 4, Therefore, for the Rayleigh's method, the maximum number of variables should not exceed 5.

5.2.2   Buckingham Pi Theorem

The other scheme for the dimensional analysis is called the *Buckingham pi theorem*, can be stated as follows: If a physical process satisfies the principle of dimensional homogeneity and involves n dimensional variables, it can be reduced to a relationship among $k = n - r$ independent dimensionless products, or 'π terms', where r is the minimum number of reference dimensions required to describe the variables.

To be specific, suppose that the physical process involves n variables, $x_1, x_2, ..., x_n$, thus the steps to perform a dimensional analysis using the method of Buckingham pi theorem are as follows:

Step 1: List all the variables that are involved in the problem.

$$f(x_1, x_2 ... x_n) = 0 \tag{5.4}$$

In this step all pertinent variables in the problem, including dimensional and nondimensional constants, should be included in the list of 'variables' to be considered for the dimensional analysis. Typically the variables will include those that are necessary to describe the *geometry* of the system (such as a pipe diameter), to define any *fluid properties* (such as a fluid viscosity), and to indicate *external effects* that influence the system (such as a driving pressure drop per unit length).

Step 2: List the dimensions of each variable in terms of basic dimensions according to $MLT\Theta$ or $FLT\Theta$ system given in Table 5-1, and determine the number of *reference dimensions*, r, and the required number of π terms, $k = n - r$. The number of reference dimensions usually corresponds to the number of the basic dimensions shown in the dimensions describing the variables involved, $r \leq 3$.

Step 3: Select *repeating variables*, of which the number required is equal to the number of reference dimensions r (usually the same as the number of basic dimensions) and is usually to be 3. Usually, pipe diameter, flow velocity and fluid density for the pipe flow, and water depth, flow velocity and fluid density for the channel flow can be selected as the repeating variables.

Step 4: Form each π term by multiplying one of the *nonrepeating variables* by the product of repeating variables each raised to an exponent that will make the combination dimensionless

$$\pi_i = x_i \cdot x_1^{a_1} \ldots x_r^{a_r} \quad (i = r + 1, \ldots, n) \tag{5.5}$$

where x_i is the nonrepeating variables; $x_1, \ldots, x_r$ are the repeating variables; $a_1, \ldots, a_r$ are the corresponding exponents of the repeating variables.

Algebraically find the exponents which make each π dimensionles sproduct according to the *principle of dimensional homogeneity*, then substitute back into Eq. 5.5 to get the expression of π_i.

Step 5: Repeat Step 4 for each of the remaining nonrepeating variables. The resulting set of π *terms* will correspond to the required number obtained from Step 2. If not, you should have made a mistake!

Step 6: Check all the resulting π terms to make sure they are dimensionless. Express the final form as a relationship among the π terms

$$f(\pi_1, \pi_2 \ldots \pi_{n-r}) = 0$$

or as a dependent π term expressed by other π terms

$$\pi_1 = g(\pi_2, \pi_3 \ldots \pi_{n-r})$$

where π_1 would contain the dependent variable in the numerator, $g()$ is the function only can be determined by the experiments.

EXAMPLE 5-3 Force on an immersed body

Repeat the development of Eq. 1 in Example 5-1 by using the *Buckingham pi theorem*.

Solution:

Step 1: Write the function and count variables:

$$f(l, U, \rho, \mu, F) = 0$$

There are five variables, $n = 5$.

Step 2: List dimensions of each variable and find the number of the reference dimensions.

From Table 5-1, we have $[F] = MLT^{-2}$, $[l] = L$, $[U] = LT^{-1}$, $[\mu] = ML^{-1}T^{-1}$, $[\rho] = ML^{-3}$. And the number of the basic dimensions equals to 3 (MLT) by inspecting the dimension combination of the variables. So is the number of the reference dimensions, $r = 3$. Thus the number of π terms is 2 ($k = n - r = 5 - 3 = 2$), that is, there are two independent dimensionless groups in this problem.

Step 3: Select a number of repeating variables. Select l, U, ρ as the repeating variables, which includes the basic dimensions M, L, T.

Step 4: Combine l, U, ρ with one additional nonrepeating variable, in sequence, to find the two π products. First add force to form π_1.

$$\pi_1 = l^{a_1} U^{b_1} \rho^{c_1} F$$

The dimensions of the above equation are

$$M^0 L^0 T^0 = (L)^{a_1}(LT^{-1})^{b_1}(ML^{-3})^{c_1}(MLT^{-2})$$

Equate exponents according to the *principle of dimensional homogeneity* to give

M: $0 = c_1 + 1$

L: $0 = a_1 + b_1 - 3c_1 + 1$

T: $0 = -b_1 - 2$

We can solve explicitly for $a_1 = -2$, $b_1 = -2$ and $c_1 = -1$. Therefore

$$\pi_1 = l^{-2} U^{-2} \rho^{-1} F = \frac{F}{\rho l^2 U^2}$$

Then, add viscosity to the combination of L, U, and ρ to find π_2:

$$\pi_2 = l^{a_2} U^{b_2} \rho^{c_2} \mu$$

The dimensions of the above equation are

$$M^0 L^0 T^0 = (L)^{a_2}(LT^{-1})^{b_2}(ML^{-3})^{c_2}(ML^{-1}T^{-1})$$

Equate exponents according to the *principle of dimensional homogeneity* to give

M: $0 = c_2 + 1$

L: $0 = a_2 + b_2 - 3c_2 - 1$

T: $0 = -b_2 - 1$

We can solve explicitly for $a_2 = b_2 = c_2 = -1$. Therefore

$$\pi_2 = l^{-1} U^{-1} \rho^{-1} \mu = \frac{\mu}{\rho U l} = \frac{1}{Re}$$

Step 5: Express the final form as a relation among the π terms

$$\frac{F}{\rho l^2 U^2} = g\left(\frac{\mu}{\rho U l}\right)$$

which is exactly Eq. 1 in Example 5-1.

DISCUSSION We can see that the dimensional analysis determines only the relevant independent dimensionless parameters of a problem, but not the exact relation between them.

EXAMPLE 5-4 Pressure drop in a pipe flow

Determine the expression of pressure drop, Δp, between two cross sections in a pipe flow on the relevant variables: pipe diameter D, uniform flow velocity U, fluid density ρ, fluid viscosity μ, the distance between two sections l and the roughness height of pipe wall Δ.

Solution:

In a pipe flow, the pressure drop, Δp, between two cross sections may depend on the following variables: D,

U, ρ, μ, l and Δ.

Step 1: Write the function and count variables:
$$f(U, D, \rho, \mu, l, \Delta, \Delta p) = 0$$
There are seven variables ($n = 7$).

Step 2: List dimensions of each variable and find the number of the reference dimensions.

From Table 5-1, we have $[\Delta p] = ML^{-1}T^{-2}$, $[l] = L$, $[U] = LT^{-1}$, $[\mu] = ML^{-1}L^{-1}$, $[\rho] = ML^{-3}$, $[\Delta] = L$ and $[D] = L$. Thus, the number of the basic dimensions equals to 3 (MLT), that is $r = 3$. The number of π terms is 4 ($k = n - r = 7 - 3 = 4$), namely there are four independent dimensionless groups in this problem.

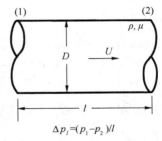

$\Delta p_l = (p_1 - p_2)/l$

Fig. E5-4

Step 3: Select a number of repeating variables. Select D, U, ρ as the repeating variables, which includes the basic dimensions MLT.

Step 4: Combine D, U, ρ with one additional nonrepeating variable, in sequence, to find the four π products:
$$\pi_1 = D^{a_1}U^{b_1}\rho^{c_1}\Delta p, \quad \pi_2 = D^{a_2}U^{b_2}\rho^{c_2}l, \quad \pi_3 = D^{a_3}U^{b_3}\rho^{c_3}\Delta, \quad \pi_4 = D^{a_4}U^{b_4}\rho^{c_4}\mu$$

For π_1:

The dimensions on both sides of π_1 are
$$M^0L^0T^0 = (L)^{a_1}(LT^{-1})^{b_1}(ML^{-3})^{c_1}(ML^{-1}T^{-2})$$
Equate exponents according to the principle of dimensional homogeneity to give

M: $0 = c_1 + 1$

L: $0 = a_1 + b_1 - 3c_1 - 1$

T: $0 = -b_1 - 2$

so $a_1 = 0$, $b_1 = -2$, $c_1 = -1$. Therefore
$$\pi_1 = \frac{\Delta p}{\rho U^2}$$

With the similar manner, we can obtain other three π terms:
$$\pi_2 = \frac{l}{D}, \quad \pi_3 = \frac{\Delta}{D}, \quad \pi_4 = \frac{\mu}{\rho DU} = \frac{1}{Re}$$

Step 5: Express the final form as a relation among the π terms
$$f\left(\frac{\Delta p}{\rho U^2}, \frac{l}{D}, \frac{\Delta}{D}, \frac{1}{Re}\right) = 0$$
or
$$\frac{\Delta p}{\rho U^2} = f_1\left(\frac{l}{D}, \frac{\Delta}{D}, Re\right)$$

In the final form $\frac{1}{Re}$ has been replaced by its reciprocal Re without changing the relation between the pi terms because Re is a dimensionless group.

The pressure drop in pipe is proportional to the pipe length, so l/D can be moved out of the bracket:
$$\Delta p = f_2\left(Re, \frac{\Delta}{D}\right)\frac{l}{D}\rho U^2$$

DISCUSSION Take the answer divided by ρg
$$\Delta h = \frac{\Delta p}{\rho g} = f_3\left(Re, \frac{\Delta}{D}\right)\frac{l}{D}\frac{U^2}{2g}$$
or
$$\Delta h = \lambda \frac{l}{D}\frac{U^2}{2g} \tag{1}$$

where
$$\lambda = f_3\left(\text{Re}, \frac{\Delta}{D}\right)$$

Equation 1 is the *Darcy-Weisbach equation* for determining the frictional head loss in a pipe flow, in which λ is the *frictional loss factor* and is the function of Reynolds number, Re, and the relative roughness of pipe wall, Δ/D.

SOME ADDITIONAL COMMENTS ABOUT DIMENSIONAL ANALYSIS

One of the most important, and difficult, steps in applying dimensional analysis to any given problem is the selection of the variables that are involved. It is vitally important that all pertinent variables be included in the list of 'variables' to be considered for the dimensional analysis. Otherwise the dimensional analysis will not be correct! Generally, it relies on a good understanding of the phenomenon involved and the governing physical laws by the experimenter. It is, therefore, imperative that sufficient time and attention be given to the first step in which the variables are determined.

It should be emphasized the relationship in terms of the π terms only can be used to describe the problem, and be clearly noted that this is as far as the experiments can go with the dimensional analysis. The actual functional relationship among the π terms must be determined by experiment.

5.3 Similitude

5.3.1 Similarity of Model and Prototype

In engineering practice, many engineering projects involve structures, aircraft, ships, rivers, harbors, dams, air and water pollution, and so on. To save time and money, frequently involve the tests performed on a geometrically scaled model, rather than on the full-scale prototype. In this case, the physical system for which the predictions are to be made is called the *prototype*, whereas the representation of a physical system used to predict the behavior of the system in some desired respect is called the *model*. Models that resemble the prototype but are generally of a different size, may involve different fluids, and often operate under different conditions (pressures, velocities, etc.). Usually a model is smaller than the prototype. Therefore, it is more easily handled in the laboratory and less expensive to construct and operate than a large prototype. Occasionally, if the prototype is very small, it may be advantageous to have a model that is larger than the prototype so that it can be more easily studied. With the successful development of a valid model, it is possible to predict the behavior of the prototype under a certain set of conditions. Therefore, it is imperative to achieve complete similarity between the model tested and the prototype to be designed.

In a general sense, *similitude* is the indication of a known relationship between a model and prototype, i.e., model tests must yield data that can be scaled to obtain the similar parameters for the prototype. A formal statement would be as follows:

Flow conditions for a model test are *completely similar* if all relevant dimensionless parameters (π terms) have the same corresponding values for the model and the prototype, that is, $\pi_{mi} = \pi_{pi}$ ($i = 1, 2, ..., k$), in which subscript 'm' and 'p' represents model and prototype respectively, k is the total number of the dimensionless parameters.

Instead of complete similarity, the engineering literature speaks of particular types of similarity, the most common being geometric, kinematic, dynamic, and thermal. In civil, coastal and hydraulic engineering, the thermal similarity is usually omitted.

5.3.2 Geometric Similarity

Geometric similarity represents a similarity of the shape of the model and the prototype. It requires that the model and the prototype be of the same shape and that *all* the linear dimensions of the model be related to corresponding dimensions of the prototype by a constant scale factor. In other words: A model and prototype are *geometrically similar* if and only if all body dimensions in all three coordinates have the same linear-scale ratio, all angles and all flow directions are preserved, that is, the orientations of model and prototype with respect to the surroundings must be identical:

$$\left.\begin{array}{l} \dfrac{l_{p1}}{l_{m1}} = \dfrac{l_{p2}}{l_{m2}} = \cdots = \dfrac{l_p}{l_m} = \lambda_1 \\ \theta_{pi} = \theta_{mi} (i = 1, 2, ..., n) \end{array}\right\} \qquad (5.6)$$

where λ_1 is the *length scale ratio*, n is the total number of all angles and all flow directions in prototype and model.

Length scales are often specified, for example, as 10:1 scale model. The meaning of this specification is that the model is one-tenth the size of the prototype, and the tacit assumption is that all relevant lengths are scaled accordingly so the model is geometrically similar to the prototype.

For area and volume, we have

Area scale ratio $\qquad\qquad\qquad \lambda_A = \dfrac{A_p}{A_m} \doteq \dfrac{l_p^2}{l_m^2} = \lambda_1^2 \qquad\qquad (5.7)$

Volume scale ratio $\qquad\qquad \lambda_V = \dfrac{V_p}{V_m} \doteq \dfrac{l_p^3}{l_m^3} = \lambda_1^3 \qquad\qquad (5.8)$

5.3.3 Kinematic Similarity

Kinematic similarity requires that the model and prototype have the same length-scale ratio and the same time-scale ratio, with the same orientations of the corresponding velocities. In a specific sense, the velocities at corresponding points are in the same direction and are related in magnitude by a constant scale factor:

$$\lambda_u = \dfrac{u_p}{u_m} = \dfrac{v_p}{v_m} \qquad\qquad (5.9)$$

where λ_u is the *velocity scale ratio*, for which the point velocity ratio equals to the ratio of averaged velocity.

Substitution of $u = l/t$ into the above equation yields

$$\lambda_u = \frac{l_p/t_p}{l_m/t_m} = \frac{\lambda_l}{\lambda_t} \tag{5.10}$$

where λ_t is the *time scale ratio*.

Equation 5.10 indicates that the kinematic similarity requires fixed ratio of the length-scale ratio to the time-scale ratio between the prototype and the model. Length-scale equivalence simply implies geometric similarity, which can guarantee the streamline patterns be related by a constant scale factor, but time-scale equivalence may require additional dynamic considerations such as equivalence of the Reynolds and Mach numbers. If viscosity, surface tension, or compressibility is important, kinematic similarity is dependent upon the achievement of dynamic similarity.

Some other useful ratios are

Acceleration scale ratio
$$\lambda_a = \frac{a_p}{a_m} \doteq \frac{u_p/t_p}{u_m/t_m} = \frac{\lambda_l}{\lambda_t^2} \tag{5.11}$$

Flow rate scale ratio
$$\lambda_Q = \frac{Q_p}{Q_m} \doteq \frac{l_p^3/t_p}{l_m^3/t_m} = \frac{\lambda_l^3}{\lambda_t} \tag{5.12}$$

5.3.4 Dynamic Similarity

When two flows have force distributions such that identical types of forces are parallel and are related in magnitude by a constant scale factor at all corresponding points, then the flows are *dynamic similar*. For the model and the prototype, the dynamic similarity exists, when both of them have same length-scale ratio, time-scale ratio and *force-scale* ratio (or *mass-scale* ratio). That is, dynamic similarity exists, simultaneous with kinematic similarity, if the model and prototype forces and *pressure coefficients* are identical. This is ensured if:

1. For compressible flow, the model and prototype Reynolds number and Mach number and *specific-heat ratio* are correspondingly equal.

2. For incompressible flow

 a. With no free surface: Reynolds numbers in model and prototype are equal.

 b. With a free surface: Reynolds number, Froude number, and (if necessary) Weber number and cavitation number in model and prototype are correspondingly equal.

The major forces influencing the flow in the fluid mechanics are the pressure force, gravity force, and the inertia force. The dynamic similarity laws listed above ensure that each of these forces will be in the same ratio and have equivalent directions between model and prototype. This *force scale ratio* is

$$\lambda_F = \frac{m_p a_p}{m_m a_m} \doteq \frac{\rho_p l_p^3 \lambda_l}{\rho_m l_m^3 \lambda_t^2} = \lambda_\rho \lambda_l^2 \left(\frac{\lambda_l}{\lambda_t}\right)^2 = \lambda_\rho \lambda_l^2 \lambda_u^2 \tag{5.13}$$

For summary, the geometric similarity is the premise and the basis of the kinematic and the dynamic similarities; the dynamic similarity is the governing factor which determines the similarities of the two fluid flows; and the kinematic similarity is the expression of the geometric and dynamic similarities. In order to have complete similarity between the model and prototype, all the similarity flow conditions must be maintained. This will automatically follow if all the important variables are included in the dimensional analysis and if all the similarity requirements based on the resulting π terms are satisfied.

5.3.5 Similarities of Initial and Boundary Conditions

Besides geometric, kinematic and dynamic similarity, the similarity of boundary conditions should be satisfied. It also includes geometric, kinematic and dynamic similarity. For example, the normal velocity on the solid boundary should be zero and the pressure on the free surface should be equal to the atmosphere pressure.

For the unsteady flow, the similarity of initial conditions between model and prototype is required.

5.4 Common Dimensionless Groups in Fluid Mechanics

5.4.1 Similitude Based on Governing Differential Equations

The most commonly used dimensionless groups in fluid mechanics can be directly developed by normalizing the differential equations governing the phenomenon of interest.

Consider a flow of an incompressible Newtonian fluid, the continuity and momentum equations with constant viscosity give

Continuity equation

$$\frac{\partial u_x}{\partial x} + \frac{\partial u_y}{\partial y} + \frac{\partial u_z}{\partial z} = 0 \tag{5.14}$$

Navier-Stokes equations

$$f_x - \frac{1}{\rho}\frac{\partial p}{\partial x} + \nu\nabla^2 u_x = \frac{du_x}{dt} = \frac{\partial u_x}{\partial t} + u_x\frac{\partial u_x}{\partial x} + u_y\frac{\partial u_x}{\partial y} + u_z\frac{\partial u_x}{\partial z} \tag{5.15a}$$

$$f_y - \frac{1}{\rho}\frac{\partial p}{\partial y} + \nu\nabla^2 u_y = \frac{du_y}{dt} = \frac{\partial u_y}{\partial t} + u_x\frac{\partial u_y}{\partial x} + u_y\frac{\partial u_y}{\partial y} + u_z\frac{\partial u_y}{\partial z} \tag{5.15b}$$

$$f_z - \frac{1}{\rho}\frac{\partial p}{\partial z} + \nu\nabla^2 u_z = \frac{du_z}{dt} = \frac{\partial u_z}{\partial t} + u_x\frac{\partial u_z}{\partial x} + u_y\frac{\partial u_z}{\partial y} + u_z\frac{\partial u_z}{\partial z} \tag{5.15c}$$

To continue the mathematical description of the problem, boundary conditions are required. For example, on all fixed solid surfaces we have $u = 0$; at inlet or outlet we have known u, p; at free surface, $z = \eta$ (where η is the elevation variation of free surface), we have $u_z = \frac{d\eta}{dt}$. In some types of problems it may be necessary to specify the pressure over some part of the boundary. For time-dependent problems, initial conditions would also have to be provided, which means that the values of all dependent variables would be given at some time (usually taken at $t = 0$).

To nondimensionalize all variables in Eqs. 5.14 and 5.15, i.e., **u**, p, x, y, z, and t, we select a reference quantity for each type of variable: a reference velocity U, a reference pressure p_0, a reference length L, and a reference time T. These reference quantities should be parameters that appear in the problem. For example, L may be a characteristic length of a body immersed in a fluid or the width of a channel through which a fluid is flowing. The velocity, U, may be the free-stream velocity or the inlet velocity. The new dimensionless (starred) variables can be expressed as

$$u_x^* = \frac{u_x}{U} \qquad u_y^* = \frac{u_y}{U} \qquad u_z^* = \frac{u_z}{U}$$

$$x^* = \frac{x}{L} \qquad y^* = \frac{y}{L} \qquad z^* = \frac{z}{L} \qquad p^* = \frac{p}{p_0} \qquad t^* = \frac{t}{T}$$

The governing equations can now be rewritten in terms of these new variables. For example,

$$\frac{\partial u_x}{\partial x} = \frac{U}{L} \frac{\partial u_x^*}{\partial x^*}$$

$$\frac{\partial^2 u_x}{\partial x^2} = \frac{U}{L} \frac{\partial}{\partial x^*}\left(\frac{\partial u_x^*}{\partial x^*}\right)\frac{\partial x^*}{\partial x} = \frac{U}{L^2} \frac{\partial^2 u_x^*}{\partial x^{*2}}$$

The other terms that appear in the equations can be expressed in a similar fashion. Thus, in terms of the new variables the governing equations become

$$\frac{\partial u_x^*}{\partial x^*} + \frac{\partial u_y^*}{\partial y^*} + \frac{\partial u_z^*}{\partial z^*} = 0 \qquad (5.16)$$

$$\left[\frac{\rho U}{T}\right]\frac{\partial u_x^*}{\partial t^*} + \left[\frac{\rho U^2}{L}\right]\left(u_x^*\frac{\partial u_x^*}{\partial x^*} + u_y^*\frac{\partial u_x^*}{\partial y^*} + u_z^*\frac{\partial u_x^*}{\partial z^*}\right) = -\left[\frac{p_0}{L}\right]\frac{\partial p^*}{\partial x^*} + \left[\frac{\mu U}{L^2}\right]\left(\frac{\partial^2 u_x^*}{\partial x^{*2}} + \frac{\partial^2 u_x^*}{\partial y^{*2}} + \frac{\partial^2 u_x^*}{\partial z^{*2}}\right)$$

$$\left[\frac{\rho U}{T}\right]\frac{\partial u_y^*}{\partial t^*} + \left[\frac{\rho U^2}{L}\right]\left(u_x^*\frac{\partial u_y^*}{\partial x^*} + u_y^*\frac{\partial u_y^*}{\partial y^*} + u_z^*\frac{\partial u_y^*}{\partial z^*}\right) = -\left[\frac{p_0}{L}\right]\frac{\partial p^*}{\partial y^*} + \left[\frac{\mu U}{L^2}\right]\left(\frac{\partial^2 u_y^*}{\partial x^{*2}} + \frac{\partial^2 u_y^*}{\partial y^{*2}} + \frac{\partial^2 u_y^*}{\partial z^{*2}}\right)$$

$$\left[\frac{\rho U}{T}\right]\frac{\partial u_z^*}{\partial t^*} + \left[\frac{\rho U^2}{L}\right]\left(u_x^*\frac{\partial u_z^*}{\partial x^*} + u_y^*\frac{\partial u_z^*}{\partial y^*} + u_z^*\frac{\partial u_z^*}{\partial z^*}\right) = -[\rho g] - \left[\frac{p_0}{L}\right]\frac{\partial p^*}{\partial z^*} + \left[\frac{\mu U}{L^2}\right]\left(\frac{\partial^2 u_z^*}{\partial x^{*2}} + \frac{\partial^2 u_z^*}{\partial y^{*2}} + \frac{\partial^2 u_z^*}{\partial z^{*2}}\right)$$

$$F_{Il} \qquad F_{Ic} \qquad\qquad\qquad F_G \qquad F_p \qquad F_v$$

The terms appearing in brackets contain the reference quantities and can be interpreted as indices of the various forces (per unit volume) that are involved: F_{Il} = inertia (local) force, F_{Ic} = inertia (convective) force, F_p = pressure force, F_G = gravitational force, and F_v = viscous force. Divided by the bracketed quantity $\rho U^2/L$, the index of the convective inertia force, the final nondimensional form of the momentum equations becomes

$$\left[\frac{L}{TU}\right]\frac{\partial u_x^*}{\partial t^*} + u_x^*\frac{\partial u_x^*}{\partial x^*} + u_y^*\frac{\partial u_x^*}{\partial y^*} + u_z^*\frac{\partial u_x^*}{\partial z^*} = -\left[\frac{p_0}{\rho U^2}\right]\frac{\partial p^*}{\partial x^*} + \left[\frac{\mu}{\rho LU}\right]\left(\frac{\partial^2 u_x^*}{\partial x^{*2}} + \frac{\partial^2 u_x^*}{\partial y^{*2}} + \frac{\partial^2 u_x^*}{\partial z^{*2}}\right) \quad (5.17a)$$

$$\left[\frac{L}{TU}\right]\frac{\partial u_y^*}{\partial t^*} + u_x^*\frac{\partial u_y^*}{\partial x^*} + u_y^*\frac{\partial u_y^*}{\partial y^*} + u_z^*\frac{\partial u_y^*}{\partial z^*} = -\left[\frac{p_0}{\rho U^2}\right]\frac{\partial p^*}{\partial y^*} + \left[\frac{\mu}{\rho LU}\right]\left(\frac{\partial^2 u_y^*}{\partial x^{*2}} + \frac{\partial^2 u_y^*}{\partial y^{*2}} + \frac{\partial^2 u_y^*}{\partial z^{*2}}\right) \quad (5.17b)$$

$$\left[\frac{L}{TU}\right]\frac{\partial u_z^*}{\partial t^*} + u_x^*\frac{\partial u_z^*}{\partial x^*} + u_y^*\frac{\partial u_z^*}{\partial y^*} + u_z^*\frac{\partial u_z^*}{\partial z^*} = -\left[\frac{gL}{U^2}\right] - \left[\frac{p_0}{\rho U^2}\right]\frac{\partial p^*}{\partial z^*} + \left[\frac{\mu}{\rho LU}\right]\left(\frac{\partial^2 u_z^*}{\partial x^{*2}} + \frac{\partial^2 u_z^*}{\partial y^{*2}} + \frac{\partial^2 u_z^*}{\partial z^{*2}}\right) \quad (5.17c)$$

The terms appearing in brackets are the standard dimensionless groups (or their reciprocals) which were developed from dimensional analysis; that is, L/TU is a form of the *Strouhal number*,

$p_0/\rho U^2$ the *Euler number*, gL/U^2 the reciprocal of the square of the *Froude number*, and $\mu/\rho LU$ the reciprocal of the *Reynolds number*. From this analysis it is now clear how each of the dimensionless groups can be interpreted as the ratio of two forces, and how these groups arise naturally from the governing equations.

For the prototype and the model systems are governed by equations of (5.16) to (5.17), then the solutions (in terms of $u_x^*, u_y^*, u_z^*, x^*, y^*, z^*, p^*,$ and t^*) will be the same if the four parameters L/TU, $p_0/\rho U^2$, U^2/gL, and $\rho LU/\mu$ are equal for the two systems. The two systems will be dynamically similar. Of course, boundary and initial conditions expressed in dimensionless form must also be equal for the two systems, and this will require complete geometric similarity. These are the same similarity requirements that would be determined by a dimensional analysis if the same variables were considered.

5.4.2 Common Dimensionless Groups

The variables that commonly arise in fluid mechanics problems are: acceleration of gravity, g; fluid bulk modulus, K; characteristic length, L; fluid density, ρ; frequency of oscillating flow, ω; pressure, p (or Δp); speed of sound, c; surface tension, σ; flow velocity, U; fluid viscosity, μ. Combinations of these variables gives *dimensionless groups* (pi terms), some of them are given in Table 5-2.

Table 5-2 Common dimensionless groups in fluid mechanics

Parameter	Definition	Interpretation (Index of force ratio indicated)	Types of Applications
Reynolds number	$Re = \dfrac{\rho LU}{\mu}$	$\dfrac{\text{Inertia force}}{\text{Viscous force}}$	Generally of importance in all types of fluid dynamics problems
Froude number	$Fr = \dfrac{U}{\sqrt{gL}}$	$\dfrac{\text{Inertia force}}{\text{Gravitational force}}$	Free-surface flow
Euler number	$Eu = \dfrac{p}{\rho U^2}$	$\dfrac{\text{Pressure force}}{\text{Inertia force}}$	Problems in which pressure, or pressure differences, are of interest, e.g. cavitation problems
Cauchy number	$Ca = \dfrac{\rho U^2}{K}$	$\dfrac{\text{Inertia force}}{\text{Compressibility force}}$	Compressible flow, i.e. water hammer in pipe system
Mach number	$Ma = \dfrac{U}{c}$	$\dfrac{\text{Inertia force}}{\text{Compressibility force}} = \dfrac{\text{Flow speed}}{\text{Sound speed}}$	Compressible flow, most of gas flow
Strouhal number	$St = \dfrac{\omega L}{U}$	$\dfrac{\text{Inertia (local) force}}{\text{Inertia (convective) force}} = \dfrac{\text{Oscillation}}{\text{Mean speed}}$	Unsteady oscillating flow
Weber number	$We = \dfrac{\rho U^2 L}{\sigma}$	$\dfrac{\text{Inertia force}}{\text{Surface tension force}}$	Film, or bubble problems in which surface tension is important

REYNOLDS NUMBER

The *Reynolds number* is undoubtedly the most famous dimensionless parameter in fluid mechanics. It is named in honor of Osborne Reynolds (1842-1912), a British engineer who first demonstrated that this combination of variables could be used as a criterion to distinguish between laminar and turbulent flow in 1883. The Reynolds number is a measure of the ratio of the inertia force on an element of fluid to the viscous force on an element and in a form of

$$\text{Re} = \frac{\rho L U}{\mu} \tag{5.18}$$

The Reynolds number is always important, with or without a free surface, and can be neglected only in flow regions away from high-velocity gradients, e. g. , away from solid surfaces, jets, or wakes. If the Reynolds number is very small ($\text{Re} \leqslant 1$), it indicates that the viscous forces are dominant compared to inertia forces, and it may be possible to neglect the inertial effects. Conversely, for large Reynolds number, viscous forces are small compared to inertial effects and flow problems are characterized as inviscid analysis.

The Reynolds number is mainly used in the fluid flow which is mainly affected by the drag of the water flow, namely, the viscous force, such as the pressure laminar flow in a pipe or tunnel, and the flow around a submerged body. Once the prototype and model are similar, we have

$$\text{Re}_p = \text{Re}_m$$

or

$$\frac{\lambda_U \lambda_L}{\lambda_\nu} = 1 \tag{5.19}$$

FROUDE NUMBER

The *Froude number*

$$\text{Fr} = \frac{U}{\sqrt{gL}} \tag{5.20}$$

is named after William Froude (1810-1879), a British civil engineer, mathematician, and naval architecter who developed the ship-model towing-tank concept and proposed similarity rules for free-surface flows (ship resistance, surface waves, open channels). The Froude number is a measure of the ratio of the inertia force on an element of fluid to the weight of the element.

The Froude number is very much significant for flows with free surface effects since gravity principally affects this type of flow and is totally unimportant if there is no free surface. The typical case is the open-channel flow, for which the characteristics length is the depth of water. It is also used to study the flow of water around ships with resulting wave motion, the flow through open conduits, the weir flows and the orifice flows. The relation of scales gives if Froude number is employed:

$$\frac{\lambda_U^2}{\lambda_g \lambda_L} = 1 \tag{5.21}$$

EULER NUMBER

The *Euler number*

$$\text{Eu} = \frac{p}{\rho U^2} \tag{5.22}$$

named after Leonhard Euler (1707-1783), a famous Swiss mathematician who pioneered work on the relationship between pressure and flow. It is rarely important unless the pressure drops low enough to cause vapor formation (*cavitation*) in a liquid. The Euler number is often written in terms of pressure differences:

$$\text{Eu} = \frac{\Delta p}{\rho U^2} \quad (5.23)$$

Also, this combination expressed as $\Delta p/(\frac{1}{2}\rho U^2)$ is called the *pressure coefficient*. Some form of the Euler number would normally be used in problems in which the pressure or the pressure difference between two points is an important variable. If Δp involves vapor pressure p_v, it is called the *cavitation number*

$$C_a = \frac{p_{at} - p_v}{\rho U^2} \quad (5.24)$$

In the similitude of prototype and model, Eu will be automatically satisfied if Re is satisfied. The relation of scales gives if Euler number is employed:

$$\frac{\lambda_p}{\lambda_\rho \lambda_U^2} = 1 \quad (5.25)$$

WEBER NUMBER

The *Weber number*

$$\text{We} = \frac{\rho U^2 L}{\sigma} \quad (5.26)$$

is named after Moritz Weber (1871-1951), a German professor of naval mechanics who was instrumental in formalizing the general use of common dimensionless groups as a basis for similitude studies. The Weber number is important only if it is of order unity or less. The Weber number is taken as an index of droplet formation and flow of thin film liquids in which there is an interface between two fluids, which typically occurs when the surface curvature is comparable in size to the liquid depth, e.g., in droplets, capillary flows, ripple waves, and very small hydraulic models. If We is large ($\text{We} \gg 1$), inertia force is dominant compared to surface tension force (e.g. flow of water in a river), while the effect of surface tension can be ignored.

The relation of scales gives if Weber number is employed:

$$\frac{\lambda_\rho \lambda_U^2 \lambda_L}{\lambda_\sigma} = 1 \quad (5.27)$$

If there is no free surface, Fr, Eu, and We drop out entirely, except for the possibility of cavitation of a liquid at very small Eu. Thus, in low-speed viscous flows with no free surface, the Reynolds number is the only important dimensionless parameter.

CAUCHY AND MACH NUMBERS

The *Mach number*

$$\text{Ma} = \frac{U}{c} = \frac{U}{\sqrt{K/\rho}} \quad (5.28)$$

and the *Cauchy number*

$$\text{Ca} = \text{Ma}^2 = \frac{\rho U^2}{K} \quad (5.29)$$

where c is the local *sound speed* in fluid, $c = \sqrt{K/\rho}$.

The Mach number is named after Ernst Mach (1838-1916), an Austrian physicist and philosopher, and the Cauchy number is named in honor of Augustin Louis de Cauchy (1789-1857), a French engineer, mathematician, and hydrodynamicist. Both the Mach number and the Cauchy number are important dimensionless groups in problems in which fluid compressibility is a significant factor.

When the Ma is relatively small (say, less than 0.3), the inertial forces induced by fluid motion are sufficiently small to cause significant change in fluid density. So, the compressibility of the fluid can be neglected. However, this number is most commonly used parameter in compressible fluid flow problems, particularly in the field of gas dynamics and aerodynamics.

STROUHAL NUMBER

The *Strouhal number*

$$\text{St} = \frac{\omega L}{U} \tag{5.30}$$

where ω is the frequency of the oscillation.

The Strouhal number is named after Vincenz Strouhal (1850-1922), a German physicist, who used this parameter in his study of 'singing wires'. It is the measure of the ratio of the inertial forces due to unsteadiness of the flow (local acceleration) to inertia forces due to changes in velocity from point to point in the flow field (convective acceleration), and is likely to be important in unsteady, oscillating flow problems. This type of unsteady flow develops when a fluid flows past a solid body (such as a wire or cable) placed in the moving stream.

For example, in a certain Reynolds number range, a periodic flow will develop downstream from a cylinder placed in a steady stream of velocity U due to a regular pattern of vortices that are shed from the body. This regular, periodic shedding is called a *von Kármán vortex street* (Fig. 5-1a), after Theodor von Kármán (1881-1963), a famous fluid mechanician, who explained it theoretically in 1912. The shedding occurs in the range $10^2 < \text{Re} < 10^7$, with an average Strouhal number $\omega d/(2\pi U) \approx 0.21$, where d is the diameter of cylinder. Resonance can occur if a vortex shedding frequency, ω, is near a body's structural vibration frequency. A striking example is the disastrous failure of the Tacoma Narrows suspension bridge in 1940, when wind-excited vortex shedding caused resonance with the natural torsional oscillations of the bridge(Fig. 5-1b).

5.5 Modelling

To have complete similarity between model and prototype, we must maintain geometric, kinematic, and dynamic similarity between the two systems. This will automatically follow if all the important variables are included in the dimensional analysis, and if all the similarity requirements based on the resulting pi terms are satisfied.

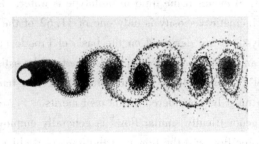 

Fig. 5-1 (a) Vortex shedding from a circular cylinder (*Courtesy of Cesareo de La Rosa Siqueira at Wikipedia*); (b) Tacoma Narrows suspension bridge distorted by wind before collapse (*courtesy of Wikipedia*)

5.5.1 Selection of Similarity

For the steady flow of an incompressible fluid without free surfaces, the dynamic and kinematic similarity will be achieved if (for geometrically similar systems) Reynolds number similarity exists. If free surfaces are involved, Froude number similarity must also be maintained. However, in the study of open-channel or free surface flows, it is usually impossible to satisfy Reynolds number similarity and Froude number similarity together. This model, for which one or more of the similar requirements are not satisfied, is called '*distorted models*'. Because, Froude number similarity, $Fr_p = Fr_m$, requires,

$$\frac{U_p}{\sqrt{g_p L_p}} = \frac{U_m}{\sqrt{g_m L_m}}$$

If the model and prototype are operated in the same gravitational field, that is, $g_p = g_m$, then the velocity scale becomes,

$$\lambda_U = \frac{U_p}{U_m} = \sqrt{\frac{L_p}{L_m}} = \sqrt{\lambda_l} \qquad (5.31)$$

For Reynolds number similarity, $Re_p = Re_m$, requires,

$$\frac{\rho_p U_p L_p}{\mu_p} = \frac{\rho_m U_m L_m}{\mu_m}$$

and the velocity scale is,

$$\lambda_U = \frac{U_p}{U_m} = \frac{\rho_m}{\rho_p} \frac{L_m}{L_p} \frac{\mu_p}{\mu_m} = \frac{\lambda_\mu}{\lambda_l \lambda_\rho} \qquad (5.32)$$

Since Reynolds number similarity and Froude number similarity should be satisfied together, the velocity scale in Eq. 5.31 must be equal to Eq. 5.32, it follows that

$$\lambda_\nu = \frac{\lambda_\mu}{\lambda_\rho} = (\lambda_l)^{3/2} \qquad (5.33)$$

Equation 5.33 requires that both model and prototype to have different kinematics viscosity scale, if both the requirements, i.e., Eqs. 5.31 and 5.32, are to be satisfied together. But practically, it is almost impossible to find a suitable model fluid for small length scale. For example, if $\lambda_l = 10$, $\lambda_v = 31.62$ according to Eq. 5.33. It means if the fluid in prototype is water, the fluid in model is difficult to be found since its kinematic viscosity is only one of 31.62 of the kinematic viscosity of water. In such cases, the systems are designed on the basis of Froude number with different Reynolds number for the model and prototype where Eq. 5.33 need not be satisfied. Such analysis will result in a 'distorted model'. Hence, there are no general rules for handling distorted models and essentially each problem must be considered on its own merits.

Thus, an incomplete similarity between geometrically similar flows is generally employed in the modeling. For example, for the pressure pipe flow and the flow past the immersed object, the viscous is dominated, thus the Reynolds number similarity is required in the model design; for the weir flow, gate flow and open channel flow, gravity is dominated, thus the Froude number similarity is required.

5.5.2 Model design

For the model design, the following procedures can be followed:
- Determine the length scale, λ_l, according to the experimental site, the modeling assemble and the measurement systems;
- Design the geometrical sizes of the model based on the length scale, and the geometrical boundaries of the model;
- Choose the dimensionless groups according to the dominated force in the flow; and
- Determine the velocity scale and flow rate scale based on the chosen dimensionless group.

The scales for Reynolds number similarity and Froude number similarity are given in Table 5-3.

Table 5-3 Model scales

Name of scale	Expression of scale			Name of scale	Expression of scale		
	Reynolds number similarity		Froude number similarity		Reynolds number similarity		Froude number similarity
	$\lambda_v = 1$	$\lambda_v \neq 1$			$\lambda_v = 1$	$\lambda_v \neq 1$	
Length scale λ_l	λ_l	λ_l	λ_l	Force scale λ_f	λ_ρ	$\lambda_v^2 \lambda_\rho$	$\lambda_l^3 \lambda_\rho$
Velocity scale λ_U	λ_l^{-1}	$\lambda_v \lambda_l^{-1}$	$\lambda_l^{1/2}$	Pressure scale λ_p	$\lambda_l^{-2} \lambda_\rho$	$\lambda_v^2 \lambda_l^{-2} \lambda_\rho$	$\lambda_l \lambda_\rho$
Acceleration scale λ_a	λ_l^{-3}	$\lambda_v^2 \lambda_l^{-3}$	λ_l^0	Work scale λ_W	$\lambda_l \lambda_\rho$	$\lambda_v^2 \lambda_l \lambda_\rho$	$\lambda_l^4 \lambda_\rho$
Flowrate scale λ_Q	λ_l	$\lambda_v \lambda_l$	$\lambda_l^{5/2}$	Power scale λ_P	$\lambda_l^{-1} \lambda_\rho$	$\lambda_v^3 \lambda_l^{-1} \lambda_\rho$	$\lambda_l^{7/2} \lambda_\rho$
Time scale λ_t	λ_l^2	$\lambda_v^{-1} \lambda_l^2$	$\lambda_l^{1/2}$				

Models are used to investigate many different types of fluid mechanics problems, and it is difficult to characterize in a general way all necessary similarity requirements, since each problem is unique. We can, however, broadly classify many of the problems on the basis of the general nature of the flow and subsequently develop some general characteristics of model designs in each of these classifications. In the following examples we will consider models for the study of (1) flow through closed conduits, (2) flow around immersed bodies, and (3) flow with a free surface.

FLOW THROUGH CLOSED CONDUITS

Common examples of this type of flow include pipe flow and flow through valves, fittings, and metering devices. Although the conduits are often circular, they could have other shapes as well and may contain expansions or contractions. Since there are no fluid interfaces or free surfaces, the dominant forces are inertial and viscous so that the Reynolds number is an important similarity parameter. For low Mach numbers (Ma < 0.3), compressibility effects are usually negligible for both the flow of liquids or gases.

Two additional points should be made with regard to modeling flows in closed conduits. First, for large Reynolds numbers, inertial forces are much larger than viscous forces, and in this case it may be possible to neglect viscous effects. The important practical consequence of this is that it would not be necessary to maintain Reynolds number similarity between model and prototype (see Section 6.6). The second point relates to the possibility of cavitation in flow through closed conduits. The use of models to study cavitation is complicated, since it is not fully understood how vapor bubbles form and grow. The initiation of bubbles seems to be influenced by the microscopic particles that exist in most liquids, and how this aspect of the problem influences model studies is not clear.

EXAMPLE 5-5 Flow through closed conduits

Model tests are to be performed to study the flow through a large valve having a 0.6-m-diameter inlet and carrying water at a flowrate of 0.84 m³/s. The working fluid in the model is water at the same temperature as that in the prototype, and the inlet diameter of valve model is 7.5cm. Assume the complete geometric similarity exists between model and prototype, determine the required flowrate in the model.

Solution:

To ensure dynamic similarity, the model tests should be run so that $Re_p = Re_m$, or $\dfrac{U_m D_m}{\nu_m} = \dfrac{U_p D_p}{\nu_p}$, where U and D correspond to the inlet velocity and diameter, respectively. Since the same fluid at same temperature is to be used in model and prototype, $\nu_p = \nu_m$, and therefore $U_m/U_p = D_p/D_m$. The discharge, Q, is equal to UA, where A is the inlet area, so

$$\frac{Q_m}{Q_p} = \frac{U_m A_m}{U_p A_p} = \frac{D_p}{D_m} \frac{\frac{\pi}{4} D_m^2}{\frac{\pi}{4} D_p^2} = \frac{D_m}{D_p} = \frac{0.075\text{m}}{0.6\text{m}} = \frac{1}{8}$$

$$Q_m = \frac{Q_p}{8} = \frac{0.84 \text{ m}^3/\text{s}}{8} = 0.105 \text{ m}^3/\text{s}$$

FLOW AROUND IMMERSED BODIES

It should be noted that these models have been widely used to study the flow characteristics associated with bodies that are completely immersed in a moving fluid. Examples include flow around aircraft, automobiles, golf balls, and buildings. Modeling laws for these problems is geometric and Reynolds number similarity is required. Since there are no fluid interfaces, surface tension (and therefore the Weber number) is not important. Also, gravity will not affect the flow patterns, so the Froude number need not be considered. The Mach number will be important for high-speed flows in which compressibility becomes an important factor, but for incompressible fluids (such as liquids or for gases at relatively low speeds) the Mach number can be omitted as a similarity requirement.

EXAMPLE 5-6 Flow around immersed bodies

The drag on an airplane cruising at 240mph in standard air is to be determined from tests on a 10:1 scale model placed in a pressurized wind tunnel. To minimize compressibility effects, the air speed in the wind tunnel is also to be 240mph. Determine the required air pressure in the tunnel (assuming the same air temperature for model and prototype), and the drag on the prototype corresponding to a measured force of 500N on the model.

Solution:

Frequently, the dependent variable of interest for this type of problem is the *drag*, F_D (see Section 6.8.3), developed on the body, and in this situation the dependent pi term would usually be expressed in the form of a drag coefficient, C_D, where $C_D = F_D/(\frac{1}{2}\rho U^2 L^2)$.

For this example, the Reynolds number similarity is to be maintained, thus $Re_p = Re_m$, or $\frac{\rho_m U_m L_m}{\mu_m} = \frac{\rho_p U_p L_p}{\mu_p}$. With the given coefficient, $U_p = U_m$ and $L_m/L_p = 1/10$, so that

$$\frac{\rho_m}{\rho_p} = \frac{\mu_m U_p L_p}{\mu_p U_m L_m} = \frac{\mu_m}{\mu_p}(1)(10) = 10\frac{\mu_m}{\mu_p}$$

This result shows that the same fluid with $\rho_p = \rho_m$ and $\mu_p = \mu_m$ cannot be used if Reynolds number similarity is to be maintained. One possibility is to pressurize the wind tunnel to increase the density of the air. We assume that an increase in pressure does not significantly change the viscosity ($\mu_m/\mu_p = 1$) so that the required increase in density is given by the relationship $\rho_m/\rho_p = 10$. For the perfect gas law for an ideal gas, $p = \rho RT$ so that $p_m/p_p = \rho_m/\rho_p$ for the same temperatures, $T_p = T_m$. Therefore, the wind tunnel would need to be pressurized to $p_m = 10p_p$. Since the prototype operates at standard atmospheric pressure, the required pressure in the wind tunnel is 10 atmospheres.

Thus, we see that a high pressure would be required and this could not be easily or inexpensively achieved. However, under these conditions Reynolds similarity would be attained and the drag could be obtained from the equation

$$\frac{F_{Dp}}{\frac{1}{2}\rho_p U_p^2 L_p^2} = \frac{F_{Dm}}{\frac{1}{2}\rho_m U_m^2 L_m^2}$$

or

$$F_{Dp} = \frac{\rho_p}{\rho_m}\left(\frac{U_p}{U_m}\right)^2\left(\frac{L_p}{L_m}\right)^2 F_{Dm} = \left(\frac{1}{10}\right)(1)^2(10)^2 F_{Dm} = 10 F_{Dm}$$

Thus, for a drag of 500N on the model the corresponding drag on the prototype is

$$F_{Dp} = 5000\text{N}$$

FLOW WITH A FREE SURFACE

Flows in canals, rivers, spillways, and stilling basins, as well as flow around ships, are all examples of flow phenomena involving a free surface. Gravitational and inertial forces are important and Froude number becomes an important similarity parameter. For large hydraulic structures, such as dam spillsways, the Reynolds numbers are large so that viscous forces are small in comparison to the forces due to gravity and inertia. In this case Reynolds number similarity is not maintained and models are designed on the basis of Froude number similarity.

EXAMPLE 5-7 Flow with a free surface

A flood spillway of a dam is 20m wide and is designed to carry 125 m³/s at flood stage. A 15:1 model is constructed to study the flow characteristics through the spillway. Determine the required model width and flowrate. What operating time for the models corresponds to a 24-hr period in the prototype? The effect of surface tension and viscosity are to be neglected.

Solution:

The width, B_m, of the model spillway is obtained from the length scale, λ_L, so that $B_p/B_m = \lambda_l$, therefore,

$$B_m = \frac{B_p}{\lambda_l} = \frac{20m}{15} = 1.33m$$

With the negligible surface tension and viscosity, the dynamics similarity will be achieved if the Froude numbers are equal between model and prototype, $Fr_p = Fr_m$, or

$$\frac{U_p}{\sqrt{g_p L_p}} = \frac{U_m}{\sqrt{g_m L_m}}$$

for $g_p = g_m$, then $U_p/U_m = \sqrt{L_p/L_m}$.

Since the flowrate is given by $Q = UA$, where A is an appropriate cross-sectional area, it follows that

$$\frac{Q_m}{Q_p} = \frac{U_m A_m}{U_p A_p} = \sqrt{\frac{L_m}{L_p}}\left(\frac{L_m}{L_p}\right)^2 = \left(\frac{L_m}{L_p}\right)^{5/2} = \frac{1}{\lambda_L^{5/2}}$$

with $A_p/A_m = (L_p/L_m)^2$.

Hence, the flowrate of the model spillway is

$$Q_m = \frac{Q_p}{\lambda_L^{5/2}} = \frac{125 \, m^3/s}{15^{5/2}} = 0.143 \, m^3/s$$

The time scale can then be obtained from the velocity scale, since $U = L/t$

$$\frac{t_m}{t_p} = \frac{U_p}{U_m}\frac{L_m}{L_p} = \frac{1}{\lambda_l^{1/2}}$$

This results indicates that times intervals in the model will be smaller than the corresponding intervals in the prototype time interval of 24 hr

$$t_m = \frac{t_p}{\lambda_l^{1/2}} = \frac{24hr}{15^{1/2}} = 6.2hr$$

Summary

There is a difference between dimensions and units: a *dimension* is a measure of a physical quantity (without numerical values), while a *unit* is a way to assign a number to that dimension. There are four *primary dimensions*: mass, length, time, temperature, in fluid mechanics. All other dimensions, *secondary dimensions*, can be formed by combination of these four primary dimensions.

In engineering practices, many problems involving fluid mechanics require experimental data for their solution, of which the physical processes satisfy the *principle of dimensional homogeneity*.

Thus, the methods of *dimensional analysis*, the *Rayleigh's method* and the *Buckingham pi theorem*, are presented and used for designing such experiments, as the aids for correlating experimental data, and as the basis for the design of physical *model*s. Alternatively, nondimensionalizing the basic equations of fluid mechanics yields the fundamental *dimensionless groups* governing flow patterns: *Reynolds number*, *Froude number*, *Strouhal number*, *Euler number*, and others. Among them the most important and common used groups are *Reynolds number*

$$\text{Re} = \frac{\rho L U}{\mu}$$

and *Froude number*

$$\text{Fr} = \frac{U}{\sqrt{gL}}$$

When all the dimensionless groups match between a model and a *prototype*, *dynamic similarity* is achieved, and we are able to directly predict prototype performance based on model experiments. However, it is not always possible to match *all* the pi groups when trying to achieve similarity between a model and a prototype. In such cases, we run the model tests under conditions of *incomplete similarity*, matching the most important pi groups as best we can, and then extrapolating the model test results to prototype conditions. In general, *Reynolds number similarity* is important for almost all flows; *Froude number similarity* is important for flows with free surfaces, such as ship resistance, open channel flows and for flows driven by the action of gravity; *Euler number similarity* is important mostly for turbomachinery flows with considerable pressure changes, for which *cavitation* may be a concern; *Mach number similarity* is important for high speed flows; *Weber number similarity* is important for problems involving interfaces between two fluids and low weight objects; *Strouhal number similarity* is important for flows with an oscillating (time periodic) flow pattern, such as *von Kármán vortices* shed from bodies.

Word Problems

W5-1 Define dimension and unit. What is the difference between a dimension and a unit? Give three examples of each.

W5-2 Define the primary dimension. List the primary dimensions commonly used in fluid mechanics.

W5-3 Define the derived dimension and the dimension formula.

W5-4 Define the dimensionless group. List some examples.

W5-5 List the three primary purposes of dimensional analysis.

W5-6 Explain the principle of dimensional homogeneity in simple terms.

W5-7 Does every empirical formula satisfy principle of dimensional homogeneity? Give several examples to explain.

W5-8 What situation can the Rayleigh's method be applicable? How about the Buckingham pi theorem?

W5-9 How to choose the repeating variables for Buckingham pi theorem?

W5-10 Why does each dimensionless group in fluid mechanics include inertia forces?

W5-11 List several examples of flows in which gravitational force and viscous force dominant respectively.

W5-12 Can a model and a prototype satisfy the Froude number and Reynolds number together?

W5-13 List and describe the necessary conditions for complete similarity between a model and a prototype.

W5-14 For each statement, choose whether the statement is true or false and discuss your answer briefly.

(a) Kinematic similarity is a necessary and sufficient condition for dynamic similarity.

(b) Geometric similarity is a necessary condition for dynamic similarity.

(c) Geometric similarity is a necessary condition for kinematic similarity.

(d) Dynamic similarity is a necessary condition for kinematic similarity.

W5-15 Think about and describe a prototype flow and a corresponding model flow that have geometric similarity, but not kinematic similarity, even though the Reynolds numbers match. Explain.

Problems

Dimension and Unit

5-1 Write the dimension of the following variables in $MLT\Theta$ system: kinematic viscosity ν; shear stress τ; surface tension σ and power $\dot{N}$.

5-2 Determine the dimension of the following quantities in $MLT\Theta$ system: (a) $\rho u \dfrac{\partial u}{\partial x}$; (b) $\int_1^2 (p - p_0)\, dA$; (c) $\rho c_p \dfrac{\partial^2 T}{\partial x \partial y}$; (d) $\iiint \rho \dfrac{\partial u}{\partial t} dx dy dz$. All quantities have their standard meanings; for example, ρ is density.

5-3 The Reynolds number, $\dfrac{\rho L U}{\mu}$, is a very important parameter in fluid mechanics. Verify that the Reynolds number is dimensionless, using both the FLT system and the MLT system for basic dimensions.

Dimensional Homogeneity

5-4 In Chap. 3 we defined the Bernoulli equation

$$z + \frac{p}{\rho g} + \frac{u^2}{2g} = H$$

Write the primary dimensions of each additive term in the equation, and verify that the equation is dimensionally homogeneous.

5-5 In Chap. 3 we defined the *Eulerian acceleration*, which is the acceleration following a fluid particle

$$\mathbf{a} = \frac{d\mathbf{u}}{dt} = \frac{\partial \mathbf{u}}{\partial t} + (\mathbf{u} \cdot \nabla)\mathbf{u}$$

(a) What are the primary dimensions of the gradient operator ∇? (b) Verify that each additive term in the equation has the same dimensions.

5-6 Group the following physical variables into the dimensionless group by dimensional analysis: (a) τ, U, ρ; (b) $\Delta p, U, \rho, g$; (c) F, U, ρ, L; (d) L, U, ρ, σ, in which σ is the surface tension.

Dimensional Analysis

5-7 The shaft power of pump $\dot{N}$ depends on the pumping flow rate Q, fluid density ρ, gravitational acceleration g and pump head H. Find the dimensionless group for this problem by Rayleigh's method.

5-8 A *weir* is an obstruction in a channel flow which can be calibrated to measure the flow rate, as in Fig. P5-8. The flow rate of the sharp-edged rectangular weir Q varies with gravity g, weir width b into the paper, and upstream water height H above the weir crest. Use the Rayleigh's method to find a unique functional relationship of the flow rate Q.

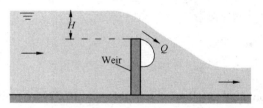

Fig. P5-8

5-9 At a sudden contraction in a pipe the diameter changes from D_1 to D_2. The pressure drop Δp, which develops across the contraction, is a function of D_1 and D_2, as well as the velocity v in the larger pipe, the fluid density ρ, and viscosity μ. Use D_1, v, and μ as repeating variables to determine a suitable set of dimensionless parameters. Why would it be incorrect to include the velocity in the smaller pipe as an additional variable?

5-10 Under laminar conditions, the volume flow Q through a small triangular-section pore of side length b and length L is a function of viscosity μ, pressure drop per unit length $\Delta p/L$, and b. Using the Buckingham pi theorem, rewrite this relation in dimensionless form. How does the volume flow change if the pore size b is doubled?

5-11 Assume the drag on an immersed sphere is a function of the sphere diameter d, the fluid viscosity μ, the fluid velocity U, and the fluid density ρ. Use the Buckingham pi theorem to find this relation in dimensionless form.

5-12 A periodic *Kármán vortex street* is formed when a uniform stream flows over a circular cylinder (Fig. P5-12). Use the pi theorem to generate a dimensionless relationship for Kármán vortex shedding frequency f_k as a function of free-stream speed U, fluid density ρ, fluid viscosity μ, and cylinder diameter D.

5-13 Consider fully developed Couette flow-flow between two infinite parallel plates separated by distance h, with the top plate moving and the bottom plate stationary as illustrated in Fig. P5-13. The flow is steady, incompressible, and two-dimensional in the xy plane. Use the pi theorem to generate a dimensionless relationship for the x component of fluid velocity u as a function of fluid viscosity μ, top plate speed U, distance h, fluid density ρ, and distance y.

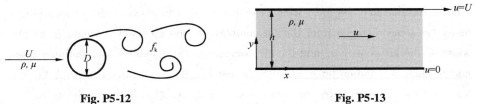

Fig. P5-12   Fig. P5-13

5-14 Consider *laminar* flow through a long section of pipe, as in Fig. P5-14. For laminar flow it turns out that wall roughness Δ is not a relevant parameter unless Δ is very large. The volume flow rate Q through the pipe is in fact a function of pipe diameter D, fluid viscosity μ, and axial pressure gradient dp/dx. If pipe diameter is doubled, all else being equal, by what factor will volume flow rate increase? Use dimensional analysis.

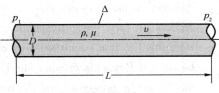

Fig. P5-14

5-15 Consider a broad-crested weir in a channel. The flow rate over the weir Q varies with gravity g, weir width b into the paper, upstream weir height P and the weir head H over the weir crest measured about $(3 \sim 4)H$ upstream of the weir. Find this relation in dimensionless form by dimensional analysis.

Similitude and Modelling

5-16 The discharge coefficient of a large Venturi meter is to be tested by a 10∶1 scale model. The identical fluid at same temperature is used in both the prototype and model. Determine the scale ratio of the flow rate if the dynamic similarity has been achieved.

5-17 The head loss of a tee-pipe in a pipe system transporting coal gas at a speed of $v = 25\,\text{m/s}$ is to be determined by model tests. The head loss is denoted as $\Delta h = \zeta \dfrac{v^2}{2g}$, in which ζ is the minor loss coefficient of tee-pipe. Take the diameter of gas pipe to be $1.2\,\text{m}$, the gas density be $40\,\text{kg/m}^3$ and the gas absolute viscosity be $0.0002\,\text{N·s/m}^2$. In the model, water at $20\,°\text{C}$ is used as the working fluid. Determine: (a) the length scale; (b) the scale ratio of head losses in terms of head and pressure.

5-18 As wind blow past buildings, complex flow patterns develop due to the flow separation and the interactions of adjacent buildings. An eight-story building is to be modeled in a low-speed environmental wind tunnel. When the air velocity in the wind tunnel is 9m/s, the measured pressures on the frontal and rear surfaces of the building are 42N/m² and −20 N/m², respectively. Determine the pressures on the frontal and rear surfaces of the building if the air velocity is increased to 12m/s. Assume the air densities and air temperatures are same in these two cases.

5-19 Consider a cross wind blowing past a train locomotive in a low-speed environmental wind tunnel, for which the typical air velocities in these tunnels are in the range of 0.1m/s to 30m/s. Assume that the local wind velocity, u, is a function of the approaching wind velocity (at some distance from the locomotive) U, the locomotive length l, height h, and width b, the air density ρ, and the air viscosity μ. (a) Establish the similarity requirements and prediction equation for a model to be used in the wind tunnel to study the air velocity u around the locomotive by the dimensional analysis. (b) If the model is to be used for cross winds gusting to $U = 25$m/s, explain why it is not practical to maintain Reynolds number similarity for a typical length scale of 50.

5-20 The drag characteristics of an airplane are to be determined by model tests in a wind tunnel operated at an absolute pressure of 1300kPa. If the prototype is to cruise in standard air at 385km/hr, and the corresponding speed of the model is not to differ by more than 20% from this (so that compressibility effects may be ignored), what range of length scales may be used if Reynolds number similarity is to be maintained? Assume the viscosity of air is unaffected by pressure, and the temperature of air in the tunnel is equal to the temperature of the air in which the airplane will fly.

5-21 In Fig. P5-21 (not to scale) water flows around a 24-m-long, 4.3-m-wide bridge pier. The water depth under the bridge is 8.2m, the corresponding sectional average flow velocity is 2.3m/s. The axis of this pier is 90m away from the abutment pier. Determine the sizes of the model and the flowrate in model if a length scale of 50 is applied.

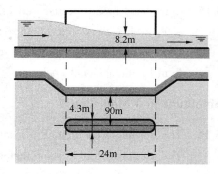

Fig. P5-21

5-22 The pressure drop in a venturi meter varies only with the fluid density, pipe approach velocity, and diameter ratio of the meter. A model venturi meter tested in water at 20℃ shows a 5kPa drop when the approach velocity is 4m/s. A geometrically similar prototype meter is used to measure gasoline at 20℃ and a flow rate of 9 m³/min. If the prototype pressure gage is most accurate at 15kPa, what should the upstream pipe diameter be?

5-23 A dam spillway is to be tested by using Froude scaling with a model with length scale of 30. The model flow has an average velocity of 0.6m/s and a volume flow of 0.05 m³/s. What will the velocity and flow of the prototype be? If the measured force on a certain part of the model is 1.5N, what will the corresponding force on the prototype be?

5-24 A prototype ship is 35m long and designed to cruise at 11m/s. Its drag is to be simulated by a 1-m-long model pulled in a tow tank. For Froude scaling find (a) the tow speed, (b) the ratio of prototype to model drag, and (c) the ratio of prototype to model power.

5-25 Air at 20℃ flows in a 600-mm-diameter smooth air duct at a velocity of 8m/s. Now it is modeled by water at 20℃ flowing in a 60-mm-diameter smooth pipe. To make sure the dynamic similarity, what will the flow velocity in the water pipe on the model be? If the model pressure drop in the water pipe is 450mmH$_2$O, what will the corresponding pressure drop on the prototype be?

5-26 The minor loss coefficient of a valve in a water supply system is measured in the same system by feeding air at 20℃ in the testing instead of water at 20℃ in the engineering practice. The pipe diameter is 50mm. If the velocity of water flow is 2.5m/s, what will the velocity of air flow in the testing be? When the pressure drop Δp of air flow is measured in the testing, what will the pressure drop of water flow be?

6 Fluid Resistance and Energy Losses

CHAPTER OPENING PHOTO: A spheroid is spinning in a right-to-left air flow. The boundary layer wrapping around it transits from smooth and orderly laminar boundary layer in the frontal section to a disorder turbulent boundary layer with evenly spaced vortices in the mid-section. Boundary layer is a thin region near the surface in which velocity varies rapidly as discussed in the internal flow and the external flow in the present chapter. (©Y. Kohama[1])

Fluid flow is classified as *external* and *internal*, depending on whether the fluid is forced to flow over a surface or in a conduit. Internal and external flows exhibit very different characteristics. In this chapter we mainly consider the fluid resistances and energy losses in the *internal flow*, which is driven primarily by a pressure difference and completely confined by the walls of conduit. In this case the viscous boundary layers grow from the sidewalls, meet downstream, and fill the entire duct.

In this chapter, we start with the definition of the energy losses and a discussion of the dimensionless Reynolds number to classify the laminar and turbulent flows. We then discuss the shear stresses, the velocity profiles and the pressure drops for both fully developed laminar and turbulent flows inside pipes, and present the frictional and minor losses with a large amount of empirical theory and formulas. Finally, we devoted to ' external' flows around bodies immersed in a fluid stream. Such a flow will have viscous (shear and no-slip) effects near the body surfaces and in its wake, but will typically be nearly inviscid far from the body.

---

[1] http://nicolesharp.com/research/stability/

6.1 Introduction to Head Losses

6.1.1 Definition of Head Losses

As water discussed in Chapter 4, when water (or any other fluid) flows, the pressure continuously drops in the streamwise direction because of friction. It is common to express this pressure drop in terms of an irreversible *head loss*, h_w, which appears in the energy equation

$$z_1 + \frac{p_1}{\rho g} + \frac{\alpha_1 v_1^2}{2g} = z_2 + \frac{p_2}{\rho g} + \frac{\alpha_2 v_2^2}{2g} + h_w$$

where h_w is the *total head loss*, the mechanical energy loss per unit weight fluid.

For flows through piping systems, open channel, and etc., there are two types of irreversible head losses: major losses and minor losses.

Major losses, or *frictional losses*, designated as h_f, are those due to viscous effect through flow sections with same shape and constant size, in which the flow is fully developed. It could happen in the long straight open channel or the pressurized straight pipe with constant diameter.

Minor losses, or *local losses*, designated as h_j, are those due to the changes of the geometric sizes, shapes of the sections in the open channel or the piping system, or due to the other 'device' other than the never changing size or shape. In the piping system, these 'device' include pipe entrance or exit, sudden expansion or contraction, bends, elbows, tees, other fittings, valves (open or partially closed), and gradual expansions or contractions, etc. The flow patterns in most of the above mentioned geometries are quite complex, see Fig. 6-1, and thus the losses commonly measured experimentally and correlated with the pipe flow parameters. As illustrated in Fig. 6-1, the minor loss occurs around the place where the geometric sizes, shapes of the sections change and last a relative long flow distance. However, for the convenience, the minor loss is summed up and considered to happen at a section.

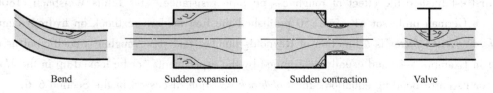

Bend   Sudden expansion   Sudden contraction   Valve

Fig. 6-1 Minor losses in piping system

Thus, the total head loss, h_w, in the energy equation can be expressed as

$$h_w = \Sigma h_f + \Sigma h_j \tag{6.1}$$

For example, the head loss in the piping system shown in Fig. 6-2 is

$$h_w = h_{f1,2} + h_{f2,3} + h_{f3,4} + h_{j1} + h_{j2} + h_{j3}$$

where $h_{f1,2}, h_{f2,3}, h_{f3,4}$ are the frictional losses occurring in the straight constant-diameter pipe sections of 1-2, 2-3 and 3-4 respectively, h_{j1}, h_{j2}, h_{j3} are the minor losses due to the sudden shape change at the pipe entrance, the converging-diverging pipe and the valve. It should be noted that the minor loss of the converging-diverging pipe, h_{j2}, has been concentratively drawn at the throat section 2 for the convenience.

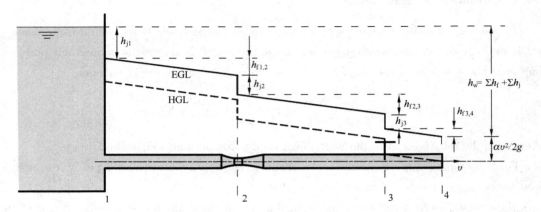

Fig. 6-2 Head losses in a simple piping system with a free jet

6.1.2 Determination of Head Losses

Major losses in piping system can be determined by

$$h_f = \lambda \frac{L}{d} \frac{v^2}{2g} \qquad (6.2)$$

where λ is the dimensionless *Darcy friction factor*, L is the length of pipe, d is the diameter of pipe, v is the sectional average velocity in the pipe.

Equation 6.2 is the *Darcy-Weisbach equation*, valid for any fully developed, steady, incompressible, laminar or turbulent pipe flows of any cross section and for laminar and turbulent flow. It is named after Henry Darcy (1803 – 1858), a French engineer whose pipe-flow experiments in 1857 first established the effect of roughness on pipe resistance, and Julius Weisbach (1806 – 1871), a German professor who in 1850 published the first modern textbook on hydrodynamics. The *Darcy friction factor* is a function of Reynolds number and pipe roughness coefficient as discussed in Example 5-4, and usually determined by the experiments, or be looked up in the *Moody chart*, or its corresponding equation, the *Colebrook equation* discussed in the Section 6.6.

For the noncircular pipe, Eq. 6.2 can be expressed as

$$h_f = \lambda \frac{L}{4R_h} \frac{v^2}{2g} \qquad (6.3)$$

where R_h is the *hydraulic radius*, defined by

$$R_h = \frac{A}{\chi} \qquad (6.4)$$

where A is the *cross-sectional flow area*, χ is the *wetted perimeter* that the fluid contacts with the solid boundary on the cross section.

Note that for the degenerate case of a circular pipe, $R_h = A/\chi = \frac{1}{4}\pi d^2/\pi d = d/4$, as expected, Eq. 6.3 is same as Eq. 6.2.

The minor losses is determined by

$$h_j = \zeta \frac{v^2}{2g} \tag{6.5}$$

where ζ is the dimensionless *minor loss coefficient*, given as a ratio of the minor head loss through the device to the velocity head $v^2/2g$ of the associated piping system and determined by the experiment.

In a system with both major and minor losses, it is important to recognize that a minor loss is defined as the *additional* loss due to the presence of the valve, elbow, etc. For example, the minor loss through an elbow does *not* include the frictional (Moody chart, see Section 6.6.2) loss through the length of the elbow. Thus, when measuring a minor loss coefficient, it is necessary to measure the length of the elbow, and to incorporate this length into the total length, L, of the pipe system.

For a piping system where the pipe diameter remains constant as in Fig. 6-2, Fqs. 6.1, 6.2 and 6.5 can be combined as follows:

$$h_w = \Sigma h_f + \Sigma h_j = \frac{v^2}{2g}\left(\lambda \frac{L}{d} + \Sigma \zeta\right) \tag{6.6}$$

In keeping with the above convention, length L represents the total length of pipe, including the centerline length through each elbow or other minor loss device.

6.2 Laminar and Turbulent Flows

6.2.1 Reynolds Experiment

Usually, the fluid flow in a pipe is streamlined at low velocities but turns chaotic as the velocity is increased above a critical value. The flow regime in the first case is said to be *laminar*, characterized by smooth streamlines and highly ordered motion, and *turbulent* in the second case, where it is characterized by velocity fluctuations and highly disordered motion. From laminar to turbulent flow the *transition flow* may occur. We can verify the existence of these laminar, transitional, and turbulent flow regimes by injecting some dye streaks into the flow in a glass pipe, as the British engineer Osborne Reynolds did over a century ago.

Fig. 6-3 shows such an experimental apparatus similar to the one of Reynolds, in which two pizeometric tubes at sections 1 and 2 are installed to measure the frictional loss with different valve opening, i.e., the flow velocity

$$h_f = \left(z_1 + \frac{p_1}{\rho g}\right) - \left(z_2 + \frac{p_2}{\rho g}\right)$$

In Fig. 6-4a correlation of the major head loss v.s. the flow velocity is plotted, which shows

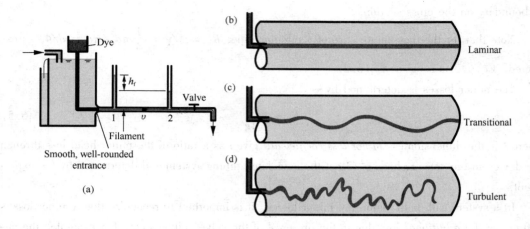

Fig. 6-3 (a) Reynolds' experiment using water in a pipe to study laminar to turbulence;
(b) Laminar flow; (c) Transitional flow and (d) Turbulent flow

that at different valve openings we can observe different flow patterns. At low velocities the dye filament is straight and smooth and unbroken along the length of the pipe (Fig. 6-3b). Increasing velocity to a critical value, v_c' in Fig. 6-4, the filament is initially smooth but become wavy at a certain distance along the pipe (Fig. 6-3c). As the velocity is increased further filament fluctuations increases until there comes a point where the dye filament suddenly mix across the tube (Fig. 6-3d). Further increase in velocity does not change the flow pattern, but the point at which mixing occurs moves closer to the inlet.

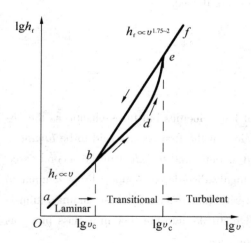

Fig. 6-4 Experimental correlation of major head loss v. s. flow velocity

Reverse the experiment procedure. Open the valve fully to make sure the turbulence occurring in the glass pipe, then gradually close the valve until the broken and mixed dye begin to congregate again and form a wavy filament. As the valve is closed further filament fluctuations becomes weak and decreases until a straight and smooth dye filament occurs. The corresponding velocity, v_c in Fig. 6.4, indicates that the flow pattern has changed from turbulence to laminar.

The critical velocity v_c' and v_c in Fig. 6-4 are called the *upper critical velocity and lower critical velocity* respectively. Obviously, the value of v_c' is larger than the value of v_c. According to the experimental observation, the value of v_c' varies in a wide range based on the disturbing level of the experimental enviroment. Laminar flow can be maintained at much higher velocity in very smooth pipes by avoiding flow disturbances and pipe vibrations. In such carefully controlled experiments, laminar flow has been maintained at Reynolds numbers of up to 100,000. However, v_c has a relative stable value. Therefore, the lower critical velocity, v_c, has been employed as the critical velocity to judge the transformation of the flow regime: laminar flow for $v < v_c$ and turbulent flow for $v > v_c$.

The relation between the major head loss and the flow velocity in laminar (line ab) and turbulence (line ef) regime can then be correlated as

$$\log h_f = \log k + \log v^n$$

or

$$h_f = k v^n \tag{6.7}$$

where k is a constant, n is the exponent of the sectional averaged velocity.

Based on the experimental data, we have $n = 1.0$ or $h_f \propto v$ for laminar flow, that is, the major loss of a fully developed laminar flow in the pipe is proportional to the averaged flow velocity; and $n = 1.75 \sim 2.0$, or $h_f \propto v^{1.75 \sim 2}$ for turbulent flow, that is, the major loss for turbulent flow is proportional to $v^{1.75 \sim 2}$.

In engineering practice, most flows encountered are turbulent. Laminar flow is encountered only when highly viscous fluids such as oil flow in small pipes or narrow passages.

6.2.2 Critical Reynolds Number

The transition from laminar to turbulent flow depends on the geometry, surface roughness, flow velocity, surface temperature, and type of fluid, among other things, and also on the degree of disturbance of the flow by surface roughness, pipe vibrations, and fluctuations in the flow. After exhaustive experiments in the 1880s, Osborne Reynolds discovered that the flow regime depends mainly on the ratio of inertia force to viscous force, that is, the *Reynolds number*. It is expressed for internal flow in a circular pipe as

$$\mathrm{Re} = \frac{\rho v d}{\mu} = \frac{v d}{\nu} \tag{6.8}$$

At large Reynolds numbers, the inertial forces, which are proportional to the fluid density and the square of the fluid velocity, are large relative to the viscous forces, and thus the viscous forces cannot prevent the random and rapid fluctuations of the fluid, *turbulent* flow occurs. At small or moderate Reynolds numbers, however, the viscous forces are large enough to suppress these fluctuations and to keep the fluid to be smooth and order, *laminar* flow occurs.

The Reynolds number at which the flow becomes turbulent is called the *critical Reynolds number* and expressed as

$$\mathrm{Re}_c = \frac{v_c d}{\nu} \tag{6.9}$$

The value of the critical Reynolds number is different for different geometries and flow conditions. For internal flow in a circular pipe, the generally accepted value of the critical Reynolds number is $Re_c = 2300$, that is, under most practical conditions, the flow in a circular pipe is laminar for $Re \leqslant 2300$, turbulent for $Re \geqslant 4000$, and transitional in between:

Laminar flow $Re \leqslant 2300$
Transitional flow $2300 \leqslant Re \leqslant 4000$
Turbulent flow $Re \leqslant 4000$

In transitional flow, the flow switches between laminar and turbulent randomly. As what discussed above, sometimes the laminar flow can be maintained at much higher Reynolds numbers, even up to 100,000.

For flow through noncircular pipes or in open channel, the Reynolds number is based on the hydraulic radius, R_h:

$$Re = \frac{vR_h}{\nu} \tag{6.10}$$

In open channel, the flow is laminar for $Re \leqslant Re_c = 575$, and turbulent flow for $Re \geqslant 575$.

EXAMPLE 6-1   Critical Reynolds number

The accepted critical Reynolds number for flow in a circular pipe is $Re_c = 2300$. For flow through a 5-cm-diameter pipe, at what velocity will this occur at 20℃ for (*a*) *air flow and* (*b*) *water flow*?

Solution:

From Tables A-2 in Appendix we take the kinematic viscosity of air at 20℃ to be $\nu = 1.516 \times 10^{-5} \, m^2/s$. From Tables A-1 we take the kinematic viscosity of water at 20℃ to be $\nu = 1.004 \times 10^{-6} \, m^2/s$.

From Eq. 6.9 we can have the critical velocity as

$$v_c = \frac{Re_c \nu}{d}$$

Thus, for air

$$v_c = \frac{Re_c \nu}{d} = \frac{(2300)(1.516 \times 10^{-5} \, m^2/s)}{0.05 \, m} = 0.7 \, m/s$$

For water

$$v_c = \frac{Re_c \nu}{d} = \frac{(2300)(1.004 \times 10^{-6} \, m^2/s)}{0.05 \, m} = 0.046 \, m/s$$

DISCUSSION   These are very low velocities, so most engineering air and water pipe flows are turbulent, not laminar. We might expect laminar pipe flow with more viscous fluids such as lubricating oils or glycerin.

EXAMPLE 6-2   Judgment of laminar and turbulent flows

Water flows in a 0.2-m-wide rectangular channel at a depth of 0.1 m and a velocity of 0.12 m/s. Take the water temperature to be 20℃. Is the flow to be laminar or turbulent?

Solution:

From Table A-1, we take the kinematic viscosity of water at 20℃ to be $\nu = 1.004 \times 10^{-6} \, m^2/s$.

For the open channel, the laminar or turbulent flow should be determined by Eq. 6.10, in which the hydraulic radius

$$R_h = \frac{A}{\chi} = \frac{bh}{b + 2h} = \frac{(0.2 \, m)(0.1 \, m)}{0.2 \, m + (2)(0.1 \, m)} = 0.05 \, m$$

Then, the Reynolds number of flow is

$$\text{Re} = \frac{vR_h}{\nu} = \frac{(0.12\text{m/s})(0.05\text{m})}{1.004 \times 10^{-6} \text{ m}^2/\text{s}} = 5976$$

Thus, the flow in the open channel is turbulent flow because $\text{Re} > \text{Re}_c = 575$.

DISCUSSION For the flow in a noncircular pipe or channel, the characteristic length in Reynolds number should be replaced by the hydraulic radius, R_h.

6.3 Fully Developed Flow and Its Shear Stress

6.3.1 Entrance Region and Fully Developed Flow

When a fluid enters a pipe at a uniform velocity as shown in Fig. 6-5, the fluid particles in the layer in contact with the pipe wall come to a complete stop because of the *no-slip condition*, which retards the fluid particles in the adjacent layers to slow down gradually as a result of friction, and accelerates the center-core flow to maintain the incompressible continuity requirement-constant flow rate through the pipe. As a result, a velocity gradient $u(r, x)$ develops along the pipe. The region of the flow in which the effects of the viscous shearing forces caused by fluid viscosity are felt is called the *boundary layer*. The hypothetical boundary surface divides the flow in a pipe into two regions: the *boundary layer region*, in which the viscous effects and the velocity changes are significant, and the *inviscid core flow region*, in which the frictional effects are negligible and the velocity remains essentially constant in the radial direction.

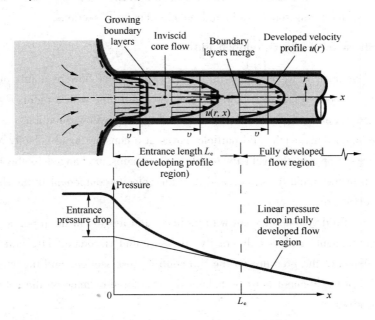

Fig. 6-5 Developing velocity profile, boundary layer and pressure changes of laminar flow in the entrance of a duct flow

At a finite distance from the entrance, the boundary layers merge and the inviscid core disappears. The pipe flow is then entirely viscous. The region from the pipe inlet to the point at which the boundary layers merge at the centerline is called the *entrance region,* and the length of this region is called the *entrance length* L_e. The region beyond the entrance region ($x > L_e$) in which the velocity profile is fully developed and remains unchanged, $u = u(r)$ only, is called the *fully developed region*. The velocity profile in the fully developed region is *parabolic* in laminar flow (Re ⩽ 2300) (see Fig. 6-5 and Section 6.4) and somewhat *flatter* (see Section 6.5) in turbulent flow. Besides, the wall shear in this region is constant due to the constant velocity profile, and the pressure drops linearly with x (Fig. 6-5) for either laminar or turbulent flow.

As with many other properties of pipe flow, the dimensionless *entrance length*, L_e/d correlates quite well with the Reynolds number. Typical entrance lengths are given by

Laminar flow $\qquad\qquad \dfrac{L_e}{d} = 0.06 \mathrm{Re}$

Turbulent flow $\qquad\qquad \dfrac{L_e}{d} = 4.4 \mathrm{Re}^{1/6}$

For very low Reynolds number flows the entrance length can be quite short ($L_e = 0.6d$ if Re = 10), whereas for large Reynolds number flows it may take a length equal to many pipe diameters before the end of the entrance region is reached ($L_e = 120d$ for Re = 2000). For many practical engineering problems, $10^4 < \mathrm{Re} < 10^5$, so that $20d < L_e < 30d$, and the pipe length would be much longer than the entrance length. Thus, the entrance effect may be neglected and a simple analysis made for fully developed flow (Sections 6.4 and 6.5). This is possible for both laminar and turbulent flows, including rough walls and noncircular cross sections.

6.3.2 Shear Stress for Fully Developed Flow

For a fully developed flow, consider the classic problem of flow in a full pipe, driven by pressure or gravity or both. Fig. 6-6a shows the geometry of a pipe with a radius R and its axis inclined to the horizontal at an angle θ. The x-axis is taken in the flow direction.

In fully developed flow, each fluid particle moves at a constant axial velocity and zero acceleration along a streamline and the velocity profile $u(r)$ remains unchanged in the flow direction. There is no motion in the radial direction, and thus the velocity component in the direction normal to flow is zero everywhere.

Now consider a fluid volume element with radius r, and length l in the pipe, as shown in Fig. 6-6b. The volume element involves only gravity, pressure and viscous effects, and thus the gravity on the whole element, the pressure on the two control cross sections and the shear force acting on the lateral sides of the element must be balanced. This force balance on the volume element in the flow direction gives

$$(p + \Delta p) A' - pA' + \rho g A' l \sin\theta - \tau \chi' l = 0$$

where A', χ' are the cross sectional area and the wetted perimeter of the r-radius fluid volume, respectively.

It can be simplified to give the expression of *pressure drop*

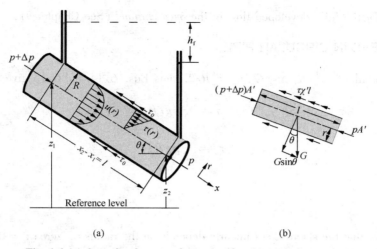

Fig. 6-6 (a) Control volume, velocity profile $u(r)$ and shear stress distribution $\tau(r)$ of steady, fully developed flow between two sections in an inclined pipe, and (b) Free-body diagram of a fluid cylinder for flow in a nonhorizontal pipe

$$\Delta p = \tau \frac{\chi' l}{A'} - \rho g l \sin\theta = \frac{\tau l}{R'_h} - \rho g l \sin\theta \qquad (6.11)$$

with $l\sin\theta = z_1 - z_2$ and $\Delta p = p_1 - p_2$ it gives

$$\left(z_1 + \frac{p_1}{\rho g}\right) - \left(z_2 + \frac{p_2}{\rho g}\right) = \frac{\tau l}{\rho g R'_h} \qquad (6.12)$$

where R'_h is the hydraulic radius of the r-radius fluid volume element, $R'_h = A'/\chi'$.

In the fully developed flow, the velocity profiles remains unchanged in the flow direction, thus we have $v_1 = v_2$ and $\alpha_1 = \alpha_2$. Thus, the application of energy equation for steady incompressible flow gives the frictional head loss

$$h_f = \left(z_1 + \frac{p_1}{\rho g}\right) - \left(z_2 + \frac{p_2}{\rho g}\right)$$

Combining Eq. 6.12 yields

$$h_f = \frac{\tau l}{\rho g R'_h} \qquad (6.13)$$

or

$$\tau = \rho g R'_h h_f / l = \rho g R'_h J \qquad (6.14)$$

where $J = h_f/l$ is the *frictional slope* of flow defined in Section 4.3.3.

At the pipe wall, $R'_h = R_h$, $\tau = \tau_0$, where τ_0 is the shear stress acting on the pipe wall, called the *wall shear stress*. Substituting into Eq. 6.14 gives

$$\tau_0 = \rho g R_h J = \rho g R_h h_f / l \qquad (6.15)$$

or

$$h_f = \frac{\tau_0 l}{\rho g R_h} \qquad (6.16)$$

Equations 6.15 and 6.16 interpret the relation of the frictional loss to the wall shear stress. It was developed from the fully developed flow in a pipe, but it is experimentally verified that it is still available for the fully developed flow in the *open channel* (See Chapter 9).

SHEAR STRESS IN CIRCULAR PIPE

For the circular flow, $R'_h = r/2$, $R_h = R/2$, thus Eqs. 6.14 and 6.15 give

$$\tau = \frac{1}{2}\rho g r J \qquad (6.17)$$

$$\tau_0 = \frac{1}{2}\rho g R J \qquad (6.18)$$

which yields

$$\tau = \frac{r}{R}\tau_0 \qquad (6.19)$$

It indicates that the shear stress linearly depends on the radius r: zero ($\tau = 0$) at the centerline of the circular pipe ($r = 0$) and maximum ($\tau = \tau_0$) at the pipe wall ($r = R$) as shown in Fig. 6-6a.

PRESSURE DROP IN HORIZONTAL CIRCULAR PIPE

For horizontal pipe, $\Delta z = l\sin\theta = 0$. Thus, the pressure drop of fully developed steady flow in a horizontal constant diameter circular pipe can be obtained from Eqs. 6.11 and 6.14

$$\Delta p = \rho g J l$$

Consider the frictional slope in Eq. 6.18 gives the expression of pressure drop related to wall shear stress in a horizontal circular pipe

$$\Delta p = \frac{2\tau_0 l}{R} = \frac{4\tau_0 l}{d} \qquad (6.20)$$

where d is the pipe diameter, $d = 2R$.

Fully developed steady flow in a constant diameter pipe may be driven by gravity and/or pressure forces. For horizontal pipe flow, $\theta = 0$, gravity has no effect in Eq. 6.11 except for a hydrostatic pressure variation across the pipe that is usually negligible. It is the pressure difference, $\Delta p = p_1 - p_2$, between one section of the horizontal pipe and another which forces the fluid through the pipe. Viscous effects provide the restraining force that exactly balances the pressure force, thereby allowing the fluid to flow through the pipe with no acceleration. If viscous effects were absent in such flows, the pressure would be constant throughout the pipe, except for the hydrostatic variation.

6.3.3 Friction Velocity

Consider the *Darcy-Weisbach equation*, Eq. 6.2, $h_f = \lambda \frac{l}{4R_h}\frac{v^2}{2g}$, and Eq. 6.16, $h_f = \frac{\tau_0 l}{\rho g R_h}$

to give

$$\tau_0 = \frac{\lambda}{8}\rho v^2 \qquad (6.21)$$

Define $u_* = \sqrt{\dfrac{\tau_0}{\rho}}$, the *friction velocity*, a quantity that has dimensions of velocity and represents the magnitude of wall shear stress, we have

$$u_* = v\sqrt{\dfrac{\lambda}{8}} \tag{6.22}$$

It indicates a relation between u_*, v and λ, which is commonly used in the analysis of energy losses in turbulent flow.

6.4 Fully Developed Laminar Flow in Pipes

In this section we consider the steady laminar flow of an incompressible fluid with constant properties in the fully developed region of a straight circular pipe. We obtain the velocity profile by combining the Newton's law of viscosity and the distribution of shear stress in circular pipe flow. Then we use it to obtain a relation for the friction factor.

6.4.1 Velocity Profile

For laminar flow of a Newtonian fluid, the shear stress is simply proportional to the velocity gradient as discussed in Section 1.4.1

$$\tau = \mu \dfrac{du}{dy}$$

For the laminar flow in a circular pipe shown in Fig. 6-7, we have $y = R - r$, thus $dy = -dr$, the above equation becomes

$$\tau = -\mu \dfrac{du}{dr} \tag{6.23}$$

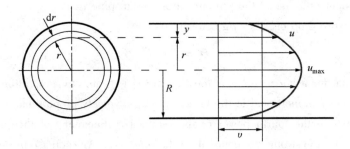

Fig. 6-7 Velocity profile of fully developed laminar flow in a circular pipe

Substitution of Eq. 6.23 into Eq. 6.17 yields

$$-\mu \dfrac{du}{dr} = \dfrac{1}{2}\rho g r J$$

or

$$du = -\dfrac{\rho g J}{2\mu} r dr$$

With a constant fluid density, viscosity and the frictional slope of flow it can be integrated to give the velocity profile as follows

$$u = -\frac{\rho g J}{4\mu}r^2 + C \tag{6.24}$$

where C is an integral constant.

Because the fluid is viscous it sticks to the pipe wall so that $u = 0$ at $r = R$. Thus, $C = \frac{\rho g J}{4\mu}R^2$. Hence, the velocity profile, Eq. 6.24, can be rewritten as

$$u = \frac{\rho g J}{4\mu}(R^2 - r^2) \tag{6.25}$$

At the pipe axis, $r = 0$, the velocity reaches a maximum

$$u_{max} = \frac{\rho g J}{4\mu}R^2 \tag{6.26}$$

Thus, Eq. 6.25 can be rewritten as the following form

$$u = u_{max}\left[1 - \left(\frac{r}{R}\right)^2\right] \tag{6.27}$$

Therefore, the velocity profile of a fully developed laminar flow in a circular pipe, plotted in Fig. 6-7, is *parabolic* in the radial coordinate, r, with a maximum velocity, u_{max}, at the pipe centerline, and a minimum velocity (zero) at the pipe wall.

Integrating Eq. 6.25 over the whole pipe section we can have the volume flow rate

$$Q = \int_A u dA = \int_0^R \frac{\rho g J}{4\mu}(R^2 - r^2)2\pi r dr = \frac{\rho g J}{8\mu}\pi R^4 \tag{6.28}$$

This equation is known as *Poiseuille's law*, and this flow is called *Hagen-Poiseuille flow* in honor of the works of Gotthilf Heinrich Ludwig Hagen (1797 – 1884), a German physicist and hydraulic engineer, and Jean Louis Marie Poiseuille (1799 – 1869), a French physician and physiologist, both of whom established the pressure-drop law in pipe flow.

With the pressure drop of $\Delta p = \rho g J l$ we have

$$Q = \frac{\pi d^4 \Delta p}{128\mu l} \tag{6.29}$$

It indicates that for a *horizontal pipe* the flowrate is (a) directly proportional to the pressure drop Δp, (b) inversely proportional to the viscosity, (c) inversely proportional to the pipe length, and (d) proportional to the fourth power of the radius (or diameter) of the pipe. The last one shows that the flow rate is strongly dependent on the pipe size. An increase in diameter by a factor of 2 will increase the flow rate by a factor of 16; or a 2% error in diameter gives an 8% error in flow rate because $\frac{\Delta Q}{Q} = \frac{4\Delta d}{d}$.

By definition, the average velocity is the flowrate divided by the cross-sectional area

$$v = \frac{Q}{A} = \frac{Q}{\pi R^2} = \frac{\rho g J}{8\mu}R^2 = \frac{u_{max}}{2} \tag{6.30}$$

Thus, the sectional average velocity in the fully developed laminar pipe flow is one half of its maximum velocity at this section. It indicates the velocity profile of the fully developed laminar flow varies a lot with a large nonuniformity, which leads to a larger kinematic correction factor

$$\alpha = \frac{\int_A u^3 \, dA}{v^3} = 2.0$$

and momentum correction factor

$$\beta = \frac{\int_A u^2 \, dA}{v^2} = \frac{4}{3}$$

which have been discussed in Section 4.3.2 and Section 4.4.2.

For *non-horizontal pipe* as in Fig. 6-6a, from the energy equation we can have the relation of the pressure drop of a steady incompressible flow in a constant diameter pipe to the frictional slope as:

$$\Delta p = \rho g J l - \rho g l \sin\theta \tag{6.31}$$

Thus, from Eq. 6.28 we have

$$Q = \frac{\pi d^4 (\Delta p + \rho g l \sin\theta)}{128 \mu l} \tag{6.32}$$

6.4.2 Frictional Loss in Circular Pipe

A quantity of interest in the analysis of pipe flow is the *pressure drop*, Δp. With $\Delta p = \rho g J l = \rho g h_f$ in a horizontal pipe Eq. 6.29 gives

$$h_f = \frac{128 \mu l Q}{\rho g \pi d^4} = \frac{32 \mu l}{\rho g d^2} v \tag{6.33}$$

or the well-known *Darcy – Weisbach equation* for the steady incompressible fully developed laminar flow

$$h_f = \frac{64}{\text{Re}} \frac{l}{d} \frac{v^2}{2g} = \lambda \frac{l}{d} \frac{v^2}{2g} \tag{6.34}$$

where the Darcy friction factor for the laminar flow is

$$\lambda = \frac{64}{\text{Re}} \tag{6.35}$$

It indicates that in the fully developed laminar flow, λ drops off with increasing Re, but independent of the roughness of pipe surface. Eq. 6.33 shows that frictional head loss in a fully developed laminar flow is proportional to the average velocity, v, as what the experimental results show in Section 6.2.1.

EXAMPLE 6-3 Fully developed laminar flow in a circular pipe

Oil at 10℃ ($\rho = 850 \text{kg/m}^3$ and $\nu = 0.18 \times 10^{-4} \text{m}^2/\text{s}$) is flowing through a 0.1-m diameter horizontal pipe steadily at an average velocity of 0.0635m/s. The flow is a fully developed laminar flow. Determine (a) the maximum velocity at the pipe centerline, (b) the velocity at the point with a radius of $r = 0.02$m, (c) the fiction factor λ, and (d) the wall shear stress and the major head loss per km length of pipe.

Solution:

(a) The average velocity of oil flow in the pipe is given as 0.0635m/s. Thus, for the fully developed laminar flow, the maximum velocity at the pipe centerline can be determined by Eq. 6.30

$$u_{max} = 2v = 2(0.0635 \text{m/s}) = 0.127 \text{m/s}$$

(b) With the maximum velocity, the velocity at the point at a radius of $r = 0.02$m can be determined by Eq. 6.27

$$u = u_{max}\left[1 - \left(\frac{r}{R}\right)^2\right] = (0.127 \text{m/s})\left[1 - \left(\frac{0.02\text{m}}{0.05\text{m}}\right)^2\right] = 0.107 \text{m/s}$$

(c) For the fully developed laminar flow in a circular pipe, the fiction factor is reversed proportional to the Reynolds number

$$\lambda = \frac{64}{Re} = \frac{64}{vd/\nu} = \frac{(64)(0.18 \times 10^{-4} \text{ m}^2/\text{s})}{(0.0635 \text{m/s})(0.1 \text{m})} = 0.181$$

(d) The wall shear stress is then determined by Eq. 6.21

$$\tau_0 = \frac{\lambda}{8}\rho v^2 = \frac{0.181}{8}(850 \text{kg/m}^3)(0.0635 \text{m/s})^2 = 0.0775 \text{N/m}^2$$

and the major head loss per km length of pipe determined by Eq. 6.34, the *Darcy-Weisbach* equation

$$h_f = \lambda \frac{l}{d}\frac{v^2}{2g} = 0.181 \frac{1000\text{m}}{0.1\text{m}} \frac{(0.0635 \text{m/s})^2}{2(9.81 \text{m/s}^2)} = 0.372 \text{m}$$

EXAMPLE 6-4 Head loss in a horizontal pipe

Water at 10℃ is steadily flowing through a 0.28-cm diameter, 10-m-long horizontal pipe involving no components at an average velocity of 0.9 m/s. Determine (a) the head loss, (b) the pressure drop, and (c) the pumping power requirement to overcome this pressure drop.

Solution:

From Table A-1 we have the fluid properties of water at 10℃: $\rho = 999.7 \text{kg/m}^3$ and $\mu = 1.307 \times 10^{-3}$ Pa·s.

(a) First we need to determine the flow regime. The Reynolds number of the flow is

$$Re = \frac{\rho vd}{\mu} = \frac{(999.7 \text{kg/m}^3)(0.9 \text{m/s})(0.0028\text{m})}{1.307 \times 10^{-3} \text{Pa·s}} = 1927$$

which is less than 2300. Therefore, the flow is laminar. Then, with the average flow velocity the head loss becomes

$$h_f = \lambda \frac{l}{d}\frac{v^2}{2g} = \frac{64}{Re}\frac{l}{d}\frac{v^2}{2g} = \frac{64}{1927}\frac{10\text{m}}{0.0028\text{m}}\frac{(0.9\text{m/s})^2}{2(9.81 \text{m/s}^2)} = 4.897 \text{m}$$

(b) Noting that the pipe is horizontal and its diameter is constant, the pressure drop in the pipe is due entirely to the frictional losses and is equivalent to the pressure loss

$$\Delta p = \rho g h_f = (999.7 \text{kg/m}^3)(9.81 \text{m/s}^2)(4.897 \text{m}) = 48.02 \text{kPa}$$

(c) The pumping power requirements is

$$\dot{N}_{pump} = \rho g Q h_f = \Delta p \frac{1}{4}\pi d^2 v = (48.02 \text{kPa})\frac{\pi}{4}(0.0028\text{m})^2(0.9 \text{m/s}) = 0.266 \text{W}$$

Therefore, power input in the amount of 0.266W is needed to overcome the frictional losses in the flow due to viscosity.

DISCUSSION The pressure rise provided by a pump is often listed by a pump manufacturer in units of head. Thus, the pump in this flow needs to provide 4.897m of water head in order to overcome the irreversible head loss.

6.5 Fully Developed Turbulent Flow in Pipes

6.5.1 Reynolds' Time-averaged Concepts

Most flows encountered in engineering practice are turbulent. However, turbulent flow is very complex process, of which the theory remains largely undeveloped.

Turbulent flow is characterized by chaotic, random and rapid fluctuations of eddies and considerable mixing of fluid throughout the flow field. Such finite-sized random mixing is an additional mechanism for momentum and energy transfer and very effective in transporting energy and mass. In laminar flow, fluid particles flow smoothly in layers, one over another, momentum and energy are transferred across streamlines by molecular diffusion. In turbulent flow, the swirling eddies transport mass, momentum, and energy to other regions of flow much more rapidly than molecular diffusion, greatly enhancing mass, momentum, and heat transfer.

Fortunately, engineer practices are toward the average or *mean* values of velocity, pressure, shear stress, etc., in a high-Reynolds-number (turbulent) flow. Fig. 6-8 shows the variation of the instantaneous velocity component u_x (in x-direction) with time at a specified location, as can be measured with a hot-wire anemometer probe or other sensitive device. We observe that the instantaneous values of the velocity fluctuate about an average value, which suggests that the velocity can be expressed as the sum of an *average value* $\overline{u_x}$ and a *fluctuating component* u'_x

$$u_x = \overline{u_x} + u'_x \qquad (6.36)$$

where the time mean $\overline{u_x}$ in a turbulent flow is defined by

$$\overline{u_x} = \frac{1}{T}\int_0^T u_x \mathrm{d}t \qquad (6.37)$$

where T is an averaging period taken to be longer than any significant period of the fluctuations themselves so that the time average levels off to a constant. For turbulent gas and water flows, an averaging period $T \approx 5\mathrm{s}$ is usually quite adequate.

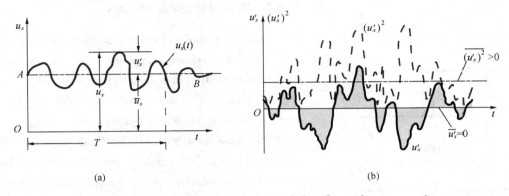

Fig. 6-8 (a) Fluctuations of velocity u_x in a turbulent flow with respect to time;
(b) Average of the fluctuations and average of the square of the fluctuations

The *fluctuation* u'_x is defined as the deviation of u_x from its average value $\overline{u_x}$

$$u'_x = u_x - \overline{u_x} \qquad (6.38)$$

also shown in Fig. 6-8a. It follows by definition that a fluctuation has zero mean value

$$\overline{u'^T_x} = \frac{1}{T}\int_0^T (u_x - \overline{u_x})\,dt = \frac{1}{T}\left[\int_0^T u_x\,dt - \overline{u_x}\int_0^T dt\right] = \frac{1}{T}(T\overline{u_x} - T\overline{u_x}) = 0 \qquad (6.39)$$

It indicates that the fluctuations are equally distributed on either side of the average as in Fig. 6-8b. However, the mean square of a fluctuation is not zero and must be positive

$$\overline{u'^2_x} = \frac{1}{T}\int_0^T u'^2_x\,dt > 0 \qquad (6.40)$$

as shown in Fig. 6-8b.

The structure and characteristics of turbulence may vary from one flow situation to another. It can be measured by the *turbulence intensity*, N, defined as the square root of the mean square of the fluctuating velocity divided by the time-averaged velocity, or

$$N = \frac{\sqrt{\overline{u'^2}}}{\overline{u}} = \frac{\sqrt{\frac{1}{3}(\overline{u'^2_x} + \overline{u'^2_y} + \overline{u'^2_z})}}{\overline{u}} \qquad (6.41)$$

where $\overline{u}$ is the mean value of velocity $u(x, y, z, t)$ at a specified point in a turbulent flow; u' is the fluctuating components of velocity $u(x, y, z, t)$; u'_x, u'_y, u'_z are the coordinate components of the fluctuating u' in the x-, y- and z-directions respectively.

The larger the turbulence intensity is, the larger the fluctuations of the velocity (and other flow parameters), which cause the larger pressure drop. Well-designed wind tunnels have typical values of $N \approx 0.01$, while values of $N \geq 0.1$ are found for the flow in the atmosphere and rivers.

For the turbulent flow, the sectional averaged velocity is defined as

$$v = \frac{1}{A}\int_A \overline{u}\,dA \qquad (6.42)$$

This is also the case for other properties to be split into mean plus fluctuating variables

$$u_y = \overline{u_y} + u'_y \qquad p = \overline{p} + p' \qquad \Theta = \overline{\Theta} + \Theta'$$

6.5.2 Reynolds Shear Stresses

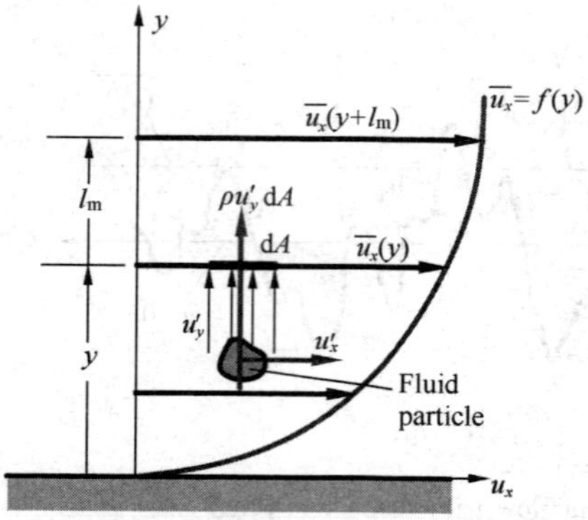

Consider turbulent flow in a plane as shown in Fig. 6-9, the random velocity components that account for the momentum transfer (hence, the shear force) are u'_x (for the x component of velocity) and u'_y (for the rate of mass transfer crossing the differential area dA).

The mass flow rate of the fluid particles rising through dA is $\rho u'_y dA$, and its net effect on the layer above dA is a reduction in its average flow velocity because of momentum transfer to the fluid particles with lower average flow velocity. Similarly, a fluid particle moving down across through dA to a layer below dA that

Fig. 6-9 Turbulent shear stress and the mixing length

is traveling with a lower x-component velocity would tend to accelerate the slower moving fluid. The x-component force acting on the fluid element above dA due to the passing of fluid particles through dA would be

$$dF = (\rho u'_y dA)(-u'_x) = -\rho u'_x u'_y dA$$

where u'_x is the negative change in x-component velocity of the fluid element above dA due to the momentum exchange; $\rho u'_y dA$ is the mass flow rate through the area dA; the negative sign provides a positive dF.

Thus, if we divide both sizes by the area dA we obtain an instantaneous *turbulent shear stress*

$$\tau_t = \frac{dF}{dA} = -\rho u'_x u'_y$$

where we know that $u'_x u'_y$ is, on the average, a negative quantity since a positive u'_y produces a negative u'_x.

This shear stress is actually a momentum exchange but since it has the same effect as a stress, we call it a shear stress. The time-averaged of this shear stress, often called the *turbulent shear stress*, or *apparent shear stress*, can be expressed as

$$\overline{\tau_t} = -\rho \overline{u'_x u'_y} \tag{6.43}$$

where $\overline{u'_x u'_y}$ is the time-averaged value of the product of the fluctuating velocity components u'_x and u'_y. Note that $\overline{u'_x u'_y} \neq 0$ even though $\overline{u'_x} = 0$ and $\overline{u'_y} = 0$ (and thus $\overline{u'_x}\,\overline{u'_y} = 0$), and experimental results show that $\overline{u'_x u'_y}$ is usually a negative quantity. The turbulent shear stresses, such as $-\rho \overline{u'_x u'_y}$ or $-\rho \overline{u'^2_x}$, are also called *turbulent stresses* or *Reynolds stresses* in honor of Osborne Reynolds who first discussed them in 1895.

Reynolds shear stress originates from the particles' inertia, and is dependent on fluid density and fluctuating intensity, and independent of the fluid viscosity.

SEMI-EMPIRICAL THEORY OF PRANDTL

The random eddy motion of groups of particles resembles the random motion of molecules in a gas—colliding with each other after traveling a certain distance and exchanging momentum in the process. Therefore, momentum transported by eddies in turbulent flows is analogous to the molecular momentum diffusion. In many of the simpler turbulence models, turbulent shear stress is expressed in an analogous manner as suggested by the French mathematician Joseph Boussinesq (1842–1929) in 1877 as

$$\overline{\tau_t} = -\rho \overline{u'_x u'_y} = \mu_t \frac{d\overline{u}}{dy} \tag{6.44}$$

where μ_t is the *eddy viscosity* or *turbulent viscosity*, which accounts for momentum transport by turbulent eddies and is independent of the fluid viscosity; $d\overline{u}/dy$ is the mean velocity gradient.

The concept of eddy viscosity is very appealing, but it is of no practical use unless its value can be determined. Several semi-empirical theories have been proposed to determine approximate values of eddy viscosity. Ludwig Prandtl (1875–1953), a German physicist and aerodynamicist, proposed that the turbulent process could be viewed as the random transport of bundles of fluid particles over a certain distance, l_m, the *mixing length* as illustrated in Fig. 6-9, from a region of one velocity to another region of a different velocity

$$\mu_t = \rho l_m \frac{d\bar{u}}{dy} \tag{6.45}$$

and expressed the turbulent shear stress as

$$\bar{\tau}_t = \mu_t \frac{d\bar{u}}{dy} = \rho l_m^2 \left(\frac{d\bar{u}}{dy}\right)^2 \tag{6.46}$$

Further considerations indicate that l_m is not a constant throughout the flow field. Near a solid surface, the turbulence is dependent on the distance from the surface, y, the mixing length l_m is proportional to the distance from the surface

$$l_m = \kappa y \tag{6-47}$$

where κ is a dimensionless constant and found to have the approximate value of 0.41 over the full range of turbulent smooth wall flows.

6.5.3 Shear Stresses in Different Flow Layers

As what discussed above, the flow properties can be split into the mean and fluctuating variables. Thus, it is convenient to think of the turbulent shear stress as consisting of two parts: the *laminar component*, which accounts for the friction between mean-averaged velocity layers in the flow direction and satisfies the well-known *Newton's law of viscosity*

$$\overline{\tau_{lam}} = \mu \frac{d\bar{u}}{dy} \tag{6.48}$$

and the *turbulent component*, which accounts for the momentum exchange between the fluctuating fluid particles and the fluid body as discussed above, and is related to the fluctuation components of velocity

$$\bar{\tau}_t = -\rho \overline{u'_x u'_y} \tag{6.49}$$

Then the *total shear stress* in turbulent flow can be expressed as

$$\bar{\tau} = \overline{\tau_{lam}} + \bar{\tau}_t = \mu \frac{d\bar{u}}{dy} - \rho \overline{u'_x u'_y} = \rho(\nu + \nu_t) \frac{d\bar{u}}{dy} \tag{6.50}$$

where ν_t is called the *kinematic eddy viscosity* or *kinematic turbulent viscosity*, $\nu_t = \mu_t/\rho$. It should be note that *kinematic eddy viscosity*, ν_t, as well as *eddy viscosity*, μ_t, is *not* a fluid property, and its value depends on flow conditions. That is, the kinematic eddy viscosity of water cannot be looked up in handbooks-its value changes from one turbulent flow condition to another and from one point in a turbulent flow to another. Kinematic eddy viscosity decreases toward the wall. Its value ranges from zero at the wall to several thousand times the value of the molecular diffusivity in the core region.

If the flow is laminar, $\overline{u'_x} = \overline{u'_y} = 0$, so that $\overline{u'_x u'_y} = 0$ and Eq. 6.45 reduces to the customary random molecule-motion-induced *laminar shear stress*, $\tau_{lam} = \mu du/dy$. For the turbulent flow, although the relative magnitude of $\overline{\tau_{lam}}$ compared to $\bar{\tau}_t$ is a complex function dependent on the specific flow involved, typical measurements indicate the structure shown in Fig. 6-10a. In a very narrow region near the wall (the *viscous sublayer* in Fig. 6-10b) , the laminar shear stress is dominant. Away from the wall (in the *outer layer*) the turbulent portion of the shear stress is dominant. There is an intermediate region, called the *overlap layer*, where both laminar and turbulent shear are important. In the outer layer $\overline{\tau_t}$ is two or three orders of magnitude (100 to 1000 times) greater than $\overline{\tau_{lam}}$, and vice versa in the wall layer.

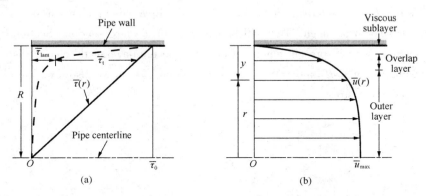

Fig. 6-10 Structure of turbulent flow in a pipe: (a) profile of shear stress and
(b) profile of velocity and the flow regimes

The viscous sublayer is usually a very thin layer adjacent to the wall. For example, for water flow in a 7.5-cm-diameter pipe with an average velocity of 3m/s, the viscous sublayer is approximately 0.05mm thick. Since the fluid motion within this thin layer is critical in terms of the overall flow (the no-slip condition and the wall shear stress occur in this layer) , it is not surprising to find that turbulent pipe flow properties can be quite dependent on the roughness of the pipe wall, unlike laminar pipe flow which is independent of roughness. Small roughness elements (scratches, rust, sand or dirt particles, etc.) can easily disturb this viscous sublayer (see Section 6.6), thereby affecting the entire flow.

6.5.4 Velocity Profile in Turbulent Flow

As is indicated in Fig. 6-10, fully developed turbulent flow in a pipe can be broken into three regions (layers) which are characterized by their distances from the wall. The *viscous* (or *laminar*) *sublayer* is very near the pipe wall, in which the viscous shear stress is dominant compared with the turbulent (or Reynolds) stress, and the random, eddying nature of the flow is essentially absent. The velocity profile in this layer is very nearly linear, and the flow is streamlined. Next to the viscous sublayer is the *overlap* (or *transition*) *layer*, in which turbulent effects are becoming significant, but still not dominant. Out of that is the *outer* (or *turbulent*) *layer* throughout the center portion of the flow in which turbulent effects dominate over molecular diffusion (viscous) effects and there is considerable mixing and randomness to the flow.

The time-average velocity profile in a pipe is quite sensitive to the magnitude of the average wall *roughness height* Δ, as sketched in Fig. 6-11. All materials are 'rough' when viewed with sufficient magnification, although glass and plastic are assumed to be smooth with $\Delta = 0$. As noted

239

in the preceding section, the laminar shear is significant only near the wall in the viscous wall layer with a thickness of δ_v, which decrease with the increasing Re. If the thickness δ_v is sufficient large, it submerges the wall roughness so that they have negligible effect on the flow, it is as if the wall were smooth. Such a condition is often referred to as being *hydraulically smooth*. If the thickness of the viscous wall layer is relatively thin, the roughness elements protrude out of this layer and the wall is *rough*. The relative roughness Δ/d and the Reynolds number can be used to determine if a pipe is smooth or rough. This will be observed in the next section, Section 6.6.

In this section, we will present empirical data for the time-averaged velocity distribution involved flows with *smooth walls* and with *rough walls*. From now on we also drop the *overbar* from the time averaged shear stress.

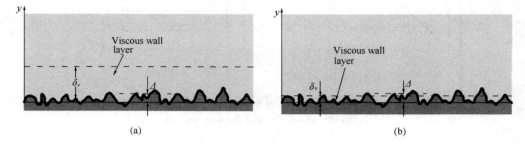

Fig. 6-11 A schematic of (a) smooth wall and (b) rough wall of a pipe

WALL REGION NEAR SMOOTH WALLS

Profile in viscous sublayer

The *thickness of viscous sublayer*, δ_v, is usually very small (typically, much less than 1 percent of the pipe diameter), but this thin layer next to the wall plays a dominant role on flow characteristics because of the large velocity gradients it involves. The wall dampens any eddy motion, and thus the flow in this layer is essentially laminar and the shear stress consists of laminar shear stress which is proportional to the fluid viscosity. Experiments confirm that in this layer the velocity profile changes from zero to nearly the core region value in a nearly *linear* manner. Then the wall shear stress can be expressed as

$$\tau_0 = \mu \frac{du}{dy} = \mu \frac{u}{y}$$

which gives

$$u = \frac{\tau_0}{\mu} y$$

or

$$\frac{u}{u_*} = \frac{y u_*}{\nu} = y^+ \qquad 0 \leq y^+ \leq 5 \qquad (6.51)$$

where u_* is the *friction velocity*, or the *shear velocity*, $u_* = \sqrt{\tau_0/\rho}$; y is the distance from the wall (note that $y = R - r$ for a circular pipe as indicated in Fig. 6-10a); y^+ is nondimensionalized distance, $y^+ = \dfrac{yu_*}{\nu}$.

Equation 6.51 represents the velocity profile in the viscous sublayer near a smooth wall, and is known as the *law of the wall*.

Profile in overlap layer

In the overlap layer, the experimental data for velocity are observed to line up on a straight line when plotted against the logarithm of distance from the wall as

$$\frac{u}{u_*} = \frac{1}{\kappa}\ln y + c \qquad (6.52)$$

where κ and c are constant and can be determined experimentally.

Equation 6.52 is the general equation of velocity profile for turbulent flow in a circular pipe and is known as the *logarithmic law*.

For the velocity at the boundary of viscous sublayer ($y = \delta_\nu$), $u = u_b$ (where u_b is the velocity at the interface between viscous sublayer and overlap layer), from Eq. 6.52 we have

$$c = \frac{u_b}{u_*} - \frac{1}{\kappa}\ln \delta_\nu$$

and from Eq. 6.51 to give

$$\delta_\nu = \frac{u_b}{u_*^2}\nu$$

Substituting c and δ_ν into Eq. 6.52 and rearranging it yields

$$\frac{u}{u_*} = \frac{1}{\kappa}\ln\frac{yu_*}{\nu} + \frac{u_b}{u_*} - \frac{1}{\kappa}\ln\frac{u_b}{u_*}$$

or

$$\frac{u}{u_*} = \frac{1}{\kappa}\ln\frac{yu_*}{\nu} + c_1$$

where κ and $c_1 \left(= \dfrac{u_b}{u_*} - \dfrac{1}{\kappa}\ln\dfrac{u_b}{u_*} \right)$ are constants whose values are determined experimentally to be about 0.40 and 5.5, respectively.

Substituting the values of the constants back to the equation, the velocity profile in overlap layer is determined to be

$$\frac{u}{u_*} = 2.5\ln\frac{yu_*}{\nu} + 5.5 \quad \text{or} \quad \frac{u}{u_*} = 5.75\lg\frac{yu_*}{\nu} + 5.5 \quad \text{for} \quad y^+ > 30,\ \frac{y}{R} < 0.15 \qquad (6.53)$$

It turns out that the logarithmic law in Eq. 6.53 satisfactorily represents experimental data for the entire flow region except for the regions very close to the wall and near the pipe center, as shown in Fig. 6-12, and thus it is viewed as a *universal velocity profile* for turbulent flow in hydraulically smooth pipes or over surfaces. In the region $5 < y^+ < 30$, the experimental data do not fit either of the curves of Eq. 6.51 and 6.53 but merge the two curves as shown in Fig. 6-12. Such a region is named the *buffer zone* from viscous sublayer to the overlap layer.

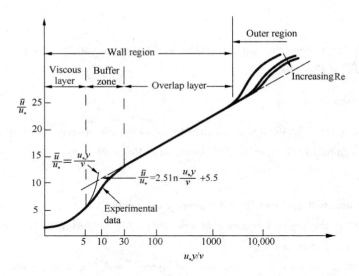

Fig. 6-12 Experimental verification of the inner-, outer-, and overlap layer laws relating velocity profiles of turbulent flow with a smooth wall (Reproduced from Fig. 7-11 in *Mechanics of Fluids* by Potter and Wiggert, 2002)

WALL REGION NEAR ROUGH WALLS

For rough pipes the viscous wall layer does not play an important role since turbulent initiates from the protruding wall elements, so only a logarithmic profile is necessary in the wall region. The boundary condition can be expressed as $y = \Delta$, $u = u_s$, in which u_s is the velocity at the interface of the pipe roughness clusters. Again, substitution into Eq. 6.52 yields

$$c = \frac{u_s}{u_*} - \frac{1}{\kappa}\ln\Delta$$

Substituting back into Eq. 6.52 and rearranging yields

$$\frac{u}{u_*} = \frac{1}{\kappa}\ln\frac{y}{\Delta} + \frac{u_s}{u_*}$$

or

$$\frac{u}{u_*} = \frac{1}{\kappa}\ln\frac{y}{\Delta} + c_2$$

where κ and $c_2\left(=\dfrac{u_s}{u_*}\right)$ are constants whose values are determined experimentally to be about 0.40 and 8.48, respectively.

Substituting the values of the constants back in the above equation, the dimensionless velocity profile in the wall region near rough wall is going to be

$$\frac{u}{u_*} = 2.5\ln\frac{y}{\Delta} + 8.48 \quad \text{or} \quad \frac{u}{u_*} = 5.75\lg\frac{y}{\Delta} + 8.48 \quad \text{for} \quad \frac{y}{R} < 0.15 \qquad (6.54)$$

OUTER REGION FOR SMOOTH AND ROUGH WALLS

In the outer region, the maximum velocity in a pipe occurs at the centerline, namely $y = R$, $u = u_{max}$. From Eq. 6.52 we have $c = \dfrac{u_{max}}{u_*} - \dfrac{1}{\kappa}\ln R$. Substituting back into Eq. 6.52 yields the empirical velocity relation for both smooth and rough pipes is

$$\dfrac{u_{max} - u}{u_*} = 2.5\ln\dfrac{R}{y} \quad \text{or} \quad \dfrac{u_{max} - u}{u_*} = 5.75\ln\dfrac{R}{y} \quad \text{for} \quad \dfrac{y}{R} \leqslant 0.15 \qquad (6.55)$$

Therefore, an additional empirical equation is needed to complete the profile for $0.15 < y/R \leqslant 1$.

Besides Eq. 6.55, an alternative, and simpler form that adequately describes the turbulent flow velocity distribution in a pipe is the *power-law velocity profile*, that is

$$\dfrac{u}{u_{max}} = \left(\dfrac{y}{R}\right)^{1/n} \quad \text{or} \quad \dfrac{u}{u_{max}} = \left(1 - \dfrac{r}{R}\right)^{1/n} \qquad (6.56)$$

where the exponent n is an integer between 5 and 10 depending on the Reynolds number and the pipe wall roughness Δ/d.

The value of n increases with increasing Reynolds number. For smooth pipes the exponents n is related to the Reynolds number as shown in Table 6-1, among which the value $n = 7$ generally approximates many flows in practice, giving rise to the term *one-seventh power-law velocity profile*.

Table 6-1　Exponents n for smooth pipes

$Re = vd/\nu$	4×10^3	2.3×10^4	1.1×10^5	1.1×10^6	$>2 \times 10^6$
n	6	6.6	7	8.8	10
u/u_{max}	0.791	0.808	0.817	0.849	0.865

Source: Data are from 刘鹤年(2004).

Various power-law velocity profiles are shown in Fig. 6-13 for $n = 6, 8$, and 10 together with the velocity profile for fully developed laminar flow for comparison. Note that the turbulent velocity profile is fuller than the laminar one with $u_{max} = (1.15 \sim 1.26)v$, while $u_{max} = 2v$ for laminar flow, where v is the averaged cross-sectional velocity. The turbulent velocity profile becomes more flat as n (and thus the Reynolds number) increases. Also note that the power-law profile cannot be used to calculate wall shear stress since it gives a velocity gradient of infinity there, and it fails to give zero slope at the centerline. But these regions of discrepancy constitute a small portion of flow, and the power-law profile gives highly accurate results for turbulent flow

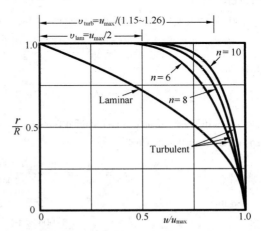

Fig. 6-13 Power-law velocity profiles for fully developed turbulent flow in a pipe for different exponents, and the fully developed laminar flow

through a pipe.

THICKNESS OF VISCOUS SUBLAYER

As discussed above, by comparing the thickness of viscous sublayer and the roughness of transitional pipe as shown in Fig. 6-11, the pipe wall could be hydraulically smooth, hydraulically transitional and hydraulically rough, among which *hydraulically transitional wall* is the wall where the values of Δ and δ_v are comparable.

If a point is located at the boundary of the viscous layer ($y = \delta_v$), the velocity at this point must satisfy both Eq. 6.51 and 6.53. Thus, the thickness of the viscous sublayer is roughly

$$\delta_v = 11.6 \frac{\nu}{u_*} \tag{6.57}$$

Introducing Eq. 6.22, $u_* = v\sqrt{\lambda/8}$, into the above equation gives

$$\delta_v = 11.6 \frac{\nu}{v\sqrt{\lambda/8}} = \frac{32.8d}{\text{Re}\sqrt{\lambda}} \tag{6.58}$$

It indicates that the thickness of the viscous sublayer is proportional to the kinematic viscosity and inversely proportional to the average flow velocity and the Reynolds number. In other words, the viscous sublayer is suppressed and gets thinner as the velocity (and thus the Reynolds number) increases. Consequently, the velocity profile becomes nearly flat and thus the velocity distribution becomes more uniform at very high Reynolds numbers.

EXAMPLE 6-5   Turbulent pipe flow properties

Water at 20℃ ($\rho = 998\text{kg/m}^3$ and $\nu = 1.004 \times 10^{-6}\text{m}^2/\text{s}$) flows through a horizontal pipe of 0.1-m diameter with a flowrate of $Q = 4 \times 10^{-2}\text{ m}^3/\text{s}$ and a pressure gradient of 2.59kPa/m. Determine (a) the approximate thickness of the viscous sublayer; (b) the approximate centerline velocity, $u_{\max}$; and (c) the ratio of the turbulent to laminar shear stress, τ_t/τ_{lam}, at a point midway between the centerline and the pipe wall (i.e., at $r = 0.025\text{m}$).

Solution:

(a) According to Eq. 6.51, the thickness of the viscous sublayer, δ_v, is approximately

$$\frac{\delta_v u_*}{\nu} = 5$$

Therefore,

$$\delta_v = 5\frac{\nu}{u_*} = 5\frac{\nu}{\sqrt{\tau_0/\rho}}$$

The wall shear stress can be obtained from the pressure drop data and Eq. 6.20, which is valid for either laminar or turbulent flows. Thus,

$$\tau_0 = \frac{d\Delta p}{4l} = \frac{(0.1\text{m})(2.59\text{kPa/m})(1\text{m})}{(4)(1\text{m})} = 64.8\text{N/m}^2$$

So that

$$\delta_v = 5\frac{\nu}{\sqrt{\tau_0/\rho}} = 5\frac{1.004 \times 10^{-6}\text{ m}^2/\text{s}}{\sqrt{(64.8\text{N/m}^2)(998\text{kg/m}^3)}} = 1.97 \times 10^{-5}\text{m} \approx 0.02\text{mm}$$

As stated previously, the viscous sublayer is very thin. Mini imperfections on the pipe wall will protrude into this sublayer and affect some of the characteristics of the flow (i.e., wall shear stress and pressure drop).

(b) The centerline velocity can be obtained from the average velocity and the assumption of a power-law velocity profile. For this flow with

$$v = \frac{Q}{A} = \frac{4 \times 10^{-2} \text{ m}^3/\text{s}}{\pi (0.1\text{m})^2/4} = 5.09 \text{m/s}$$

The Reynolds number is

$$\text{Re} = \frac{vd}{\nu} = \frac{(5.09\text{m/s})(0.1\text{m})}{1.004 \times 10^{-6} \text{ m}^2/\text{s}} = 5.07 \times 10^5$$

Thus, from Table 6-1, $n = 7.72$ so that

$$\frac{u}{u_{max}} = \left(1 - \frac{r}{R}\right)^{1/7.72} \tag{1}$$

To determine the centerline velocity, u_{max}, we must know the relationship between the average velocity v and u_{max}. This can be obtained by integration of the power-law velocity profile as

$$Q = \int u dA = u_{max} \int_{r=0}^{r=R} \left(1 - \frac{r}{R}\right)^{1/n} 2\pi r dr$$

which can be integrated to give

$$Q = 2\pi R^2 u_{max} \frac{n^2}{(n+1)(2n+1)}$$

With $Q = Av = \pi R^2 v$, we have

$$u_{max} = \frac{(n+1)(2n+1)}{2n^2} v = 1.203v = (1.203)(5.09\text{m/s}) = 6.12\text{m/s}$$

(c) From Eq. 6.19, which is valid for laminar or turbulent flow, the shear stress at $r = 0.025$m is

$$\tau = \frac{r}{R} \tau_0 = \frac{0.025\text{m}}{0.05\text{m}} (64.8\text{N/m}^2) = 32.4\text{N/m}^2$$

With the power-law velocity profile, Eq. 1, the laminar component of the total shear stress can be obtained from the Newtonian law of viscosity

$$\tau_{lam} = \mu \frac{du}{dy} = -\mu \frac{du}{dr} = \rho \nu \frac{u_{max}}{nR}\left(1 - \frac{r}{R}\right)^{\frac{1-n}{n}}$$

$$= (998\text{kg/m}^3)(1.004 \times 10^{-6} \text{ m}^2/\text{s}) \frac{6.12\text{m/s}}{(7.72)(0.05\text{m})}\left(1 - \frac{0.025\text{m}}{0.05\text{m}}\right)^{\frac{1-7.72}{7.72}} = 0.029\text{N/m}^2$$

Thus, the ratio of turbulent to laminar shear stress is given by

$$\frac{\tau_t}{\tau_{lam}} = \frac{\tau - \tau_{lam}}{\tau_{lam}} = \frac{32.4\text{N/m}^2 - 0.029\text{N/m}^2}{0.029\text{N/m}^2} = 1116$$

DISCUSSION As expected, most of the shear stress at this location in the turbulent flow is due to the turbulent shear stress.

6.6 Frictional Losses in Turbulent Flow

6.6.1 Nikuradse's Experiments

As what discussed in Section 6.1, the major losses can be determined by the *Darcy-Weisbach equation*

$$h_f = \lambda \frac{L}{d} \frac{v^2}{2g} \tag{6.2}$$

The friction factor λ in fully developed laminar pipe flow only depends on the Reynolds number, $\lambda = 64/\mathrm{Re}$ as indicated in Eq. 6.35. However, for the fully developed turbulent pipe flow, it may depend on both the Reynolds number and the *relative roughness*, Δ/d, which is the ratio of the mean height of roughness of the pipe, Δ, to the pipe diameter

$$\lambda = f(\mathrm{Re}, \Delta/d)$$

The functional form of this dependence cannot be obtained from a theoretical analysis, and all available results are obtained from painstaking experiments using artificially roughened surfaces (usually by gluing sand grains of a known size on the inner surfaces of the pipes). Most such experiments were conducted by Prandtl's student Johann Nikuradse (1894 – 1979), a German engineer and physicist, in 1933, followed by the works of others. The friction factor was calculated from the measurements of the flow rate and the pressure drop at the corresponding Reynolds number and relative roughness, as indicated in Fig. 6-14, from which the following characteristics are observed:

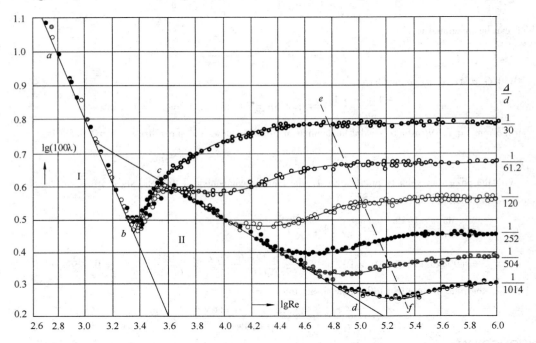

Fig. 6-14 Experiments with sand-grain roughness by Nikuradse show a systematic increase of the turbulent friction factor with the roughness ratio

- Line ab: *laminar zone*, $\mathrm{Re} < 2300$. All experimental data with different relative roughness distribute around line ab, which indicates λ only depends on Re and is independent of relative roughness, $\lambda = f(\mathrm{Re})$.

- Curve bc: *transition zone*, $\mathrm{Re} = 2300 \sim 4000$. All experimental data with different relative roughness stay around the curve bc, which indicates $\lambda = f(\mathrm{Re})$.

- Line cd: *hydraulically smooth* zone, $\mathrm{Re} > 4000$. All experimental data with different relative roughness distribute along line cd, $\lambda = f(\mathrm{Re})$. For the flow in a pipe with larger relative roughness Δ/d, the data leave this line at smaller Re, otherwise leaves with a relative larger Re.

- Cluster curves between line cd and ef: *hydraulically transitional zone* with moderate values of Re, the friction factor is indeed dependent on both the Reynolds number and relative roughness, $\lambda = f(\text{Re}, \Delta/d)$.

- Cluster lines on the right side of line ef: *hydraulically rough zone* with very large Reynolds numbers. Experimental data with different relative roughness form different horizontal line, $\lambda = f(\Delta/d)$.

From Fig. 6-14 we also see that laminar friction is unaffected, but turbulent friction, after an *onset* point, increases monotonically with the roughness ratio Δ/d. For any given Δ/d, the friction factor becomes constant (fully rough) at high Reynolds numbers. These points of change are certain values of $\text{Re}_* = \dfrac{u_* \Delta}{\nu}$. With Eq. 6.57, we can have

$$\frac{\Delta}{\delta_s} = \frac{1}{11.6} \frac{u_* \Delta}{\nu} = \frac{1}{11.6} \text{Re}_* \tag{6.59}$$

Thus, the points of change are:

$\dfrac{u_* \Delta}{\nu} < 5$: *hydraulically smooth walls*, no effect of roughness on friction; for which $\delta_s > 2.32\Delta$, $\lambda = f(\text{Re})$;

$5 \leqslant \dfrac{u_* \Delta}{\nu} \leqslant 70$: *hydraulically transitional walls*, transitional roughness, moderate Reynolds-number effect; for which $0.17\Delta \leqslant \delta_s \leqslant 2.32\Delta$, $\lambda = f(\text{Re}, \Delta/d)$;

$\dfrac{u_* \Delta}{\nu} > 70$: *hydraulically fully rough walls*, sublayer totally broken up and friction independent of Reynolds number; for which $\delta_s < 0.17\Delta$, $\lambda = f(\Delta/d)$.

6.6.2 Losses in Developed Pipe Flow

SEMI-EMPIRICAL FORMULAS

Computing the mean velocity from the logarithmic-law correlation, with the frictional velocity, Eq. 6.22 and the velocity profiles for smooth pipe and fully rough zone, Eq. 6.53 and Eq. 6.54, then adjusted the constants slightly to fit friction data better, we obtain a relation between the friction factor and Reynolds number for turbulent pipe flow

Smooth wall zone: $\dfrac{1}{\sqrt{\lambda}} = 0.86 \ln \text{Re} \sqrt{\lambda} - 0.8$ or $\dfrac{1}{\sqrt{\lambda}} = 2\log \dfrac{\text{Re}\sqrt{\lambda}}{2.51}$ (6.60)

Fully rough zone: $\dfrac{1}{\sqrt{\lambda}} = -0.86 \ln \dfrac{\Delta/d}{3.7}$ or $\dfrac{1}{\sqrt{\lambda}} = -2\log \dfrac{\Delta/d}{3.7}$ (6.61)

COLEBROOK EQUATION

In 1939, Cyril F. Colebrook (1910-1997) combined the available data for transitional and turbulent flow in smooth as well as rough pipes into the following implicit relation known as the *Colebrook equation*:

$$\frac{1}{\sqrt{\lambda}} = -2\log\left(\frac{\Delta/d}{3.7} + \frac{2.51}{\text{Re}\sqrt{\lambda}}\right) \qquad (6.62)$$

It actually just couples the smooth pipe equation to the completely rough zone equation. It would be Eq. 6.60 with $\Delta = 0$, and Eq. 6.61 with $\text{Re} = \infty$.

The Colebrook equation is implicit in λ, and thus the determination of the friction factor requires some iteration unless an equation solver is used. An alternative explicit formula for λ given by S. E. Haaland in 1983 as

$$\frac{1}{\sqrt{\lambda}} \cong -1.8\log\left[\left(\frac{\Delta/d}{3.7}\right)^{1.11} + \frac{6.9}{\text{Re}}\right] \qquad (6.63)$$

varies less than 2 percent from the Colebrook equation.

MOODY CHART

Colebrook equation is the accepted design formula for turbulent friction. It was plotted in 1944 by Lewis F. Moody (1880-1953) into what is now called the *Moody chart* for pipe friction (Fig. 6-15). It presents the Darcy friction factor for pipe flow as a function of the Reynolds number and Δ/d over a wide range. It is probably one of the most widely accepted and used charts in engineering. Although it is developed for circular pipes, it can also be used for noncircular pipes by replacing the diameter by the hydraulic diameter.

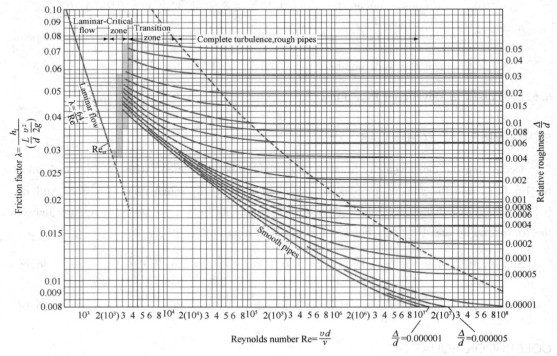

Fig. 6-15 The Moody chart for pipe friction with smooth and rough walls. This chart is identical to Eq. 6.62 for turbulent flow. Reproduced from Fig. 6-13 in *Fluid Mechanics* by Frank M. White(1998)

We make the following observations from the Moody chart:

- For laminar flow, $Re < 2300$, $\lambda = 64/Re$ the friction factor decreases with increasing Reynolds number, and it is independent of surface roughness.

- The transition region from the laminar to turbulent regime ($2300 < Re < 4000$) is indicated by the shaded area in the Moody chart (Fig. 6-15). The flow in this region may be laminar or turbulent, depending on flow disturbances, or it may alternate between laminar and turbulent, and thus the friction factor may also alternate between the values for laminar and turbulent flow. The data in this range are the least reliable. At small relative roughnesses, the friction factor increases in the transition region and approaches the value for smooth pipes.

- The friction factor is a minimum for a smooth pipe (but still not zero because of the no-slip condition) and increases with roughness. The Colebrook equation, Eq. 6.62 in this case ($\Delta = 0$) reduces to *Prandtl equation* expressed as Eq. 6.60. In this zone, the pipe roughness is considerably less than the viscous sublayer thickness.

- At very large Reynolds numbers (to the right of the dashed line on the chart), $\lambda = f(\Delta/d)$, the friction factor curves corresponding to specified relative roughness curves are nearly horizontal, and thus the friction factors are independent of the Reynolds number (Fig. 6-15). The flow in that region is called *fully rough turbulent flow* or just *fully rough flow* because the thickness of the viscous sublayer decreases with increasing Reynolds number, and it becomes so thin that it is negligibly small compared to the surface roughness height. The viscous effects in this case are produced in the main flow primarily by the protruding roughness elements, and the contribution of the laminar sublayer is negligible. The Colebrook equation in the *fully rough* zone ($Re \to \infty$) reduces to the *von Kármán equation* expressed as Eq. 6.61 which is explicit in λ. The friction loss of the flow is found to be proportional to v^2 from Eq. 6.2. Such a zone is also called the *resistance square zone*.

EQUIVALENT ROUGHNESS HEIGHT

In commercially available pipes the roughness is not as uniform and well defined as in the artificially roughened pipes used by Nikuradse. However, it is possible to obtain a measure of the effective relative roughness, the equivalent roughness height, of typical pipes and thus to obtain the friction factor.

Equivalent roughness height is defined as the assumed roughness height of the artificially roughened pipe with uniform roughness when its friction factor λ is equal to the friction factor λ in the rough zone of a commercial pipe with same diameter. Some typical equivalent roughness values for some commercial pipes are given in Table 6-2 as well as on the Moody chart. But it should be kept in mind that these values are for *new* pipes, and the relative roughness of pipes may increase with use as a result of corrosion, scale buildup, and precipitation. As a result, the friction factor may increase by a factor of 5 to 10. Also, the Moody chart and its equivalent Colebrook equation involve several uncertainties (the roughness size, experimental error, curve fitting of data, etc.), and thus the results obtained should not be treated as 'exact.' It is usually considered to be accurate to ± 15 percent over the entire range in the figure.

Table 6-2 Equivalent roughness values for new commercial ducts

Material	Condition	Roughness, Δ (mm)	Material	Condition	Roughness, Δ (mm)
Steel	Sheet metal, new	0.05	Iron	Asphalted cast	0.12
	Stainless, new	0.002	Brass	Drawn, new	0.002
	Commercial, new	0.046	Plastic	Drawn tubing	0.0015
	Riveted	3	Glass	—	smooth
	Rusted	2	Concrete	Smoothed	0.04
Iron	Cast, new	0.26		Rough	2
	Wrought, new	0.046	Rubber	Smoothed	0.01
	Galvanized, new	0.15	Wood	Stave	0.5

Source: Data are from Table 6.1 in *Fluid Mechanics* (4th Edition) by Frank W. White (1998). The uncertainty in these values can be as much as ±60 percent.

OTHER EMPIRICAL FORMULA

Blasius formula for smooth wall pipe:

$$\lambda = \frac{0.3164}{\mathrm{Re}^{0.25}} \qquad 4000 < \mathrm{Re} < 10^5 \tag{6.64}$$

Щифринсон *formula* for fully rough zone:

$$\lambda = 0.11 \left(\frac{\Delta}{d}\right)^{0.25} \tag{6.65}$$

Chézy formula for fully rough flow in ducts and open channel:

$$v = \sqrt{\frac{8g}{\lambda}} \sqrt{R_h J} = C\sqrt{R_h J} \tag{6.66}$$

which can be deduced from Eq. 6.3 with the definition of friction slope, $J = h_f/l$.

Equation 6.66 is called the *Chézy formula*, first developed by the French engineer, Antoine Chézy(1718-1798), in conjunction with his experiments on the Seine River and the Courpalet Canal in 1769. The quantity C, called the *Chézy coefficient*, has a unit of $\sqrt{m}$/s and is defined as

$$C = \sqrt{\frac{8g}{\lambda}} \tag{6.67}$$

can be used to present the frictional losses as the friction factor, λ. Over the past century a great deal of hydraulics research has been devoted to the correlation of the Chézy coefficient with the roughness, shape, and slope of various open channels. Among them, in 1890, Robert Manning (1816-1897), an Irish engineer, proposed a simple formula based on the conclusion that the Chézy coefficient C increased approximately as the sixth root of the channel size found in tests with real channels

$$C = \frac{1}{n} R_h^{1/6} \tag{6.68}$$

where n is *Manning resistance coefficient* for a given surface condition for the walls and bottom of the channel. Its value is dependent on the surface material of the channel's wetted perimeter and is obtained from experiments. It is not dimensionless, having the unit of $s/m^{1/3}$.

Equaton 6.68 is called the *Manning equation*, which has a good agreement for the fully rough flow in the ducts and small channel with $n < 0.02$ and $R_h < 0.5\text{m}$. It is widely and commonly used in the engineering practice worldwide.

It should be noted that Eq. 6.66 can be used in the turbulent flow from the smooth region to the fully rough region. However, if Eq. 6.68 is used to determine the *Chézy coefficient* in Eq. 6.66, which only depends on the n and R_h, and nothing to do with Re. In this situation, the *Chézy* equation only can be used for the fully rough turbulent flow.

Typical values of the Manning coefficient are indicated in Table 6-3. As expected, the rougher the wetted perimeter, the larger the value of n. For example, the roughness of floodplain surfaces increases from pasture to brush to tree conditions. So does the corresponding value of the Manning coefficient. Precise values of n are often difficult to obtain. Except for artificially lined channel surfaces like those found in new canals or flumes, the channel surface structure may be quite complex and variable. It should be noted that the values of n given in Table 6-3 are valid only for water as the flowing fluid.

EXAMPLE 6-6 Friction loss in turbulent flow

Oil, with $\rho = 900\text{kg/m}^3$ and $\nu = 10 \times 10^{-6}\,\text{m}^2/\text{s}$, through a 300-m-long of 200-mm-diameter cast-iron pipe flows at a rate of $0.2\text{m}^3/\text{s}$. Determine (a) the frictional head loss in the pipe and (b) the pressure drop if the pipe slopes down at 5° in the flow direction.

Solution:

(a) First compute the velocity from the known flow rate

$$v = \frac{Q}{\pi d^2/4} = \frac{0.2\text{m}^3/\text{s}}{\pi(0.2\text{m})^2/4} = 6.37\text{m/s}$$

Then the Reynolds number is

$$\text{Re} = \frac{vd}{\nu} = \frac{(6.37\text{m/s})(0.2\text{m})}{10 \times 10^{-6}\,\text{m}^2/\text{s}} = 127400$$

From Table 6-2, we have $\Delta = 0.26\text{mm}$ for cast-iron pipe. Then

$$\frac{\Delta}{d} = \frac{0.26}{200} = 0.0013$$

From Moody chart with $\Delta/d = 0.0013$ and Re = 127400 to read $\lambda \approx 0.0228$. Then the head loss in the pipe can be obtained by the Darcy-Weisbach equation

$$h_f = \lambda\,\frac{L}{d}\,\frac{v^2}{2g} = 0.0228\,\frac{300\text{m}}{0.2\text{m}}\,\frac{(6.37\text{m/s})^2}{(2)(9.81\text{m}^2/\text{s})} = 70.73\text{m}$$

(b) The energy equation in an inclined constant-diameter pipe

$$z_1 + \frac{p_1}{\rho g} = z_2 + \frac{p_2}{\rho g} + h_f$$

with $z_1 - z_2 = L\sin 5°$, $p_1 - p_2 = \Delta p$ yields

$$\Delta p = \rho g(h_f - L\sin 5°) = (900\text{kg/m}^3)(9.81\text{m/s}^2)[70.73\text{m} - (300\text{m})(\sin 5°)] = 393.6\text{kPa}$$

Table 6-3 Experimental mean values of Manning coefficient n

	Wetted Perimeter	Manning's n		Wetted Perimeter	Manning's n
A.	*Natural channels*			Corrugated metal	0.022
	Clean and straight	0.030		Rubble masonry	0.025
	Sluggish with deep pools	0.040		Smooth masonry	0.017
	Major rivers	0.035	E.	*Ducts*	
	Mountain streams	0.050		Polyethylene PE - Corrugated with smooth inner walls	0.009-0.015
B.	*Floodplains*				
	Pasture, farmland	0.035		Polyethylene PE - Corrugated with corrugated inner walls	0.018-0.025
	Light brush	0.050			
	Heavy brush	0.075		Polyvinyl Chloride PVC - with smooth inner walls	0.009-0.011
	Trees	0.150			
C.	*Excavated earth channels*			Uncoated cast iron	0.013
	Clean	0.022		Coated cast iron	0.012
	Gravelly	0.025		Commercial wrought iron-black	0.013
	Weedy	0.030		Commercial wrought iron galvanized	0.014
	Stony, cobbles	0.035			
D.	*Artificially lined channels*			Concrete pipe	0.013
	Glass	0.010		Smooth brass and glass	0.010
	Brass	0.011		Riveted and spiral steel pipe	0.015
	Steel, smooth	0.012		Lockbar and welded steel pipe	0.012
	Steel, painted	0.014		Common clay drainage tile	0.012
	Steel, riveted	0.015		Brick in cement mortar, brick sewers	0.013
	Cast iron	0.013			
	Concrete, finished	0.012		Cement mortar surfaces	0.012
	Concrete, unfinished	0.014		Wood stave pipe	0.011
	Wood, planed	0.012		Corrugated metal	0.022*
	Wood, unplaned	0.013		Vitrified sewer pipe	0.013
	Clay tile	0.014		Glazed brickwork	0.012
	Brickwork	0.015		Neat cement surfaces	0.011
	Asphalt	0.016		Smooth lockbar and welded "OD"	0.011

* Corrugated metal pipe n value can vary significantly with pipe diameter and type of corrugations (values can range from 0.012 to 0.033) -AISI (1980).

Source: Data for channels are from Ven Te Chow (1959), *Open Channel Hydraulics*. Data for ducts are from Metcalf and Eddy (1981), *Wastewater Engineering: Collection and Pumping of Wastewater*, and AISI (1980), *Modern Sewer Design*.

DISCUSSION The friction factor also can be obtained from the Colebrook equation, Eq. 6.62. However, it needs to be iterated. Alternatively, we can explicitly obtain the friction factor from Haaland equation, Eq. 6.63, $\lambda \approx 0.0226$, which is slightly less than what we obtain from the Moody chart.

EXAMPLE 6-7 Fully developed rough flow

Water flows through 400-mm-diameter, 100-m-long new water supply pipe with a Manning coefficient of $n = 0.011$. The flow is fully developed rough flow, and total frictional head loss in the pipe is 0.4m. Determine the flowrate in the pipe.

Solution:

Compute the cross sectional area, hydraulic radius with the given parameters

$$A = \frac{1}{4}\pi d^2 = \frac{1}{4}\pi(0.4\text{m})^2 = 0.126\text{m}^2$$

$$R_h = \frac{d}{4} = \frac{0.4\text{m}}{4} = 0.1\text{m}$$

Then the Chézy coefficient can be determined by Eq. 6.68

$$C = \frac{1}{n}R_h^{1/6} = \frac{1}{0.011}(0.1)^{1/6} = 61.94 \text{ m}^{1/2}/\text{s}$$

So the flowrate can be determined by the *Chézy formula*, Eq. 6.66

$$Q = vA = AC\sqrt{R_h J} = AC\sqrt{R_h \frac{h_f}{L}} = (0.126\text{m}^2)(61.94\text{m}^{1/2}/\text{s})\sqrt{(0.1\text{m})\frac{0.4\text{m}}{100\text{m}}} = 0.156\text{m}^3/\text{s}$$

6.6.3 Losses in Non-circular Ducts

If the duct is noncircular, the analysis of fully developed flow follows that of the circular pipe but is more complicated algebraically. For laminar flow, one can solve the exact equations of continuity and momentum. For turbulent flow, the logarithm law velocity profile can be used by introducing the hydraulic diameter.

For noncircular ducts, the frictional losses in a noncircular is also determined by Darcy-Weisbach equation but with the form of Eq. 6.3

$$h_f = \lambda \frac{L}{4R_h}\frac{v^2}{2g} \quad (6.3)$$

This is equivalent to Eq. 6.1 for pipe flow except that d is replaced by $4R_h$. Therefore we customarily define the *hydraulic diameter* for the noncircular duct as

$$D_h = \frac{4A}{\chi} = \frac{4 \times \text{area}}{\text{wetted perimeter}} = 4R_h \quad (6.69)$$

We should stress that the wetted perimeter includes all surfaces acted upon by the shear stress.

We would therefore expect by dimensional analysis that this friction factor λ, based upon hydraulic diameter as in Eq. 6.3, would correlate with the Reynolds number and roughness ratio based upon the hydraulic diameter

$$\lambda = f(\text{Re}_h, \Delta/D_h)$$

where $\text{Re}_h = vR_h/\nu$.

But we should not necessarily expect the Moody chart (Fig. 6-15) to hold exactly in terms of this new length scale. And it does not, but it is surprisingly accurate:

$$\lambda \approx \begin{cases} 64/\text{Re}_h & \pm 40\% \quad \text{Laminar flow} \\ \lambda_{\text{Moody}}(\text{Re}_h, \Delta/D_h) & \pm 15\% \quad \text{Turbulent flow} \end{cases} \qquad (6.70)$$

6.7 Minor Losses

6.7.1 Minor loss coefficient

As discussed in the previous section, the head loss in long, straight sections of pipe, the major losses, can be calculated by use of the friction factor obtained from either the Moody chart or the Colebrook equation. Most pipe systems, however, consist of considerably components more than straight pipes. These additional components, e.g., pipe entrance or exit; sudden expansion or contraction; bends, elbows, tees, and other fittings; valves, open or partially closed and gradual expansions or contractions, add to the overall head loss of the system. Such losses are generally termed *minor losses*, or *local losses*, denoted as h_j.

The minor loss is usually given as the dimensionless *minor loss coefficient* ζ (also called the *resistance coefficient*), and defined as the ratio of the head loss h_j through the device to the velocity head $v^2/2g$ of the associated piping system

$$\zeta = \frac{h_j}{v^2/2g} \qquad (6.71)$$

Thus, when the loss coefficient for a component is available, the head loss for that component is determined from Eq. 6.5: $h_j = \zeta v^2/2g$.

Although ζ is dimensionless, it unfortunately is not correlated in the literature with the Reynolds number and roughness ratio but strongly dependent on the geometry of the component considered

$$\zeta = f(\text{Geometry}) \qquad (6.72)$$

and almost all data of ζ in the literature are reported for turbulent-flow conditions. The following are some typical minor losses in the piping system.

6.7.2 Sudden Expansion and Contraction

Losses occur because of a change in pipe diameter, i.e. sudden or gradual expansion or contraction. The losses are usually much greater in the case of sudden expansion (or wide-angle expansion) because of flow separation. By combining the conservation of mass, momentum, and energy equations, the loss coefficient for the case of sudden expansion can be theoretical obtained.

As the sudden expansion indicated in Fig. 6-16a, the flow velocity decreases and so the pressure increases from section 1 to 2. Turbulent eddies at section give rise to the local loss. Experiments confirm that the shear stress in the corner separated flow, or dead water region, is negligible. Thus, a control-volume analysis between the expansion section 1 and the end of the separation zone 2 gives a theoretical loss.

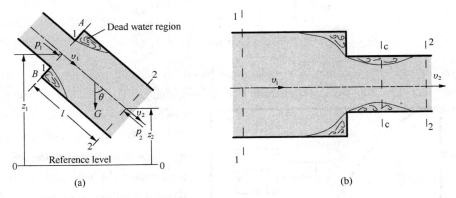

Fig. 6-16 (a) Sudden expansion and (b) sudden contraction

Due to the short length of the sudden expansion, the major loss between section 1-1 and 2-2 is negligible. Apply the energy equation between sections 1-1 and 2-2 to give the minor loss as

$$h_j = \left(z_1 + \frac{p_1}{\rho g}\right) - \left(z_2 + \frac{p_2}{\rho g}\right) + \frac{\alpha_1 v_1^2 - \alpha_2 v_2^2}{2g} \quad (6.73)$$

Now take the volume enclosed by sections 1-1 and 2-2 and the pipe wall to be the control volume, apply the momentum equation between section A-B and 2-2 to give:

$$\Sigma F = \rho Q(\beta_2 v_2 - \beta_1 v_1)$$

where ΣF includes: the pressure force on section A-B, $p_1 A_2$, based on the observation that the shear stress in the corner separated flow is negligible and the pressure in this section satisfy the principle of hydrostatic distribution; the pressure force on section 2-2, $p_2 A_2$; the component of gravity in the flow direction, $G\cos\theta = \rho g A_2(z_1 - z_2)$; the frictional force on the pipe wall, neglected. Thus, the above equation becomes

$$p_1 A_2 - p_2 A_2 + \rho g A_2(z_1 - z_2)$$
$$= \rho Q(\beta_2 v_2 - \beta_1 v_1)$$

Divided by $\rho g A_2$ and considering the continuity equation, $Q = v_2 A_2$ gives

$$\left(z_1 + \frac{p_1}{\rho g}\right) - \left(z_2 + \frac{p_2}{\rho g}\right) = \frac{v_2}{g}(\beta_2 v_2 - \beta_1 v_1) \quad (6.74)$$

Take $\alpha_1 = \alpha_2 \approx 1.0$, $\beta_1 = \beta_2 \approx 1.0$, the combination of Eqs. 6.73 and 6.74 gives

$$h_j = \frac{(v_1 - v_2)^2}{2g} \quad (6.75)$$

This is called *Borda equation*, which indicates that the minor loss in the sudden expansion equals to the velocity head computed by the difference of average velocities before and after the expansion. Equation 6.75 is in excellent agreement with experiment.

Substituting the continuity equation, $v_2 = v_1 A_1/A_2$ or $v_1 = v_2 A_2/A_1$ into Eq. 6.75, we can have the expression of the minor loss coefficient

$$\zeta_1 = \left(1 - \frac{A_1}{A_2}\right)^2 \quad \text{for} \quad h_j = \zeta_1 \frac{v_1^2}{2g} \tag{6.76}$$

$$\zeta_2 = \left(\frac{A_2}{A_1} - 1\right)^2 \quad \text{for} \quad h_j = \zeta_2 \frac{v_2^2}{2g} \tag{6.77}$$

The loss for sudden expansion of circular pipe based on velocity head in the small pipe is graphed in Fig. 6-17. Note that $\zeta = 0$ when there is no area change ($D = d$) and $\zeta = 1$ when a pipe discharges into a reservoir ($D \gg d$).

For the sudden contraction, however, flow separation in the downstream pipe causes the main stream to contract through a minimum diameter, called the *vena contracta*, c-c as sketched in Fig. 6.16b. Because the theory of the vena contracta is not well developed, the loss coefficient in the figure for sudden contraction is experimental. It fits the empirical formula

$$\zeta = 0.5\left(1 - \frac{A_2}{A_1}\right) \quad \text{for} \quad h_j = \zeta \frac{v_2^2}{2g} \tag{6.78}$$

which is also graphed in Fig. 6-17.

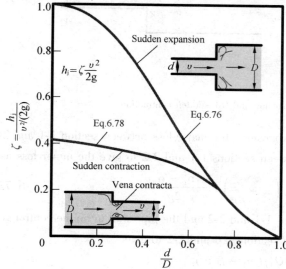

Fig. 6-17 Losses in sudden expansion and sudden contraction. Note that the loss is based on the velocity head in the small pipe

6.7.3 Pipe Inlet and Exit

When the fluid flows from a large tank into a pipe (Fig. 6-18b), $A_1 \gg A_2$, $A_2/A_1 \approx 0$, so Eq. 6.78 gives $\zeta = 0.5$ for the sharp-edged entrance. That is, a sharp edged inlet causes half of the velocity head to be lost as the fluid enters the pipe. This is because the fluid cannot make sharp 90° turns easily, especially at high velocities. A *vena contracta* is formed and separated flow occurs as indicated in Fig. 6-19. The velocity increases in the vena contracta region (and the pressure decreases) because of the reduced effective flow area and then decreases as the flow fills the entire cross section of the pipe, along which the viscous dissipation caused by the intense mixing and the turbulent eddies convert part of the kinetic energy into frictional heating and result in an irreversible pressure drop, the head loss.

The head loss at the inlet of pipe is highly dependent upon the entrance geometry. Even slight rounding of the edges can result in significant reduction of ζ, as shown in Fig. 6-18c and d. Sharp edges or protrusions in the entrance cause large zones of flow separation and large losses. A little rounding goes a long way, and a well rounded entrance ($r = 0.2d$) has a nearly negligible loss, $\zeta = 0.04$.

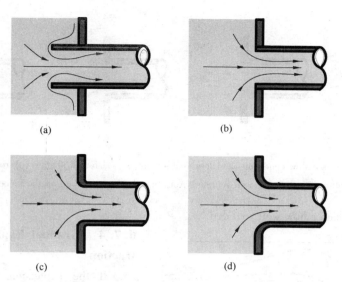

Fig. 6-18 Entrance flow conditions and loss coefficients. (a) Reentrant, $\zeta = 0.8$; (b) sharp-edged, $\zeta = 0.5$; (c) slightly rounded, $\zeta = 0.2$, and (d) well-rounded, $\zeta = 0.04$. Note that the loss is based on the velocity in the pipe

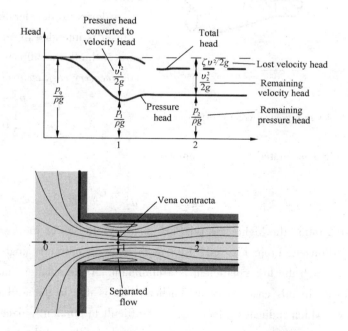

Fig. 6-19 Graphical representation of flow contraction and the associated head loss at a sharp-edged pipe inlet

When a pipe expands into a large tank (Fig. 6-20), $A_1 \ll A_2$, $A_1/A_2 \approx 0$, so Eq. 6.76 gives $\zeta = 1$. That is, the head loss is equal to the velocity head just before the expansion into the tank. It means that at a submerged exit, the flow simply passes out of the pipe into the large downstream reservoir and loses its all velocity head due to viscous dissipation. Therefore $\zeta = 1$ for all *sub-*

257

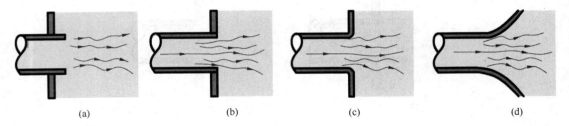

Fig. 6-20 Submerged exit flow conditions and loss coefficients. $\zeta = 1$ for all shapes: (a) reentrant; (b) sharp-edged; (c) slightly rounded; (d) well-rounded. Note that the loss is based on the velocity in the pipe

merged exits, no matter how well rounded.

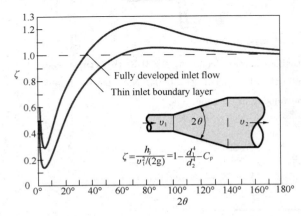

Fig. 6-21 Flow losses in a gradual conical expansion

6.7.4 Gradual Expansion and Contraction

If the expansion or contraction is gradual, the losses are quite different. Fig. 6-21 shows the loss through a gradual conical expansion, usually called a *diffuser*, a device to decelerate a fluid. Since a diffuser is intended to raise the static pressure of the flow, diffuser data list the *pressure-recovery coefficient* of the flow

$$C_p = \frac{p_2 - p_1}{\frac{1}{2}\rho v^2} \quad (6.79)$$

The loss coefficient is related to this parameter by

$$\zeta = \frac{h_j}{v^2/(2g)} = 1 - \frac{d_1^4}{d_2^4} - C_p \quad (6.80)$$

For a given area ratio, the higher the pressure recovery, the lower the loss; hence large C_p means a successful diffuser. From Fig. 6-21 we see there is an optimum angle ($2\theta \approx 8°$ for the case illustrated) for which the loss coefficient is a minimum. Angle smaller than this gives a excessively long diffuser. In this case, most of the head loss is due to the wall shear stress as in fully developed flow, which indicates a fact that it is difficult to efficiently decelerate a fluid.

For moderate or large angles, the flow separates from the walls and the losses are due mainly to a dissipation of the kinetic energy of the jet leaving the smaller diameter pipe. In fact, for moderate or large values of θ (i.e., $\theta > 35°$ for the case shown in Fig. 6-21), the conical diffuser is, perhaps unexpectedly, less efficient than a sharp-edged expansion which has $\zeta = (1 - A_1/A_2)^2$. It would actually be better to use a sudden expansion.

Flow in a conical contraction (a nozzle; reverse the flow direction shown in Fig. 6-21) is less complex than that in a conical expansion. Typical loss coefficients based on the velocity in the smaller pipe can be quite small (Fig. 6-22).

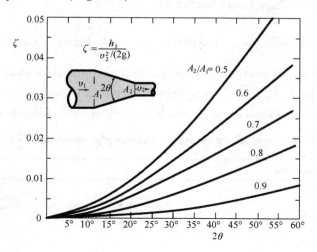

Fig. 6-22 Flow losses in a gradual conical contraction

6.7.5 Pipe Bend

Bends or curves in pipes always produce a greater head loss than the simple Moody friction loss when the pipe is straight. The losses are due to flow separation at the region near the inside of the bend and the swirling *secondary flow* arising from the imbalance centripetal forces as a result of the curvature of the pipe centerline. The loss coefficients ζ in Fig. 6-23 are for this additional bend loss for large Reynolds number flows through a 90° bend. The Moody loss due to the axial length of the bend must be computed separately, i.e., the bend length should be added to the pipe length.

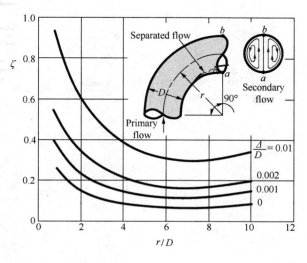

Fig. 6-23 Character of the flow in a 90° bend and the associated loss coefficient

6.7.6 Other Typical Minor Loss Coefficients

Loss coefficients for typical pipe components $\left(h_j = \zeta \dfrac{v^2}{2g}\right)$ are given in Table 6-4. These typical components are designed more for ease of manufacturing and costs than for reduction of the head losses that they produce. As with many system components, the head loss in valves is mainly a result of the dissipation of kinetic energy of a high-speed portion of the flow.

Table 6-4 Loss coefficients for typical circular pipe components $\left(h_j = \zeta \dfrac{v^2}{2g}\right)$

Component	ζ
a. Elbows	
Regular 90°, flanged	0.3
Regular 90°, threaded	1.5
Long radius 90°, flanged	0.2
Long radius 90°, threaded	0.7
Long radius 45°, flanged	0.2
Regular 45°, threaded	0.4
b. 180° return bends	
180° return bends, flanged	0.2
180° return bends, threaded	1.5
c. Tees	
Line flow, flanged	0.2
Line flow, threaded	0.9
Branch flow, flanged	1.0
Branch flow, threaded	2.0
d. Union, threaded	0.08

e. Mitered bend (without vanes)

θ	30°	40°	50°	60°	70°	80°	90°	90° (with vanes)
ζ	0.20	0.30	0.40	0.55	0.70	0.90	1.10	0.20

Component	ζ
f. Valve	
Globe, fully open	10
Angle, fully open	2
Gate:	
fully open	0.15
1/4 open	0.26
1/2 open	2.1
3/4 open	17
Swing check:	
forward flow	2
backward flow	∞
Ball valve:	
fully open	0.05
1/3 close	5.5
2/3 close	210

6.7.7 Disturbance between Components

It should keep in mind that the values of minor loss coefficients for different pipe components listed above are representative values for loss coefficients. Actual values strongly depend on the design and manufacture of the components and may differ from the given values considerably (especially for valves). Therefore, the actual manufacturer's data should be consulted in the final design of piping systems rather than relying on the representative values in handbooks.

Also, these representative values of components are obtained individually. Once some of those components are connected in sequent in a close distance, the velocity fields and their fluctuations will disturb each other, which leads to a 0.5 to 3.0 times variation of the values of each individual loss coefficients. It indicates that the total minor loss coefficients of those related components are not the simple sum of the every individual loss coefficient, which may have a much large decrease or increase of values.

EXAMPLE 6-8 Minor head loss

Water, $\rho = 998.2 \text{kg/m}^3$ and $\nu = 1.011 \times 10^{-6} \text{ m}^2/\text{s}$, is pumped between two reservoirs at $0.006 \text{m}^3/\text{s}$ through a 120-m-long, 5-cm-diameter pipe and several minor losses, as shown in Fig. E6-8. The roughness ratio is $\Delta/d = 0.001$. Compute the pump horsepower required.

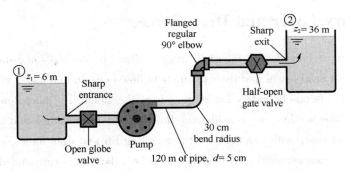

Fig. E6-8

Solution:

Take the computational surfaces to be the free surfaces of reservoirs, write the steady flow energy equation between the two reservoir surfaces:

$$z_1 + \frac{p_1}{\rho g} + \frac{\alpha_1 v_1^2}{2g} + h_p = z_2 + \frac{p_2}{\rho g} + \frac{\alpha_2 v_2^2}{2g} + h_f + \Sigma h_j$$

where h_p is the head increase across the pump. With the given parameters $z_1 = 6\text{m}$, $z_2 = 36\text{m}$, $p_1 = p_2 = p_{at}$, $v_1 = v_2 \approx 0$ and assuming $\alpha_1 = \alpha_2 \approx 1.0$, solve for the pump head

$$h_p = z_2 - z_1 + h_f + \Sigma h_j = 36\text{m} - 6\text{m} + \frac{v^2}{2g}\left(\lambda \frac{l}{d} + \Sigma \zeta\right) \qquad (1)$$

Now with the flow rate known, calculate the average velocity in the pipe

$$v = \frac{Q}{A} = \frac{0.006 \text{m}^3/\text{s}}{\pi(0.05\text{m})^2/4} = 3.06 \text{m/s}$$

List and sum the minor loss coefficients:

Pipe components	ζ
Sharp entrance (Fig. 6-18)	0.5
Open globe valve (Table 6-4)	10
30-cm radius bend (Fig. 6-23)	0.13
Regular 90° elbow (Table 6-4)	0.3
Half-closed gate valve (Table 6-4)	2.1
Sharp exit (Fig. 6-20)	1.0
	$\Sigma\zeta = 14.03$

Calculate the Reynolds number and the friction factor of pipe

$$\text{Re} = \frac{vd}{\nu} = \frac{(3.06\,\text{m/s})(0.05\,\text{m})}{1.011 \times 10^{-6}\,\text{m}^2/\text{s}} = 151335$$

For $\Delta/d = 0.001$, from the Moody chart read $\lambda = 0.0218$. Substitute into Eq. 1

$$h_p = 30\,\text{m} + \frac{(3.06\,\text{m/s})^2}{(2)(9.81\,\text{m/s}^2)}\left(0.0218\,\frac{120\,\text{m}}{0.05\,\text{m}} + 14.03\right) = 61.66\,\text{m}$$

The pump must provide a power to the water of

$$N = \rho g Q h_p = (998.2\,\text{kg/m}^3)(9.81\,\text{m/s}^2)(0.006\,\text{m}^3/\text{s})(61.66\,\text{m}) = 3622.77\,W = 4.86\,\text{hp}$$

DISCUSSION Allowing for an efficiency of 70 to 80 percent, a pump is needed with an input of about 7 hp.

6.8 Boundary Layer and Drag Force

The above sections considered 'internal' flows confined by the walls of a duct, with emphasis on pressure drop and head losses and their relations to flow rate. In that case, viscous shear is the dominant effect, the viscous boundary layers grow from the sidewalls, meet downstream, and fill the entire duct. In this section, we consider the flow of fluids over bodies that are immersed in a fluid, called *external flow*, with emphasis on the resulting lift and drag forces.

External flow is characterized by a freely growing boundary layer surrounded by an outer flow region that involves small velocity and temperature gradients. Such immersed-body flows are commonly encountered in engineering studies: *aerodynamics* (airplanes, rockets, projectiles), *hydrodynamics* (ships, submarines, torpedos), *transportation* (automobiles, trucks, cycles), *wind engineering* (buildings, bridges, water towers, wind turbines), and *ocean engineering* (buoys, breakwaters, pilings, cables, moored instruments).

Much of the information about external flows comes from experiments carried out, for the most part, on scale models of the actual objects. Such testing includes the obvious wind tunnel testing of model airplanes, buildings, and even entire cities. The use of water tunnels and towing tanks also provides useful information about the flow around ships and other objects.

6.8.1 Boundary Layer Structure on Flat Plate

In this section we consider the simplest situation of boundary layer, one in which the *boundary layer* is formed on an infinitely long flat plate along which flows a viscous, incompressible fluid as is shown in Fig. 6-24. If the surface were curved (i.e., a circular cylinder or an airfoil), the boundary layer structure would be more complex. Such flows are discussed in next section.

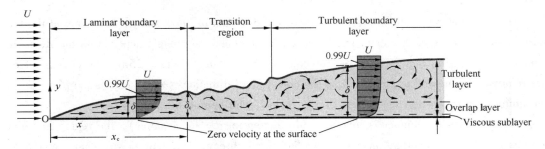

Fig. 6-24 Boundary layer structure on a flat plate (Not to scale)

Consider a thin flat plate of infinite length, infinite width, and negligible thickness, which lies in the xy plane, and whose edges correspond to $x = 0$ and $y = 0$ as in Fig. 6-24, the original of the xOy coordinate system. Suppose that the plate immersed in a uniform steady stream of viscous fluid, whose undisturbed velocity U is perpendicular to the sharp leading edge and parallel to the plate surface.

As any given volume of fluid sweeps over the plate from the leading edge, the effect of the viscous retardation due to the plate surface extends progressively into the fluid column. For immediately downstream of the leading edge only those elements immediately adjacent to the solid surface are perceptibly retarded by it and $u_x = 0$; but as the fluid moves on, these slower-moving elements exert a viscous drag on their neighbors in the stream further from the plate surface, and the frictional retardation is propagated into the fluid by a process akin to diffusion. The layer of fluid significantly affected by this viscous effect with large gradient (du_x/dy) is known as the *boundary layer* forming above and below the plate as in Fig. 6-25. Outside this boundary layer, the viscous effects are negligible, the fluid flow in this region can be considered as the ideal fluid flow.

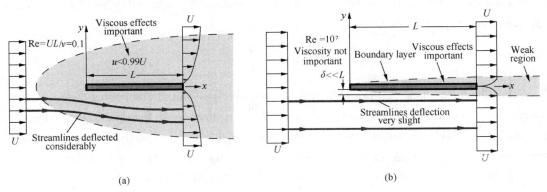

Fig. 6-25 Character of the steady, viscous flow past a flat plate parallel to the upstream velocity:
(a) low Reynolds number flow; (b) large Reynolds number flow

We define the *boundary layer thickness*, δ, to be the distance in y-direction from the plate to the locus of points at which the fluid velocity parallel to the plate reaches the 99% of the upstream velocity (see Fig. 6-24), namely

$$\delta = y \quad \text{where } u_x = 0.99U \tag{6.81}$$

The boundary layer thickness increases in the direction of flow, starting from zero at the forward or leading edge of the plate. The boundary layer thickness strongly depends on the Reynolds number, as shown in Fig. 6-25. If the Reynolds number is small (Fig. 6-25a), the viscous effects are relatively strong and are felt far from the object in all directions, which causes the plate affects the uniform upstream flow far ahead, above, below, and behind the plate.

As the Reynolds number is increased (by increasing U, for example), the region in which viscous effects are important becomes smaller in all directions except downstream. When the Reynolds number is large (but not infinite) (Fig. 6-25b), the flow is dominated by inertial effects and the viscous effects are negligible everywhere except in a region very close to the plate and in the relatively thin *wake region* behind the plate. Since the fluid viscosity is not zero ($Re < \infty$), it follows that the fluid must stick to the solid surface (the no-slip boundary condition). In this thin boundary layer region next to the plate, $\delta = \delta(x) \ll L$ (i.e., thin relative to the length of the plate L), the fluid velocity changes from the upstream value of $u = U$ to zero velocity on the plate.

In 1904, Ludwig Prandtl (1875-1953), a German physicist and aerodynamicist, proposed an idea of the boundary layer in 1904, a thin region on the surface of a body in which viscous effects are very important and outside of which the fluid behaves essentially as if it were inviscid. That is, the flow past an object can be treated as the combination of viscous flow in the boundary layer and inviscid flow elsewhere if the Reynolds number is large enough, even though the viscosity is not zero. By using such a hypothesis it is possible to simplify the analysis of large Reynolds number flows, thereby allowing solution to external flow problems that are otherwise still unsolvable.

The flow within the boundary layer may be *laminar* or *turbulent*, depending on various parameters involved as shown in Fig. 6-24. For an infinitely long plate, we use x, the coordinate distance along the plate from the leading edge, as the characteristic length and define the *local Reynolds number* in the boundary layer as

$$\text{Re}_x = \frac{Ux}{\nu} \tag{6.82}$$

This definition shows that this Reynolds number increases with the increasing distance downstream from the leading edge. Therefore, at some distance downstream from the leading edge, the boundary layer flow becomes turbulent and the fluid particles become greatly distorted because of the random, irregular nature of the turbulence. The transition from a *laminar boundary layer* to a *turbulent boundary layer* occurs at a critical value of the Reynolds number

$$\text{Re}_{x_c} = \frac{U_0 x_c}{\nu} \tag{6.83}$$

on the order of 2×10^5 to 3×10^6, depending on the roughness of the surface and the amount of turbulence in the upstream flow. Assume the finite length of plate is L. If $L < x_c$, the flow in the boundary layer will be laminar because $\text{Re}_x = \dfrac{UL}{\nu} < \text{Re}_{x_c}$; If $L > x_c$, it will be laminar before the point at $x_c(x < x_c)$ and turbulent after the point at $x_c(x_c \leqslant x \leqslant L)$. Viscous layer, overlap layer also occur in the turbulent boundary layer (Fig. 6-24). The turbulent boundary layer profile on a flat plate closely resembles the boundary layer profile in fully developed turbulent pipe flow discussed in Section 6.5.4.

The *boundary layer thickness* on a flat plate is determined by

Laminar boundary layer: $\qquad \delta = \dfrac{5x}{\text{Re}_x^{1/2}} \qquad\qquad$ (6.84)

Turbulent boundary layer: $\qquad \delta = \dfrac{0.377x}{\text{Re}_x^{1/5}} \qquad\qquad$ (6.85)

In all case, the boundary layers are so thin that their displacement effect on the outer inviscid layer is negligible. For example, the air at 20℃ flows around a flat plate at a velocity of $U = 10\text{m/s}$, the thickness $\delta \approx 1.8\text{mm}$ at $x = 1.0\text{m}$, and $\delta \approx 2.5\text{mm}$ at $x = 2.0\text{m}$.

6.8.2 Boundary Layer Separation

As with the flow past the flat plate described above, the flow past a blunt object (such as a circular cylinder or spheres) also varies with Reynolds number, of which the characteristic length for a circular cylinder or sphere is taken to be the external diameter D, that is $\text{Re} = UD/\nu$. As usual, the larger the Reynolds number, the smaller the region of the flow field in which viscous effects are important.

Cross-flow over a cylinder exhibits complex flow patterns, as shown in Fig. 6-26. The fluid approaching the cylinder branches out and encircles the cylinder, forming a boundary layer that wraps around the cylinder. The fluid particles on the midplane strike the cylinder at the stagnation point, bringing the fluid to a complete stop and thus raising the pressure at that point. The pressure decreases in the flow direction while the fluid velocity increases.

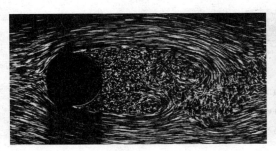

Fig 6-26 Flow past a cylinder at (a) Re = 2000[*Photograph courtesy Werle and Gallon (ONERA)*] and (b) Re = 10000[Photograph courtesy Thomas Corke and Hasan Najib(*Illinois Institute of Technology, Chicago*)] ❶

For a flow with a very low Reynolds number ($\text{Re} \leqslant 1$), as the case of Re = 0.1 in Fig.

---

❶ http://nptel.ac.in/courses/Webcourse-contents/IIT-KANPUR/FLUID-MECHANICS/lecture-31/31-3_mechanics.htm

6-27a, the viscous effects are important in all direction from the cylinder, and the fluid completely wraps around the cylinder and follows the curvature of the cylinder. A somewhat surprising characteristic of this flow is that the streamlines in regular orders are essentially symmetric about the center of the cylinder—the streamline pattern is the same in front of the cylinder as it is behind the cylinder.

At larger Reynolds number, as the case of $Re = 10^5$ in Fig. 6-27b, the fluid still hugs the cylinder on the frontal side, but with the loss of energy due to the viscous, residual kinetic energy of the fluid particle (left after overcoming viscous forces) may be insufficient to allow the flow to proceed into regions of increasing pressure. As a result, the fluid particle comes to a stop at the *separation location S*, at which point the boundary layer detaches from (lifts off) the surface, forming an irregular, unsteady (perhaps turbulent) *wake region* that extends far downstream of the cylinder (Fig. 6-27b). This is termed *boundary layer separation*. Flow in the wake region is characterized by periodic vortex formation, see Fig. 6-26b, and some of the fluid in the separation bubble behind the cylinder is actually flowing upstream, against the direction of the upstream flow. The mainstream is deflected by this wake.

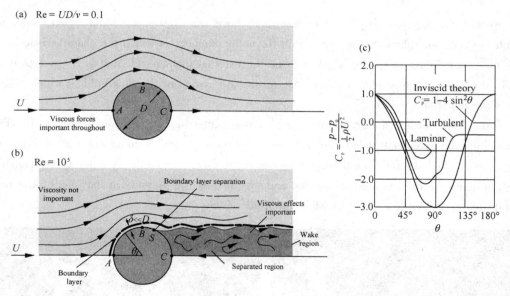

Fig. 6-27 Character of the steady, viscous flow past a circular cylinder: (a) low Reynolds number flow; (b) large Reynolds number flow; (c) surface pressure distributions for inviscid flow and boundary layer flow.

Prandtl showed that separation like that in Fig. 6-27b is caused by excessive momentum loss near the wall in a boundary layer trying to move downstream against increasing pressure, $\partial p/\partial x > 0$, which is called an *adverse pressure gradient*. The opposite case of decreasing pressure, $\partial p/\partial x < 0$, is called a *favorable pressure gradient*, where flow separation can never occur.

For the inviscid flow, a fluid particle traveling from the front to the back of the cylinder coasts down the 'pressure hill' from $\theta = 0$ to $\theta = 90°$ in Fig. 6-27c (from front point A to shoulder

B in Fig. 6-27a) and then back up the hill to $\theta = 180°$ (from shoulder B and to rear C) without any loss of energy as indicated in Fig. 6-27c, in which p_0 and U are the pressure and velocity, respectively, in the free stream. There is an exchange between kinetic and pressure energy, but there are no energy losses.

For a viscous flow, because of the viscous effects involved, a fluid particle within the boundary layer flowing from front A to shoulder B and to rear C experiences an additional loss of energy as it flows along. In the front half of cylinder from A to B, the fluid particle moves along with the positive pressure gradient, $\partial p/\partial x > 0$, and has residual kinetic energy after overcoming the surface friction force. But in the rear half of cylinder from B to C the fluid particle in boundary layer encounters not only the surface friction force, but also the *adverse* pressure gradient, $\partial p/\partial x < 0$. Both can retard the fluid particle and slow down its velocity rapidly. Finally, the fluid particle comes to a stop and *boundary layer separation* occurs. At the separation location S (velocity profile shown in Fig. 6-28), the velocity gradient at wall and the wall shear stress are zero:

$$\left.\frac{\partial u}{\partial y}\right|_{y=0} = 0 \quad \text{and} \quad \tau_0 = \mu\left(\frac{\partial u}{\partial y}\right)_{y=0} = 0 \tag{6.86}$$

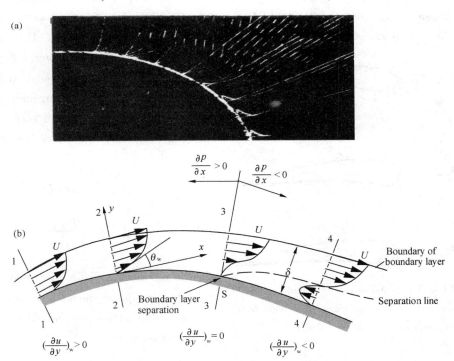

Fig. 6-28 Separation of boundary layer on a curved surface. (a) Flow visualization (© from *Aerodynamics of an Airplane* by C. B. Millikan ❶); (b) Velocity profile in boundary layer

Beyond that location (from S to C in Fig. 6-27b) there is a reverse flow in the boundary layer

---

❶ http://people.rit.edu/pnveme/MECE356/drag/boundary_layer.html

(see Fig. 6-28a). The region downstream of the object enclosing by the separation lines is termed as *wake flow*. Typical velocity profile is shown at section 4-4 in Fig. 6-28b.

As is indicated in Fig. 6-27c, because of the boundary layer separation, the average pressure on the rear half of the cylinder is considerably less than that on the front half. Thus, a large pressure drag (discussed in next section) is developed, even though (because of small viscosity) the viscous shear drag may be quite small. However, the turbulent boundary layer can flow further around the cylinder (further up the pressure hill) before it separates than the laminar boundary layer. Fig. 6-27c shows that the laminar flow separates at about 80° ($C_D = 1.2$, where C_D is the drag coefficient defined in next section), while the turbulent flow separates at 120°($C_D = 0.3$) (see also Fig. 6-30b: D, E). It is why golf balls are deliberately dimpled: to induce a turbulent boundary layer and delay the boundary layer separation, thus lower drag ($C_D = 0.2$ for turbulent boundary layer and $C_D = 0.5$ for laminar).

6.8.3 Drag and Lift Forces

When any body moves through a fluid, an interaction between the body and the fluid occurs. This effect can be described in terms of the forces at the fluid-body interface: the wall shear stresses, τ_0, due to viscous effects, and the normal stresses due to the pressure, p. Their resultant forces are termed the *drag force*, F_D, in the direction of the upstream velocity, and the *lift force*, F_L, normal to the upstream velocity as indicated in Fig. 6-29. The *lift coefficient*, C_L, and *drag coefficient*, C_D, are defined as

$$C_L = \frac{F_L}{\frac{1}{2}\rho U^2 A} \quad (6.87)$$

and

$$C_D = \frac{F_D}{\frac{1}{2}\rho U^2 A} \quad (6.88)$$

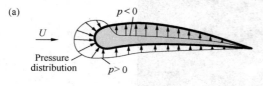

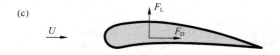

Fig. 6-29 Forces from the surrounding fluid on a two-dimensional object: (a) pressure force; (b) viscous force; (c) resultant force (lift and drag forces).

where A is a characteristic area of the object, typically taken to be *frontal area*, the projected area at the section perpendicular to the upstream velocity, U.

DRAG FORCE

The drag force on a body could be divided into two components: friction drag and pressure drag. *Friction drag*, F_{Df}, is due directly to the shear stress on the object. *Pressure drag*, F_{Dp}, often referred to as *form drag* because of its strong dependency on the shape or form of the object, is due directly to the pressure on an object.

Most of the information pertaining to drag on objects is a result of numerous experiments with wind tunnels, water tunnels, towing tanks, and other ingenious devices that are used to measure the drag on scale models. Typically, drag coefficient, C_D is the function of object's shape and some dimensionless parameters such as Reynolds number Re, Mach number Ma, Froude number Fr, and relative roughness of the surface, Δ/L (where L is some *characteristic length* of the object):

$$C_D = f(\text{shape, Re, Fr, Ma, } \Delta/L)$$

Some typical values of C_D are shown in Fig. 6-30, Tables 6-5 and 6-6.

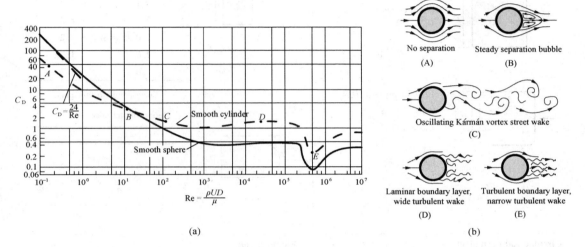

Fig. 6-30 (a) Drag coefficient as a function of Reynolds number for a smooth circular cylinder and a smooth sphere; (b) Typical flow patterns for flow past a circular cylinder at various Reynolds numbers as indicated in (a).

As indicated in Fig. 6-30a, the curves exhibit different behaviors in different ranges of Reynolds numbers:

• For very small Reynolds number flows ($Re \leqslant 1$), drag coefficient decreases with increasing Reynolds number. For a sphere, the drag coefficient varies inversely with the Reynolds number $C_D = 24/Re$. There is no flow separation in this regime.

• At about $Re = 10$, separation starts occurring on the rear of the body with vortex shedding starting at about $Re \cong 90$. The region of separation increases and the drag coefficient decreases with increasing Reynolds number up to about $Re = 10^3$. At this point, the drag is mostly (about 95 percent) due to pressure drag.

• For moderate Reynolds number, the drag coefficients producing by flows past blunt bodies are relatively constant, for the spheres and circular cylinders this character appears in the range $10^3 < Re < 10^5$. The flow in the boundary layer is laminar in this range, but the flow in the separated region past the cylinder or sphere is highly turbulent with a wide turbulent wake.

• For large Reynolds number, the boundary layer becomes turbulent, it indicates a later separation of boundary layer, a thinner wake and smaller pressure drag, which induces a sudden decrease in C_D for $10^5 < Re < 10^6$ (usually, at about 2×10^5).

Table 6-5 Typical drag coefficients for regular two-dimensional objects

Drag coefficients C_D of various two-dimensional bodies for Re $> 10^4$ based on the frontal area $A = bD$, where b is the length in direction normal to the page (for use in the drag force relation $F_D = C_D \rho U^2 A/2$, where U is the upstream velocity)

Square rod		**Rectangular rod**		
Sharp corners: $C_D = 2.2$		Sharp corners:		
		L/D		C_D
		0.0*		1.9
		0.1		1.9
		0.5		2.5
		1.0		2.2
		2.0		1.7
		3.0		1.3
Round corners ($r/D = 0.2$): $C_D = 1.2$		*Corresponds to thin plate		
		Round frontal edge:		
		L/D		C_D
		0.5		1.2
		1.0		0.9
		2.0		0.7
		4.0		0.7

Circular rod (cylinder)	**Elliptical rod**		
Laminar: $C_D = 1.2$	L/D	C_D Laminar	C_D Turbulent
Turbulent: $C_D = 0.3$	2	0.60	0.20
	4	0.35	0.15
	8	0.25	0.10

Equilateral triangular rod	**Semicircular shell**	**Semicircular rod**
$C_D = 1.5$	$C_D = 2.3$	$C_D = 1.2$
$C_D = 2.0$	$C_D = 1.2$	$C_D = 1.7$

Source: Data are from Table 11-1 in *Fluid Mechanics: Fundamentals and Applications* by Çengel and Cimbala (2006).

Table 6-6 Typical drag coefficients for regular three-dimensional objects

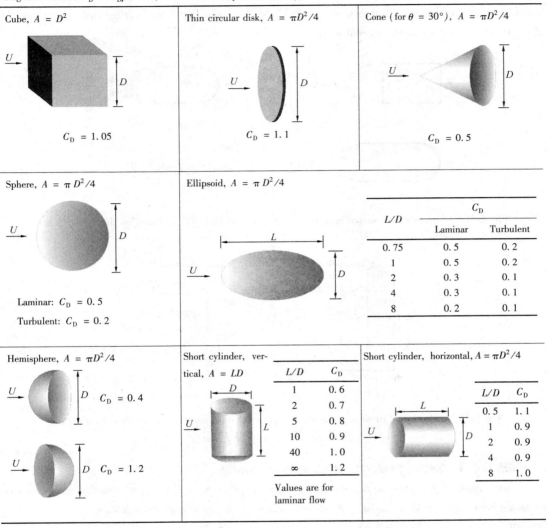

Representative drag coefficients C_D for various three-dimensional bodies for Re $> 10^4$ based on the frontal area (for use in the drag force relation $F_D = C_D \rho U^2 A/2$, where U is the upstream velocity)

Source: Data are from Table 11-1 in *Fluid Mechanics: Fundamentals and Applications* by Çengel and Cimbala (2006).

The structure of the flow field at selected Reynolds numbers indicated in Fig. 6-30a is shown in Fig. 6-30b. It indicates that for a given object there is a wide variety of flow situations depending on the Reynolds number involved.

In general, we cannot overstress the importance of body streamlining to reduce drag at Reynolds numbers above about 100. This is illustrated in Fig. 6-31. The rectangular cylinder (Fig. 6-31a) has rampant separation at all sharp corners and very high drag. Rounding its nose (Fig. 6-31b) reduces drag by about 45 percent, but C_D is still high. Streamlining its rear to a sharp trailing edge (Fig. 6-31c) reduces its drag another 85 percent to a practical minimum for the given thickness. As a dramatic contrast, the circular cylinder (Fig. 6-31d) has one-eighth the thickness and one-threehundredth of the cross section in Fig. 6-31c, yet it has the same drag as in Fig. 6-31c. For high-performance vehicles and other moving bodies, the name of the game is drag reduction. Better streamlining of car shapes has resulted over the years in a large decrease in the automobile drag coefficient, as shown in Fig. 6-32. Modern cars have an average drag coefficient of about 0.35, based upon the frontal area.

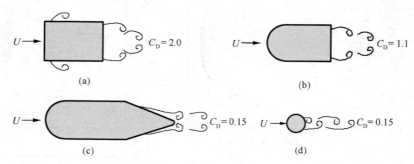

Fig. 6-31 The important of streamlining in reducing drag of body (C_D is based on the frontal area)

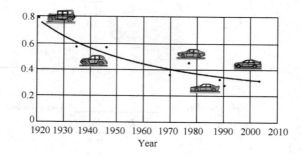

Fig. 6-32 The historical trend of streamlining automobiles to reduce their aerodynamic drag and increase their miles per gallon

EXAMPLE 6-9 Settling velocity of sand in still water

A small grain of sand, diameter $D = 0.10$mm and specific gravity $S_G = 2.3$, settles to the bottom of a lake after having been stirred up by a passing boat. Determine how fast it falls through the still water.

Solution:

A free-body diagram of the particle (relative to the moving particle) is shown in Fig. E6-9a. The particle moves downward with a constant velocity U that is governed by a balance between the weight of the particle, G, the buoyancy force of the surrounding water, F_B, and the drag of the water on the particle, F_D.

From the free-body diagram, we obtain

$$G = F_B + F_D \tag{1}$$

where

$$G = \rho_{sand} g \forall = S_G \rho_{H_2O} g \frac{\pi D^3}{6} \tag{2}$$

$$F_B = \rho_{H_2O} g \forall = \rho_{H_2O} g \frac{\pi D^3}{6} \tag{3}$$

We assume (because of the smallness of the object) that the flow will be creeping flow (Re < 1) with $C_D =$

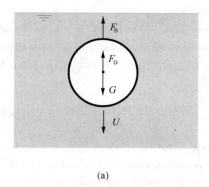
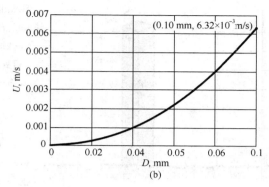

(a)   (b)

Fig. E6-9

$24/Re$ (see Fig. 6-30a) so that

$$F_D = \frac{1}{2}\rho_{H_2O}U^2 A C_D = \frac{1}{2}\rho_{H_2O}U^2 \frac{\pi D^2}{4} \frac{24}{\rho_{H_2O}UD/\mu_{H_2O}}$$

or
$$F_D = 3\pi\mu_{H_2O}UD \tag{4}$$

We must eventually check to determine if this assumption is valid or not. Equation 4 is called *Stokes law* in honor of Sir George Gabriel Stokes (1819-1903), a British mathematician and physicist. Substituting Eqs. 2, 3, and 4 into Eq. 1 yields

$$S_G \rho_{H_2O} g \frac{\pi D^3}{6} = \rho_{H_2O} g \frac{\pi D^3}{6} + 3\pi\mu_{H_2O}UD$$

or
$$U = \frac{(S_G - 1)gD^2}{18\,\nu_{H_2O}} \tag{5}$$

From Table A-1 for water at 20°C we have $\nu_{H_2O} = 1.004 \times 10^{-6}$ m²/s. Thus, from Eq. 5 we obtain

$$U = \frac{(2.3 - 1)(9.81 \text{m/s}^2)(0.0001\text{m})^2}{(18)(1.004 \times 10^{-6}\text{m}^2/\text{s})} = 7.06 \times 10^{-3} \text{m/s}$$

Since
$$Re = \frac{UD}{\nu_{H_2O}} = \frac{(7.06 \times 10^{-3}\text{m/s})(0.0001\text{m})}{1.004 \times 10^{-6}\text{m}^2/\text{s}} = 0.703$$

We see that Re < 1, and the form of the drag coefficient used is valid.

By repeating the calculations for various particle diameters, D, the results shown in Fig. E6-9b are obtained. Note that very small particles fall extremely slowly. Thus, it can take considerable time for silt to settle to the bottom of a river or lake.

DISCUSSION   Note that if the density of the particle were the same as the surrounding water (i.e., $S_G = 1$), from Eq. 5 we would obtain $U = 0$. This is reasonable since the particle would be neutrally buoyant and there would be no force to overcome the motion-induced drag. Note also that we have assumed that the particle falls at its *steady terminal velocity*. That is, we have neglected the acceleration of the particle from rest to its terminal velocity. Since the terminal velocity is small, this acceleration time is quite small. For faster objects (such as a free-falling sky diver) it may be important to consider the acceleration portion of the fall.

EXAMPLE 6-10   Bending moment on a pile by drag force

A square 15cm piling is acted on by a water flow of 1.5m/s that is 6m deep, as shown in Fig. E6-10. Estimate the maximum bending stress exerted by the flow on the bottom of the piling.

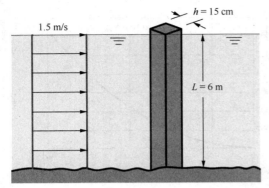

Fig. E6-10

Solution:
From Table A-3 at 15.6℃ we take the density and the kinematic viscosity of seawater to be $\rho = 1030 \text{kg/m}^3$ and $\nu = 1.17 \times 10^{-6} \text{m}^2/\text{s}$. With a piling width of 15cm, we have

$$\text{Re}_h = \frac{Ul}{\nu} = \frac{(1.5 \text{m/s})(0.15 \text{m})}{1.17 \times 10^{-6} \text{m}^2/\text{s}} = 1.92 \times 10^5$$

This is the range where Table 6-5 applies. The worst case occurs when the flow strikes the flat side of the piling, $C_D = 2.2$. The frontal area is $A = Ll = (6\text{m})(0.15\text{m}) = 0.9\text{m}^2$. The drag is estimated by

$$F_D = \frac{1}{2}\rho U^2 A C_D = \frac{1}{2}(1030\text{kg/m}^3)(1.5\text{m/s})^2(0.9\text{m}^2)(2.2) = 2294\text{N}$$

If the flow is uniform, the center of this force should be at approximately middepth. Therefore the bottom bending moment is

$$M_0 = F_D \frac{L}{2} = (2294\text{N}) \frac{6\text{m}}{2} = 6882\text{N}\cdot\text{m}$$

According to the flexure formula from strength of materials, the bending stress at the bottom would be

$$S = \frac{M_0 y}{I} = \frac{(6882\text{N}\cdot\text{m})\left(\frac{0.15\text{m}}{2}\right)}{\frac{1}{12}(0.15\text{m})^4} = 12.23\text{MPa}$$

to be multiplied, of course, by the stress-concentration factor due to the built-in end conditions.

6.8.4 Measurement of Drag Force

Wind Tunnel

A *wind tunnel* is a research tool used in aerodynamic research to study the effects of air moving past solid objects. A wind tunnel consists of a tubular passage with the object under test mounted in the middle. Air is drawn or blown by mechanical means in order to achieve a specified speed and predetermined flow pattern at a given instant. The flow so achieved can be observed from outside the wind tunnel through transparent windows that enclose the test section and flow characteristics are measurable by using specialized instruments. The test object, often called a *wind tunnel model*, or some full-scale engineering structure, typically a vehicle, or part of it, can be immersed into the established flow, thereby disturbing it. The objectives of the immersion include being able to simulate, visualize, observe, and/or measure how the flow around the immersed object affects the immersed object.

There are many different kinds of wind tunnels, the most typical kinds are low-speed wind tunnel, high-speed wind tunnel, supersonic wind tunnel, hypersonic wind tunnel and subsonic and transonic wind tunnel.

In civil and hydraulic engineering, the most commonly used wind tunnel is the *low-speed wind tunnel*. It is used for operations at very low Mach number, with speeds in the test section up to 480 km/h (~ 134 m/s, $Ma = 0.4$). They may be of *open-return type* (also known as the *Eiffel type*, see Fig. 6-33a), or *closed-return flow* (also known as the *Prandtl type*, see Fig. 6-33b) with air moved by a propulsion system usually consisting of large axial fans that increase the dynamic pressure to overcome the viscous losses. In a return-flow wind tunnel the return duct must be properly designed to reduce the pressure losses and to ensure smooth flow in the test section. In the field of civil engineering, the low-speed wind tunnel is usually applied to analyze the physical modelling based external wind effect on complex large scale civil structures, i.e., tall and super-tall buildings, large span roof structures and sports stadia, long span bridges, special structures and airports. Fig. 6-34 shows the wind tunnel testing on the tall buildings and a gymnasium in a low-speed wind tunnel at Zhejiang University.

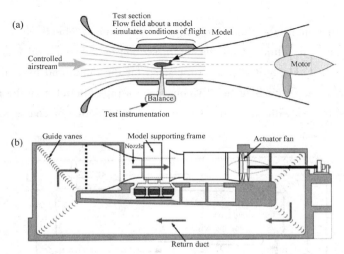

Fig. 6-33 Schematics of (a) open wind tunnel and (b) closed wind tunnel

Fig. 6-34 Wind tunnel testing on (a) the interference effects of tall buildings and (b) external wind effect on the gymnasium (*Courtesy of Zhejiang University*)

Water Tunnel

A *water tunnel* is an experimental facility used for testing the hydrodynamic behavior of submerged bodies in flowing water, i.e., the cavitation, elastic, turbulence of fluid and the boundary layer around the submerged bodies. Few water tunnels are straight and open to return water. Most water tunnels are closed-return flow (see Fig. 6-35), which is very similar to a recirculating low-speed wind tunnel but with water as the working fluid, and the related phenomena, i. e., the forces on scale models of submarines, lift and drag on hydrofoils, are investgated and measured. Water tunnels are sometimes used in place of wind tunnels to perform measurements because techniques like *particle image velocimetry* (PIV) are easier to implement in water. For many cases as long as the Reynolds number is equivalent, the results are valid, whether a submerged water vehicle model is tested in air or an aerial vehicle is tested in water. For low Reynolds number flows, tunnels can be made to run oil instead of water. The advantage is that the increased kinematic viscosity will allow the flow to be a faster speed (and thus easier to maintain stably) for a lower Reynolds number.

In water and oil tunnels the fluid is circulated with pumps, effectively using a net pressure head difference to move the fluid. Thus the return section of water and oil tunnels does not need any flow management; typically it is just a pipe sized for the pump and desired flow speeds. The upstream section of a water tunnels generally consists of a pipe (outlet from the pump) with several holes along its side and with the end open followed by a series of coarse and fine screens to

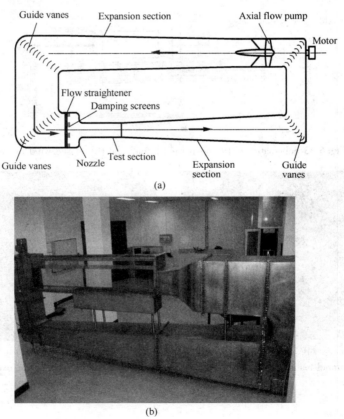

Fig. 6-35 (a) Schematic of water tunnel and (b) a photo of water tunnel (*Courtesy of Jiannong Gu, Naval University of Engineering, PLA*)

even the flow before the contraction into the test section.

Additionally, many water tunnels are sealed and can reduce or increase the internal static pressure, to perform cavitation studies. These are referred to as *cavitation tunnels*.

Summary

Fluid flow is classified as *external* and *internal*, depending on whether the fluid is forced to flow over a surface or in a conduit. This chapter deals with those two flows, especially on the internal flow in pipes and ducts, the most common problems encountered in engineering fluid mechanics.

The *internal flows* are very sensitive to the *Reynolds number* (in pipes)

$$\text{Re} = \frac{\rho \, vd}{\mu} = \frac{vd}{\nu}$$

and change from *laminar* to *transitional* to *turbulent* flow as the Reynolds number increases:
- *Laminar flow*, characterized by smooth streamlines and highly ordered motion at Re < 2300 in pipe;
- *Turbulent flow*, characterized by velocity fluctuations and highly disordered motion at Re > 4000 in pipe; and
- *Transitional flow* in between.

In a *fully developed flow*, the *shear stress* linearly depends on the radius r: zero at the centerline of the circular pipe and maximum at the pipe wall:

$$\tau = \frac{r}{R} \tau_0$$

with the *wall shear stress* is determined by

$$\tau_0 = \rho g R_h J = \frac{\lambda}{8} \rho v^2$$

In a *fully developed laminar pipe flow*, the velocity profile is *parabolic*

$$u = u_{\max}\left[1 - \left(\frac{r}{R}\right)^2\right]$$

with a maximum velocity occurs at the centerline of pipe and two times of the cross-sectional averaged velocity, $u_{\max} = 2v$, and the *major loss* is determined by the Darcy-Weisbach equation

$$h_f = \lambda \frac{L}{d} \frac{v^2}{2g}$$

in which the *Darcy friction coefficient* is a constant and determined by $\lambda = \dfrac{64}{\text{Re}}$.

In a *fully developed turbulent pipe flow*, the flow properties should be *time-averaged*. The flow regime is classified into three layers: *viscous sublayer*, *overlap layer* and *outer* (or *turbulent*) *layer* characterized by their distances from the wall. In the viscous sublayer, the laminar shear stress is dominant, and the velocity profile satisfies the *law of the wall*. Away from the wall, in the outer layer, the *turbulent (Reynolds) shear stress* is dominant, the velocity profile can be taken to satisfy the *logarithmic law* or *power-law*. In the overlap layer, where both laminar and turbulent shears are important, the velocity profile satisfies the *logarithmic law*. Anyway, the shear stress in a turbulent flow includes both laminar and Reynolds shear stresses and is defined as

$$\overline{\tau} = \overline{\tau_{lam}} + \overline{\tau_t} = \mu\frac{d\overline{u}}{dy} - \rho\overline{u'_x u'_y}$$

The *Darcy friction factor* in a turbulent flow is a function of the Reynolds number and/or the relative roughness depending on the different wall types of pipe, $\lambda = f(\text{Re}, \Delta/d)$. For the transition and turbulent flows in smooth as well as rough pipes, the friction factor is given by the *Moody chart*, or *Colebrook equation*

$$\frac{1}{\sqrt{\lambda}} = -2\log\left(\frac{\Delta/d}{3.7} + \frac{2.51}{\text{Re}\sqrt{\lambda}}\right)$$

Besides the friction losses due to the viscous effect, the addition of *minor losses* due to the geometry change, e. g. sudden/gradual expansion and contraction, bending, valve and other devices, is presented in the form of loss coefficients

$$\zeta = \frac{h_j}{v^2/2g}$$

When all the loss coefficients are available, the total *head loss* in a piping system is determined from

$$h_w = \Sigma h_f + \Sigma h_j = \Sigma \lambda \frac{L}{d}\frac{v^2}{2g} + \Sigma \zeta \frac{v^2}{2g}$$

Finally, the *external flows* are presented by begins with a discussion of the flat-plate boundary layer, together with the *laminar* and *turbulent boundary layers*. Then the *boundary layer separation*, where the boundary layer breaks away from the surface and forms a broad and low-pressure *wake*, caused by the pressure-gradient effects are briefly introduced. The drag and lift forces of immersed bodies in fluids are characterized by the *lift coefficient* and *drag coefficient*

$$C_L = \frac{F_L}{\frac{1}{2}\rho U^2 A} \qquad C_D = \frac{F_D}{\frac{1}{2}\rho U^2 A}$$

This chapter ends with a brief discussion of the measurement of drag forces: wind tunnels and water tunnels.

Word Problems

W6-1 Define major loss and minor loss. What cause these losses?
W6-2 Define the hydraulic radius. How is it related to diameter d for a circular pipe?
W6-3 Define laminar flow and turbulent flow. What are their differences?
W6-4 What is the physical significance of the Reynolds number? How is it defined for (a) flow in a circular pipe with an internal diameter of d and (b) flow in a rectangular duct with a cross section of $a \times b$?
W6-5 The lower critical Reynolds number, instead of upper critical Reynolds number, is used

to classify the laminar and turbulent flows. Why?

W6-6 How does the Reynolds number changes with (a) an increasing flow rate in a constant diameter pipe, and (b) a constant flow rate in a diffuser?

W6-7 Consider a person walking first in air and then in water at the same speed. For which motion will the Reynolds number be higher?

W6-8 What is the generally accepted value of the Reynolds number above which the flow in smooth pipe is turbulent?

W6-9 Consider the flow of air and water in pipes with identical diameter, at the same temperature, and at the same mean velocity. Which flow is more likely to be turbulent? Why?

W6-10 How do the shear stress and the velocity vary in a cross section in the fully developed laminar and turbulent flows?

W6-11 How is the hydrodynamic entry length defined for flow in a pipe? Is the entry length longer in laminar or turbulent flow?

W6-12 How does the wall shear stress τ_0 vary along the flow direction in the fully developed region in (a) laminar flow and (b) turbulent flow?

W6-13 What fluid property is responsible for the development of the velocity boundary layer? For what kinds of fluids will there be no velocity boundary layer in a pipe?

W6-14 In the fully developed region of flow in a circular pipe, will the velocity profile change in the flow direction?

W6-15 Someone claims that the shear stress at the center of a circular pipe during fully developed laminar flow is zero. Do you agree with this claim? Explain.

W6-16 Someone claims that in fully developed turbulent flow in a pipe, the shear stress is a maximum at the pipe surface. Do you agree with this claim? Explain.

W6-17 How to compute the friction factor in a fully developed laminar pipe flow? Does it can be used in the entrance region of pipe? Explain.

W6-18 The frictional loss in a fully developed laminar pipe flow is proportional to the flow velocity. Verify it from the Darcy-Weisbach equation for circular pipe flow: $h_f = \lambda \dfrac{L}{d} \dfrac{v^2}{2g}$.

W6-19 The volume flow rate in a circular pipe with laminar flow can be determined by measuring the velocity at the centerline in the fully developed region, multiplying it by the cross-sectional area, and dividing the result by 2. Is it true? Explain.

W6-20 Can the average velocity of a fully developed laminar flow in a circular pipe be determined by simply measuring the velocity at $r_0/2$ (midway between the wall surface and the centerline)? Explain.

W6-21 Consider fully developed laminar flow in a circular pipe. If the diameter of the pipe is reduced by half while the flow rate and the pipe length are held constant, the head loss will (a) double, (b) triple, (c) quadruple, (d) increase by a factor of 8, or (e) increase by a factor of 16.

W6-22 How is head loss related to pressure loss? For a given fluid, explain how you would convert head loss to pressure loss.

W6-23 Consider laminar flow of air in a circular pipe with perfectly smooth surfaces. Do you think the friction factor for this flow will be zero? Explain.

W6-24 Why should the time-averaged properties be introduced in the analysis of turbulent flow? How to define the steady and unsteady turbulent flow?

W6-25 Define the instant velocity, fluctuation velocity, time-average velocity and sectional averaged velocity? How are they related to each other?

W6-26 In fully developed turbulent flow, the shear stress includes laminar and turbulent shear stress. What are the causes and the related properties for each of them? In what part of pipe does each of them dominate?

W6-27 Define the viscous sublayer, overlap layer and outer layer? How do they classified?

W6-28 How to define the hydraulically smooth, transitional and rough zone? How does the friction factor vary with the Reynolds number and relative roughness in those zones?

W6-29 In fully rough zone, the frictional loss $h_f \propto v^2$. Explain it by the Darcy-Weisbach equation.

W6-30 How does surface roughness affect the pressure drop in a pipe if the flow is turbulent? What would your response be if the flow were laminar?

W6-31 For the fully developed flow in a circular pipe with negligible entrance effects, someone claims that the head loss will double if the length of the pipe is doubled. Do you agree? Explain.

W6-32 Explain why the friction factor is independent of the Reynolds number at very large Reynolds numbers.

W6-33 From the Moody chart, rough surfaces, such as sand grains or ragged machining, do not affect laminar flow. Why? They *do* affect turbulent flow, why?

W6-34 What is the minor loss in pipe flow? How is the minor loss coefficient ζ defined?

W6-35 Water flows in a pipe with a sudden change of diameter. If the flow direction is changed to the opposite, do the minor losses at the sudden change of pipe diameter in those two cases equal? Explain.

W6-36 The effect of rounding of a pipe inlet on the loss coefficient is (a) negligible, (b) somewhat significant, or (c) very significant.

W6-37 The effect of rounding of a pipe exit on the loss coefficient is (a) negligible, (b) somewhat significant, or (c) very significant.

W6-38 Which has a greater minor loss coefficient for a pipe flow in (a) gradual expansion, or (b) gradual contraction? Why?

W6-39 A piping system involves sharp turns, and thus large minor head losses. Give two ways to reduce the head loss.

W6-40 During a retrofitting project of a fluid flow system to reduce the pumping power, it is proposed to install vanes into the miter elbows or to replace the sharp turns in 90° miter elbows by smooth curved bends. Which approach will result in a greater reduction in pumping power requirements?

W6-41 What is the boundary layer? How to define its thickness?

W6-42 What are the laminar and turbulent boundary layer?

W6-43 What is flow separation? What causes it? How the flow separation affect the drag coefficient?

W6-44 What is the difference between streamlined and blunt bodies? Is a tennis ball a streamlined or blunt body?

W6-45 Why is flow separation in flow over cylinders delayed in turbulent flow?

W6-46 What is drag? What causes it? Why do we usually try to minimize it?

W6-47 What is lift? What causes it? Does wall shear contribute to the lift?

W6-48 Define the frontal area of a body subjected to external flow. When is it appropriate to use the frontal area in drag and lift calculations?

W6-49 Which car is more likely to be more fuel-efficient: the one with sharp corners or the one that is contoured to resemble an ellipse? Why?

W6-50 Which bicyclist is more likely to go faster: the one who keeps his head and his body in the most upright position or the one who leans down and brings his body closer to his knees? Why?

W6-51 In flow over cylinders, why does the drag coefficient suddenly drop when the flow becomes turbulent? Isn't turbulence supposed to increase the drag coefficient instead of decreasing it?

Problems

Laminar and Turbulent Flows

6-1 Blue and yellow streams of paint at 15.6℃ (each with a density of $824 kg/m^3$ and a viscosity 1000 times greater than water) enter a pipe with an average velocity of 1.2 m/s as shown in Fig. P6-1. Would you expect the paint to exit the pipe as green paint or separate streams of blue and yellow paint? Explain. Repeat the problem if the paint were 'thinned' so that it is only 10 times more viscous than water. Assume the density remains the same.

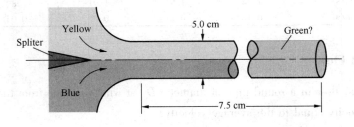

Fig. P6-1

6-2 A soft drink with the properties of 10℃ water is sucked through a 4-mm-diameter, 0.25-m-long straw at a rate of 4 cm³/s. Is the flow at the outlet of the straw laminar? Is it fully developed? Explain.

6-3 Oil with $\nu = 2.2 \times 10^{-4}\,\mathrm{m^2/s}$ is flowing through a pipe at a flowrate of $0.02\,\mathrm{m^3/s}$. Determine the maximum pipe diameter if the oil flow is laminar flow.

6-4 Air at 30 °C with a kinematic viscosity of $16.07 \times 10^{-6}\,\mathrm{m^2/s}$ is passing through a 0.25-m-diameter pipe at a velocity of 3 m/s. (a) Is the flow laminar or turbulent flow? (b) Determine the maximum flow rate if the air flow is restricted to be laminar flow.

6-5 Water is flowing through a 0.25-m-wide rectangular channel at a flowrate of $0.01\,\mathrm{m^3/s}$ at a depth of 0.3 m. Let the water temperature be 20 °C. Is the flow to be laminar or turbulent?

6-6 Water at 10 °C is flowing through a 10-mm-diameter pipe at a velocity of 0.2 m/s. (a) Is the flow laminar or turbulent flow? (b) If the flow velocity and water temperature are same, determine the minimum pipe diameter to make the flow in pipe be turbulent. (c) If the pipe diameter were changed into 30 mm, is the flow laminar or turbulent?

6-7 Oil, which has a viscosity of 0.29 Pa·s and a density of 851 kg/m³, is flowing through a 200-mm-diameter pipe. Determine the minimum flow rate in pipe to make sure the flow in pipe be turbulent?

6-8 Water flows uniformly in a 0.02-m-diameter and 100-m-long pipe with a frictional slope of $J = 0.008$. Determine: (a) the wall shear stress; (b) the shear stress at $r = 0.005\,\mathrm{m}$; (c) the head loss.

6-9 Oil with density of 890 kg/m³ and a viscosity of 0.07 kg/m·s, is to be pumped through 15-m-long straight horizontal pipe with a power input of 1 hp. (a) What is the maximum possible mass flow rate, and corresponding pipe diameter, if laminar flow is to be maintained? (b) What is the wall shear stress?

6-10 The following table lists the velocities of the fixed point in an open channel at a time interval of 0.5 s measured by laser Doppler anemometer. Determine: (a) the fluctuation intensity of turbulent flow; and (b) the Reynolds shear stress.

No. of time interval	1	2	3	4	5
u_x (m/s)	1.88	2.05	2.34	2.30	2.17
u_y (m/s)	0.10	-0.06	-0.21	0.19	0.12
No. of time interval	6	7	8	9	10
u_x (m/s)	1.74	1.91	1.91	1.98	2.19
u_y (m/s)	0.18	0.21	0.06	-0.04	-0.10

Fully Developed Flow

6-11 For laminar flow in a round pipe of diameter D, at what distance from the centerline is the actual velocity equal to the average velocity?

6-12 A fluid with a density of 1000 kg/m³ and a viscosity of 0.30 N·s/m² flows steadily down a vertical 0.10-m-diameter pipe and exits as a free jet from the lower end. Determine the maximum pressure allowed in the pipe at a location 10 m above the pipe exit if the flow is to be laminar.

6-13 Water at 20℃ flows through a horizontal 1-mm-diameter tube to which are attached two pressure taps a distance 1m apart. (a) What is the maximum pressure drop allowed if the flow is to be laminar? (b) Assume the manufacturing tolerance on the tube diameter is $D = 1.0 \pm 0.1$ mm. Given this uncertainty in the tube diameter, what is the maximum pressure drop allowed if it must be assured that the flow is laminar?

6-14 Water at 10℃ is flowing steadily in a 0.20-cm-diameter, 15-cm-long pipe at an average velocity of 1.2m/s. Determine (a) the pressure drop, (b) the head loss, and (c) the pumping power requirement to overcome this pressure drop.

6-15 In fully developed laminar flow in a circular pipe, the velocity at $R/2$ (midway between the wall surface and the centerline) is measured to be 6m/s. Determine the velocity at the center of the pipe.

6-16 The velocity profile in fully developed laminar flow in a circular pipe of inner radius $R = 2$ cm, in m/s, is given by $u(r) = 4\left(1 - \dfrac{r^2}{R^2}\right)$. Determine the average and maximum velocities in the pipe and the volume flow rate.

6-17 Consider an air solar collector that is 1m wide and 5m long and has a constant spacing of 3cm between the glass cover and the collector plate. Air flows at an average temperature of 45℃ at a rate of 0.15 m³/s through the 1-m-wide edge of the collector along the 5-m-long passageway. Disregarding the entrance and roughness effects, determine the pressure drop in the collector.

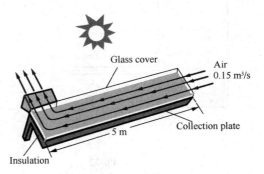

Fig. P6-17

6-18 Oil with a density of 876kg/m³ and a viscosity of 0.24kg/m·s is flowing through a 1.5-cm-diameter pipe that discharges into the atmosphere at 88kPa. The absolute pressure 15m before the exit is measured to be 135kPa. Determine the flow rate of oil through the pipe if the pipe is (a) horizontal, (b) inclined 8° upward from the horizontal, and (c) inclined 8° downward from the horizontal.

6-19 Light oil with a density of 880kg/m³ and a viscosity of 0.015kg/m·s is flowing down a vertical 6-mm-diameter tube due to gravity only. Estimate the volume flow rate if (a) $L = 1$m and (b) $L = 2$m. (c) Verify that the flow is laminar.

6-20 Verify that the velocity at a depth of $y' = 0.632h$ (where h is the total water depth in river, y' is the water depth above the computational point A) in a very wide rectangular river as shown in Fig. P6-20 equals to the average velocity of this

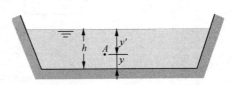

Fig. P6-20

cross section by using Eq. 6.52: $\dfrac{u}{u_*} = \dfrac{1}{\kappa}\ln y + c$, where y is the vertical distance from the river bed to the computational point A. Hint: $\int_0^h \ln\dfrac{y}{h}d\left(\dfrac{y}{h}\right) = -1$.

Frictional Losses and Minor Losses

6-21 Ethanol at 20℃ flows at $0.00789 \text{m}^3/\text{s}$ through a horizontal 12-m-long and 5-cm-diameter cast-iron pip. Neglecting entrance effects, estimate (a) the pressure gradient, dp/dx; (b) the wall shear stress, τ_0; and (c) the percent reduction in friction factor if the pipewalls are polished to a smooth surface.

6-22 The tank-pipe system in Fig. P6-22 is to deliver at least $11 \text{ m}^3/\text{h}$ of water at 20℃ to the reservoir. What is the maximum roughness height Δ allowable for the pipe?

6-23 The pipe flow in Fig. P6-23 is driven by pressurized air in the tank. What gage pressure p_1 is needed to provide a 20℃ water flow rate of $60\text{m}^3/\text{h}$?

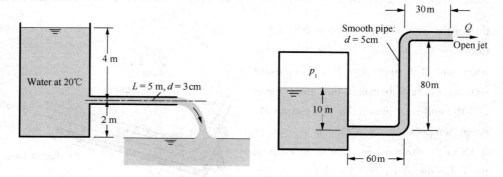

Fig. P6-22   Fig. P6-23

6-24 In Fig. P6-23 suppose $p_1 = 700\text{kPa}$ and the fluid specific gravity is 0.68. If the flow rate is $27\text{m}^3/\text{h}$, estimate the viscosity of the fluid. What fluid in Table A-3 is the likely suspect?

6-25 The reservoirs in Fig. P6-25 contain water at 20℃. If the pipe is smooth with $L = 4500\text{m}$ and $d = 4\text{cm}$, what will the flow rate in m^3/h be for $\Delta z = 100\text{m}$?

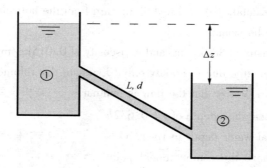

Fig. P6-25

6-26 Water flows in a 200-mm-diameter pipe at a flow rate of $0.094 \text{m}^3/\text{s}$. The frictional slope of this pipe flow is 0.046. Determine the frictional factor in this pipe.

6-27 Consider a uniform flow in a trapezoidal-section clean-earth channel with a 2-m-wide base. This channel has a bottom slope of 0.0004 in flow direction, a side slope of bank of 1.5, and a Manning's coefficient of 0.0225. If the water depth in channel is 1.5m, determine: (a) the average velocity and (b) the flow rate in the channel.

6-28 Water at 20℃ is flowing through a 250-mm-diameter, 700-m-long iron pipe at a flow rate of $0.056 \text{ m}^3/\text{s}$. The Manning's coefficient of pipe is 0.013. Let the kinematic viscosity of water be $1.01 \times 10^{-6} \text{m}^2/\text{s}$. (a) Is the flow in pipe a laminar or turbulent? (b) Determine the frictional loss in this pipe.

6-29 The pressure in a trunk pipe in urban water supply system is 196.2kPa. A 250-mm-diameter branch pipe is horizontally connected to this main pipe. If the roughness height of the branch pipe is 0.5mm, and the flow rate in it is $0.05 \text{ m}^3/\text{s}$, how far can water be transported? Let the water temperature be 20℃.

6-30 Water at 10℃ is flowing through a 300-mm-diameter, 1000-m-long iron pipe, which has a roughness height of 1.2mm. If the friction loss in this pipe is 7.05m, determine the flow rate in the pipe.

6-31 The following data were obtained for flow of 20°C water at 20 m^3/h through a badly corroded 5-cm-diameter pipe which slopes downward at an angle of 8°: $p_1 = 420\text{kPa}$, $z_1 = 12\text{m}$, $p_2 = 250\text{kPa}$, $z_2 = 3\text{m}$. Estimate (a) the roughness height of the pipe and (b) the percent change in head loss if the pipe were smooth and the flow rate the same.

6-32 The small turbine in Fig. P6-32 extracts 400W of power from the water flow. Both pipes are wrought iron ($\Delta = 0.046\text{mm}$). Compute the flow rate Q m^3/h. Sketch the EGL and HGL accurately.

6-33 In Fig. P6-33 the connecting pipe is commercial steel 6cm in diameter. Estimate the flow rate, in m^3/h, if the fluid is water at 20℃. Which way is the flow?

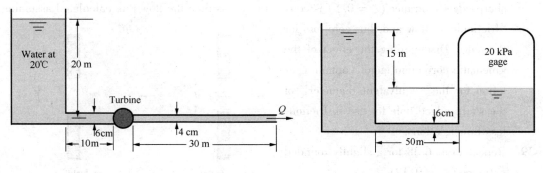

Fig. P6-32 Fig. P6-33

6-34 A large room uses a fan to draw in atmospheric air at 20℃ through a 30-cm by 30-cm commercial-steel duct of 12m-long, as in Fig. P6-34. Estimate (a) the air flow rate in m^3/h if the room pressure is 10Pa vacuum and (b) the room pressure if the flow rate is 1200m^3/h.

Neglect minor losses.

6-35 In Fig 6-35 is a cross section of the concrete prismatic open channel. Water is flowing through it at a flow rate of 7.05 m³/s with a frictional slope of 1/800. Determine: (a) the average frictional velocity at the cross section; (b) the Manning's roughness of channel.

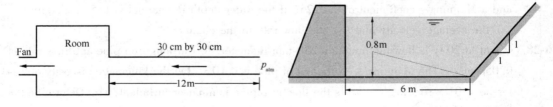

Fig. P6-34 Fig. P6-35

6-36 Water is to be withdrawn from a 3-m-high water reservoir by drilling a 1.5-cm-diameter hole at the bottom surface. Disregarding the effect of the kinematic correction factor, determine the flow rate of water through the hole if (a) the entrance of the hole is well-rounded and (b) the entrance is sharp-edged.

6-37 A horizontal pipe has an abrupt expansion from 8cm to 16cm. The water velocity in the smaller section is 10m/s and the flow is turbulent. The pressure in the smaller section is 300kPa. Taking the kinematic correction factor to be 1.06 at both the inlet and the outlet, determine the pressure in the larger section, and estimate the error that would have occurred if Bernoulli's equation had been used.

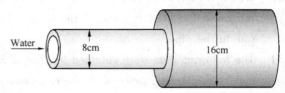

Fig. P6-37

6-38 Consider flow from a water reservoir through a circular hole of diameter d at the side wall at a vertical distance H from the free surface. The flow rate through an actual hole with a sharp-edged entrance ($\zeta = 0.5$) is considerably less than the flow rate calculated assuming 'frictionless' flow and thus zero loss for the hole. Disregarding the effect of the kinematic correction factor, obtain a relation for the 'equivalent diameter' of the sharp-edged hole for use in frictionless flow relations.

6-39 Repeat Prob. 6-38 for a slightly rounded entrance ($\zeta = 0.12$).

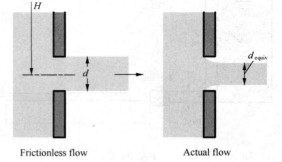

Fig. P6-38

6-40 In Fig. P6-40 water is flowing through a 500-mm-diameter 1200-m-long pipe at a flow rate of 0.2 m³/s. The pipe has a frictional factor of 0.015, a minor loss coefficient at pipe entrance of 0.5, at the location of valve 0.2, and for each bend 0.8. Determine the required water head above the axis at pipe exit H.

6-41 The system in Fig. P6-41 consists of 1200-m-long 5-cm-diameter cast iron pipe, two 45° and four 90° flanged long-radius elbows, a fully open flanged globe valve, and a sharp exit into a reservoir. Determine the required gage pressure at point 1 to deliver 0.005 m³/s of water at 20℃ into the reservoir.

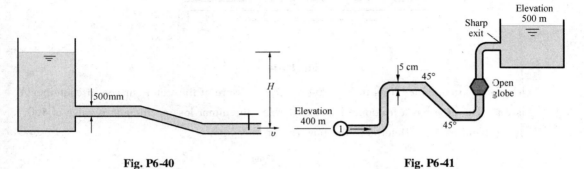

Fig. P6-40   Fig. P6-41

6-42 In Fig. P6-42 the pipe is galvanized iron. The minor loss coefficient for the reentrant inlet with $l/D = 1.2$ is about 1.0, for the butterfly valve at 30° is about 80 and at fully open is 0.3. Estimate the percentage increase in the flow rate (a) if the pipe entrance is cut off flush with the wall and (b) if the butterfly valve is opened wide.

6-43 The shower head in Fig. P6-43 delivers water at 50°C. An orifice-type flow reducer is to be installed. The upstream pressure is constant at 400 kPa. What flow rate results without the reducer? What reducer orifice diameter would decrease the flow by 40 percent?

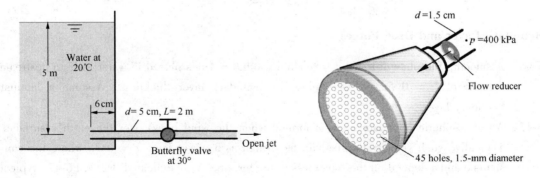

Fig. P6-42   Fig. P6-43

6-44 Consider a simple apparatus to measure the minor loss coefficient of a valve as in Fig. P6-44. To eliminate the friction effects, four piezometer tubes have been assembled to a 50-mm-diameter pipe, among which two are prior to and two are after the valve. The distances between the tubes are l_1 and l_2, respectively. When the average velocity in the pipe is 1.2 m/s, the water levels in piezometer tubes are as shown. Determine the minor loss coefficient of the valve.

6-45 Water discharged from the closed container A is flowing through a 25-mm-diameter 10-m-

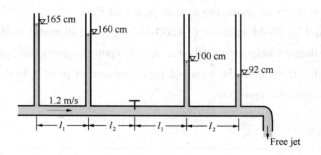

Fig. P6-44

long pipe to container B as in Fig. P6-45. The pressure at the water surface in container A is 2at. The pipe has a frictional factor of 0.025, the minor loss coefficient at valve of 4.0, at the bend of 0.3. Determine the flow rate in the pipe.

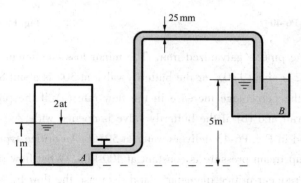

Fig. P6-45

Boundary Layer and Drag Forces

6-46 A smooth, flat plate of length $l = 6m$ and width $b = 4m$ is placed in water with an upstream velocity of $U = 0.5 m/s$. Determine the boundary layer thickness. Assume a laminar boundary layer.

6-47 An atmospheric boundary layer is formed when the wind blows over the earth's surface. Typically, such velocity profiles can be written as a power law: $u = ay^n$, where the constants a and n depend on the roughness of the terrain. As is indicated in Fig. P6-47, typical values are $n = 0.40$ for urban areas, $n = 0.28$ for woodland or suburban areas, and $n = 0.16$ for flat open country. (a) If the velocity is 6m/s at the bottom of the sail on your boat ($y = 1.2m$), what is the velocity at the top of the mast ($y = 9m$)? (b) If the average velocity is 10km/h on the tenth floor of an urban building, what is the average velocity on the sixtieth floor?

6-48 A 38.1-mm-diameter, 0.0245-N-weight table tennis ball is released from the bottom of a swimming pool. With what velocity does it rise to the surface? Assume it has reached its terminal velocity.

6-49 A 500-N cube of specific gravity $S_G = 1.8$ falls through water at a constant speed of U. De-

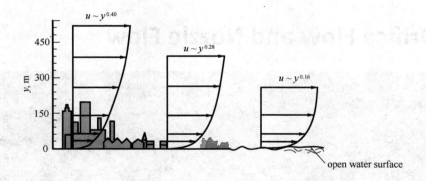

Fig. P6-47

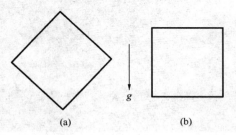

Fig. P6-49

termine U if the cube falls (a) as oriented in Fig. P6-49a, (b) as oriented in Fig. P6-49b.

6-50 If for a given vehicle it takes 20 hp to overcome aerodynamic drag while being driven at 65 mph, estimate the horsepower required at 75 mph.

7 Orifice Flow and Nozzle Flow

CHAPTER OPENING PHOTO: The musical fountain of Xihu Lake in Hangzhou, China, with pivoting nozzles to vary the patterns of the water, is controlled by computers and accompanied by music. The design of the heights and the flowrates out of the nozzles are based on the principles in the present chapter and Chapter 4. (ⓒ三颗桃子[1])

This chapter deals with the basic theories of orifice flow and nozzle flow, the common phenomena in engineering practice. For example, the flows through the sluice gate and the orifice plate meter are orifice flows; the flows through the culver under highway, spillway tunnel in a dam, hydraulic soil excavator nozzle, jetter, turbine draft tube, artificial rain device, nozzle meter, etc., can be considered as the nozzle flows.

7.1 Introduction

An *orifice* is any opening, mouth, hole or vent, as of a pipe, plate. In engineering practice, the sluice gate, the hole used to discharge liquid from containers, the orifice plate meter can be considered as the orifices. The hydraulic phenomenon of fluid flowing through the orifice is termed the *orifice flow*, for example, the fluid flows through a hole located at the vessel wall as shown in Fig. 7-1 is a typical orifice flow.

Considered a well-contoured horizontal orifice as indicated in Fig. 7-1, the fluid is unable to turn the sharp 90° corner indicated by the dotted lines in Fig. 7-1 but to be a smooth curve, which

---

[1] http://www.19lou.com/forum-138-thread-14801352897328862-1-1.html

indicates that the fluid flowing through the orifice only contacts with the orifice perimeter line, and the thickness of the orifice wall has no influence on the discharge. This orifice is termed the *sharp-edged orifice*. The jet through the orifice continues to contract till to the section about $d/2$ away from the orifice, section c-c in Fig. 7-1, where the streamlines are nearly parallel to each other, and the diameter, d_c, is less than the diameter of the orifice, d. This section is termed the *vena contracta*. Its effect is designated by the *contraction coefficient*, ε

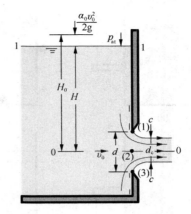

$$\varepsilon = \frac{A_c}{A} \quad (7.1)$$

Fig. 7-1 Steady free jet through sharp-edged small orifice

where A_c and A are the area of the jet at the vena contracta and the area of the orifice, respectively.

For the orifice flow, i.e. the one in Fig. 7-1, the velocity at the centerline will be slightly greater than that at the top, and slightly less than that at the bottom due to the differences in elevation. The highest pressure occurs along the centerline at (2) and the lowest pressure, $p_1 = p_3 = 0$, is at the edge of the jet. However, if $d/H < 0.1$, we can safely use the centerline velocity as a reasonable 'average velocity' and assume a uniform velocity with straight streamlines and constant pressure ($p = 0$) in the plane of the vena contracta, section c-c. Such an orifice is termed *small orifice*. On contrary, if $d/H > 0.1$, the assumption of constant pressure at the plane of vena contracta is no longer valid, the velocity and pressure varies vertically along the plane. Such an orifice is termed *big orifice*.

A *nozzle* is a device designed to control the direction or characteristics of a fluid flow (especially to increase velocity) as it exits (or enters) an enclosed chamber or pipe. A nozzle is often a pipe or tube of varying cross sectional area with 3 to 4 times of orifice diameter long connected to the orifice. Fig. 7-2 shows several typical nozzles commonly used in the engineering practice.

Nozzles are frequently used to control the rate of flow, speed, direction, mass, shape, and/or the pressure of the stream that emerges from them. In nozzle, velocity of fluid increases on the expense of its pressure energy. The hydraulic phenomenon that a liquid flows through a nozzle and fully fills the nozzle outlet section is termed *nozzle flow* (Fig. 7-3).

When the fluid is freely discharged into atmosphere through the orifice, nozzle, or pipe outlet, the pressure on the vena contracta can be considered as atmospheric pressure, that is $p_c = p_{at}$. Such an outflow is termed the *free discharge*, or *free jet*. Otherwise, if the orifice or nozzle is immersed in the downstream liquid body, it is termed the *submerged discharge*, or *submerged jet*. If the water level over the orifice, nozzle or the pipe outlet is kept unchanged, the flow through the orifice, nozzle or the pipe outlet would be *steady*; otherwise, it is *unsteady*.

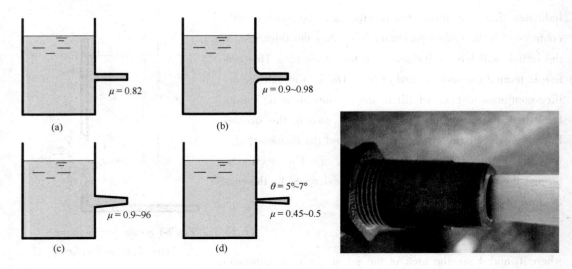

Fig. 7-2 Commonly used nozzles: (a) Cylindrical outer nozzle; (b) Streamlining outer nozzle; (c) Converging cone nozzle; (d) Diverging cone nozzle.

Fig. 7-3 Water nozzle (*Courtesy of Wikipedia*)

7.2 Steady Flow through Sharp-edged Orifice

7.2.1 Free Jet through Small Orifice

For the sharp-edged small orifice, the velocity and pressure at the plane of vena contracta, section c-c in Fig. 7-1, can be considered to be distributed uniformly. Take the flow to be steady and the jet be free, the energy equation between free surface in tank and the vena contracta with a datum plane at the orifice centerline gives

$$H + \frac{p_{at}}{\rho g} + \frac{\alpha_0 v_0^2}{2g} = 0 + \frac{p_c}{\rho g} + \frac{\alpha_c v_c^2}{2g} + h_w \tag{7.2}$$

Since the frictional loss in the tank is negligible, the head loss only includes the minor loss through the orifice, that is

$$h_w = h_j = \zeta_o \frac{v_c^2}{2g} \tag{7.3}$$

We also have $p_c = p_{at}$ since it is the free jet from the sharp-edged small orifice. Take $\alpha_c = 1.0$ and define $H_0 = H + \frac{\alpha_0 v_0^2}{2g}$, Eq. 7.2 becomes

$$H_0 = (\alpha_c + \zeta_o) \frac{v_c^2}{2g} = (1 + \zeta_o) \frac{v_c^2}{2g}$$

which gives the velocity at the vena contracta and the flowrate through orifice

$$v_c = \frac{1}{\sqrt{1 + \zeta_o}} \sqrt{2gH_0} = \varphi \sqrt{2gH_0} \tag{7.4}$$

$$Q = v_c A_c = \varepsilon A \varphi \sqrt{2gH_0} = \mu A \sqrt{2gH_0} \qquad (7.5)$$

where

H_0 is the *acting head* above the orifice centerline, if $v_0 \approx 0$, $H_0 = H$;

ζ_o is the *minor loss coefficient* for flow through orifice;

φ is the *coefficient of velocity*, which is the ratio of actual velocity at the vena contracta, v_c, to the ideal fluid velocity, $\sqrt{2gH_0}$, $\varphi = \dfrac{1}{\sqrt{\alpha_c + \zeta_o}} \approx \dfrac{1}{\sqrt{1 + \zeta_o}}$. According to the experimental data, $\varphi = 0.97$ for sharp-edged small circular orifice, thus the minor loss coefficient for sharp-edged small circular orifice is

$$\zeta_o = \frac{1}{\varphi^2} - 1 = \frac{1}{0.97^2} - 1 = 0.06.$$

ε is the *coefficient of contraction* defined in Eq. 7.1, which depends on the orifice geometry and the location on the tank wall. For sharp-edged small circular orifice, the flow has been completely contracted in all directions at the plane of vena contracta, the diameter at the vena contracta is $d_c = 0.8d$, thus $\varepsilon = 0.64$. However, if the orifice is located at tank wall with any one of the distance to the tank edges or the free surface is less than $3d$, the value of ε would be different and should be determined by experiments. This will be discussed in Section 7.2.3.

μ is the *discharge coefficient of orifice*, $\mu = \varepsilon \varphi$. For sharp-edged small circular orifice, $\mu = \varepsilon \varphi = 0.62$.

7.2.2 Submerged Jet through Orifice

For a steady orifice flow submerged in the downstream liquid body as shown in Fig. 7-4, the energy equation between sections 1-1 and 2-2 with a datum located at the orifice centerline gives

$$H_1 + \frac{p_{at}}{\rho g} + \frac{\alpha_1 v_1^2}{2g} = H_2 + \frac{p_{at}}{\rho g} + \frac{\alpha_2 v_2^2}{2g} + h_w \qquad (7.6)$$

The head loss, h_w, in the submerged jet includes the minor loss fluid flowing through the orifice and the sudden expansion starting from the vena contractra in the downstream liquid body

$$h_w = (\zeta_o + \zeta_{se}) \frac{v_c^2}{2g} \qquad (7.7)$$

Define $H_0 = H_1 - H_2 + \dfrac{\alpha_1 v_1^2}{2g} - \dfrac{\alpha_2 v_2^2}{2g}$, Eq. 7.6 becomes

$$H_0 = h_w = (\zeta_o + \zeta_{se}) \frac{v_c^2}{2g}$$

which gives the velocity at the vena contracta and the

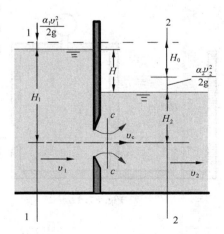

Fig. 7-4 Steady submerged jet through sharp-edged small orifice

flowrate through orifice

$$v_c = \frac{1}{\sqrt{\zeta_o + \zeta_{se}}} \sqrt{2gH_0} = \varphi \sqrt{2gH_0} \qquad (7.8)$$

$$Q = v_c A_c = \varepsilon A \varphi \sqrt{2gH_0} = \mu A \sqrt{2gH_0} \qquad (7.9)$$

where H_0 is the *acting head* above the orifice centerline, if $v_1 \approx 0, v_2 \approx 0$, $H_0 = H_1 - H_2 = H$; ζ_o is the *minor loss coefficient* for flow through orifice same as the one for free jet; ζ_{se} is the *minor loss coefficient* when jet gets a sudden expansion from vena contracta c-c to section 2-2, if $A_2 \gg A_c$, $\zeta_{se} \approx 1.0$ based on Eq. 6.76 for the sudden expansion; φ is the *velocity coefficient* for submerged jet, $\varphi = \frac{1}{\sqrt{\zeta_o + \zeta_{se}}} \approx \frac{1}{\sqrt{1 + \zeta_o}}$; ε is the *coefficient of contraction*, $\varepsilon = 0.64$; μ is the *discharge coefficient* of submerged orifice, $\mu = \varepsilon \varphi$. For sharp-edged small orifice, $\varphi \approx 0.97$, thus we have $\mu = 0.62$, which equals to the value of μ for the free jet flow.

It should be noted that Eqs. 7.8 and 7.9 can also be used for the submerged jet through both the big and small orifices, because from Eqs. 7.8 and 7.9 we can see that the velocity and flow rate of submerged jet only depend on the difference of water levels and have nothing to do with the water depth above the orifice center line.

The comparison of Eqs. 7.5 and 7.9 shows that we have the exactly same expression and the same values of discharge coefficients to determine the flow rates of free jet and submerged jet through small sharp-edged orifices. However, the acting heads differ from each other. For free jet, the head H is the water depth above the orifice centroid; for submerged jet, the head H is the difference of water levels between upstream and downstream tanks if the velocities in the tanks are negligible.

EXAMPLE 7-1 Free jet and submerged jet

There are two small sharp-edged circular orifices with same diameters located on the tank walls at the same height from the tank bottom as shown in Fig. E7.1. If the water levels in tanks are steady, determine the pressure reading of the gage. Assume water is at 20℃.

Solution:

Assume the flow is incompressible and the velocities in tanks are negligible. Take the water density at 20℃ to be 998.2 kg/m³.

The flowrate for the submerged jet and free jet can be given as

$$Q_1 = \mu_1 A_1 \sqrt{2g \left[H_1 - \left(H_2 + \frac{p}{\rho g} \right) \right]}$$

and

$$Q_2 = \mu_2 A_2 \sqrt{2g \left(H_2 + \frac{p}{\rho g} \right)}$$

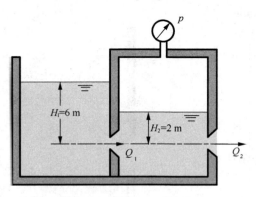

Fig. E7.1

where $A_1 = A_2$, $\mu_1 = \mu_2$ because both orifices are small sharp-edged circular orifices and have same diameters.

For the steady flow, we have $Q_1 = Q_2$, which gives

$$H_1 - \left(H_2 + \frac{p}{\rho g}\right) = H_2 + \frac{p}{\rho g}$$

or

$$p = \frac{1}{2}\rho g(H_1 - 2H_2) = \frac{1}{2}(998.2\text{kg/m}^3)(9.81\text{m/s}^2)[6\text{m} - (2)(2\text{m})] = 9792\text{Pa}$$

DISCUSSION When the tank is closed as indicated in this example, the pressure at the water surface in the enclosed tank is not atmospheric pressure anymore, which should be taken into consideration in the formula computing the flowrate through orifice.

7.2.3 Free Jet through Big Orifice

For big orifice, the distributions of velocity and pressure are no longer uniform at the plane of vena contracta. Thus, the flowrate through big orifice has to be added up by separating big orifice along its vertical size into several small orifices as indicated in Fig. 7-5

$$Q = \int dQ = \int_{H_1}^{H_2} \mu\sqrt{2gh}b\,dh = \frac{2}{3}\mu b\sqrt{2g}(H_2^{3/2} - H_1^{3/2}) \tag{7.10}$$

where b is the width of orifice; H_1 and H_2 are the water depths above the upper and lower edges of the big orifice, respectively; μ is the *discharge coefficient* of big orifice.

It should be noted that in engineering practice, the discharge formula for sharp-edged small orifice, Eq. 7.5, can also be used to estimate the flowrate through a shape-edged big orifice by using water depth above the centerline of big orifice and with $\mu = 0.62$. However, the discharge coefficient of big orifice is usually bigger than the one of small orifice because of the larger contraction coefficient of the big orifice as listed in Table 7-1, in which the *perfect contraction* and *non-perfect contraction* can be illustrated in Fig. 7-6. When the distance from the orifice edge to the corresponding adjacent tank edge in all directions is three times larger than the orifice size in the same direction, that is $l>3a$ and $l>3b$, the flow contraction has nothing to do with the distance away from the tank edge. Such a flow contraction is termed the *perfect contraction*, i.e. the orifice A in Fig. 7-6. When $l>3a$ or $l>3b$ in any direction cannot be satisfied, the flow will not contract completely in this direction. Such a flow contraction is termed the *non-perfect contraction*, i.e. the orifices B, C and D in Fig. 7-6. It should be noted that the coefficient of discharge for non-perfect con-

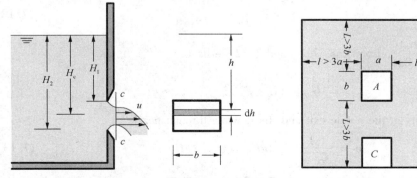

Fig. 7-5 Steady free jet through big orifice   **Fig. 7-6** Schematic of orifice locations at the tank wall

traction orifice is always bigger than that for perfect contraction orifice, as indicated in Table 7-1.

Table 7-1   The discharge coefficient of big orifice

Contraction of flow	μ
Non-perfect contraction in all direction	0.70
No contraction in the bottom, but with moderate contractions in other directions (orifice C in Fig. 7-6)	0.66 ~ 0.70
No contraction in the bottom, but with very small contractions in other directions	0.70 ~ 0.75
No contraction in the bottom, but with extreme small contractions in other directions	0.80 ~ 0.90

7.3   Steady Nozzle Flow

7.3.1   Cylindrical Outer Nozzle Flow

Fig. 7-7 shows a steady flow through a cylindrical outer nozzle with a constant water level.

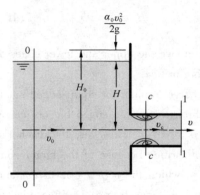

Fig. 7-7 Steady free jet through cylindrical outer nozzle

Due to the inability to turn 90° sharp turn, the flow through the nozzle entrance contracts continuously till to the section of *vena contracta*, c-c, then gradually expands to the full section of nozzle before it exits the nozzle. A separation zone occurs just downstream of the nozzle entrance due to the flow contraction.

Neglect the frictional loss in the nozzle because its value in such a short nozzle is negligible compared to the minor loss in nozzle. Thus, the energy equation between section 0-0 and the nozzle exit plane 1-1 with a datum at the orifice centerline gives

$$H + \frac{\alpha_0 v_0^2}{2g} = 0 + \frac{p_{at}}{\rho g} + \frac{\alpha v^2}{2g} + h_j \quad (7.11)$$

where h_j is the minor loss caused by fluid flowing through the nozzle entrance and the sudden enlargement after the vena contracta c-c and defined as

$$h_j = \zeta_n \frac{v^2}{2g} \quad (7.12)$$

Take $H_0 = H + \frac{\alpha_0 v_0^2}{2g}$, from Eqs. 7.11 and 7.12 we have

$$H_0 = (\alpha + \zeta_n) \frac{v^2}{2g}$$

which gives the velocity at the nozzle exit and the flowrate through nozzle

$$v = \frac{1}{\sqrt{\alpha + \zeta_n}} \sqrt{2gH_0} = \varphi_n \sqrt{2gH_0} \quad (7.13)$$

$$Q = vA = A\varphi_n \sqrt{2gH_0} = \mu_n A \sqrt{2gH_0} \quad (7.14)$$

where H_0 is the *acting head* above the nozzle centerline, if $v_0 \approx 0$, $H_0 = H$; ζ_n is the *minor loss coefficient* when fluid flows through the nozzle, its value is near to the minor loss coefficient when fluid flows through sharp-edged pipe entrance, $\zeta_n \approx 0.5$; φ_n is the *velocity coefficient of nozzle*, $\varphi_n = \dfrac{1}{\sqrt{\alpha + \zeta_n}} = \dfrac{1}{\sqrt{1 + 0.5}} = 0.82$; μ_n is the *discharge coefficient* of nozzle, $\mu_n = \varphi_n = 0.82$ because there is no contraction at the nozzle outlet.

The discharge formula of orifice flow and nozzle flow, Eq. 7.5 and Eq. 7.14, have the same mathematical expression. However, the discharge coefficient $\mu_n = 1.32\mu$, which indicates that the nozzle discharge is 1.32 times of the orifice discharge if their acting heads and their orifice areas are the same.

7.3.2 Vacuum in Cylindrical Outer Nozzle

The nozzle is a short pipe connected to the orifice, which increase the resistance force when fluid flows through the nozzle. However, the nozzle discharge is greater than the orifice discharge as discussed above, which is contributed by the vacuum occuring in the cylindrical outer nozzle.

The energy equation between the section of vena contracta *c-c* and the nozzle exit section 1-1 in Fig. 7-7 gives

$$\frac{p_c}{\rho g} + \frac{\alpha_c v_c^2}{2g} = \frac{p_{at}}{\rho g} + \frac{\alpha v^2}{2g} + h_j \tag{7.15}$$

where the head loss, h_j, is caused by the sudden enlargement of flow after the vena contracta, $h_j = \zeta_{se} \dfrac{v^2}{2g}$, where ζ_{se} is the minor loss coefficient of sudden expansion, $\zeta_{se} = \left(\dfrac{A}{A_c} - 1\right)^2 = \left(\dfrac{1}{\varepsilon} - 1\right)^2$.

Apply the continuity equation between the sections of vena contracta *c-c* and the nozzle exit to give $A_c v_c = Av$, which yields

$$v_c = \frac{A}{A_c} v = \frac{1}{\varepsilon} v$$

The velocity at nozzle exit v can be determined by Eq. 7.13: $v = \varphi_n \sqrt{2gH_0}$. Thus, Eq. 7.15 can be rearranged as

$$\frac{p_c - p_{at}}{\rho g} = \frac{\alpha v^2}{2g} - \frac{\alpha_c v_c^2}{2g} + \zeta_{se} \frac{v^2}{2g} = \left[\alpha - \frac{\alpha_c}{\varepsilon^2} + \left(\frac{1}{\varepsilon} - 1\right)^2\right] \frac{v^2}{2g} = \left[\alpha - \frac{\alpha_c}{\varepsilon^2} + \left(\frac{1}{\varepsilon} - 1\right)^2\right] \varphi_n^2 H_0$$

With the given parameters of $\varepsilon = 0.64$, $\varphi_n = 0.82$ and $\alpha = \alpha_c = 1.0$ gives the pressure at the vena contracta

$$\frac{p_c - p_{at}}{\rho g} = -0.75 H_0 \tag{7.16}$$

It indicates that vacuum occurs at the plane of vena contracta when fluid flows through the cylindrical outer nozzle. Its *vacuum height* is

$$\frac{p_v}{\rho g} = \frac{p_{at} - p_c}{\rho g} = 0.75 H_0 \tag{7.17}$$

For the orifice flow, the contraction of flow occurs in the air. However for the nozzle flow, the contraction of flow occurs within the nozzle accompanied by a vacuum about $0.75H_0$. It indicates the nozzle flow achieves about 75% extra head than the orifice flow even though they have the same acting head, which leads to a larger flowrate through the nozzle than through the orifice.

Equation 7.17 also indicates that the larger the acting head H_0, the larger the vacuum on the vena contracta. When the vacuum height on the section of vena contracta is greater than 7.0 m, the air may be sucked into the nozzle from the nozzle exit, which may destroy the vacuum at the vena contracta. Once the vacuum disappears, the nozzle exit section can no longer be fully filled by the liquid. The nozzle flow turns into the orifice flow. Therefore, to make sure the vacuum height at the plane of vena contracta less than $0.75H_0$, that is $\frac{p_{at}}{\rho g} \leqslant 0.75H_0$, the acting head of nozzle should be less than a limited value $[H_0] = 7\text{m}/0.75 = 9$ m.

The length of nozzle also has to be limited. If the length of nozzle is too short, $l < (3\text{-}4)d$, the fluid clusters cannot be expanded to whole section before the nozzle exit, the vacuum cannot form on the section of vena contracta, the flow acts as orifice flow. If the length of nozzle is too long, the percent of frictional loss increases, the nozzle flow turns into a flow in short pipe.

Therefore, the conditions for nozzle flow occurring in a cylindrical outer nozzle are: the acting head $H_0 \leqslant 9$m; the length of nozzle $l = (3\text{-}4)d$; the nozzle exit section is fully occupied.

7.3.3 Typical Nozzles

In the engineering practice, there are many kinds of nozzle. Fig. 7-2 shows several typical nozzles commonly used in the engineering practice:

- *Cylindrical outer nozzle*, with a sharp-edged entrance, the flow first contracts then enlarges to fill the whole cross section of nozzle.
- *Streamlining outer nozzle*, with a streamlined entrance, has the smallest loss coefficient because the flow has no contraction and enlargement in the nozzle.
- *Diverging cone nozzle*, expanding from a smaller diameter to a larger one and accelerating subsonic fluids, is usually applied as jetter, turbine draft tube and artificial rain device due to its bigger discharge and lower velocity.
- *Converging cone nozzle*, arrowing down from a wide diameter to a smaller diameter and slowing fluid in the direction of the flow, is usually applied as hydraulic soil excavator nozzle, and nozzle in fire control because of the larger velocity at the exit.

All the nozzles have the same formula, Eq. 7.14, to determine the discharge from the nozzle. However, they have the different discharge coefficient as indicated in Fig. 7-2.

7.4 Unsteady Flow through Orifice

7.4.1 Empty Time of Tank

Figure 7-8 shows an orifice flow from a tank with a constant cross sectional area. Once the water in the tank cannot be implemented simultaneously, the water level in the tank changes with the time, so does the flowrate through the orifice. Assume at the moment t, the head above the orifice centerline is h as indicated in Fig. 7-8, thus the fluid volume through the orifice during the time interval, dt, can be determined by Eq. 7.5

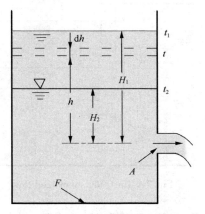

Fig. 7-8 Unsteady free jet

$$d\mathcal{V} = Qdt = \mu A\sqrt{2gh}\,dt$$

which equals to the decreased volume in the tank during the time interval dt

$$d\mathcal{V} = -Fdh$$

where A is the area of orifice; F is the cross sectional area of tank.

Thus, we have

$$\mu A\sqrt{2gh}\,dt = -Fdh$$

$$dt = -\frac{F}{\mu A\sqrt{2g}}\frac{dh}{\sqrt{h}}$$

Integration of the above equation yields

$$t = \int_{H_1}^{H_2} -\frac{F}{\mu A\sqrt{2g}}\frac{dh}{\sqrt{h}} = \frac{2F}{\mu A\sqrt{2g}}(\sqrt{H_1} - \sqrt{H_2}) \qquad (7.18)$$

If $H_2 = 0$, we can get the *empty time* of the tank from the initial water level to the orifice centerline

$$T = \frac{2F\sqrt{H_1}}{\mu A\sqrt{2g}} = \frac{2FH_1}{\mu A\sqrt{2gH_1}} = \frac{2\mathcal{V}}{Q_{max}} \qquad (7.19)$$

where $\mathcal{V} = FH_1$ is the total liquid volume released from the tank when water level falls from the initial water level to the orifice centerline; $Q_{max} = \mu A\sqrt{2gH_1}$ is the maximum flowrate discharged into the air through the orifice at the very beginning.

Eq. 7.19 indicates that for the variable water level in the tank, the empty time of constant sectional area tank is twice of the empty time for discharging same liquid volume from same tank under the constant water level H_1 at the very beginning.

7.4.2 Filling Time of Tank

Two tanks are separated by a wall with a submersed orifice as shown in Fig. 7-9. The water level in the upstream tank is constant, while the water level in downstream tank changes with the volume of fed liquid from the upstream tank through the orifice. The downstream tank is taken to be fully filled when the water levels in downstream and upstream tanks equals.

To determine the filling time of the downstream tank, the similar manner for empty time computation can be used. Thus we have

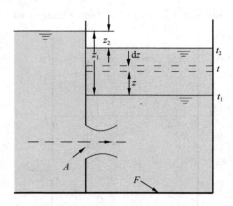

Fig. 7-9 Unsteady submerged flow

$$T = \frac{2V}{\mu A \sqrt{2gz_1}} = \frac{2V}{Q_{max}} \quad (7.20)$$

where $V = Fz_1$ is the total volume filled into the downstream tank during the time interval T; $Q_{max} = \mu A \sqrt{2gz_1}$ is the maximum flowrate discharged into the downstream tank through the orifice at the very beginning; A is the area of orifice; F is the cross sectional area of downstream tank; z_1 is the water level difference at the very beginning.

Equation 7.20 indicates that for the variable water level in the downstream tank, the *filling time* of constant sectional area tank is twice of the filling time for feeding same liquid volume into the same tank under the constant difference of water levels z_1 at the very beginning.

EXAMPLE 7-2 Emptying time from a tank

A 1.2-m-high, 0.9-m-diameter cylindrical water tank whose top is open to the atmosphere is initially filled with water. Now the discharge plug near the bottom of the tank is pulled out, and a water jet whose diameter is 1.27-cm streams out (Fig. E7-2). Determine how long it will take for the water level in the tank to drop to 0.6m from the bottom.

Solution:

Assume water is an incompressible substance and the distance between the bottom of the tank and the center of the hole is negligible compared to the total water height. Take the discharge coefficient of orifice to be 0.62.

Thus, from Eq. 7.18, we have the empty time taken for the water level in the tank to drop to 0.6m from the bottom

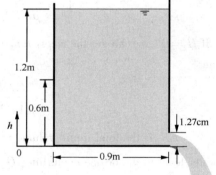

Fig. E7-2

$$t = \frac{2F}{\mu A \sqrt{2g}} (\sqrt{H_1} - \sqrt{H_2})$$

$$= \frac{(2)\left(\frac{\pi}{4}\right)(0.9\text{m})^2}{(0.62)\left(\frac{\pi}{4}\right)(0.0127\text{m})^2 \sqrt{2(9.81\text{m/s}^2)}} (\sqrt{1.2\text{m}} - \sqrt{0.6\text{m}})$$

$$= 1173\text{s} = 19.55\text{min}$$

Therefore, half of the tank will be emptied in 19.55 min after the discharge hole is unplugged.

DISCUSSION   Using the same relation with $h_2 = 0$ gives $t = 66.77$min for the discharge of the entire amount of water in the tank. Therefore, emptying the bottom half of the tank takes much longer than emptying the top half. This is due to the decrease in the average discharge velocity of water with decreasing h. On the other hand, if we neglect the loss through the orifice, that is, the discharge coefficient is taken to be $\mu = 1$. The average velocity of the jet is then given by $v = \sqrt{2gh}$. Thus, the time emptying the top half and the whole tank are 12.12min and 41.4min, which are much short than the time considering the loss.

EXAMPLE 7-3   Filling time of tank

As indicated in Fig. E7-3, two tanks is connected by a short pipe with an area of a. The discharge coefficient of this short pipe is $\mu = 0.6$. The initial difference between free surfaces in the tanks is $H = 4$m. The relations of cross sectional areas of tanks to the area of pipe are $A_1 = 20a$ and $A_2 = 10a$ respectively. Once the valve in the middle of short pipe is opened, how long does it take to make the water levels in two tanks leveling with each other?

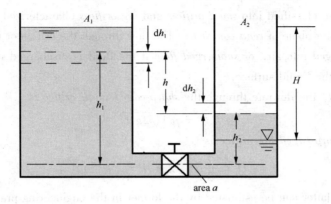

Fig. E7-3

Solution:

Assume the water to be incompressible.

Assume at time t, the water levels in two tanks are h_1 and h_2 respectively. The water level difference is

$$h = h_1 - h_2 \tag{1}$$

After an time interval dt, the water level in upstream tank drops dh_1, while the water level in downstream tank rises dh_2. According to the conservation law of mass, the decreased volume in upstream tank for an incompressible fluid must equal to the increased volume in downstream tank, that is

$$-A_1 dh_1 = A_2 dh_2 \tag{2}$$

Combining Eqs. 1 and 2 gives

$$dh_1 = \frac{dh}{1 + \dfrac{A_1}{A_2}} \tag{3}$$

During the time interval dt, the water volume passing the short pipe equals to the decreasing water volume

$$\mu a \sqrt{2gh}\, dt = -A_1 dh_1 \tag{4}$$

Thus we have

$$dt = \frac{-A_1 dh}{\mu a \sqrt{2gh}\left(1 + \dfrac{A_1}{A_2}\right)}$$

Integrating it from $h = H$ to $h = 0$ yields

$$\int_0^T dt = \int_H^0 \frac{-A_1}{\mu a \sqrt{2g}\left(1 + \dfrac{A_1}{A_2}\right)} \frac{dh}{\sqrt{h}}$$

or

$$T = \frac{2A_1}{\mu a \sqrt{2g}\left(1 + \dfrac{A_1}{A_2}\right)} \sqrt{H} = \frac{(2)(20a)}{(0.6a)\sqrt{(2)(9.81\text{m/s}^2)}\left(1 + \dfrac{20a}{10a}\right)} \sqrt{4\text{m}} = 10.03\text{s}$$

Summary

This chapter is mainly concerned with the *orifice flow* and the *nozzle flow*, the common hydraulic phenomena in engineering practice.

The orifices are classified into *small orifices* and *big orifices* characterized by the variations of fluid properties at the plane of *vena contracta*. The flow through the orifice or nozzle is a *free jet* if the fluid is discharged into air, or *submerged flow* if the fluid is discharged into the downstream liquid body under the liquid surface.

For the *free jet*, the flowrate through the *sharp-edged small orifice* is

$$Q = \mu A \sqrt{2gH_0}$$

and through the *sharp-edged big orifice* is

$$Q = \frac{2}{3}\mu b \sqrt{2g}(H_2^{3/2} - H_1^{3/2})$$

Actually, the latter can be estimated by the former in the engineering practice, even through it may be 5% ~ 10% less.

For the *submerged jet*, the flowrate through the sharp-edged both small and big orifices are determined by

$$Q = \mu A \sqrt{2gH_0}$$

which has an identical expression with the free jet through sharp-edged small orifice, but a different *acting head* H_0.

Nozzle is a short pipe or tube connected to the orifice. The flowrate through an outer cylindrical nozzle is determined by

$$Q = \mu_n A \sqrt{2gH_0}$$

in which $\mu_n = 0.82$. It indicates the flowrate discharged by an outer cylindrical nozzle with same acting head and same cross-sectional area is 1.32 times that of the sharp-edged small orifice. It is because a *vacuum* with a height of $0.75H_0$ occurs at the *vena contracta* in the nozzle, which results in the restricted condition for a nozzle: the acting head $H_0 \leq 9$m; the length of nozzle $l = (3-4)d$; the nozzle exit section is fully occupied.

Finally the *emptying* and *filling time* of a tank with a variable water level (unsteady flow) are discussed. Either the emptying or filling time of a tank is twice of the corresponding time of a steady emptying or filling flow under the constant head at the very beginning.

Word Problems

W7-1 Define small orifice and big orifice. How does the pressure and velocity distribute along the vertical plane of these two orifices?

W7-2 How does the discharge formula for the small orifice differ from the big orifice?

W7-3 Assume all the orifices in Fig. W7-3 are sharp-edged small orifices with perfect contractions and identical orifice-diameter, and the flows are steady. Determine the relation of the flowrates through the orifices in each case.

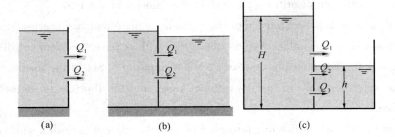

Fig. W7-3

W7-4 Can Eq. 7.5 for small orifice be used to determine the flowrate through the big orifice? Explain.

W7-5 Define nozzle flow. In Fig. W7-5 which flow is nozzle flow? Explain.

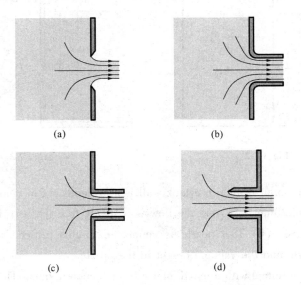

Fig. W7-5

W7-6 How to make sure the cylindrical outer nozzle work normally? Explain.

W7-7 In engineering practice, nozzle, instead of the simple hole, is more commonly used to release liquid from tank, reservoir or other containers. Why?

Problems

Orifice and Nozzle Flow

7-1 A tank has a sharp-edged circular orifice with a diameter of 10mm on the wall. The orifice center is at a depth of 2m. The diameter at the vena contracta of the steady free jet through this orifice is 8mm. If water volume through the orifice is 0.01m^3 in 32s, determine the contraction coefficient ε, the velocity coefficient φ, the orifice discharge coefficient μ, and the minor loss coefficient of orifice ζ_o. Neglect the velocity in the tank.

7-2 In Fig. P7-2 a tank is separated into two parts by a plate and two sharp-edged circular orifices are made on the walls. Assume those two orifices have the identical discharge coefficient, $\mu = 0.62$. For a steady flow with a water depth of 4.8m in the upstream tank, determine the water depth H_2 in the downstream tank, and the flowrate to the open air through the orifice.

7-3 A reservoir discharge its flooding through a rectangular big orifice with a width of 0.7m and a height of 1.5m as shown in Fig. P7-3. Assume the flow is steady and the discharge coefficient of the orifice is 0.62, determine the flowrate through this big orifice by the discharge formula of small and big orifices respectively.

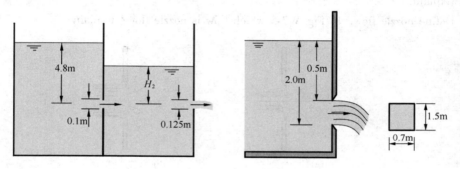

Fig. P7-2   Fig. P7-3

7-4 A sharp-edged circular orifice in a tank wall is at a depth of 2m below the free surface and has a diameter of 20mm. Assume the flow is a steady free jet. (a) Determine the flowrate through the orifice; (b) If a nozzle is connected to this orifice, determine the flowrate through the nozzle and the vacuum height in the nozzle.

7-5 Fig. P7-5 shows a tunnel with a length of 4m in the concrete dam. The steady flowrate in the tunnel is $10\text{m}^3/\text{s}$. Determine the diameter of the tunnel and the vacuum at the vena contracta in the tunnel. Take the discharge coefficient of the tunnel to be 0.82.

7-6 In Fig. P7-6 there are three tanks, among which tank A and tank B are connected by a short pipe with a valve, tank B and tank C is separated by a plate. A contraction nozzle is connected to the bottom of tank C. With constant water levels in the tanks, water flows steadily from tank A to tank B, then to tank C and finally to the air. In this system, the pipe, orifice and the exit section of nozzle have an identical diameter of 60 mm, the minor loss coefficients of valve, inlet and outlet in the short pipe are 3.0, 0.5 and 1.0 respectively, and the discharge coefficient for the orifice on the plate is 0.6, for the nozzle 0.96. Determine (a) the water depth H_B in tank B and H_C in tank C, (b) the flowrate in the short pipe. Disregarding the major loss in the short pipe.

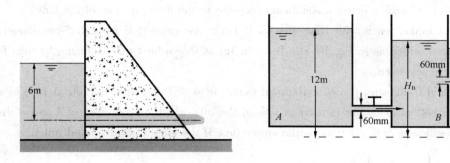

Fig. P7-5 **Fig. P7-6**

7-7 In Fig. P7-7 a tank is separated into part A and part B by a plate with an orifice of 4-cm-diameter. An cylindrical outer nozzle with a diameter of 3cm is connected to the bottom of part B. When the flow is steady, determine: (a) h_1 and h_2; (b) the flowrate out of the tank through the nozzle.

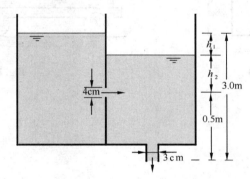

Fig. P7-7

Unsteady Orifice and Nozzle Flow

7-8 A boat has a flat bottom with a area of $8m^2$, a 0.5-m-height boardside and weights 9.8kN. But now the flat bottom of boat is cracked by an unexpected rock and a hole with a diameter of about 10cm appears at the flat bottom. Water is poured in through the hole. Assume the boat is empty before it is cracked, how long does it take the boat to be sunk?

7-9 A 3-m-diameter tank is initially filled with water 2m above the center of a sharp-edged 10-cm-diameter orifice. The tank water surface is open to the atmosphere, and the orifice drains to the atmosphere. Neglecting the effect of the kinematic correction factor, calculate (a) the initial velocity from the tank and (b) the time required to empty the tank. Does the loss

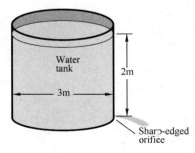

Fig. P7-9

coefficient of the orifice cause a significant increase in the draining time of the tank?

7-10 Two water tanks, each with base area of $0.1 m^2$, are connected by a 1.5-cm-diameter sharp-edged orifice as in Fig. P7-10. If $h = 0.3 m$ as shown for $t = 0$, estimate the time for two water levels even.

7-11 A l-long and D-diameter vented cylindrical tanker is to be filled with fuel oil. A hole located at the bottom of tanker is used to unload the oil, which has an area of A and a discharge coefficient of μ. Determine the empty time if the tanker is fully filled initially.

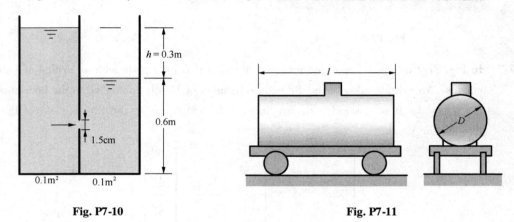

Fig. P7-10　　　　　　　　　　Fig. P7-11

8 Pipe Flow

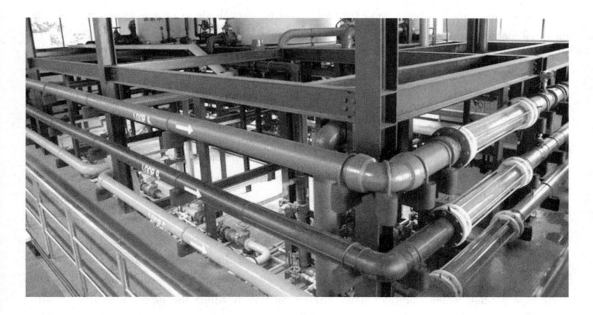

CHAPTER OPENING PHOTO: Large-scale model of drink water piping system focused on the water quality and water age in the laboratory at Zhejiang University, China. The principles for its design and problem solving are discussed in the present chapter. (*Courtesy of Institution of Municipal Engineering, Zhejiang University*)

The transport of a fluid (liquid or gas) in a fully filled closed conduit (commonly called a *pipe* if it is of round cross section or a *duct* if it is not round), driven primarily by a pressure difference, are encountered in almost every engineering design and thus have been studied extensively. Thus, this chapter is completely devoted to this important practical fluids engineering problem. It begins with applying the basic governing principles concerning mass, momentum, and energy to the steady flow of viscous, incompressible fluids in simple short pipes, complex piping systems and pipe network. Then, unsteady flow in a pipe, water hammer, is briefly presented. Finally, we present an overview of flow measurement devices.

8.1 Introduction

Piping system, or *pipeline system*, is a network of pipes connected by using pipe fittings (elbow, bend, return, tee, cross, reducer, end cap, plug, etc.), valves and other special components (expansion joints, orifice plates, restriction orifices, etc.) to perform the required mode of transferring fluids (liquids, gas or slurry) from one location to another location.

Pipeline system used in water distribution, industrial application and in many engineering systems may range from simple arrangement to extremely complex one. The plumbing network supplying water into our home and the sewerage network collecting the wastewater out of our homes are common examples of pipeline system. Oil is often transferred from their source by pressure pipelines to refineries while gas is conveyed by pipelines into a distribution network for supply. Numerous hoses and pipes carry hydraulic fluid or other fluids to various components of vehicles and machines. The air quality within our buildings is maintained at comfortable levels by the distribution of conditioned (heated, cooled, humidified/dehumidified) air through a maze of pipes and ducts. Thus, it can be seen that the fluid flow in conduits is of immense practical significance in civil engineering. Although all of these systems are different, the fluid mechanics principles governing the fluid motions are common. The purpose of this chapter is to understand the basic processes involved in such flows.

The pipeline system can be classified into *simple pipeline* and *complex pipeline*. The former is a pipe system separated by pipe fittings or components into several pipe elements with same diameters connected, through which the flow rate is constant. The latter includes two or more pipe elements with different diameters, or branches. If the pipe elements form to loops, this system is termed the *pipe network*, i.e, the water supply system.

For all flows involved in this chapter, we assume that the pipe is completely filled with the fluid being transported. Such a flow in pipe is called *pipe flow*. Thus, we will not consider a concrete pipe through which rainwater flows without completely filling the pipe. Such flows, called *open-channel flow*, are treated in Chapter 9. The difference between open-channel flow and the pipe flow of this chapter is in the fundamental mechanism that drives the flow. For open-channel flow, gravity alone is the driving force--the water flows down a hill. For pipe flow, gravity may be important (the pipe need not be horizontal), but the main driving force is likely to be a pressure gradient along the pipe. If the pipe is not full, it is not possible to maintain this pressure drop along the pipe.

Pipe flow is a *pressure flow*, for which the pressure in the pipe is usually greater or less than the atmospheric pressure. When fluid flows through the pipe, both major and minor head losses would occur. A pipe, through which the minor head loss and the velocity head of flow are comparable with the major loss and not negligible, is termed the *short pipe*, such as the suction pipe of pump, siphon pipe, railway culvert, etc. However, a pipe, through which the major head loss is dominated, the minor head loss and the velocity head are negligible compared to the major head loss (less than 5% ~ 10%) or computed as a fixed percent of the major head loss, is termed the *long pipe*. For example, the pipe elements in the water supply system are usually considered as a long pipe.

Problems regarding pipelines are usually tackled by the use of continuity and energy equations. The main idea involved is to apply the energy equation between appropriate locations within the flow system, with the head loss written in terms of the friction factor and the minor loss coefficients. We will consider two classes of pipe systems: those containing a single pipe (whose length may be interrupted by various components), and those containing multiple pipes in parallel, series, or network configurations.

8.2 Steady Flow in Simple Short Pipeline

8.2.1 Fundamental Formula in Pipe Flow

For a steady pressure flow in a short pipe with a free jet at the pipe exit as indicated in Fig. 8-1, both the minor loss and the friction loss are not negligible. The minor losses can be found at the section of pipe entrance, orifice plate and valve as indicated by the energy gradient line and piezometric gradient line. Thus, the energy equation between section 1-1 and the pipe exit section 2-2 with a horizontal datum passing center point of pipe exit gives

$$H + \frac{p_{at}}{\rho g} + \frac{\alpha_0 v_0^2}{2g} = 0 + \frac{p_{at}}{\rho g} + \frac{\alpha v^2}{2g} + h_{w1-2} \tag{8.1}$$

or

$$H_0 = \frac{\alpha v^2}{2g} + h_{w1-2} \tag{8.2}$$

where $H_0 = H + \frac{\alpha_0 v_0^2}{2g}$ is the initial total head of pipe system, also termed the *acting head*; H is the elevation difference between the free surface in tank and the center point of pipe exit; v_0 is the sectional average velocity at section 1-1, termed the *approaching velocity*; v is the sectional average velocity in the simple pipe system; h_{w1-2} is the total head loss from sections 1-1 to 2-2.

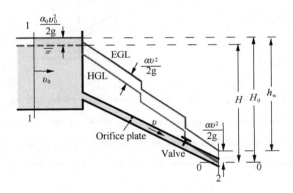

Fig. 8-1 Flow in a simple short pipeline with free jet

Equation 8.2 indicates that all the total head in the tank has been consumed and transferred into the frictional and minor head losses, excepting the parts transferring into the velocity head of free jet at the pipe exit.

For simple short pipeline, the diameters for all pipe elements are identical, thus the head loss in the system can be expressed as

$$h_{w1-2} = \Sigma h_f + \Sigma h_j = \left(\Sigma \lambda \frac{l}{d} + \Sigma \zeta\right)\frac{v^2}{2g}$$

Substitution into Eq. 8.2 yields the velocity in the pipe

$$v = \frac{1}{\sqrt{\alpha + \Sigma \lambda \dfrac{l}{d} + \Sigma \zeta}} \sqrt{2gH_0} \qquad (8.3)$$

and the flow rate

$$Q = vA = \mu A \sqrt{2gH_0} \qquad (8.4)$$

where $\mu = \dfrac{1}{\sqrt{\alpha + \Sigma \lambda \dfrac{l}{d} + \Sigma \zeta}}$ is the *discharge coefficient* for flow in a simple short pipeline with a free jet.

Equation 8.4 is the fundamental equation to determine the flow rate in a simple short pipeline. If the cross sectional area is large enough in the upstream tank, the approaching velocity head is negligible, the acting head in the above equation can be simplified as $H_0 \approx H$.

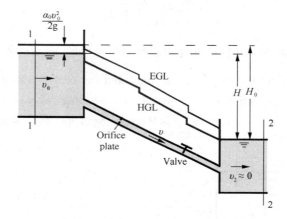

Fig. 8-2 Flow in simple short pipeline with a submerged jet

If the pipe exit is fully submerged under the water level in the downstream tank as indicated in Fig. 8-2, the fluid first contracts then expands in the downstream tank. Assume fluid has completely expanded at section 2-2, thus the energy equation between sections 1-1 to 2-2 gives

$$H + \frac{p_{at}}{\rho g} + \frac{\alpha_0 v_0^2}{2g} = 0 + \frac{p_{at}}{\rho g} + \frac{\alpha_2 v_2^2}{2g} + h_{w1-2} \qquad (8.5)$$

or

$$H_0 = h_{w1-2} \qquad (8.6)$$

where $H_0 = H + \dfrac{\alpha_0 v_0^2}{2g} - \dfrac{\alpha_2 v_2^2}{2g}$ is the total head difference between upstream and downstream tanks; H_0 equals to the difference of water levels in the upstream and downstream tanks, $H_0 = H$, if the velocity in tanks are negligible; h_{w1-2} is the total head loss from sections 1-1 to 2-2 and can also be expressed as

$$h_{w1-2} = \Sigma h_f + \Sigma h_j = \left(\Sigma \lambda \frac{l}{d} + \Sigma \zeta \right) \frac{v^2}{2g}$$

Substitution into Eq. 8.6 yields the velocity and the flow rate in pipe

$$v = \frac{1}{\sqrt{\Sigma \lambda \dfrac{l}{d} + \Sigma \zeta}} \sqrt{2gH_0} \qquad (8.7)$$

$$Q = vA = \mu A \sqrt{2gH_0} \qquad (8.8)$$

where $\mu = \dfrac{1}{\sqrt{\Sigma \lambda \dfrac{l}{d} + \Sigma \zeta}}$ is the *discharge coefficient* for flow in a simple short pipeline with a submerged jet.

Even though the expressions of the discharge coefficients in Eqs. 8.4 and 8.8 are different but their values are same for the two systems only with different exits. In the expressions of μ for both flows, the minor losses at the pipe entrance, orifice plate and valve are included. The only difference is α included in μ for free jet and the minor loss at pipe exit into downstream tank, ζ_{exit}, included in μ for submerged jet. Generally, the pipe flow in engineering practice is turbulent, for which the value of kinematic factor can be approximately taken to be 1.0, that is $\alpha \approx$ 1.0, because $\alpha \approx 1.04 \sim 1.11$ for fully developed turbulent pipe flow. The minor loss at pipe exit is nearly to $\zeta_{exit} = 1.0$ as indicated in Fig. 6-20. Therefore, the discharge coefficients for both flows are approximate equal.

8.2.2 Fundamental Problems in Simple Short Pipe

In the design and analysis of simple short pipe systems, we can encounter three fundamental problems (the fluid and the roughness of the pipe are assumed to be specified in all cases):

(1) Determining the *acting head* of simple short pipeline when the pipe length and diameter are given for a specified flow rate (or velocity). It can be directly computed by the above discussed equations.

(2) Determining the *flow rate* when the pipe length and diameter are given for a specified pressure drop (or head loss). It also can be directly computed by the above discussed equations.

(3) Determining the *pipe diameter* when the pipe length and flow rate are given for a specified pressure drop (or head loss).

Problems of the *first type* are straightforward and can be solved directly by using the Moody chart. Problems of the *second type* and *third type* are commonly encountered in engineering design (in the selection of pipe diameter, for example, that minimizes the sum of the construction and pumping costs), but the use of the Moody chart with such problems requires an iterative approach unless an equation solver is used.

In problems of the *second type*, the diameter is given but the flow rate is unknown. A good guess for the friction factor in that case is obtained from the completely turbulent flow region for the given roughness. This is true for large Reynolds numbers, which is often the case in practice. Once the flow rate is obtained, the friction factor can be corrected using the Moody chart or the Colebrook equation, and the process is repeated until the solution converges. Typically only a few iterations are required for convergence to three or four digits of precision.

In problems of the *third type*, the diameter is not known and thus the Reynolds number and the relative roughness cannot be calculated. Therefore, we start calculations by assuming a pipe diameter. The pressure drop calculated for the assumed diameter is then compared to the specified pressure drop, and calculations are repeated with another pipe diameter in an iterative fashion until convergence.

EXAMPLE 8-1 Effect of flushing on flow rate from a shower

The bathroom plumbing of a building consists of 1.5-cm-diameter copper pipes ($\Delta = 1.5 \times 10^{-6}$ m) with threaded connectors, as shown in Fig. E8-1. (a) If the gage pressure at the inlet of the system is 200kPa during a shower and the toilet reservoir is full (no flow in that branch), determine the flow rate of water through the shower head. (b) Determine the effect of flushing of the toilet on the flow rate through the shower head. Take the loss coefficients of the shower head and the reservoir to be 12 and 14, respectively. Assume the flow is a steady and incompressible, fully developed turbulent flow; the reservoir is open to the atmosphere and the velocity heads are negligible. The water temperature is at 20℃.

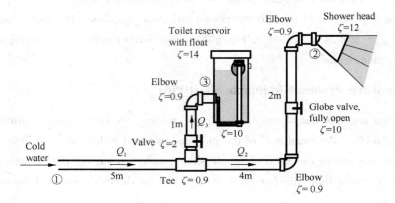

Fig. E8-1

Solution:

From Table A-1 we have the properties of water at 20℃ to be $\rho = 998 \text{kg/m}^3$, $\nu = 1.004 \times 10^{-6} \text{m}^2/\text{s}$. The datum plane is taken at the axis of horizontal inlet pipe.

This is a problem of the second type since it involves the determination of the flow rate for a specified pipe diameter and pressure drop. The solution involves an iterative approach since the flow rate (and thus the flow velocity) is not known.

(a) The piping system of the shower alone involves $l_{1-2} = (5 + 4 + 2)\text{m} = 11\text{m}$ of piping, a tee with line flow ($\zeta_t = 0.9$), two standard elbows ($\zeta_e = 0.9$ each), a fully open globe valve ($\zeta_{gv} = 10$), and a shower head ($\zeta_s = 12$). Therefore, $\Sigma\zeta_{1-2} = 0.9 + 2 \times 0.9 + 10 + 12 = 24.7$.

The energy equation between points 1 and 2 gives

$$z_1 + \frac{p_1}{\rho g} + \frac{\alpha_1 v_1^2}{2g} = z_2 + \frac{p_{at}}{\rho g} + \frac{\alpha_2 v_2^2}{2g} + h_{wl-2}$$

Noting that the shower head is open to the atmosphere, and the velocity heads are negligible. Then, with the given values: $p_1 = 200\text{kPa}$, $z_2 - z_1 = 2\text{m}$, the head loss between points 1 to 2 can be determined by

$$h_{wl-2} = \frac{p_1}{\rho g} - (z_2 - z_1) = \frac{200\text{kPa}}{(998\text{kg/m}^3)(9.81\text{m/s}^2)} - 2\text{m} = 18.43\text{m}$$

also
$$h_{wl-2} = \left(\lambda \frac{l_{1-2}}{d} + \Sigma \zeta_{1-2}\right)\frac{v^2}{2g}$$

$$18.43\text{m} = \left(\lambda \frac{11\text{m}}{0.015\text{m}} + 24.7\right)\frac{v^2}{(2)(9.81\text{m/s}^2)}$$

Since the diameter of the piping system is constant, the average velocity in the pipe and the friction factor are the same in each individual pipe. Assume the friction factor in pipes $\lambda = 0.0218$, thus we can get the velocity from the above equation $v = 2.98\text{m/s}$. Thus, the Reynolds number and the relative roughness height are

$$\text{Re} = \frac{vd}{\nu} = \frac{(2.98\text{m/s})(0.015\text{m})}{1.004 \times 10^{-6}\text{m}^2/\text{s}} = 44522$$

$$\frac{\Delta}{d} = \frac{1.5 \times 10^{-6}\text{m}}{0.015\text{m}} = 0.0001$$

which satisfies the Colebrook equation

$$\frac{1}{\sqrt{\lambda}} = -2\log\left(\frac{\Delta/d}{3.7} + \frac{2.51}{\text{Re}\sqrt{\lambda}}\right)$$

It indicates the assumption of friction factor $\lambda = 0.0218$ is correct. Thus the flow rate of water through the shower head is

$$Q_2 = vA = v\frac{1}{4}\pi d^2 = (2.98\text{m/s})\frac{1}{4}(3.14)(0.015\text{m})^2 = 0.00053 \text{ m}^3/\text{s}$$

(b) When the toilet is flushed, the float moves and opens the valve. The discharged water starts to refill the reservoir, resulting in parallel flow after the tee connection. The head loss and minor loss coefficients for the shower branch were determined in (a) to be $h_{wl-2} = 18.43\text{m}$ and $\Sigma \zeta_{1-2} = 24.7$, respectively. The corresponding quantities for the reservoir branch can be determined similarly to be

$$h_{wl-3} = \frac{p_1}{\rho g} - (z_3 - z_1) = \frac{200\text{kPa}}{(998\text{kg/m}^3)(9.81\text{m/s}^2)} - 1\text{m} = 19.43\text{m}$$

$$\Sigma \zeta_{1-3} = 2 + 0.9 + 10 + 14 = 26.9$$

The continuity equation at the tee, the average velocities, the friction factors and the Reynolds numbers in the pipe systems are:

$$Q_1 = Q_2 + Q_3$$

$$h_{wl-2} = \lambda_1 \frac{5\text{m}}{0.015\text{m}}\frac{v_1^2}{(2)(9.81\text{m/s}^2)} + \left(\lambda_2 \frac{4\text{m}+2\text{m}}{0.015\text{m}} + 24.7\right)\frac{v_2^2}{(2)(9.81\text{m/s}^2)} = 18.43\text{m}$$

$$h_{wl-3} = \lambda_1 \frac{5\text{m}}{0.015\text{m}}\frac{v_1^2}{(2)(9.81\text{m/s}^2)} + \left(\lambda_3 \frac{1\text{m}}{0.015\text{m}} + 26.9\right)\frac{v_3^2}{(2)(9.81\text{m/s}^2)} = 19.43\text{m}$$

$$v_1 = \frac{Q_1}{\frac{1}{4}(3.14)(0.015\text{m})^2} \quad v_2 = \frac{Q_2}{\frac{1}{4}(3.14)(0.015\text{m})^2} \quad v_3 = \frac{Q_3}{\frac{1}{4}(3.14)(0.015\text{m})^2}$$

$$\text{Re}_1 = \frac{v_1(0.015\text{m})}{1.004 \times 10^{-6}\text{m}^2/\text{s}} \quad \text{Re}_2 = \frac{v_2(0.015\text{m})}{1.004 \times 10^{-6}\text{m}^2/\text{s}} \quad \text{Re}_3 = \frac{v_3(0.015\text{m})}{1.004 \times 10^{-6}\text{m}^2/\text{s}}$$

$$\frac{1}{\sqrt{\lambda_1}} = -2\log\left(\frac{0.0001}{3.7} + \frac{2.51}{\text{Re}_1\sqrt{\lambda_1}}\right)$$

$$\frac{1}{\sqrt{\lambda_2}} = -2\log\left(\frac{0.0001}{3.7} + \frac{2.51}{\text{Re}_2\sqrt{\lambda_2}}\right)$$

$$\frac{1}{\sqrt{\lambda_3}} = -2\log\left(\frac{0.0001}{3.7} + \frac{2.51}{\text{Re}_3\sqrt{\lambda_3}}\right)$$

Solving these 12 equations in 12 unknowns simultaneously, the flow rates are determined to be
$$Q_1 = 0.00090 \text{m}^3/\text{s}, \quad Q_2 = 0.00042 \text{m}^3/\text{s}, \quad Q_3 = 0.00048 \text{m}^3/\text{s}$$
Therefore, the flushing of the toilet reduces the flow rate of cold water through the shower by 21 percent from 0.53 to 0.42L/s, causing the shower water to suddenly get very hot.

DISCUSSION   If the velocity heads were considered, the flow rate through the shower would be 0.43 instead of 0.42L/s. Therefore, the assumption of negligible velocity heads is reasonable in this case. Note that a leak in a piping system will cause the same effect, and thus an unexplained drop in flow rate at an end point may signal a leak in the system.

Following some commonly used simple short pipes in practice are presented.

FLOW IN SIPHON

Siphon can be described as a pipe or tube (small diameter pipe) fashioned or deployed in an inverted *U* shape and filled until atmospheric pressure is sufficient to force a liquid from a reservoir in one end of the tube over a barrier higher than the reservoir and out the other end at a lower level than the reservoir as shown in Fig. 8-3. Siphon allows liquids to flow uphill, above the surface of the tank or reservoir, without pumps. Liquids flow down the tube under the pull of gravity. The liquid is discharged at a level lower than the surface of the reservoir or the tank. The pull on the surface of liquids from a higher to lower altitude is known as *siphoning*.

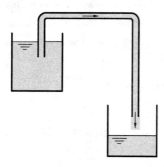

Fig. 8-3 Schematic of a siphon

The working principle of a siphon is first because of gravity then the vacuum on the top. To begin with a siphoning, a siphon is firstly primed by sucking the tube or the pipe like a straw until it is filled with the liquid to form the hydrostatic pressure in it, then one end of the siphon is inserted into the reservoir and the other end is left free or settled inside a container collecting the falling liquid. Thus, the siphon is started. The gravity pulling down on the taller column of liquid causes less pressure (less than the atmospheric pressure) in the tube or pipe segment higher than the reservoir. However, the pressure at the intake of a siphon is always greater than the atmospheric pressure, which forces the liquid entering the tube and flowing upward, passing the top barrier and then flowing downwards and leaving the tube. It continues to work until the outlet of a siphon or the level of the container equals the level of the reservoir, and the largest vacuum height at the section of siphon with highest elevation must be less than 7-8m.

Siphon is used in many day-to-day applications. Siphoning can prevent impurities from being transferred to a fresh container in the fermentation of wine and beer. A siphon can be used to clean a fish tank or aquarium, and to remove accumulated water in flooded homes or cellars, etc. Siphon is also used in many engineering practices, such as the siphon filter in water supply plant, siphon spillway in hydraulic engineering and the siphon transporting water from nearby ditches or channels into fields in irrigation engineering, etc. Large siphons used in municipal waterworks and industries are controlled by using valve inlet and valve outlet. These siphons are generally primed using a manual or electric pump to start the process.

The hydraulic problems for a siphon are to determine the flow rate in a siphon or the permissive mounted height (level difference between the free surfaces of the reservoirs and the highest level of the siphon).

EXAMPLE 8-2 Flow rate and mounted height in a siphon

An 0.2-m-diameter riveted steel pipe is used as a siphon to transport water from river to an irrigation ditch as indicated in Fig. E8-2, for which the mounted height h_B = 4.5m. The level difference between the water levels in river and ditch is $H = 1.6$m. The lengths of siphon upwards and downwards are l_{AB} = 30m and l_{BC} = 40m, respectively. The minor loss coefficients at the siphon intake, the first bend and the second bend of siphon are ζ_1 = 0.5, ζ_2 = 0.3 and ζ_3 = 0.4 respectively. The siphon has a free jet at the exit. (a) Determine the flow rate in the siphon, and (b) check the available of the mounted height, h_B. The water in siphon is at 20℃.

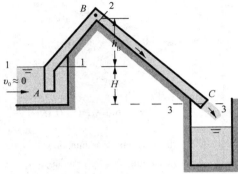

Fig. E8-2

Solution:

Assume the flow in siphon is steady and incompressible.

(a) Siphon can be considered as a simple short pipe. Take the approaching velocity in the river to be negligible, $v_0 \approx 0$, thus $H_0 = H + \dfrac{\alpha_0 v_0^2}{2g} \approx H$.

From Table 6-3 we have the manning roughness of a riveted steel pipe as $n = 0.019$. Thus, the *Chézy coefficient* can be obtained from Eq. 6.68

$$C = \frac{1}{n} R_h^{1/6} = \frac{1}{n}\left(\frac{d}{4}\right)^{1/6} = \frac{1}{0.019}\left(\frac{0.2\text{m}}{4}\right)^{1/6} = 31.94 \sqrt{\text{m}}/\text{s}$$

and the friction factor of steel pipe from Eq. 6.67

$$\lambda = \frac{8g}{C^2} = \frac{(8)(9.81 \text{ m/s}^2)}{(31.94\sqrt{\text{m}}/\text{s})^2} = 0.0769$$

The total minor loss coefficient in the siphon system

$$\Sigma \zeta = \zeta_1 + \zeta_2 + \zeta_3 = 0.5 + 0.3 + 0.4 = 1.2$$

Thus, the velocity in the siphon can be obtained from Eq. 8.3

$$v = \mu\sqrt{2gH} = \frac{\sqrt{2gH}}{\sqrt{1 + \Sigma\lambda\dfrac{l}{d} + \Sigma\zeta}} = \sqrt{\frac{(2)(9.81\text{m/s}^2)(1.6\text{m})}{1 + 0.0769\dfrac{30\text{m} + 40\text{m}}{0.2\text{m}} + 1.2}} = 1.038\text{m/s}$$

and the flowrate in siphon is

$$Q = \frac{1}{4}\pi d^2 v = \frac{1}{4}(3.14)(0.2\text{m})^2(1.038\text{m/s}) = 0.0326 \text{ m}^3/\text{s}$$

From Table A-1 we can get the kinematic viscosity $\nu = 1.004 \times 10^{-6}$ m²/s when the water temperature is 20℃. Check the Reynolds number

$$Re = \frac{vd}{\nu} = \frac{(1.038\text{m/s})(0.2\text{m})}{1.004 \times 10^{-6} \text{ m}^2/\text{s}} = 206773$$

and the roughness height for a riveted pipe from Table 6-2, 2 mm, gives $\dfrac{\Delta}{d} = \dfrac{3\text{mm}}{0.2\text{m}} = 0.015$. From Moody chart shows the flow is in the fully rough region. Besides, we have $n = 0.019 < 0.02$ and $R_h = \dfrac{d}{4} = 0.05\text{m} < 0.5\text{m}$. Thus, Eq. 6.67 and 6.68 are valid in this problem.

(b) To check the mounted height, the vacuum height at the summit of siphon should be determined first. Take the free surface in the river to be reference level, the energy equation between the free surface in the river and the section 2-2 at the siphon summit gives

$$0 + \frac{p_{at}}{\rho g} + \frac{\alpha_0 v_0^2}{2g} = h_B + \frac{p_2}{\rho g} + \frac{\alpha_2 v_2^2}{2g} + h_{w1-2}$$

Thus, with $v_0 \approx 0$, $v_2 = v$ and $\alpha_2 \approx 1.0$, the vacuum height at section 2-2 is

$$h_v = \frac{p_{at} - p_2}{\rho g} = h_B + \frac{\alpha v^2}{2g} + \left(\lambda \frac{l_{AB}}{d} + \Sigma \zeta\right)\frac{v^2}{2g}$$

$$= 4.5\text{m} + \left(1 + 0.0769 \frac{30\text{m}}{0.2\text{m}} + 1.2\right)\frac{(1.038\text{m/s})^2}{(2)(9.81\text{ m/s}^2)} = 5.254\text{m}$$

It indicates the maximum vacuum height in the siphon is 5.254m, which is less than the limited vacuum height of siphon in the engineering practice, $[h_v] = (7 \sim 8)$m. Therefore, the siphon can work normally at this mounted height.

FLOW IN INVERTED SIPHON

Inverted siphons (sometimes called *sag culverts* or *sag lines*) are used to convey water by gravity under roads, railroads, other structures, various types of drainage channels and depressions. An inverted siphon is a closed conduit designed to run full and under pressure. The structure should operate without excess head when flowing at design capacity.

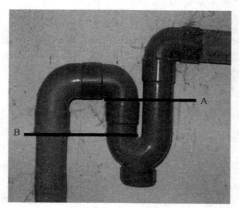

Fig. 8-4 Plumbing 'trap', water seal under a sink. Inverted siphoning occurs below the line A (*Courtesy of McGeddon at Wikimedia*)

Inverted siphons are commonly called *traps* for their function in preventing smelly sewer gases from coming back out of drains (see Fig. 8-4). Liquid flowing in one end simply forces liquid up and out the other end, but solids like sand will accumulate. This is especially important in sewage systems or culverts which must be routed under rivers or other deep obstructions where the better term is '*depressed sewer*'. Large inverted siphons are used to convey water being carried in canals or flumes across valleys, for irrigation or mining.

The purpose in the hydraulic problem of inverted siphon is to determine the diameter of the inverted siphon.

EXAMPLE 8-3 Diameter of inverted siphon

A 200-m-lomg reinforced concrete circular inverted siphon is used to connect the rivers on the both sides of a river dike as indicated in Fig. E8-3. The roughness height of this inverted siphon is 2.3mm. The water level difference between upstream and downstream rivers is $H = 8$m. The minor loss coefficients at entrance A, bend B, bend C and exit D are $\zeta_A = 0.5$, $\zeta_B = 0.1$, $\zeta_C = 0.1$ and $\zeta_D = 1.0$ respectively. Assume the water flow rate through the inverted siphon is 25 m³/s, determine the diameter of the inverted siphon. The water is at 20℃.

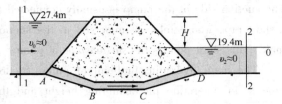

Fig. E8-3

Solution:

Assume the flow in inverted siphon is steady and incompressible. Neglect the velocities in the upstream and downstream river.

The out end of inverted siphon is submerged in the downstream water body. Take the free surface of downstream river as the reference level, the energy equation between sections 1-1 and 2-2 gives

$$H + \frac{p_{at}}{\rho g} + \frac{\alpha_0 v_0^2}{2g} = 0 + \frac{p_{at}}{\rho g} + \frac{\alpha_2 v_2^2}{2g} + h_w$$

with $v_0 \approx 0$, $v_2 \approx 0$, we have

$$H = h_w = \left(\lambda \frac{l}{d} + \zeta_A + \zeta_B + \zeta_C + \zeta_D\right)\frac{1}{2g}\left(\frac{Q}{\pi d^2/4}\right)^2$$

Assume the flow is in fully rough zone and $\lambda = 0.02$, it follows

$$8\text{m} = \left(0.02\frac{200\text{m}}{d} + 0.5 + 0.1 + 0.1 + 1.0\right)\frac{1}{(2)(9.81\text{m/s}^2)}\left[\frac{(4)(25\text{ m}^3/\text{s})}{3.14\ d^2}\right]^2$$

$$d^5 - 10.984d - 25.847 = 0$$

After several rounds of trial and error, we get the diameter of the inverted siphon $d = 2.185$m.

Check the assumption of $\lambda = 0.02$. From Table A-1 in Appendix A, we have $\nu = 1.004 \times 10^{-6}$ m²/s when water temperature is 20℃. With the given parameters, the relative roughness height is $\frac{\Delta}{d} = \frac{2.3\text{mm}}{2.185\text{m}} = 0.00105$.

Thus the velocity and Reynolds number are

$$v = \frac{Q}{A} = \frac{25\text{ m}^3/\text{s}}{\frac{1}{4}(3.14)(2.185\text{m})^2} = 6.67\text{m/s}$$

$$Re = \frac{vd}{\nu} = \frac{(6.67\text{m/s})(2.185\text{m})}{1.004 \times 10^{-6}\text{ m}^2/\text{s}} = 1.45 \times 10^7$$

From Moody chart, we can get $\lambda \approx 0.02$. Thus, the diameter, $d = 2.185$m, is the result required. However, in the engineering practice, the diameter may be determined to be 2.25m to meet the nominal diameter of the commercial pipe.

SUCTION PIPE OF PUMP

Pumps typically have an inlet where the fluid enters the pump and an outlet where the fluid comes out (see Fig. 8-5). The inlet location is said to be at the suction side of the pump. The outlet location is said to be at the discharge side of the pump. Operation of the pump creates suction (a lower pressure) at the suction side so that fluid can enter the pump through the inlet.

Pump operation also causes higher pressure at the discharge side by forcing the fluid out at the outlet. The suction side of the pump is usually connected to a short pipe whose inlet is immersed in the suction tank. This short pipe is the *suction pipe* of pump.

The main hydraulic problem of pump suction pipe is to determine the mounting height and the head of pump.

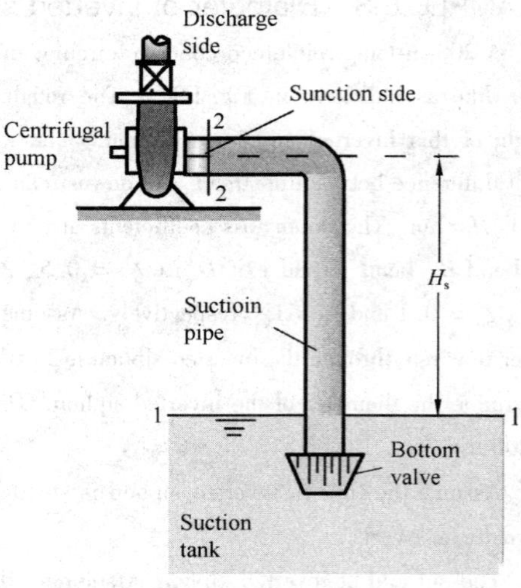

Fig. 8-5 Schematic of centrifugal pump

EXAMPLE 8-4 Mounting height of a pump

A centrifugal pump has a 0.3-m-diameter and 12-m long suction pipe as indicated in Fig. 8-5. Its design flow rate is 306m³/h. The frictional factor of the suction pipe is 0.016. The minor loss coefficients at intakes and bend are ζ_1 = 5.5 and ζ_2 = 0.3 respectively. The permission vacuum height in suction pipe is $[h_v]$ = 6m. Determine the permission mounting height of this pump.

Solution:

Assume the flow in the suction pipe is steady and incompressible.

Let the free surface of tank to be the reference level. The energy equation between free surface in tank and the inlet section of pump 2-2 gives

$$0 + \frac{p_{at}}{\rho g} + \frac{\alpha_1 v_1^2}{2g} = H_s + \frac{p_2}{\rho g} + \frac{\alpha_2 v_2^2}{2g} + h_w$$

with $v_1 \approx 0$, $v_2 \approx v$, $\alpha_2 \approx 1$, we have

$$\frac{p_{at}}{\rho g} = H_s + \frac{p_2}{\rho g} + \frac{\alpha v^2}{2g} + h_w$$

where

$$v = \frac{Q}{A} = \frac{Q}{\frac{1}{4}\pi d^2} = \frac{(306/3600)\ \text{m}^3/\text{s}}{\frac{1}{4}(3.14)(0.3\text{m})^2} = 1.20\ \text{m/s}$$

$$h_w = \left(\lambda \frac{l}{d} + \zeta_1 + \zeta_2\right)\frac{v^2}{2g} = \left(0.016\frac{12\text{m}}{0.3\text{m}} + 5.5 + 0.3\right)\frac{(1.20\text{m/s})^2}{(2)(9.81\ \text{m/s}^2)} = 0.47\text{m}$$

$$\frac{p_{at} - p_2}{\rho g} = [h_v] = 6\text{m}$$

Substituting the values back into the energy equation yields

$$H_s = \frac{p_{at} - p_2}{\rho g} - \frac{\alpha v^2}{2g} - h_w = 6\text{m} - \frac{(1.20\text{m/s})^2}{(2)(9.81\ \text{m/s}^2)} - 0.47\text{m} = 5.45\text{m}$$

8.3 Steady Flow in Simple Long Pipeline

Long pipe is a *simple* pipeline in which the transported fluid fully fills. The flow in long pipe is a pressure flow with negligible minor losses and velocity head. Thus, the energy equation between sections 1-1 to 1-2 in Fig. 8-6 gives

$$H = h_f = \lambda \frac{l}{d} \frac{v^2}{2g} \qquad (8.9)$$

Equation 8.9 indicates that the acting head above the axis point of long pipe exit will be completely consumed by the frictional losses. The energy line drops continuously in the flow direction, and coincides with the hydraulic gradient line, as indicated in Fig. 8-6.

In the environment and civil engineering, flow rate is often used to replace the velocity. With $v = \frac{4Q}{\pi d^2}$ Eq. 8.9 becomes

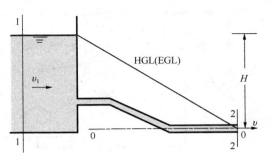

Fig. 8-6 Simple long pipe

$$h_f = \lambda \frac{l}{d} \frac{v^2}{2g} = \frac{8\lambda}{g\pi^2 d^5} lQ^2 = SlQ^2 \qquad (8.10)$$

where $S = \frac{8\lambda}{g\pi^2 d^5}$ is the head loss per unit flow rate passing through per unit length pipe, called the *pipe specific resistance*, the dimension is $T^2 L^{-6}$.

In the practices of civil engineering, the flow is always a fully developed turbulent flow in the fully rough zone, Manning equation (Eq. 6.68), $C = \frac{1}{n} R_h^{1/6}$, and Eq. 6.67, $\lambda = \frac{8g}{C^2}$, are valid in this case. Substituting into $S = \frac{8\lambda}{g\pi^2 d^5}$ gives

$$S = \frac{10.3\, n^2}{d^{5.333}} \qquad (8.11)$$

In the hydraulic and transportation engineering, the flow is also a fully developed turbulent flow in the fully rough zone, the Chézy equation, $Q = AC\sqrt{RJ} = K\sqrt{h_f/l}$, is always introduced to give the major loss in the system as

$$h_f = \frac{l}{K^2} Q^2 \qquad (8.12)$$

where $K = AC\sqrt{R}$ is the *pipe discharge modulus* with a dimension of $L^3 T^{-1}$, which represents the combined effects of sectional shape, size and roughness on the carrying capacity of pipe.

In pipe system, an alternate, and less desirable, procedure is to report the minor loss as if it were an *equivalent length, l'*, of pipe, namely, the head loss caused by the equivalent long component (such as the angle valve) is equivalent to the head loss caused by a section of the pipe, satisfying the Darcy friction-factor relation

$$h_j = \Sigma \zeta \frac{v^2}{2g} = \lambda \frac{l'}{d} \frac{v^2}{2g}$$

$$l' = \frac{d}{\lambda} \Sigma \zeta \qquad (8.13)$$

Although the equivalent length should take some of the variability out of the loss data, it is an artificial concept.

EXAMPLE 8-5 Simple long pipe

A 0.3-m-diameter and 3500-m-long cast iron pipe with a manning roughness coefficient $n = 0.011$ is used to supply water from a reservoir to a factory as indicated in Fig. E8-5. The elevation of ground is $z_2 = 110.0$m. The service head, H_z, in the factory is 25m. If the required flowrate in the pipe system is $0.085 \text{m}^3/\text{s}$, determine the water level in the reservoir, z_1.

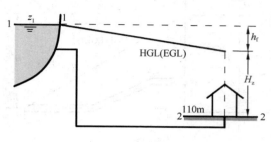

Fig. E8-5

Solution:

Assume the flow in the pipe is steady and incompressible.

From Eq. 8.11 we have the pipe specific resistance

$$S = \frac{10.3 \, n^2}{d^{5.333}} = \frac{(10.3)(0.011^2)}{0.3^{5.333}} = 0.766 \text{ s}^2/\text{m}^6$$

From Eq. 8.10 gives

$$h_f = SlQ^2 = (0.766 \text{ s}^2/\text{m}^6)(3500\text{m})(0.085 \text{ m}^3/\text{s})^2 = 19.37\text{m}$$

Thus the water level in the reservoir can be determined by

$$z_1 = z_2 + H_z + h_f = 110.0\text{m} + 25\text{m} + 19.37\text{m} = 154.37\text{m}$$

DISCUSSION   Take the minor loss coefficients at the pipe inlet to be $\zeta_1 = 0.5$, $\zeta_2 = 0.3$ for each of three bends, $\zeta_3 = 2.5$ at the outlet. Thus, the minor head loss is

$$h_j = (\zeta_1 + 3\zeta_2 + \zeta_3) \frac{\left(\frac{4Q}{\pi d^2}\right)^2}{2g} = (0.5 + 3(0.3) + 2.5) \frac{\left[\frac{(4)(0.085 \text{ m}^3/\text{s})}{(3.14)(0.3\text{m})^2}\right]^2}{(2)(9.81\text{m/s}^2)} = 0.288\text{m}$$

which is about 1.49% of the major head loss. It indicates that neglecting minor loss is reasonable in this problem.

8.4 Steady Flow in Complex Long Pipeline

Most piping systems encountered in practice such as the water distribution systems in cities or commercial or residential establishments involve numerous parallel and series connections as well as several sources (supply of fluid into the system) and loads (discharges of fluid from the system). A piping project may involve the design of a new system or the expansion of an existing system. The engineering objective in such projects is to design a piping system that will deliver the specified flow rates at specified pressures reliably at minimum total cost (initial plus operating and maintenance). Once the layout of the system is prepared, the determination of the pipe diameters and the pressures throughout the system, while remaining within the budget constraints, typically requires solving the system repeatedly until the optimal solution is reached.

Piping systems typically involve several pipes connected to each other in series and/or in parallel as discussed follows.

8.4.1 Serial Piping

The pipes (three or more) connected in series, as shown in Fig. 8-7, are termed *serial piping*.

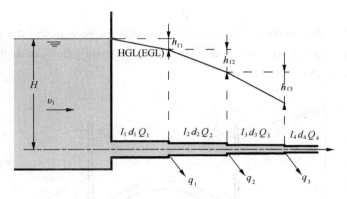

Fig. 8-7 Pipes in series

The flow rate in each individual pipe can be determined by the continuity equation

$$Q_{i+1} = Q_i - q_i \tag{8.14}$$

where i is the sequential number of pipes; q_i is the discharge flowrate from the end of each i th-individual pipe.

If $q_i = 0$, the flow rate through the entire system remains constant regardless of the diameters of the individual pipes in the system: $Q_1 = Q_2 = \cdots = Q_i$. This is a natural consequence of the conservation of mass principle for steady incompressible flow.

When each individual pipe is considered as a long pipe, the minor loss and the velocity head are negligible, and the major loss is determined by Eq. 8.10

$$h_{fi} = S_i l_i Q_i^2 \tag{8.15}$$

Then the total head loss in the system is the sum of the major losses in each individual pipe in the system

$$H = \Sigma h_{fi} = \Sigma S_i l_i Q_i^2 \tag{8.16}$$

Combination of Eqs. 8.14 to 8.16 can give the flow rates, pipe diameters or the necessary head H for the serial piping.

For a long pipe as in Fig. 8-7, the energy gradient line is coincided with the hydraulic gradient line because the velocity head and the minor losses in long pipe are considered to be negligible. Thus, the energy gradient line for serial piping is a connected line of serial lines with different slopes caused by the different diameters of each individual pipe (see Fig. 8-7), the smaller the diameter pipe, the larger the velocity, and the larger the slope of EGL.

8.4.2 Parallel Piping

A pipe, which branches out into two (or more) parallel pipes and then those pipes rejoin again at a junction downstream, is termed the *parallel piping*.

The flow rates at junction A and B in a parallel piping system shown in Fig. 8-8 should be balanced according to the continuity equation for steady incompressible flow, thus

Junction A: $\qquad Q_1 = q_A + Q_2 + Q_3 + Q_4 \qquad$ (8.17)
Junction B: $\qquad Q_2 + Q_3 + Q_4 = q_B + Q_5 \qquad$ (8.18)
or $\qquad \Sigma Q_i = 0$

It indicated that the continuity equation requires that for each *node* (the junction of two or more pipes) the net flowrate is zero. What flows into a node must flow out at the same rate for a steady incompressible flow.

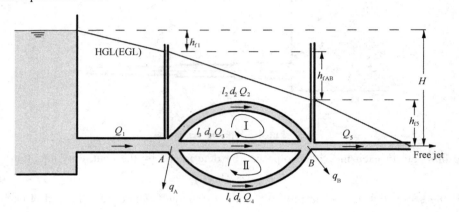

Fig. 8-8 Pipes in parallel

Consider the head loss in each individual pipe connected in parallel, write the energy equation between section A and B for each individual pipe with negligible minor loss and velocity head as

$$h_{fi} = h_{wi} = \left(z_A + \frac{p_A}{\rho g}\right)_i - \left(z_B + \frac{p_B}{\rho g}\right)_i$$

where subscript 'i' presents the No. of the parallel pipes, i.e., $i = 2, 3, 4$ in Fig. 8-8.

Because the junction A and B are the common points for all the individual pipes in parallel, the junction elevations (z_{Ai} and z_{Bi}) and the junction pressures (p_{Ai} and p_{Bi}) are identical for all the individual pipes. Thus, we have

$$h_{fAB} = h_{f2} = h_{f3} = h_{f4} \qquad (8.19)$$

or

$$S_2 l_2 Q_2^2 = S_3 l_3 Q_3^2 = S_4 l_4 Q_4^2 \qquad (8.20)$$

or consider two loops as shown in the figure and take the head loss along the loop direction to be positive and counter to be negative, in loop I we have

$$\Sigma h_f = h_{f2} - h_{f3} = 0$$

and in loop II

$$\Sigma h_f = h_{f3} - h_{f4} = 0$$

It indicated that the net head loss, or the net pressure difference completely around a *loop* (starting at one location in a pipe and returning to that location) must be zero.

With the given total flow rate and the discharged flow rates, q_A and q_B, diameters and lengths of all the individual pipes and the manning roughness efficient n, the flow rates in each individual pipe, Q_i, and the major head loss, h_{fi}, can be determined by combining Eqs. 8.17, 8.18 and 8.20 and the energy equation from the reservoir to the outlet of piping system:

$$H = h_{f1} + h_{fAB} + h_{f5}$$

EXAMPLE 8-6 Serial and parallel pipes

In Fig. E8-6 is a piping system including serial and parallel pipes with a free jet at the end point D. All the pipe parameters are listed in the following table. The flowrate in pipe AB is $Q_0 = 0.2 \text{m}^3/\text{s}$, the flowrates out of junction B and C are $q_1 = 0.0295 \text{m}^3/\text{s}$ and $q_2 = 0.0705 \text{m}^3/\text{s}$ respectively. The flows in all individual pipes are fully developed turbulent flow. The manning roughness coefficients for all individual pipes are $n = 0.013$. Determine (a) the flowrates in each individual pipe of parallel pipes, Q_1, Q_2, Q_3; (b) the head loss between junction B and C, h_{fBC}; and (c) the required total head at the inlet of the piping system, H.

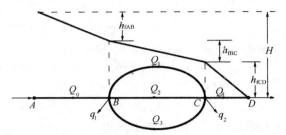

Fig. E8-6

	Pipe AB	Pipe BC: No. 1	Pipe BC: No. 2	Pipe BC: No. 3	Pipe CD
Diameter (m)	0.35	0.25	0.20	0.20	0.25
Length (m)	500	800	1000	600	300

Solution:

Assume the flow in the pipe system is steady and incompressible. In this piping system all individual pipes are assumed to be long pipes.

(a) The head losses for the parallel pipes in the system satisfy Eq. 8.20

$$S_1 l_1 Q_1^2 = S_2 l_2 Q_2^2 = S_3 l_3 Q_3^2 \tag{1}$$

in which S_i can be determined by Eq. 8.11

$$S_1 = \frac{10.3 \, n^2}{d_1^{5.333}} = \frac{(10.3)(0.013^2)}{0.25^{5.333}} \text{ s}^2/\text{m}^6 = 2.83 \text{ s}^2/\text{m}^6$$

$$S_2 = S_3 = \frac{10.3 \, n^2}{d_2^{5.333}} = \frac{(10.3)(0.013^2)}{0.2^{5.333}} \text{ s}^2/\text{m}^6 = 9.30 \text{ s}^2/\text{m}^6$$

Thus Eq. 1 gives

$$Q_2 = Q_1 \left(\frac{S_1 l_1}{S_2 l_2}\right)^{1/2} = Q_1 \left[\frac{(2.83 \text{ s}^2/\text{m}^6)(800\text{m})}{(9.30 \text{ s}^2/\text{m}^6)(1000\text{m})}\right]^{1/2} = 0.493 \, Q_1 \tag{2}$$

$$Q_3 = Q_1 \left(\frac{S_1 l_1}{S_3 l_3}\right)^{1/2} = Q_1 \left[\frac{(2.83 \text{ s}^2/\text{m}^6)(800\text{m})}{(9.30 \text{ s}^2/\text{m}^6)(600\text{m})}\right]^{1/2} = 0.637 \, Q_1 \tag{3}$$

The continuity equation at junction A gives

$$Q_0 = q_1 + Q_1 + Q_2 + Q_3 \tag{4}$$

Combining Eqs. 2 to 4 with the given values of Q_0, q_1 yields the flow rates in each individual pipe of parallel pipes

$$Q_1 = 0.08 \text{ m}^3/\text{s}, \quad Q_2 = 0.0395 \text{ m}^3/\text{s}, \quad Q_3 = 0.0510 \text{ m}^3/\text{s}$$

(b) The head loss between junction B and C can then be determined directly by

$$h_{fBC} = S_1 l_1 Q_1^2 = (2.83 \text{ s}^2/\text{m}^6)(800\text{m})(0.08 \text{ m}^3/\text{s})^2 = 14.49\text{m}$$

(c) The total head of the piping system, H, is determined by the energy equation between point A and D

$$H = h_{fAB} + h_{fBC} + h_{fCD} = S_0 l_0 Q_0^2 + h_{fBC} + S_4 l_4 Q_4^2$$

in which $S_4 = S_1$ because of $d_4 = d_1$,

$$S_0 = \frac{10.3 \, n^2}{d_0^{5.333}} = \frac{(10.3)(0.013^2)}{0.35^{5.333}} \text{ s}^2/\text{m}^6 = 0.47 \text{ s}^2/\text{m}^6$$

$$Q_4 = Q_0 - q_1 - q_2 = (0.2 - 0.0295 - 0.0705) \text{ m}^3/\text{s} = 0.1 \text{ m}^3/\text{s}$$

With the given parameters in the table yields

$$H = (0.47 \text{ s}^2/\text{m}^6)(500\text{m})(0.2 \text{ m}^3/\text{s})^2 + 14.49 \text{ m} + (2.83 \text{ s}^2/\text{m}^6)(300\text{m})(0.1 \text{ m}^3/\text{s})^2$$
$$= 32.38\text{m}$$

EXAMPLE 8-7 Parallel piping

Water is to be pumped from a reservoir A to reservoir B at a higher elevation through two 25-m-long plastic pipes connected in parallel as indicated in Fig. E8-7. The diameters of the two pipes are $d_1 = 3$cm and $d_2 = 5$cm respectively, and the corresponding Darcy friction factors are $\lambda_1 = 0.0164$ and $\lambda_2 = 0.0139$ respectively. Water is to be pumped by a 68%-efficient motor-pump unit that draws 7kW of electric power during operation. The minor and major losses in the pipes that connect the parallel pipes to the two reservoirs are considered to be negligible. Determine (a) the total flowrate between the reservoirs, and (b) the flowrates through each of the parallel pipes.

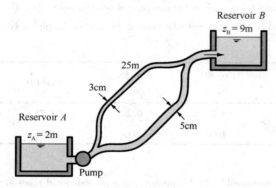

Fig. E8-7

Solution:

Take the density of water to be $\rho = 1000\text{kg/m}^3$. Assume the flow is steady and incompressible.
Based on the given parameters, we can determine the head provided by pump

$$h_p = \frac{\eta \dot{N}}{\rho g Q} = \frac{(0.68)(7000\text{W})}{(1000\text{kg/m}^3)(9.81\text{m/s}^2)(Q \text{ m}^3/\text{s})} = \frac{0.485}{Q}\text{m} \quad (1)$$

The energy equation between the free surfaces in reservoir A and B gives

$$z_A + \frac{p_{at}}{\rho g} + \frac{\alpha_A v_A^2}{2g} + h_p = z_B + \frac{p_{at}}{\rho g} + \frac{\alpha_B v_B^2}{2g} + h_w \quad (2)$$

where $\alpha_A = \alpha_B = 1.0$, $v_A = v_B = 0$, $z_A = 2$m, $z_B = 9$m. Thus, from Eq. 2 we have

$$h_p = z_B - z_A + h_w = \left(7 + \lambda_2 \frac{l_2}{d_2} \frac{v_2^2}{2g}\right)\text{m} = (7 + 0.354 \, v_2^2)\text{m} \quad (3)$$

The parallel pipes give

$$h_w = h_{w1} = h_{w2} \quad (4)$$

which gives

$$\lambda_1 \frac{l_1}{d_1} \frac{v_1^2}{2g} = \lambda_2 \frac{l_2}{d_2} \frac{v_2^2}{2g}$$

Thus

$$v_1 = v_2 \sqrt{\frac{\lambda_2 l_2 d_1}{\lambda_1 l_1 d_2}} = v_2 \sqrt{\frac{0.0139}{0.0164} \frac{25\text{m}}{25\text{m}} \frac{3\text{cm}}{5\text{cm}}} = 0.713 v_2 \quad (5)$$

The continuity equation at the junction is

$$Q = Q_1 + Q_2 \quad (6)$$

where

$$Q_1 = \frac{1}{4}\pi d_1^2 v_1 = \frac{1}{4}(3.14)(0.03)^2 v_1 = (7.065 v_1 \times 10^{-4})\ \text{m}^3/\text{s} \quad (7)$$

$$Q_2 = \frac{1}{4}\pi d_2^2 v_2 = \frac{1}{4}(3.14)(0.05)^2 v_2 = (19.625 v_2 \times 10^{-4})\ \text{m}^3/\text{s} \quad (8)$$

Combining Eqs. 5 to 8 yields:

$$Q = (7.065 v_1 + 19.625 v_2) \times 10^{-4}\ \text{m}^3/\text{s} = 24.662 v_2 \times 10^{-4}\ \text{m}^3/\text{s} \quad (9)$$

Combining Eqs. 1, 3 and 9 yields

$$v_2 = 7.42\ \text{m/s}$$

Thus,

$$Q = (24.662 v_2 \times 10^{-4})\ \text{m}^3/\text{s} = 0.0183\ \text{m}^3/\text{s}$$

$$Q_2 = (19.625 v_2 \times 10^{-4})\ \text{m}^3/\text{s} = 0.0146\ \text{m}^3/\text{s}$$

$$Q_1 = Q - Q_2 = (0.0183 - 0.0146)\ \text{m}^3/\text{s} = 0.0037\ \text{m}^3/\text{s}$$

8.4.3 Pipeline with Uniform Draw-off

In the supply of water for domestic, commercial and irrigation purposes, it is common for water to be taken from the pipe line in many places along the length. As a result, the pressure of the fluid at various points along the pipeline will vary. Thus, if fluid in a pipe is drawn off at a uniform discharge per unit length q, this pipeline is called *pipeline with uniform draw-off* (Fig. 8-9). The discharge drawn off continuously along the pipe is called *the draw-off discharge*.

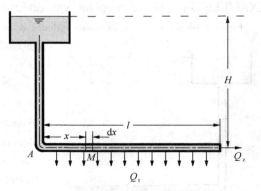

Fig. 8-9 Pipeline with uniform draw-off

Consider a pipe with uniform draw-off compared with a pipe of the same diameter and length but discharging at the end. Let the *end flowrate* at the end of pipe with draw-off to be Q_z, the uniform draw-off discharge per unit length to be q, the total draw-off discharge in a l-long pipe with draw-off to be $Q_t = ql$. Consider a short length of pipe dx at a distance x from the entry point A of pipeline with uniform draw-off, the flow rate at this section is

$$Q_x = Q_z + Q_t - qx = Q_z + Q_t - \frac{Q_t}{l}x$$

Thus, the head loss in this short pipe is

$$dh_f = SQ_x^2 dx = S\left(Q_z + Q_t - \frac{Q_t}{l}x\right)^2 dx$$

Assume the roughness and diameter of pipe elements are constant, and the flow is a fully developed turbulent flow, the specific resistance S is a constant. Thus, the total head loss in this pipeline with a length of l can be obtained by integrating the above equation

$$h_f = \int_0^l S\left(Q_z + Q_t - \frac{Q_t}{l}x\right)^2 dx = Sl\left(Q_z^2 + Q_zQ_t + \frac{1}{3}Q_t^2\right) \quad (8.21)$$

which can be approximately simplified as

$$h_f = Sl(Q_z + 0.55Q_t)^2 = SlQ_c^2 \quad (8.22)$$

where Q_c is the computational flowrate in the pipeline with uniform draw-off, termed the *equivalent discharge*

$$Q_c = Q_z + 0.55Q_t \quad (8.23)$$

Equation 8.22 indicates the total head loss in a pipe with uniform draw-off can also be computed by Eq. 8.10 by replacing the flowrate in Eq. 8.10 by the equivalent discharge.

If the end flowrate in the system is zero, that is $Q_z = 0$, Eq. 8.22 gives

$$h_f \approx \frac{1}{3}SlQ_t^2 \quad (8.24)$$

It indicates that the total head loss in a pipe with uniform draw-off and no end discharge is only one third of the head loss when the fluid flows in a pipeline with no draw-off discharge at a constant flow rate of the total draw-off discharge.

EXAMPLE 8-8 Serial pipeline with uniform draw-off discharge

Figure E8-8 presents a water piping system supplied by a water tank. The main truck is composed of three individual cast iron pipe elements with diameters and lengths as: $d_1 = 0.2$m, $d_2 = 0.15$m, $d_3 = 0.125$m and $l_1 = 500$m, $l_2 = 150$m, $l_3 = 200$m. The middle one in the serial pipeline is a uniform draw-off pipe with a total draw-off discharge $Q_t = 0.015$ m^3/s. The flowrate taken from the junction B is $q_B = 0.01$ m^3/s, the end flowrate of trunk pipe is $Q_z = 0.02$ m^3/s. Determine the mounting elevation of the water tank, H, if the water depth in tank is 4m.

Solution:

Assume the flow in the system is steady and incompressible.

According to the continuity equation and Eq. 8.24 the flow rates in each individual pipes are

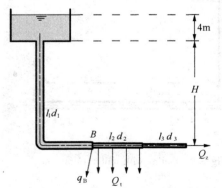

Fig. E8-8

$$Q_1 = Q_z + Q_t + q_B = (0.02 + 0.015 + 0.01) \text{ m}^3/\text{s} = 0.045 \text{ m}^3/\text{s}$$

$$Q_2 = Q_c = Q_z + 0.55 Q_t = 0.02 \text{ m}^3/\text{s} + (0.55)(0.015 \text{ m}^3/\text{s}) = 0.028 \text{ m}^3/\text{s}$$

$$Q_3 = Q_z = 0.02 \text{ m}^3/\text{s}$$

Take the axis of horizontal pipe as the reference level, the mounting elevation of water tank, H, can then be determined by the energy equation:

$$H + h = h_{f1} + h_{f2} + h_{f3} = S_1 l_1 Q_1^2 + S_2 l_2 Q_2^2 + S_3 l_3 Q_3^2 \tag{1}$$

With the manning coefficient of cast iron pipe $n = 0.013$ from Table 6-4, and the pipe specific resistance S from Eq. 8.11, Eq. 1 yields

$$H = 10.3 n^2 \left(\frac{l_1 Q_1^2}{d_1^{5.333}} + \frac{l_2 Q_2^2}{d_2^{5.333}} + \frac{l_3 Q_3^2}{d_3^{5.333}} \right) - h$$

$$= (10.3)(0.013^2) \left[\frac{(500)(0.045)^2}{(0.2)^{5.333}} + \frac{(150)(0.028)^2}{(0.15)^{5.333}} + \frac{(200)(0.02)^2}{(0.125)^{5.333}} \right] \text{m} - 4\text{m}$$

$$= 19.60 \text{m}$$

DISCUSSION Assume the middle pipe is an pipe with no uniform draw-off and the flowrates in the first and third pipe elements are unchanged, $Q_1 = 0.045 \text{ m}^3/\text{s}$, $Q_3 = Q_z = 0.02 \text{ m}^3/\text{s}$, thus the flowrates through the middle pipe and taken at junction B are

$$Q_2 = Q_3 = 0.02 \text{ m}^3/\text{s}$$
$$q_B = Q_1 - Q_2 = 0.045 \text{ m}^3/\text{s} - 0.02 \text{ m}^3/\text{s} = 0.025 \text{ m}^3/\text{s}$$

The mounting elevation of water tank is

$$H = 10.3 n^2 \left(\frac{l_1 Q_1^2}{d_1^{5.333}} + \frac{l_2 Q_2^2}{d_2^{5.333}} + \frac{l_3 Q_3^2}{d_3^{5.333}} \right) - h = 17.12 \text{m}$$

It indicates that the mounting elevation of water tank now is lower than the one in the original problem. It is because a smaller flowrate in the second pipe now (0.02 m³/s compared to the original 0.028 m³/s) leads to a smaller loss of 2.58m compared to the original 5.06m.

8.5 Steady Flow in Pipe Networks

8.5.1 Introduction to Pipe Network

In order to direct water to many individuals in a municipal water supply, many times the water is routed through a water supply network. From the layout point of view, these water distribution systems are either branched or looped topology, or a combination of both.

Branched systems are those that convey water from a distribution main to different consumption points, following a treelike pattern; all their branches finish in dead-ends (Fig. 8-10a). Their design is straightforward but has a main disadvantage in the fact that it causes stagnant water pockets in all dead-ends. If repairs are necessary, large areas must be cut off from service. Head losses, due to heavy local demands or during a fire, may be excessive unless the pipes are quite large.

Looped network systems usually have a ring main to which secondary pipes may be connected, that is pipes are interconnected throughout the system such that the flow to a demand node can be supplied through several connected pipes (Fig. 8-10b). Their design is much more complicated;

with them the possibility of stagnant water is reduced. If part of the pipeline needs cleaning or repair, it may be isolated from the rest of the system (with appropriate valves); all watering points outside of it may continue to be supplied. The flow direction in a looped system can change based on spatial or temporal variation in water demand. Thus, the flow directions in looped system pipes are not unique.

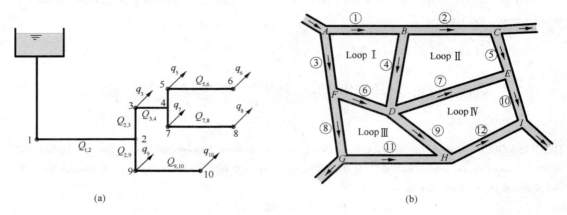

Fig. 8-10 (a) Branched pipe network; (b) Looped pipe network

The solution for pipe network problems is often carried out by use of node and loop equations similar in many ways to that done in electrical circuits. For example, as what presented in Section 8.4.2, the continuity equation requires that for each *node* the net flowrate is zero, the net head loss completely around a *loop* must be zero. By combining these ideas with the usual head loss and pipe flow equations, the flow throughout the entire network can be obtained. Of course, trial-and-error solutions are usually required because the directions of flows and the friction factors may not be known.

In the analysis of the pipe network, all individual pipes are considered as *long* pipes.

8.5.2 Branched Pipe Network

TRUNK

Branched pipe network consists of trunk and branches (subbranches). In general, the *trunk* is taken as the pipeline from the source to the most farest locus with the highest altitude and the largest required flowrate. It consists of serial pipes with different diameter, in which water flows at identical or different flowrates. Thus, the hydraulic design of the trunk is to determine the diameters of the trunk mains, the flowrates in them and the mounting height of water tank. It should satisfy the basic principles: the continuity equation, the superposition of the head losses and the energy equation.

The continuity equation. The net flowrate at any nodes of the trunk mains must be zero, $\Sigma Q_i + \Sigma q_j = 0$, where Q_i is the flow rates in the trunk pipes connected to this junction; q_j is the flow rate taken from this node by the branch pipes or the consumers.

The superposition of the head losses. The total head loss in the trunk is the sum of head losses in each individual trunk pipes which are connected in series

$$h_f = \Sigma h_{fi} = \Sigma S_i l_i Q_i^2 \qquad (8.25)$$

where i is the No. of the truck mains.

The energy equation. To make sure the fluid supply at the most difficult point and overcome the resistance in the trunk pipeline, a *residual head* (also called *free head*), H_z, at the end of trunk pipeline must be remained. Thus, the mounting elevation of water tank can be determined by energy equation

$$H = H_z + \Sigma h_{fi} + (z_t - z_0) \qquad (8.26)$$

where H is the total head from the referred ground level to the water level in water tank; z_0 is the referred ground level where water tank is mounted; z_t is the ground level at the end of trunk pipeline.

If the diameters of each individual trunk pipes are given, we can determine the mounting elevation of water tank by Eq. 8.26. In the engineering practices, the pipe diameters are usually determined by the *economic velocity*, designated as v_c, recommended by engineering manuals. For example, the *maximum permissive velocity* in water supply system is 2.5-3.0m/s for avoiding the high pressure surge caused by the water hammer, the *minimum permissive velocity* is 0.6 m/s for avoiding the silting in the network. Thus, the pipe diameter can be estimated by

$$d_i = \sqrt{\frac{4Q_i}{\pi v_c}} \qquad (8.27)$$

BRANCHES

Now, we can start to determine the diameters of branches, the pipes connecting the trunk and the subbranches or the consumers. For branches, the pressures at the start and end ends are known, also the pipe length. Thus, we can have the averaged frictional slopes for each branch

$$\overline{J}_{jk} = \frac{H_j - H_k}{l_{jk}} \qquad (8.28)$$

where subscript j and k present the reference numbers of the start and the end of the branch respectively; $\overline{J}_{jk}$ is the averaged frictional slopes for this branch; l_{jk} is the length of this branch; H_j and H_k are the total heads at the start and the end of this branch, respectively.

Thus the pipe specific resistance S_{jk} can be determined by

$$S_{jk} = \frac{h_{fjk}}{l_{jk} q_{jk}^2} = \frac{\overline{J}_{jk}}{q_{jk}^2} \qquad (8.29)$$

Finally, the diameter of branch by Eq. 8.11 gives

$$d_{jk} = \sqrt[5.333]{\frac{10.3 \, n_{jk}^2}{S_{jk}}} \qquad (8.30)$$

and is rounded to a nominal diameter used in engineering practices. In general, the check calculation of whole network with the designed nominal diameters is necessary to make sure the free head is not less than the required value.

EXAMPLE 8-9 Branched water supply system

Figure E8-9 gives a topology of branched water supply system in a new community. The water level in the suction tank of pump station is $z_0 = 150$m. The elevations at junctions 2, 4 and 6 are $z_2 = 153$m, $z_4 = 155$m and $z_6 = 154.5$m respectively. The required service head at junction 4 and 6 are $H_{z4} = 20$m and $H_{z6} = 12$m, respectively. The pipe lengths and the flow rates in pipes are listed in the table. Assume all pipes have the same Manning coefficients, that is $n = 0.013$, and the head loss in the suction pipe of pump is negligible. Determine the pump head and the diameters of each individual pipes.

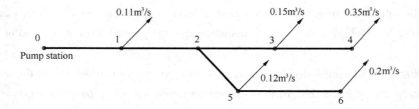

Fig. E8-9

Pipes	Given parameters			Calculated parameters		
	End No. of pipes	Pipe length l (m)	Flow rate Q (m³/s)	Pipe diameter d (mm)	Pipe specific resistance S (s²m⁻⁶)	Head loss h_f (m)
Trunks	0,1	1000	0.93	1000	0.00174	1.505
	1,2	800	0.82	900	0.00305	1.642
	2,3	500	0.50	700	0.01165	1.456
	3,4	1000	0.35	600	0.0265	3.245
Branches	2,5	500	0.32	450	0.1172	
	5,6	600	0.20	400	0.30	

Solution:

(1) Determine the pump head and the diameters of trunk pipes.

Take the serial pipes 0-1-2-3-4 as the trunk to determine the pipe diameters and the head losses. Assume the flows in pipes are fully developed steady incompressible turbulent flows.

Take 1.2m/s to be the economic velocity in the trunk pipe 0-1. Thus, the diameter of trunk pipe 0-1 is

$$d_{0,1} = \sqrt{\frac{4Q_{0,1}}{\pi v_c}} = \sqrt{\frac{(4)(0.93\text{m}^3/\text{s})}{(3.14)(1.2\text{m/s})}} = 0.993\text{m}$$

For the nominal diameter it would be taken as $d_{0,1} = 1.0$m. Determine the pipe specific resistance and head loss:

$$S_{0,1} = \frac{10.3\, n_{0,1}^2}{d_{0,1}^{5.333}} = \frac{(10.3)(0.013^2)}{1}\, \text{s}^2/\text{m}^6 = 0.00174\, \text{s}^2/\text{m}^6$$

$$h_{f0,1} = S_{0,1} l_{0,1} Q_{0,1}^2 = (0.00174\, \text{s}^2/\text{m}^6)(1000\text{m})(0.93\, \text{m}^3/\text{s})^2 = 1.505\text{m}$$

With the exactly similar manner, we can determine the diameters of trunk pipes and head losses in the trunk pipes. The results are listed in the above table. Thus, the energy equation gives the pump head

$$H = H_{z4} + \Sigma h_{fjk} + (z_4 - z_0)$$
$$= 20\text{m} + (1.505 + 1.642 + 1.456 + 3.245)\text{ m} + (155\text{m} - 150\text{m}) = 32.85\text{m}$$

(2) Determine the diameters of branches.

The head at the start end (junction 2) of branch (2, 5) is

$$H = H_{z4} + z_4 + h_{f2,3} + h_{f3,4} = 20\text{m} + 155\text{m} + 1.456\text{m} + 3.245\text{ m} = 179.701\text{m}$$

The head at junction 6 is

$$H = H_{z6} + z_6 = 12\text{m} + 154.5\text{m} = 166.5\text{m}$$

The average frictional slope of branch (2, 6) is

$$\overline{J}_{2,6} = \frac{H_2 - H_6}{l_{2,6}} = \frac{179.701\text{m} - 166.5\text{m}}{500\text{m} + 600\text{m}} = 0.012$$

The pipe specific resistance of branch (2, 5) is

$$S_{2,5} = \frac{\overline{J}_{2,5}}{q_{2,5}^2} = \frac{0.012}{(0.32 \text{ m}^3/\text{s})^2} = 0.1172 \text{ s}^2/\text{m}^6$$

Thus, the diameter of branch (2,5) can be determined by Eq. 8.30

$$d_{2,5} = \sqrt[5.333]{\frac{10.3 \, n_{2,5}^2}{S_{2,5}}} = \sqrt[5.333]{\frac{(10.3)(0.013^2)}{0.1172}} \text{ m} = 0.454\text{m}$$

For the nominal diameter it would be taken as $d_{2,5} = 0.45\text{m}$. With the exactly similar manner, the diameters of branch (5, 6) could be $d_{5,6} = 0.40\text{m}$ as listed in the above table.

8.5.3 Looped pipe network

Most of water supply system in modern city are composed by the looped pipe network as indicated in Fig. 8-10b. It might be a water supply system for a community or subdivision or even a city.

In looped pipe network pattern, all the pipes are interconnected with no dead-ends. In such a system, water can reach any point from more than one direction. Thus, at the time of fires, by manipulating the cut-off valves, plenty of water supply may be diverted and concentrated for firefighting. However, the looped pipe network cost much more because relatively more length of pipes and more valves are required. Moreover, the calculation of pipe sizes is more complicated.

Even though the looped network is quite complex algebraically, but follows basic rules, some of them are same as for the branched network.

Continuity equation. The net flow into any node must be zero

$$\Sigma Q_i = 0 \tag{8.31}$$

where Q_i is the flow rates in pipes connected to this node.

For a network with n_j nodes, we can get only $n_j - 1$ independent continuity equations. It is because the No. n_j equation can be induced from the $n_j - 1$ continuity equations. It is dependent.

Energy relationship. The net head loss around any closed loop must be zero as indicated in Section 8.4.2. In other words, the HGL at each node must have one and only one elevation, that is

$$\Sigma h_{fi} = \Sigma S_i l_i Q_i^2 = 0 \tag{8.32}$$

For a network with n_l loops, we can get n_l equations.

Based on the above two rules, we can get $n_j + n_l - 1$ equations, which can be used to solve n_s flow rates in the whole network. Because the total number of pipes, n_p, in a looped network with n_j nodes and n_l loops is exactly equal to $n_j + n_l - 1$.

By supplying these rules to each node and independent loop in the network, one obtains a set of simultaneous equations for the flow rates in each pipe leg and the HGL (or pressure) at each node. Solution may then be obtained by numerical iteration, as first developed in a hand-calculation technique by Prof. Hardy Cross (1885-1959), a structural engineering professor at the University of Illinois at Urbana-Champaign, in 1936. Computer solution of pipe-network problems is now quite common and covered in at least one specialized text. Network analysis is quite useful for real water distribution systems if well calibrated with the actual system head-loss data.

The *Hardy Cross method* is an iterative method for determining the flow in pipe network systems where the inputs and outputs are known, but the flow inside the network is unknown. Hardy Cross developed two methods for solving flow networks. One of them is the *method of balancing heads*. It uses an initial guess that satisfies continuity equation of flow at each node and then balances the flow energy relationships until continuity of potential is also achieved over each loop in the system. The detail is as follows.

1. Guess the flow rates in each pipe, make sure that the total inflow is equal to the total outflow at each node. Take the inflow to be positive and outflow be negative.

2. Determine each closed loop in the system.

3. For each loop, determine the clockwise head losses and counter clockwise head losses. Head loss in each pipe are calculated using $h_{fi} = S_i l_i Q_i^2$. Take the counter-clockwise head losses to be positive and clockwise head losses be negative.

4. Determine the total head loss in the loop, Σh_{fi}, by adding all head losses in the loop.

5. If the initial guess of flow rates in each pipe is correct, the change in head loss over a loop in the system, Δh_{fi}, would equal to zero. However, if the initial guess is not correct, then the change in head will be non-zero and a change in flow rate, ΔQ, must be applied. Thus, the head loss in each pipe should be

$$h_{fi} + \Delta h_{fi} = S_i l_i (Q_i + \Delta Q)^2 = S_i l_i [Q_i^2 + 2Q_i \Delta Q + (\Delta Q)^2]$$

For a small $(\Delta Q)^2$ compared to ΔQ, the third term in the bracket vanishes, leaving

$$h_{fi} + \Delta h_{fi} = S_i l_i Q_i^2 + 2 S_i l_i Q_i \Delta Q$$

yields

$$\Delta h_{fi} = 2 S_i l_i Q_i \Delta Q$$

For a loop, the total head loss should be zero, $\Sigma h_{fi} = 0$, thus

$$\Sigma(h_{fi} + \Delta h_{fi}) = \Sigma h_{fi} + \Sigma \Delta h_{fi} = \Sigma S_i l_i Q_i^2 + 2 \Delta Q \Sigma S_i l_i Q_i = 0$$

which yields

$$\Delta Q = -\frac{\sum S_i l_i Q_i^2}{2 \sum S_i l_i Q_i} = -\frac{\sum h_{fi}}{2\sum \dfrac{S_i l_i Q_i^2}{Q_i}} = -\frac{\sum h_{fi}}{2\sum \dfrac{h_{fi}}{Q_i}} \qquad (8.33)$$

6. If the change in flow rate is positive, add it to all the flow rate in counter clockwise direction in the loop and subtract it to all the flow rate in clockwise direction in the loop. If the change in flow rate is negative, add it to the flow rate in clockwise direction and subtract it to the flow rate in counter clockwise direction.

7. Repeat from step 3 to step 6 until the change in flow rate is within a satisfactory range.

8. Determine the head of each node according to the calculated head losses, then determine the pump head or mounting elevation of water tank.

The Hardy Cross method is useful because it relies on only simple math, circumventing the need to solve a system of equations. It can iteratively correct the mistakes in the initial guess used to solve the problem, and the subsequent mistakes in calculation. If the method is followed correctly, the proper flow rate in each pipe can still be found if small mathematical errors are consistently made in the process. As long as the last few iterations are done with attention to detail, the solution will still be correct.

8.6　Unsteady Flow in Pipeline

8.6.1　Introduction to Transition Flow

In the above sections, steady pressure flows in pipelines have been discussed thoroughly. In this section, an unsteady pressure flow, or *transition flow*, in the pipe will be discussed.

Transient flow is a transition from one steady state to another steady state in a fluid flow system. It occurs in all fluids, confined and unconfined. The most common transition flows in pipeline are water hammer and surge.

Water hammer (or *hydraulic shock*) is the momentary increase in pressure, which occurs in a liquid system when there is a sudden change of direction or velocity of the liquid. For example, when a rapidly closed valve suddenly stops water flowing in a pipeline, pressure energy is transferred to the valve and pipe wall. Shock waves are set up within the system. Pressure waves travel backward until encountering the next solid obstacle, then forward, then back again. The pressure wave's velocity is equal to the speed of the sound; therefore it 'bangs' as it travels back and forth, until dissipated by friction losses. Anyone who has lived in an older house is familiar with the 'bangs' that resounds through the pipes when a faucet is suddenly closed. This is an effect of water hammer.

A less severe form of hammer is called *surge*, a slow motion of mass oscillation of liquid caused by internal pressure fluctuations in the system. This can be pictured as a slower 'wave' of pressure building within the system.

Water hammer is a common but serious problem in residential plumbing systems. If not controlled, both water hammer and surge produce potential damage to pipes, fittings, and valves, cause leaks and shorten the life of the system. Neither the pipe nor the water will compress to absorb the shock. For example, the dramatic pressure rise can cause pipes to rupture. Accompanying the high pressure wave, there is a negative wave, which is often overlooked, can cause very low pressures leading to the possibility of contaminant intrusion.

8 6.2 Processes of Water Hammer

If the valve is suddenly closed, the flowing water will be obstructed and momentum will be destroyed and consequently a wave of high pressure will be created which travels back and forth starting at the valve, traveling to the reservoir, and returning back to the valve and so on. This wave of high pressure not only has a very high speed (called *celerity* of *pressure wave*, c) which may reach the speed of sound wave and may create noise called *knocking*, but also has the effect of hammering action on the walls of the pipe and hence is commonly known as the *water hammer phenomenon*.

Consider a tank-pipe-valve system as shown in Fig. 8-11a, that is, a horizontal pipeline AB with a constant internal diameter and a length of l fed by a reservoir at end B with a constant water level and controlled by a valve at the downstream end A. Let v_0 be the initial mean flow velocity in the pipe, A be the cross-sectional area of pipe, Δp be the instantaneous rise in pressure due to sudden closure of valve and t be the actual closure time of valve. Thus, the whole process of water hammer in a single-conduit, frictionless pipeline following its sudden closure can be interpreted by Fig. 8-11.

1. Initial status. At $t = 0$, the pressure profile is steady, which is shown by the pressure head ($H = p/\rho g$) line running horizontally because of the assumed lack of friction. Under steady-state conditions, the flow velocity is v_0 (Fig. 8-11a).

2. Process one ($0 < t \leq l/c$). The sudden closure of the gate valve at the downstream end of the pipeline causes a pulse of high pressure Δp, and the pipe wall is stretched. The pressure wave generated runs in the opposite direction to the steady-state direction of the flow at the speed of sound, c, accompanied by a reduction of the flow velocity to $v = 0$ in the high pressure ($p + \Delta p$) zone. While the zone pressure wave untouched still has a pressure p and a velocity v_0. The process takes place in a period of time $0 < t < l/c$ (Fig. 8-11b). At $t = l/c$, the pressure wave arrives at the reservoir (Fig. 8-11c), the fluid in whole pipe has a higher pressure, $p + \Delta p$, and zero velocity, $v = 0$.

3. Process two ($l/c < t \leq 2l/c$). At $t = l/c$, there is an unbalanced condition at point B because the reservoir pressure p = constant and the pressure in the pipe is $p + \Delta p$. With a change of sign, the pressure wave is reflected in the opposite direction. A relief wave with a head of $-\Delta p$ travels downstream towards the gate valve, while the flow velocity changes sign ($-v_0$) and is now headed in the direction to the reservoir. The pressure and velocity in the zone pressure wave passing are p and $-v_0$ respectively, leaving the untouched zone $p + \Delta p$ and $v = 0$ (Fig. 8-11d). At time $t = 2l/c$, the relief wave arrives at the closed gate valve (Fig. 8-11e) and the pressure re-

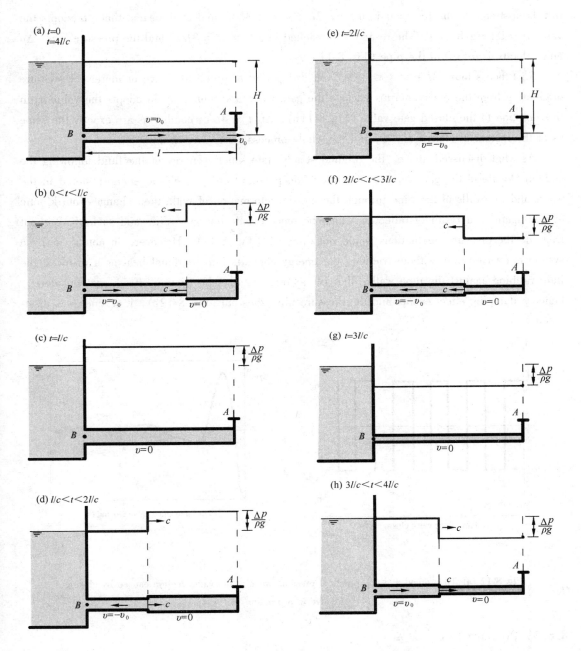

Fig. 8-11 Pressure and velocity waves in a single, frictionless constant-diameter pipe following its sudden closure. The expansion and contraction of the pipeline as a result of rising and falling pressure levels, respectively, are shown(Not to scale)

sumes the reservoir's pressure head, p, and the flow velocity in pipe, $-v_0$.

4. Process three ($2l/c < t \leqslant 3l/c$). Upon arrival at the gate valve, the fluid continues to head in the direction of the reservoir due to the inertia and the velocity changes from $-v_0$ to $v = 0$. This causes a sudden negative change in pressure of $-\Delta p$ at the gate. The low pressure wave $-\Delta p$

335

travels upstream to the reservoir in a time $2l/c < t < 3l/c$, and at the same time, v adopts the value $v = 0$ (Fig. 8-11f). The reservoir is reached in a time $t = 3l/c$, and the pressure is $p - \Delta p$ and velocity is $v = 0$ in the pipe (Fig. 8-11g).

5. Process four ($3l/c < t \leqslant 4l/c$). In this period of time, the wave of increased pressure originating from the reservoir runs back to the gate valve and v once again adopts the value v_0 in the direction to the closed gate valve (Fig. 8-11h). At $t = 4l/c$, conditions are exactly the same as at the instant of closure $t = 0$, and the whole process starts over again.

As what discussed above, the original steady-state kinetic energy of the fluid following the sudden closure of the gate valve is converted into potential energy (elastic energy) stored in the water and the walls of the pipe through the elastic deformation of both, then changes into kinetic energy again as a result of reflection, then becomes elastic energy again, and so forth. Without friction, the pressure fluctuations would not diminish (Fig. 8-12). However, in actual fact, no system is ever entirely without friction, the energy converts into frictional heat as a result of the fluid rubbing against the pipe walls, that is, in real practice the friction effects are considered, hence a damping effect occurs and the pressure wave dies out (Fig. 8-12b), i. e. energy is dissipated.

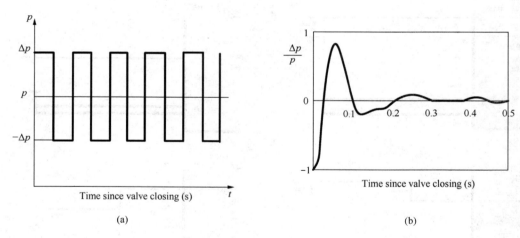

Fig. 8-12 (a) Theoretical and (b) practical pressure wave at the valve section caused by closing a valve in a pipeline

8.6.3 Pressure Rise

In the whole process of water hammer, the time required for the pressure wave generated due to the closure or open of valve to travel once from the point of origin to reservoir over the entire length of the pipeline and back to the point of origination is termed the *critical time*, which is designated T_r. It is the *reflection time* of the pipe, and has a value $T_r = 2l/c$.

Let T_c to be the time of valve totally closed, then the transient flow is referred to as *gradual closure* when $T_c > T_r$, and *sudden closure* when $T_c < T_r$. Then, the pressure rise due to water hammer can be estimated, which mainly depends upon: (a) The initial velocity of the flow of liquid in the pipe, (b) The length of pipe, (c) Time taken to close the valve and (d) Elastic properties of the material of the pipe.

SUDDEN VALVE CLOSURE

For sudden valve closure and compressible fluid in pipe, *Joukowsky's formula*, which originates from Newton's laws of motion, can be used to describe the pressure change that results from a rapid change in velocity:

$$\Delta p = \rho c v_0 \tag{8.34}$$

or

$$\Delta h = \frac{\Delta p}{\rho g} = \frac{c v_0}{g} \tag{8.35}$$

By analyzing the formula, it is clear that the larger the magnitude of the velocity change and the larger the magnitude of the wave speed is, the greater the change in pressure will be.

GRADUAL VALVE CLOSURE

When the valve is closed slowly compared to the critical time for a pressure wave to travel the length of the pipe to reservoir and return to the valve, the wave of reduced pressure reflected by the tank has arrived at the gate valve after T_r has elapsed, and evens out some of the pressure increase at the valve. In this case, the pressure rise at valve is determined by

$$\Delta p = \rho c v_0 \frac{T_r}{T_c} = \rho v_0 \frac{2l}{T_c} \tag{8.36}$$

or

$$\Delta h = \frac{\Delta p}{\rho g} = \frac{v_0}{g} \frac{2l}{T_c} \tag{8.37}$$

CELERITY OF PRESSURE WAVE

The pressure rise in water hammer phenomena is proportional to the celerity of pressure wave as discussed above. The *celerity of pressure wave* can be determined by

$$c = \frac{c_0}{\sqrt{1 + \frac{Kd}{E\delta}}} \tag{8.38}$$

where c_0 is the *speed of sound* in water, $c_0 = 1435 \text{m/s}$ in fresh water at 10℃ and of pressure at 1-25 atm; K is the *bulk modulus of elasticity* of water, $K = 2.1 \times 10^9 \text{N/m}^2$; E is the *elastic modulus* of the pipe wall, N/m^2; d is the internal diameter of pipe, m; δ is the thickness of pipe wall, m.

For a regular steel pipe, $d/\delta \approx 100$, $K/E \approx 0.01$, substituting into Eq. 8.38 yields $c \approx 1014 \text{m/s}$. If water travels at $v_0 = 1.0 \text{m/s}$ before the closure of valve, thus the pressure rise for a sudden closure of valve determined by Eq. 8.35 is about $\Delta h \approx 103 \text{m}$, which may cause a great

337

damage to the pipe.

8.6.4 Practical Solutions to Water Hammer

A water transport system's operating conditions are almost never at a steady state. Pressures and flows change continually as pumps start and stop, demand fluctuates, and tank levels change. The causes of water hammer are varied. There are, however, four common events that typically induce large changes in pressure: pump startup, pump power failure, valve opening and closing, and improper operation or incorporation of surge protection devices.

Water hammer has caused accidents and fatalities, but usually damage is limited to breakage of pipes or appendages. An engineer should always assess the risk of a pipeline burst. Pipelines transporting hazardous liquids or gases warrant special care in design, construction, and operation. Hydroelectric power plants especially must be carefully designed and maintained because the water hammer can cause water pipes to fail catastrophically. Thus, the following characteristics can be applied to reduce or eliminate water hammer:

(a) Lower the fluid velocities. To keep water hammer low, pipe-sizing charts for the water distribution system recommend flow velocity at or below 3.0m/s.

(b) Fit slowly closing or opening valves to prolong the operation time of valve, i.e. toilet fill valves are available in a quiet fill type that closes quietly, or a good pipeline control (start-up and shut-down procedures).

(c) Shorten the lengths of straight pipe, i.e. add elbows, expansion loops, and adopt the pipe with smaller elastic modulus. Water hammer is related to the speed of sound in the fluid. Smaller elastic modulus of pipe material reduces the speed of pressure wave. With looped piping, a branch can be served by flows from both sides of a loop with lower velocities.

(d) In long pipelines (such as the pipelines or tunnels in hydroelectric generating stations), surge can be relieved with a tank of water directly connected to the pipeline called a '*surge tank.*' When surge is encountered, the tank will act to relieve the pressure, and can store excess liquid, giving the flow alternative storage better than that provided by expansion of the pipe wall and compression of the fluid. Surge tanks can serve for both positive and negative pressure fluctuations.

(e) *Water towers* (used in many drinking water systems) help maintain steady flow rates and trap large pressure fluctuations.

(f) *Air vessels* work in much the same way as water towers, but are pressurized. They typically have an air cushion above the fluid level in the vessel, which may be regulated or separated by a bladder. Sizes of air vessels may be up to hundreds of cubic meters on large pipelines. They come in many shapes, sizes and configurations. Such vessels often are called *accumulators* (Fig. 8-13b) or *expansion tanks*.

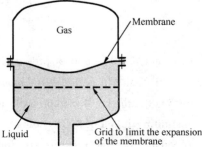

Fig. 8-13 Schematic of accumulator

(g) Install *hydropneumatic device* between the water pipe and the machine to absorb the shock and stop the banging, or install flywheel on pump.

(h) Install air valves at high points in the pipeline to remediate low pressures.

8.7 Flow Measurements in Pipes

Almost all practical fluids engineering problems are associated with the need for an accurate flow measurement. There is a need to measure *local* properties (velocity, pressure, temperature, density, viscosity, turbulent intensity), *integrated* properties (mass flow and volume flow), and *global* properties (visualization of the entire flow field). We have discussed pressure measurement in Section 2.3.3, the flow visualization technology in Section 3.3.1, and will discuss the flow measurement schemes suitable for open-channel and other free-surface flows in Chapter 10. In this section, we concentrate on the velocity and volume-flow measurements in pipe flow.

Flowmeters range widely in their level of sophistication, size, cost, accuracy, versatility, capacity, pressure drop, and the operating principle. We give an overview of the meters commonly used to measure the flowrate of liquids and gases flowing through pipes or ducts, and limit our consideration to incompressible flow.

8.7.1 Local-Velocity Measurements

Velocity averaged over a small region, or point, can be measured by several different physical principles, listed in order of increasing complexity and sophistication:

1. Trajectory of floats or neutrally buoyant particles
2. Rotating mechanical devices: propeller meter; turbine meter
3. Pitot-static tube (Fig. 3-29)
4. Electromagnetic current meter
5. Thermal (hot wires and hot films) anemometer
6. Laser-doppler anemometer (LDA)

ROTATING SENSORS

The rotating devices can be used in either gases or liquids, and their rotation rate is approximately proportional to the flow velocity. The deduced-propeller (turbine) and free-propeller meters must be aligned with the flow parallel to their axis of rotation. They can sense reverse flow because they will then rotate in the opposite direction. All these rotating sensors can be attached to counters or sensed by electromagnetic or slip-ring devices for either a continuous or a digital reading of flow velocity. All have the disadvantage of being relatively large and thus not representing a 'point.'

PITOT-STATIC TUBE

A *Pitot-static tube* is a slender tube aligning with the flow (Fig. 3-29) to measure local velocity by means of a pressure difference, which has been discussed in detailed in Section 3.6.5. The velocity can be determined by Eq. 3.64.

ELECTROMAGNETIC METER

A *full-flow electromagnetic flowmeter* is a nonintrusive device that consists of a magnetic coil that encircles the pipe, and two electrodes drilled into the pipe along a diameter flush with the inner surface of the pipe so that the electrodes are in contact with the fluid but do not interfere with the flow and thus do not cause any head loss. The electrodes are connected to a voltmeter. The coils generate a magnetic field when subjected to electric current, and the voltmeter measures the electric potential difference between the electrodes. This potential difference is proportional to the flow velocity of the conducting fluid, and thus the flow velocity can be calculated by relating it to the voltage generated. Electromagnetic flowmeters measure flow velocity indirectly, and thus careful calibration is important during installation. Electromagnetic flowmeters are well-suited for measuring flow velocities of liquid metals such as mercury, sodium, and potassium that are used in some nuclear reactors. They can also be used for liquids that are poor conductors, such as water, blood and seawater, provided that they contain an adequate amount of charged particles. Electromagnetic flowmeters can also be used to measure the flow rates of chemicals, pharmaceuticals, cosmetics, corrosive liquids, beverages, fertilizers, and numerous slurries and sludges, provided that the substances have high enough electrical conductivities. Their use is limited by their relatively high cost, power consumption, and the restrictions on the types of suitable fluids with which they can be used.

THERMAL ANEMOMETERS

Thermal anemometers were introduced in the late 1950s and have been in common use since then in fluid research facilities and labs. As the name implies, thermal anemometers in volve an electrically heated sensor, as shown in Fig. 8-14, and utilize a thermal effect to measure flow velocity. A thermal anemometer is called a *hot-wire anemometer* if the sensing element is a wire, and a *hot-film anemometer* if the sensor is a thin metallic film (less than 0.1 μm thick) mounted usually on a relatively thick ceramic support having a diameter of about 50 μm. The hot-wire anemometer is characterized by its very small sensor wire-usually a few microns in diameter and a couple of millimeters in length. The sensor is usually made of platinum, tungsten,

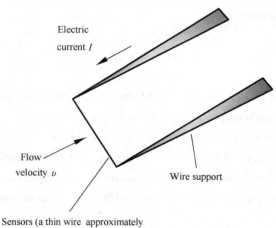

Fig. 8-14 The electrically heated sensor and its support of a hot-wire probe

or platinum-iridium alloys, and it is attached to the probe through holders. The fine wire sensor of a hot-wire anemometer is very fragile because of its small size and can easily break if the liquid or gas contains excessive amounts of contaminants or particulate matter; thus it is not always suitable for studying the fine details of turbulent flow.

Thermal anemometers have extremely small sensors, and thus they can be used to measure the instantaneous velocity at any point in the flow without appreciably disturbing the flow. They can take thousands of velocity measurements per second with excellent spatial and temporal resolution, and thus they can be used to study the details of fluctuations in turbulent flow. They can measure velocities in liquids and gases accurately over a wide range-from a few centimeters to over a hundred meters per second. Thermal anemometers can be used to measure two- or three-dimensional velocity components simultaneously by using probes with two or three sensors respectively.

LASER DOPPLER VELOCIMETRY (LDV)

Laser Doppler velocimetry (LDV), also called *laser velocimetry* (LV) or *laser Doppler anemometry* (LDA), is an optical technique to measure flow velocity at any desired point without disturbing the flow. It can accurately measure velocity at a very small volume, and thus it can also be used to study the details of flow at a locality, including turbulent fluctuations, and it can be traversed through the entire flow field without intrusion.

In the LDA a laser beam provides highly focused, coherent monochromatic light which is passed through the flow (see Fig. 8-15). When this light is scattered from a moving particle in the flow, a stationary observer can detect a change, or *doppler shift*, in the frequency of the scattered light. The shift Δf is proportional to the velocity of the particle.

$$v = \frac{\lambda \Delta f}{2\sin(\theta/2)} \tag{8.39}$$

where λ is the wave length of the laser light, θ is the angle of two beams.

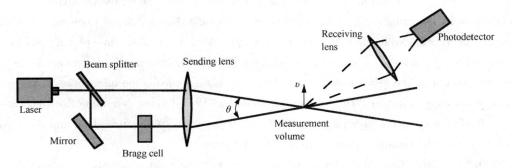

Fig. 8-15 A dual-beam LDV system in forward scatter mode

Multiple components of velocity can be detected by using more than one photodetector and other operating modes. Either liquids or gases can be measured as long as scattering particles are present. In liquids, normal impurities serve as scatters, but gases may have to be seeded by smoke or with particles made of latex, oil, or other materials. By using three laser beam pairs at different wavelengths, the LDV system is also used to obtain all three velocity components at any point in the flow.

The LDA has following advantages: No disturbance of the flow; High spatial resolution of the flow field; Velocity data that are independent of the fluid thermodynamic properties; An output voltage that is linear with velocity; No need for calibration. The disadvantages are that both the apparatus and the fluid must be transparent to light and that the cost is high.

8.7.2 Whole-field Velocity Measurement

There is a whole-field velocity measurement technique: Particle Image Velocimetry.

Particle image velocimetry (PIV) is a double-pulsed laser technique used to measure the instantaneous velocity distribution in a plane of flow by photographically determining the displacement of particles in the plane during a very short time interval. Unlike methods like hot-wire anemometry and LDV that measure velocity at a point, PIV combines the accuracy of LDV with the capability of flow visualization and provides instantaneous flow field mapping throughout an entire cross section, and thus it is a whole-field technique. PIV provides the accurate *quantitative* description of various flow quantities such as the velocity field, and thus the capability to analyze the flow numerically using the velocity data provided. Because of its whole-field capability, PIV is also used to validate computational fluid dynamics (CFD) codes.

The PIV technique has been used since the mid-1980s. The accuracy, flexibility, and versatility of PIV systems with their ability to capture whole-field images with submicrosecond exposure time have made them extremely valuable tools in the study of supersonic flows, explosions, flame propagation, bubble growth and collapse, turbulence, and unsteady flow.

The PIV technique for velocity measurement consists of two main steps: visualization and image processing. The first step is to seed the flow with suitable particles (markers) to trace the fluid motion, slicing a pulse of laser light sheet at the desired plane, record the positions of particles in that plane by detecting the light scattered by particles on a digital video or photographic camera positioned at right angles to the light sheet (Fig. 8-16). After a very short time period, record new positions of the particles by a second pulse of laser light sheet, and finally determine the magnitude and direction of the velocity of the particles in the plane by using the information on these two superimposed camera images. The built-in algorithms of PIV systems determine the velocities at thousands of area elements called *interrogation regions* throughout the entire plane and display the velocity field on the computer monitor in any desired form.

With PIV, other flow properties such as vorticity and strain rates can also be obtained, and the details of turbulence can be studied. Recent advances in PIV technology have made it possible to obtain three-dimensional velocity profiles at a cross section of a flow using two cameras. This is done by recording the images of the target plane simultaneously by both cameras at different angles, processing the information to produce two separate two-dimensional velocity maps, and combining these two maps to generate the instantaneous three-dimensional velocity field.

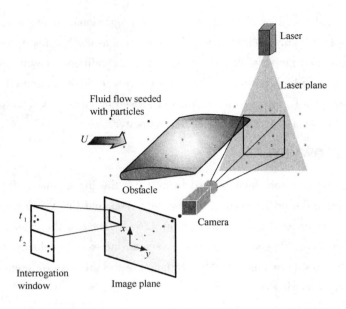

Fig. 8-16 Schematic image showing a typical 2D PIV arrangement

8.7.3 Volume-Flow Measurements

It is often desirable to measure the integrated mass, or volume flow, passing through a duct, especially the accurate measurement of water, gas, oil in billing customers. Generally, there are two types of devices available to make this measurement: mechanical instruments and head-loss instruments.

The *mechanical instruments* measure actual mass or volume of fluid by trapping it and counting it. The main types of measurement are: (a) Mass measurement: weighing tanks, tilting traps; (b) Volume measurement: volume tanks, reciprocating pistons, rotating slotted rings, nutating disk, sliding vanes, gear or lobed impellers, reciprocating bellows and sealed-drum compartments, among which the last three are suitable for gas flow measurement.

The *head-loss devices* obstruct the flow and cause a pressure drop which is a measure of flux. The main types of devices are: (a) *Obstruction flowmeters*, also called the *Bernoulli-type devices*: thin-plate orifice, flow nozzle and Venturi tube; (b) Friction-loss devices: capillary tube and porous plug. The friction-loss meters cause a large non-recoverable head loss and obstruct the flow too much to be generally useful.

Six other widely used meters operating on different physical principles are: turbine meter, vortex meter, ultrasonic flowmeter, rotameter, Coriolis mass flowmeter and laminar flow element.

TURBINE FLOWMETER

Turbine flowmeter consists of a cylindrical flow section that houses a turbine (a vaned rotor) that is free to rotate, additional stationary vanes at the inlet to straighten the flow, and a sensor that generates a pulse each time a marked point on the turbine passes by to determine the rate of rotation. The rotational speed of the turbine is nearly proportional to the flow rate of the fluid. Turbine flowmeters give highly accurate results (as accurate as 0.25 percent) over a wide range of flow rates when calibrated properly for the anticipated flow conditions. Paddlewheel flowmeters are low-cost alternatives to turbine flowmeters for flows where very high accuracy is not required. All these rotating sensors can be attached to counters or sensed by electromagnetic or slip-ring devices for either a continuous or a digital reading of flow velocity.

OBSTRUCTION FLOWMETERS

Three of the most common devices used to measure the instantaneous flowrate in pipes are Bernoulli-type devices: the orifice meter, the nozzle meter, and the Venturi meter. Just like the flow in the Venturi meter as discussed in Section 3.6.5, each of these meters operates on the principle that a decrease in flow area in a pipe causes an increase in velocity that is accompanied by a decrease in pressure (see Fig. 8-17). Correlation of the pressure difference with the velocity provides a means of measuring the flowrate as Eq. 3.65.

$$Q = A_2 \sqrt{\frac{2(p_1 - p_2)}{\rho[1 - (A_2/A_1)^2]}} = A_2 \sqrt{\frac{2(p_1 - p_2)}{\rho(1 - \beta^4)}}$$

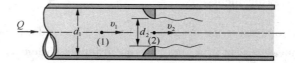

Fig. 8-17 Schematic of typical configuration of obstruction flowmeter

However, the actual measured volume flowrate will be less than the theoretical value because of the energy loss in the viscous flow. Thus, a discharge coefficient, C_d, is used to account for this difference by multiplying the ideal volume flowrate:

$$Q = C_d A_2 \sqrt{\frac{2(p_1 - p_2)}{\rho(1 - \beta^4)}} \tag{8.40}$$

where C_d is called the *discharge coefficient* of the orifice meter, the nozzle meter, or the Venturi meter; A_2 is the throat area, $A_2 = \frac{1}{4}\pi d_2^2$; d_2 is the diameter of the throat; β is the *throat-to-pipe diameter ratio*, $\beta = d_2/d_1$; d_1 is the diameter before the converging.

The typical *orifice meter* is constructed by inserting between two flanges of a pipe a flat plate with a hole, as shown in Fig. 8-18. The contraction after the orifice, the swirling flow and turbulent motion near the orifice plate are accounted by the *orifice discharge coefficient*, C_o. Thus, Eq. 8.40 becomes

$$Q = C_o A_o \sqrt{\frac{2(p_1 - p_2)}{\rho(1 - \beta^4)}} \tag{8.41}$$

where $A_o = \pi d^2/4$ is the area of the hole in the orifice plate. The value of C_o is a function of $\beta = d/D$ and the Reynolds number $Re = \rho v D/\mu$, where $v = Q/A = 4Q/(\pi D^2)$. Typical values of C_o

obtained from experiments are given in Fig. 8-18.

Nozzle meter typically places a contoured nozzle between flanges of pipe sections rather than a simple plate with a hole as in an orifice meter (Fig. 8-19). These constructions can lead to a slighter vena contracta and less severe secondary flow separation, but still with viscous effect. Thus, *nozzle discharge coefficient*, C_n, is used to account these effects

$$Q = C_n A_n \sqrt{\frac{2(p_1 - p_2)}{\rho(1 - \beta^4)}} \qquad (8.42)$$

where $A_n = \pi d^2/4$. The value of C_n is also the function of $\beta = d/D$ and the Reynolds number Re $= \rho v D/\mu$ as in Fig. 8-19. Note that $C_n > C_o$; the nozzle meter is more efficient (less energy dissipated) than the orifice meter.

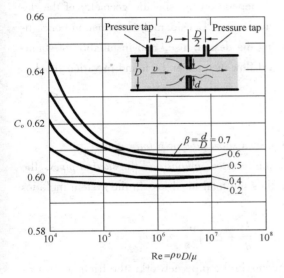 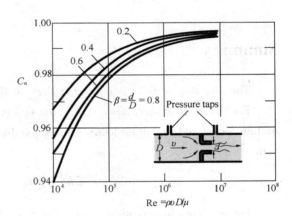

Fig. 8-18 Orifice meter discharge coefficient   **Fig. 8-19** Nozzle meter discharge coefficient

The most precise and most expensive of the three obstruction-type flow meters is the Venturi meter (Fig. 8-20), which has been discussed in Section 3.6.5. Its geometry is designed to reduce head losses to a minimum by providing a relatively streamlined contraction (which eliminates separation ahead of the throat) and a very gradual expansion downstream of the throat (which eliminates separation in this decelerating portion of the device). Most of the head loss that occurs in a well-designed Venturi meter is due to friction losses along the walls rather than losses associated with separated flows and the inefficient mixing motion that accompanies such flow.

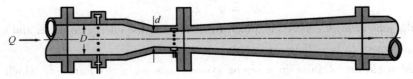

Fig. 8-20 Typical Venturi meter construction

As discussed in Section 3.6.5, the flowrate through a Venturi meter is given by

$$Q = C_v A_T \sqrt{\frac{2(p_1 - p_2)}{\rho(1 - \beta^4)}} \qquad (8.43)$$

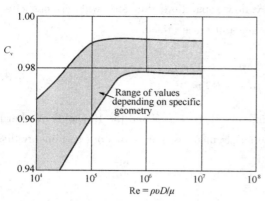

Fig. 8-21 Venturi meter discharge coefficient

where $A_T = \pi d^2/4$ is the throat area. The range of values of C_v, the *Venturi discharge coefficient*, is given in Fig. 8-21. The throat-to-pipe diameter ratio ($\beta = d/D$), the Reynolds number, and the shape of the converging and diverging sections of the meter are among the parameters that affect the value of C_v.

Again, the precise values of C_o, C_n, and C_v depend on the specific geometry of the devices used. Considerable information concerning the design, use, and installation of standard flow meters can be found in various books.

Summary

The *pipe flow* of fluid is a *pressure flow*. It flows in a *short pipe* or a *long pipe*.

For the flows in *short pipes*, i.e., the *siphons*, the *inverted siphons* and the *suction pipes*, the friction losses and the minor losses are comparable, thus the total head loss in the system includes both

$$h_w = \Sigma h_f + \Sigma h_j = \left(\Sigma \lambda \frac{l}{d} + \Sigma \zeta\right)\frac{v^2}{2g}$$

For the flows in *long pipes*, i.e., the pipe elements in the pipe network, the frictional losses dominate, the minor losses and the velocity heads are negligible, thus the total head loss in the system only account the frictional loss

$$h_w = \Sigma h_f = \Sigma SlQ^2$$

in which S is the *pipe specific resistance*, $S = \dfrac{10.3\, n^2}{d^{5.333}}$.

When the pipes are connected in *series* and no flow is taken from the nodes, the flowrate through the entire system remains constant regardless of the diameters of each individual pipes. For the *parallel pipelines*, the continuity equation at nodes should be satisfied and the net head loss in a loop should be zero:

$$\Sigma Q_i = 0 \qquad \Sigma h_{fi} = 0$$

Thus, the *pipe network* can be computed by combining the energy equation and take each individual pipe element to be a long pipe.

The sudden closure of valve in a piping system causes a *water hammer*, which induces the pressure wave propagating forward and backward in the pipe with an additional pressure surplus or decrement of

$$\Delta p = \rho c v_0$$

for *sudden valve closure*, and

$$\Delta p = \rho c v_0 \frac{T_r}{T_c} = \rho v_0 \frac{2l}{T_c}$$

for *gradual valve closure*, in which

$$c = \frac{c_0}{\sqrt{1 + \frac{Kd}{E\delta}}}$$

is the *celerity of pressure wave*.

Water hammer is a common but serious problem in residential plumbing systems, several measures to control the pressure rise are presented in this chapter.

This chapter ends with a brief present of *flowmeters* for the velocity and volume flowrates measurement in pipe, especially *PIV*, *LDV* and the *Bernoulli-obstruction-type meters*. Flowmeters always requires careful experimental calibration.

Word Problems

W8-1 Define short pipe and long pipe. Do they have the practical meaning?

W8-2 Do the discharge coefficients in Eqs. 8.4 and 8.8 equal if the two pipe systems only differ from each other at the pipe exit as shown in Figs. 8-1 and 8-2? Explain.

W8-3 Sketch the energy gradient line if the downstream tank connected by pipeline in Fig. 8-2 is a large reservoir. How does the HGL change if the downstream tank is a small channel?

W8-4 For a siphon illustrated in Fig. W8-4, list the influent factors on the permission mounted height above the water level of upstream tank at point B. Explain.

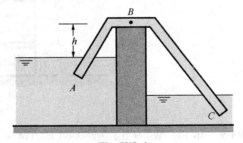

Fig. W8-4

W8-5 Explain how and why a siphon works. Someone proposes siphoning cold water over a 7-m-high wall. Is this feasible? Explain.

W8-6 Water flows steadily out a reservoir through a pipe into the air (Fig. W8-6a) or into a downstream reservoir (Fig. W8-6b). Assume two cases have the identical acting head H, pipe length l, pipe diameter d and friction factor λ. (a) Do the flowrates in pipes equal? Explain. (b) Do the pressures at the corresponding points in the pipes equal? Explain.

W8-7 Water is pumped from a large lower reservoir to a higher reservoir. Someone claims that if the head loss is negligible, the required pump head is equal to the elevation difference between the free surfaces of the two reservoirs. Do you agree?

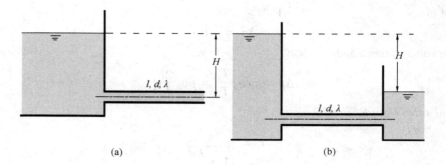

(a)   (b)

Fig. W8-6

W8-8 Two water transportation systems in a concrete gravity dam for the electricity generation are proposed as 1 and 2 shown in Fig. W8-8. Assume the diameters, materials, and roughnesses of pipes, the flowrates through the turbine in two systems are identical, also the frictional losses in the pipeline prior to and after the turbine, h_{f1} and h_{f2}. How would you compare the (a) turbine heads and (b) the pressure drops before and after the turbine in these two pipeline systems?

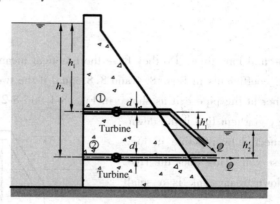

Fig. W8-8

W8-9 In Fig. W8-9 the diversion system of hydropower station has a d-diameter and l-long pipeline with a friction factor of λ and a total minor loss coefficient ζ. Determine (a) the pressure at the turbine inlet section A-A as a function of the water level difference H between upper and lower reservoirs, and (b) the pressure at the inlet of draft tube B-B as a function of its height over the water level in the lower reservoir, z.

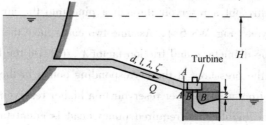

Fig. W8-9

W8-10 How do the HGL and EGL for the short pipe differ from the one for the long pipe?

W8-11 A piping system involves three pipes of different lengths and diameters (but of identical roughness) connected in parallel. Do the friction losses in these pipes equal? Explain.

W8-12 A piping system involves two pipes made of two different materials (but of identical length, diameter and the flowrate) connected in series. The ratio of Manning's roughness coefficient is 0.8. How would you compare the friction losses in these two pipes?

W8-13 A piping system involves three pipes connected in series as shown in Fig. W8-13a, while the system shown in Fig. W8-13b is the inverse connection of these three pipes. Take the acting heads H in these two systems to be equal. How would you compare the flow rates in these two systems if (a) all losses in both systems are neglected, (b) only minor losses are neglected, and (c) all losses are needed to be considered.

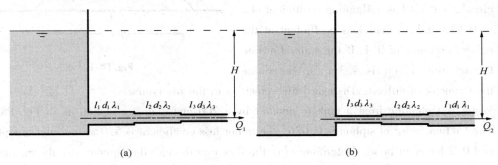

Fig. W8-13

W8-14 In Fig. W8-14 there is a piping system involving two branch pipes A and B. Take the flowrates in pipes A and B to be Q_1 and Q_2 respectively. Explain what would happen to the flowrates Q_1 and Q_2 (a) if a horizontal pipe (dash line shown in the figure) is connected to pipe B, and (b) if pipe B is connected by another pipe (dash line shown in the figure) along its flow direction.

W8-15 A piping system involves two pipes of identical diameter and friction factor, but with different lengths of $l_2 = 3l_1$, connected in parallel. How would you compare the flowrates in these two pipes?

W8-16 The flow rates in pipes illustrated in Fig. W8-16 are Q_0, Q_1, Q_2 respectively when the valve K is fully opened. How do the flow rates change when the valve is partly closed? Assume the other related parameters are unchanged.

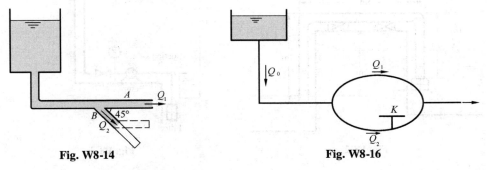

Fig. W8-14   **Fig. W8-16**

W8-17 A piping system involves two pipes of different diameters (but of identical length, material, and roughness) connected in parallel. How would you compare the (a) flow rates and (b) pressure drops in these two pipes?

W8-18 A piping system involves two pipes of identical diameters but of different lengths connected in parallel. How would you compare the pressure drops in these two pipes?

Problems

Simple pipelines

8-1 In Fig. P8-1 the 20-m-long concrete pressure circular culvert has a Manning coefficient of $n = 0.013$ and a minor loss coefficient at the culvert entrance of 0.1. If the water flowrate through the culvert is $4.3 \text{m}^3/\text{s}$, determine the diameter of culvert. Disregard the velocities in the reservoirs.

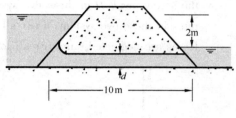

Fig. P8-1

8-2 Water is siphoning from a tank to another by a 200-mm-diameter siphon as in Fig. P8-2. The friction factor of siphon is 0.026, the minor loss coefficient is 5.0 at the inlet of siphon and 0.2 for each bend. Determine: (a) the flow rate through the siphon; (b) the maximum vacuum height in the siphon and its location.

8-3 A enclosed water tank is to be filled from a lower reservoir by a pump with a mounted height of 2m as in Fig. P8-3. The 100-mm-diameter, 6-m-long suction pipe of pump has a friction factor of 0.025, and the minor coefficient for the filter at the entrance of suction pipe be 7.0. The 80-mm-diameter, 60-m-long pressure side pipe of pump has a friction factor of 0.028, and the minor coefficient for the valve be 8.0, for each bend be 0.5 and at the exit of pressure pipe be 1.0. If the gage pressure at the exit section of pump is $p_m = 245.2\text{kPa}$, in the tank is $p_0 = 117.7\text{kPa}$, determine: (a) the pumping flow rate and (b) the pump head h_P.

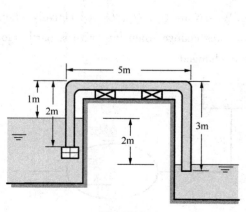

Fig. P8-2

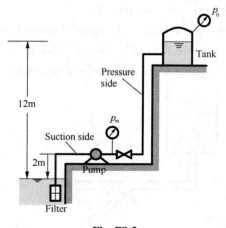

Fig. P8-3

8-4 In Fig. P8-4 water is flowing through a 50-m-long pressure inverted siphon under a highway. The siphon has a friction factor of 0.023, and the minor coefficient at the entrance be 0.5, for each bend be 0.15 and at exit be 1.0. Determine the diameter of the inverted siphon if its required transporting capacity is $3 m^3/s$. Disregard the flow velocities in the rivers.

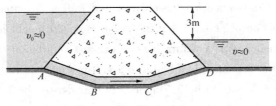

Fig. P8-4

8-5 In Fig. P8-5 water is discharged into air from a reservoir through a 1-m-diameter pipeline consisted of a 50-m-long tunnel and a 200-m-long pipe. The entrance of tunnel and the exit of pipe are lower than the water level in the reservoir by 15m and 49m respectively. Both the tunnel and pipe have an identical friction factor of 0.02. The minor coefficients at both entrance and bend are 0.5. Determine: (a) the flowrate in the pipe; (b) the minimum pressure in pipeline and its location.

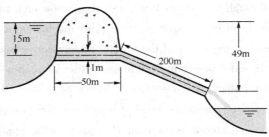

Fig. P8-5

8-6 In Fig. P8-6 water flows gravitationally from pool to the suction well of pump through a 50-

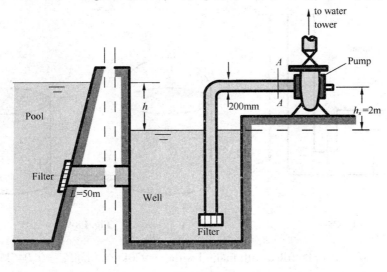

Fig. P8-6

351

m-long simple short pipe, then is pumped to the water tower by a pump, which has a 200-mm-diameter 6-m-long suction pipe. The minor loss coefficients for both the filters are 6.0, for the bend is 0.3. Both the short pipe and the pump suction pipe have a friction factor of 0.02. When steadily pumped flowrate is $0.064 \text{m}^3/\text{s}$, determine: (a) the diameter of short pipe if the elevation difference h between pool and well is less than 2m; (b) the absolute pressure at inlet section of pump A-A if the pump mounted height is $h_s = 2\text{m}$.

8-7 In Fig. P8-7 is a boiler with a natural smoke exhaust ventilation system, which evacuates smoke from the outlet section 1-1 of boiler to a 1.0-m-diameter vertical chimney then to the air. Take the smoke density to be 0.7kg/m^3, the air density be 1.209kg/m^3, and the minor loss through the stove be $4.8v^2/2g$, where v is the velocity in the chimney. The sectional area at section 1-1 equals the sectional area of chimney. When the flowrate through the stove is $8.33\text{m}^3/\text{s}$, the vacuum at section 1-1 is $10\text{mmH}_2\text{O}$, determine the required height of chimney, H.

8-8 A water tank filled with solar-heated water at 40°C is to be used for showers in a field using gravity-driven flow. The system includes 20m of 1.5-cm-diameter galvanized iron piping with four miter bends (90°) without vanes and a wide-open globe valve. If water is to flow at a rate of 0.7L/s through the shower head, determine how high the water level in the tank must be from the exit level of the shower. Disregard the losses at the entrance and at the shower head, and neglect the effect of the kinetic energy correction factor.

8-9 Two water reservoirs A and B are connected to each other through a 40-m-long, 2-cm-diameter cast iron pipe with a sharp-edged entrance. The pipe also involves a swing check valve and a fully open gate valve. The water level in both reservoirs is the same, but reservoir A is pressurized by compressed air while reservoir B is open to the atmosphere at 88kPa. If the initial flow rate through the pipe is 1.2L/s, determine the absolute air pressure on top of reservoir A. Take the water temperature to be 10°C.

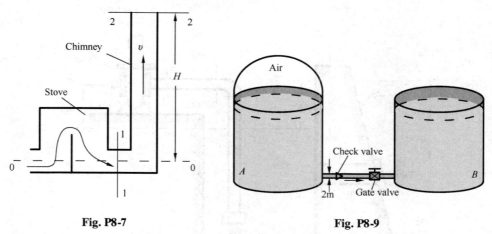

Fig. P8-7　　　　　　　　　　Fig. P8-9

8-10 A vented tanker is to be filled with fuel oil with $\rho = 920\text{kg/m}^3$ and $\mu = 0.045\text{kg/m·s}$ from an underground reservoir using a 20-m-long, 5-cm-diameter plastic hose with a slightly rounded entrance and two 90° smooth bends. The elevation difference between the oil level in the reservoir and the top of the tanker where the hose is discharged is 5 m. The capacity of the tanker is 18m³ and the filling time is 30min. Taking the kinetic energy correction factor at hose discharge to be 1.05 and assuming an overall pump efficiency of 82 percent, determine the required power input to the pump.

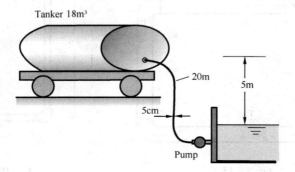

Fig. P8-10

8-11 In Fig. P8-11 a 125-m-long, 5-cm-diameter cast iron pipe with sharp-edged entrance transports water from a small reservoir at a head of 40m to a small turbine for generation. If the flow rate is 0.004m³/s, what power is extracted by the turbine?

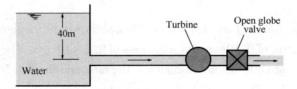

Fig. P8-11

8-12 In Fig. P8-12 the turbine is fed by a 0.3-m-diameter, 120-m-long cast-iron pipe and develops 400kW electricity. Determine the flowrate if (a) head losses are negligible or (b) head loss due to friction in the pipe is considered. Assume the cast-iron pipe has a frictional factor of 0.02. *Note*: There may be more than one solution or there may be no solution to this problem.

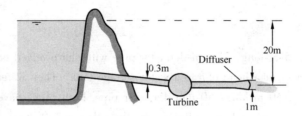

Fig. P8-12

8-13 A fan is to produce a constant air speed of 40m/s throughout the pipe loop in a building shown in Fig. P8-13. The 3-m-diameter pipes are smooth, and each of the four 90° elbows has a loss coefficient of 0.30. Determine the power that the fan adds to the air.

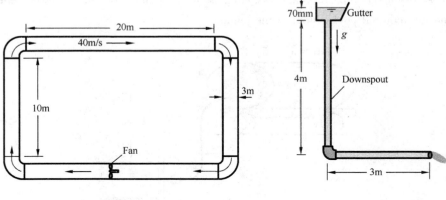

Fig. P8-13   Fig. P8-14

8-14 Rainwater flows through the galvanized iron downspout shown in Fig. P8-14 at a rate of $0.006 m^3/s$. Neglect the velocity of the water in the gutter at the free surface and the head loss associated with the elbow. Determine the size of the downspout cross section if it is (a) a rectangle with an aspect ratio of 1.7 to 1 and (b) a circular, which is completely filled with water.

8-15 Water flows from a large tank that sits on frictionless wheels as shown in Fig. P8-15. The pipe has a diameter of 0.50m and a roughness of 9.2×10^{-5}m. The loss coefficient for the filter is 8; other minor losses are negligible. The tank and the first 50-m section of the pipe are bolted to the last 75-m section of the pipe which is clamped firmly to the floor. Determine the tension in the bolts.

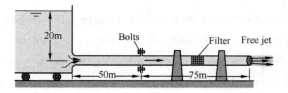

Fig. P8-15

Complicated pipelines

8-16 In Fig. P8-16 a 25-m-long, 75-mm-diameter pipe with sharp-edged entrance transports water from a reservoir at a head of h to a water storage tank. Then water is transported to the users by another 150-m-long, 50-mm-diameter pipe. Both pipes have an identical friction factor of 0.03. The minor loss coefficient for the valve is 3.0. (a) Determine the flowrate in the pipes and the level difference h of free surfaces between reservoir and tank; (b) Sketch the energy gradient line and the hydraulic gradient line.

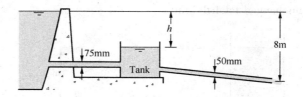

Fig. P8-16

8-17 In Fig. P8-17 there are 37.5-m of 5-cm pipe, 22.5-m of 15-cm pipe, and 45-m of 7.5-cm pipe, all cast iron. There are three 90° elbows (ζ_b = 0.95) and an open globe valve (ζ_v = 6.3), all flanged. The friction factors in all pipes are identical of 0.0287. The minor coefficient for the sudden expansion is ζ_{ex} = 0.79. If the exit elevation is zero, what horsepower is extracted by the turbine when the flow rate is 4.53L/s of water at 20℃?

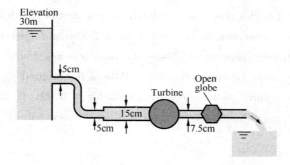

Fig. P8-17

8-18 In Fig. P8-18 two tanks are connected by serial pipes with an identical friction factor of 0.02. The minor loss coefficient for the sudden contraction between those two pipes is 0.6, which is related to the velocity in the pipe with smaller diameter. Consider all the losses in the system, determine the flow rate in the pipe and quantitatively plot EGL and HGL along the pipeline.

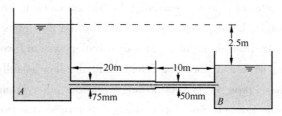

Fig. P8-18

8-19 In Fig. P8-19 two tanks are connected by two parallel cast-iron pipes of identical length (l_1 = l_2 = 30m), which are assembled at the same elevation. The diameter of one pipe is 50mm and the another is 100mm. Both pipes have an identical friction factor of 0.02, and identical total minor loss coefficient of $\Sigma\zeta$ = 0.5. Determine: (a) the flowrate in each pipe; (b) the pipe diameter if those two parallel pipes are replaced by only one pipe with identical length and unchanged total flowrate.

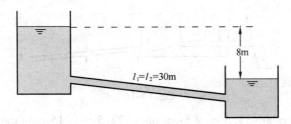

Fig. P8-19

8-20 One galvanized pipe and one cast iron pipe are connected in parallel. The galvanized pipe has a diameter of 150mm, a length of 1500m and the Manning's coefficient of 0.011. The cast iron pipe has a diameter of 200mm and the Manning's coefficient of 0.013. If the flowrates in these parallel pipes are identical, determine the length of the cast iron pipe.

8-21 In Fig. P8-21 two reservoirs is connected by a 300-mm-diameter, 3000-m-long pipe. Then, a new 300-mm-diameter, 1500-m-long pipe was assembled parallel to the second half of the original pipe by connecting the middle of the original pipe and the downstream reservoir. All pipes have an identical friction factor of 0.02. If all the minor losses are negligible, determine the flow rate increment when the new parallel pipe is added.

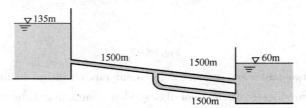

Fig. P8-21

8-22 Two pipes of identical length and material are connected in parallel. The diameter of pipe A is twice the diameter of pipe B. Assuming the friction factor to be the same in both cases and disregarding minor losses, determine the ratio of the flow rates in the two pipes.

8-23 A certain part of cast iron piping of a water distribution system involves a parallel section. Both parallel pipes have a diameter of 30cm, and the flow is fully developed turbulent. One of the branches (pipe A) is 1000m long while the other branch (pipe B) is 3000m long. If the flow rate through pipe A is $0.4 \text{m}^3/\text{s}$, determine the flow rate through pipe B. Disregard minor losses and assume the water temperature to be 15℃.

8-24 Repeat Prob. 8-23 assuming pipe A has a halfway closed gate valve ($\zeta = 2.1$) while pipe B has a fully open globe valve ($\zeta = 10$), and the other minor losses are negligible. Assume the flow to be fully turbulent.

8-25 In large buildings, hot water in a water tank is circulated through a loop so that the user doesn't have to wait for all the water in long piping to drain before hot water starts coming out. A certain recirculating loop involves 40-m-long, 1.2-cm-diameter cast iron pipes with six 90° threaded smooth bends and two fully open gate valves. If the average flow velocity through the loop is 2.5 m/s, determine the required power input for the recirculating pump. Take the average water temperature to be 60°C and the efficiency of the pump to be 70 percent.

8-26 A pipeline that transports oil at 40°C at a rate of $3 \text{m}^3/\text{s}$ branches out into two parallel pipes made of commercial steel that reconnect downstream as shown in Fig. P8-26. Pipe A is 500m long and has a diameter of 30cm while pipe B is 800m long and has a diameter of 45cm. The minor losses are considered to be negligible. Determine the flow rate through each of the parallel pipes.

8-27 The parallel galvanized-iron pipe system in Fig. P8-27 delivers gasoline at 20°C with a total flowrate of $0.036 \text{m}^3/\text{s}$. If the pump is wide open and not running with a minor loss coefficient of 1.5, determine (a) the flow rate in each pipe and (b) the overall pressure drop.

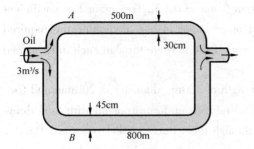

Fig. P8-26

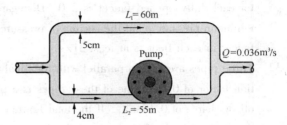

Fig. P8-27

8-28 For the series-parallel system in Fig. P8-28, all pipes are 8-cm-diameter asphalted cast iron. If the total pressure drop $p_1 - p_2 = 750\text{kPa}$, find the resulting flowrate $Q \text{m}^3/\text{hr}$ for water at 20°C. Neglect minor losses.

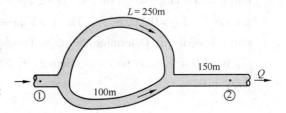

Fig. P8-28

8-29 Three cast-iron pipes are laid in parallel with these dimensions: $l_1 = 800\text{m}$, $l_2 = 600\text{m}$, $l_3 = 900\text{m}$; $d_1 = 12\text{cm}$, $d_2 = 8\text{cm}$, $d_3 = 10\text{cm}$. The total flowrate is $200 \text{m}^3/\text{hr}$ of water at 20°C. Determine (a) the flow rate in each pipe and (b) the pressure drop across the system.

8-30 In Fig. P8-30 water is supplied from a water tower to users by a serial-parallel piping system with these dimensions: $d_1 = d_4 = 200\text{mm}$, $d_2 = d_3 = 150\text{mm}$, $l_1 = l_4 = 100\text{m}$, $l_2 = 50\text{m}$, $l_3 = 200\text{m}$. The friction factors for all pipes are identical of 0.02. The flowrate out of tank is $Q = 0.1 \text{m}^3/\text{s}$. Disregard the minor losses. Determine (a) the flowrates in the parallel pipes, Q_2 and Q_3, and (b) the elevation of free surface in water tower, H.

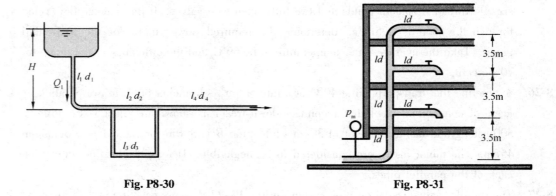

Fig. P8-30 **Fig. P8-31**

8-31 In Fig. P8-31 shows a water distribution system for a three-floor building with a pressure gage located at the main pipe. For each floor both the vertical pipe and horizontal branch pipe has a diameter of $d = 60$mm and a length of $l = 4$m. The elevation difference between faucets is 3.5m. All branch pipes have an identical friction factor of 0.03. The minor loss coefficient for each fully opened faucet is 3.0. Disregard other minor losses, determine the required minimum pressure p_m at the seciton of pressure gage if the flowrate through each fully opened faucet in each floor is at least 3L/s.

8-32 Three pipes are laid in parallel with identical length of 300m, diameter of 200mm and friction factor of 0.02. One of these pipes is a pipeline with uniform draw-off with total draw-off discharge of $0.15\text{m}^3/\text{s}$. If the total flowrate through these three parallel pipes is $1.0\text{m}^3/\text{s}$, determine (a) the flowrate in each pipe and (b) the head loss in the parallel pipe.

8-33 In Fig. P8-33 water is flowing through a 200-mm-diameter, 1000-m-long truck pipe with an end flowrate of $Q_z = 0.04\text{m}^3/\text{s}$. Along the length of the truck, water is taken every 50m at a flowrate of $q = 0.002\text{m}^3/\text{s}$. The truck pipe has a friction factor of 0.025. Disregard all minor losses. (a) Determine the major loss h_f in the truck; (b) Determine the required acting head H if there is no water taken every 50m; (c) Determine the required acting head H if $Q_z = 0$.

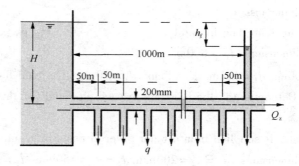

Fig. P8-33

8-34 In Fig. P8-34 water is supplied from a water tower to users by three serial pipes. The friction factors for all pipes are 0.03. The flowrate out of tank is $0.04\text{m}^3/\text{s}$. The second pipe with l_2-length is a pipe with uniform draw-off, which discharges half of the total flowrate. Disregard all the minor losses. Determine the required acting head over the axis of horizontal pipes, H.

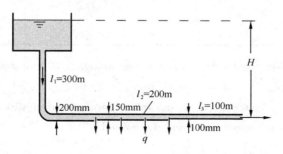

Fig. P8-34

Pipe Network

8-35 In Fig. P8-35 all pipes are 8-cm-diameter cast iron. Determine the flowrate out of the reservoir (1) if the valve at pipe C is (a) closed and (b) open with a minor coefficient of 0.5.

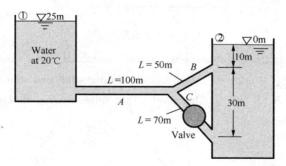

Fig. P8-35

8-36 In Fig. P8-36 water in reservoirs A and B is discharged into air at point C. The parameters of the connected pipes are given in the following Table. Determine the flow rates in each pipe.

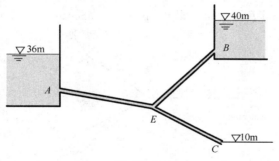

Fig. P8-36

No. of pipes	Pipe length (m)	Pipe diameter (mm)	Friction factor of pipe λ
AE	300	300	0.024
BE	450	375	0.020
CE	60	450	0.024

8-37 The water of a factory is fed by a water tower through a 75-mm-diameter, 140-m-long uncoated cast iron pipe at a flow rate of $36 \text{m}^3/\text{hr}$ as in Fig. P8-37. If the required service head in the factory should be at least 12m, determine the elevation difference H between the water level in the water tower and the ground.

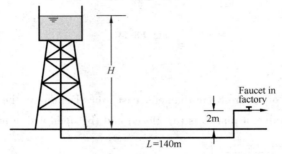

Fig. P 8-37

8-38 In Fig. P8-38 a branched water supply system assembled at the same elevation in a factory is fed by the pump at point A. All the pipes are uncoated cast iron pipe. Water at 20°C is taken from nodes A, B and C. Disregard all the minor losses, determine the required head at the exit of pump.

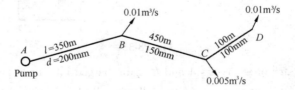

Fig. P8-38

8-39 In Fig. P8-39 the branched water supply network has a ground level at the water tower of 15m, at the pipe end node C of 20m and end node D of 15m. The required service heads at both nodes C and D are 5m. All the pipes are uncoated cast iron pipes. Determine (a)

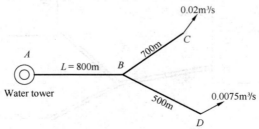

Fig. P8-39

the acting head at the water tower, and (b) the diameters of pipes AB, BC and BD.

8-40 In the five-pipe horizontal network of Fig. P8-40, assume that all pipes have a friction factor of 0.025. For the given inlet and exit flow rate of $0.0566 \text{m}^3/\text{s}$ of water at 20°C, determine the flow rate and direction in all pipes. If $p_A = 587 \text{kPa}$ gage, determine the pressures at points B, C, and D.

8-41 In Fig. P8-41 there is a rectangular looped network ABCD and a triangle looped network ADE, which are connected by pipe AD. The inflow rates into node E and all outflow rates at nodes B, C and D are in m^3/s, the pipe discharge modulus $K(=Sl)$ are in s^2/m^5. Determine the flowrates in all pipes and draw the flow directions of all pipes in the figure.
Hints: $h_f = SlQ^2 = KQ^2$.

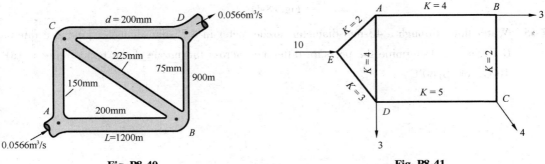

Fig. P8-40   Fig. P8-41

Flowmeters

8-42 In Fig. P8-42 a Pitot-static probe is arranged to measure the air flow. Its manometer fluid is water at 20°C. Estimate (a) the centerline velocity, (b) the pipe volume flow, and (c) the (smooth) wall shear stress.

8-43 An engineer who took college fluid mechanics on a passfail basis has placed the static pressure hole far upstream of the stagnation probe, as in Fig. P8-43, thus contaminating the pitot measurement ridiculously with pipe friction loss. If the pipe flow is air at 20°C and 1atm and the manometer fluid is Meriam red oil ($S_G = 0.827$), estimate the air centerline velocity for the given manometer reading of 16cm. Assume a smooth-walled tube.

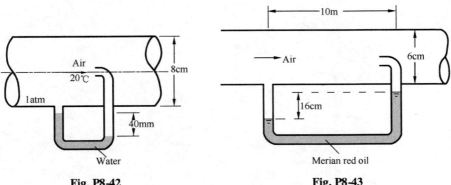

Fig. P8-42   Fig. P8-43

8-44 Water at 20°C is flowing through the orifice in Fig. P8-44, which is monitored by a mercury manometer. Determine (a) the manometer reading h if the flowrate is $20 m^3/hr$, and (b) the flowrate in the pipe in m^3/hr if the reading is $h = 58$ cm.

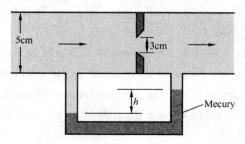

Fig. P8-44

8-45 Water flows through a 40-mm-diameter nozzle meter in a 75-mm-diameter pipe at a rate of $0.015 m^3/s$. Determine the pressure difference across the nozzle if the temperature is (a) 10°C, or (b) 80°C.

9 Open Channel Flow

CHAPTER OPENING PHOTO: Tidal bore with a single breaking wavefront with a roller at the Qiantang river, Hangzhou, China on August 7, 2009. Qiantang Bore is known as the world's largest tidal bore up to 9 meters high and traveling at up to 40 kilometers per hour, and is a rough turbulent flow in a natural open channel. The present chapter discusses the theory about the open channel flow including the wave speed of the tidal bore. (*Courtesy of Wikipedia*)

Open-channel flow involves the flow of a liquid (usually water) in a channel or conduit that is not completely filled and open to the gas (usually air, which is at atmospheric pressure). Open channel flow has a free surface. The main driving force for such flows is the fluid weight: gravity forces the fluid to flow downhill, and achieves a dynamic balance with friction. For the steady and fully developed flow in open channel, the pressure in the vertical direction varies hydrostatically.

Practical open-channel problems are generally turbulent, due to its large scale and small kinematic viscosity, and are three-dimensional, sometimes unsteady, and often surprisingly complex due to geometric effects. In this chapter we mainly present the basic principles of uniform and nonuniform open-channel flows and the associated correlations for steady flow in straight channels of common cross sections, and a brief discussion of the theories of a rapidly varied flow, hydraulic jump, and the determination of the water surface profile of gradually varied flow.

9.1 Introduction

9.1.1 Channel Geometry

An *open channel* is a type of landform consisting of the outline of a path of relatively shallow and narrow body of fluid opened to the atmosphere. Open channels can be either natural or artificial. *Natural channels*, such as streams, rivers, valleys, estuaries, floodplains, etc. are generally irregular in shape, alignment and roughness of the surface. *Artificial channels* are built for some specific purpose, such as flumes, spillways, canals, weirs, drainage ditches, uncovered culverts, etc., in the irrigation, water supply, wastewater, water power development, and rain collection systems. These are generally regular in shape and alignment with uniform roughness of the boundary surface.

An open channel can be prismatic or non-prismatic based on the geometries of the cross sections. *Prismatic channel* is a long straight runs of channel with same geometry shape and constant sizes of cross section along the flow direction. When either the shape or the sizes of the cross section of channel (or both) change, the channel is referred to as *non-prismatic channel*. It is obvious that only artificial channel can be prismatic.

The most common shapes of prismatic channels are rectangular, triangular, trapezoidal and circular as shown in Fig. 9-1.

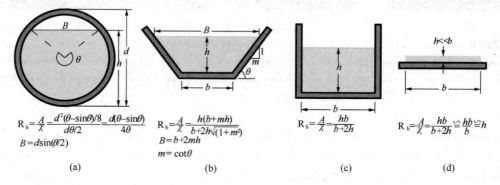

Fig. 9-1 Most common shapes of prismatic channel: (a) Circular channel (θ in rad); (b) Trapezoidal channel; (c) Rectangular channel; (d) Liquid film of thickness h

Flow in open channels is gravity driven, and thus a typical channel is slightly sloped down as in Fig. 9-2. The *bed slope*, or the *bottom slope* of the channel is defined as

$$i = \frac{\Delta z}{l} = \sin\theta \tag{9.1}$$

where θ is the angle the channel bottom making with the horizontal line; Δz is the elevation drop between two sections and l is the distance between those two sections as in Fig. 9-2a.

In general, the bed slope of channel i is very small, and thus the channel bed is nearly horizontal. Therefore, $l \approx l_x$, where l_x is the horizontal distance between two sections as indicated in

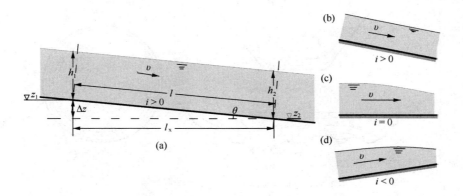

Fig. 9-2 (a) Definition of channel bed slope; (b) downhill slope; (c) horizontal slope and (d) adverse slope

Fig. 9-2a. Also, the flow depth, which is measured normal to the channel bed, can be taken to be the depth in the vertical direction, h, with negligible error. Thus, Eq. 9.1 gives

$$i = \frac{z_1 - z_2}{l_x} = \frac{\Delta z}{l_x} = \tan\theta \qquad (9.2)$$

Based on the elevation change of channel bed in the flow direction (see Fig. 9-2), channel bed slope could be a *downhill slope*, or *positive slope* ($i > 0$) if the channel bed elevation falls in the flow direction, that is $\Delta z = z_1 - z_2 > 0$; the *adverse slope*, or *negative slope* ($i < 0$) if the channel bed elevation rises, $\Delta z < 0$; or the *horizontal slope* ($i = 0$) if channel bed elevation does not change in the flow direction.

9.1.2 Classification of Open Channel Flow

Open-channel flow is the flow of a liquid in a channel with a free surface opened to the air, which coincides with the hydraulic grade line (HGL). The pressure at the free surface is a constant: atmospheric pressure. But the height of free surface from the channel bottom and thus all dimensions of the flow cross-section along the channel is not known a priori: it changes along with average flow velocity. Open channel flow is a *gravity flow* because the gravity is the power of flow. The channel bed slope affects the depth and flow velocity. The sudden variation of the boundary can also affect the flow statues in channel.

In an open channel, the flow velocity is zero at the side and bottom surfaces because of the *no-slip condition*, and maximum occurring in the midplane about 20 percent below the free surface as shown in Fig. 9-3. In very broad shallow channels the maximum velocity is near the surface, and the velocity profile is nearly logarithmic from the bottom to the free surface. Furthermore, flow velocity also varies in the flow direction in most cases. Therefore, the velocity distribution (and thus flow) in open channels is, in general, three-dimensional. However, the one-dimensional equations in terms of the average velocity at a cross section of the channel still can provide remarkably accurate results. Therefore, the one-dimensional-flow approximation is commonly used in practical engineering because of the simplicity.

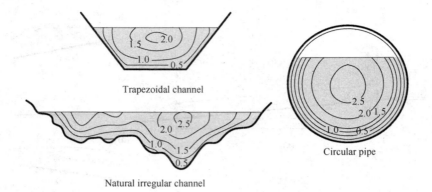

Fig. 9-3 Measured isovelocity contours in typical straight open channel flows. (From Ven Te Chow, 1959)

Open-channel flow can be classified and described in various ways based on the change in flow depth with respect to time and space.

BY TEMPORAL VARIATION

The representative quantity in open-channel flows is the flow depth (or alternately, the average velocity), which may vary along the channel. The flow is said to be *steady* if the flow depth does not vary with time at any given location along the channel (although it may vary from one location to another), or if it can be assumed to be constant during the time interval under consideration. Otherwise, the flow is *unsteady*.

BY REYNOLDS NUMBER

The flow in open channel can be either laminar or turbulent depend on the value of Reynolds number. Considering that open channels come with rather irregular cross sections, the *hydraulic radius*, $R_h (= A/\chi)$, serves as the characteristic dimension and brings uniformity to the treatment of open channels. Thus, the Reynolds number in channel defined as

$$\text{Re} = \frac{\rho v R_h}{\mu} = \frac{v R_h}{\nu} \qquad (9.3)$$

The open-channel flow comes to be *laminar* for $\text{Re} \lesssim 475$, *turbulent* for $\text{Re} \gtrsim 2500$ and *transitional* for $475 \lesssim \text{Re} \lesssim 2500$. In practice, however, the laminar flow occurs very rarely. It can be encountered when a thin layer of water (such as the rainwater draining off a road or parking lot) flows at a low velocity. The engineer is concerned mainly with turbulent flow.

Note that the wetted perimeter χ in hydraulic radius includes the sides and the bottom of the channel in contact with the liquid—it does not include the free surface and, of course, not the parts of the sides above the water level. For example, if a rectangular channel is b wide and H high and contains water to depth h, its wetted perimeter is $\chi = b + 2h$, not $b + 2H$. For the most common shapes, the width of free surface, the area and the wetted perimeter are shown in Fig. 9-1.

Also, the Reynolds number is constant for the entire uniform flow section of an open channel.

BY DEPTH VARIATION

Flow in open channels is also classified as being uniform or nonuniform (also called *varied*), depending on how the flow depth h (the distance of free surface from the bottom of the channel measured in the *vertical* direction) varies along the channel. The flow in a channel is said to be *uniform* if the flow depth (and thus the average velocity) remains constant at every section of the channel ($dh/ds = 0$). Uniform flow (UF) can be steady or unsteady, depending on whether or not the depth changes with time (although unsteady uniform flow is rare). A channel in uniform flow is said to be moving at its *normal depth*, h_0, which is an important characteristic parameter and is a function of flow rate, channel geometry and bottom slope.

If the channel slope or cross section changes or there is an obstruction in the flow, such as a gate, then the flow depth varies, thus the flow is said to be *nonuniform* or *varied*, indicating that the flow depth varies with distance in the flow direction ($dh/ds \neq 0$). Such varied flows are common in both natural and human-made open channels such as rivers, irrigation systems, and sewer lines. The varied flow is called *rapidly varied flow* (RVF) ($dh/ds \sim 1$) if the flow depth changes markedly over a relatively short distance in the flow direction (such as the flow of water past a partially open gate or over a falls), and *gradually varied flow* (GVF) if the flow depth changes gradually over a long distance along the channel ($dh/ds \ll 1$). A gradually varied flow region typically occurs between rapidly varied and uniform flow regions, as shown in Fig. 9-4. The relative importance of the various types of forces involved (pressure, weight, shear, inertia) is different for the different types of flows.

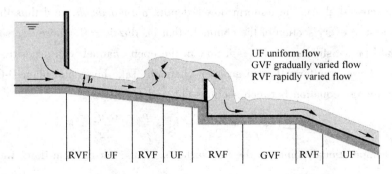

Fig. 9-4 Open-channel flow classified by regions of RVF, GVF, and UF depth profiles.

Thus, the flow classified by the depth variation could be:
(1) Uniform flow (constant depth and slope)
(2) Varied flow, or nonuniform flow
   a. Gradually varied (one-dimensional)
   b. Rapidly varied (multi-dimensional)

Typically uniform flow is separated from rapidly varied flow by a region of gradually varied flow. Gradually varied flow can be worked with the one-dimensional average velocity. For a known discharge rate, the profile of the free surface in a gradually varied flow region in a specified open channel can be determined in a step-by-step manner by starting the analysis at a cross section where the flow conditions are known, and evaluating head loss, elevation drop, and then the average velocity for each step, which will be discussed detailed in Section 9.7. But rapidly varied flow usually requires experimentation or three dimensional potential theory.

BY FROUDE NUMBER

Open-channel flow is also classified as *tranquil*, *critical*, or *rapid*, depending on the value of the dimensionless Froude number discussed in Chapter 5 and defined as

$$F_r = \frac{v}{\sqrt{gh}} \tag{9.4}$$

where h is the flow depth of liquid in channel, v is the average liquid velocity at the cross section.

Thus, the channel flow can be classified as

$F_r < 1.0$ *Tranquil flow*, or *subcritical flow*
$F_r = 1.0$ *Critical flow*
$F_r > 1.0$ *Rapid flow*, or *supercritical flow*

It will be discussed in detail in Section 9.4.

9.2 Uniform Flow in Open Channel

9.2.1 Conditions

As what discussed above, the uniform flow's depth, *normal depth*, and thus the average velocity, are constant at every section of the channel, that is, $dh/ds = 0$. However, some restriction conditions should be satisfied even if it is a flow in the open channel with a positive bed slope.

Figure 9-5 shows a uniform flow in an open channel. Take horizontal plane 0-0 as the reference level, the energy equation between sections (1) and (2) gives

$$(h_1 + \Delta z) + \frac{p_1}{\rho g} + \frac{\alpha_1 v_1^2}{2g} = h_2 + \frac{p_2}{\rho g} + \frac{\alpha_2 v_2^2}{2g} + h_w$$

Take the computational points on the free surfaces. Thus, for uniform flow, we have $p_1 = p_2$

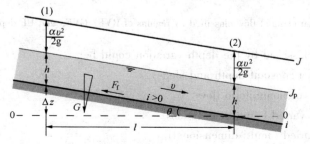

Fig. 9-5 Uniform flow in open channel

$= p_{at}$, $h_1 = h_2 = h$, $v_1 = v_2 = v$, $\alpha_1 = \alpha_2$ and $h_w = h_f$. Thus the energy equation becomes
$$\Delta z = h_f \tag{9.5}$$

Divided by the length l yields
$$i = J \tag{9.6}$$
where l is the horizontal distance between sections 1 and 2; $J = h_f/l$ is the *friction slope* as defined in Section 4.3.3.

Equation 9.6 shows that for the uniform flow, the kinematic energy remains constant along the channel if the drop of the bed elevation equals to the frictional loss due to the frictional effects. Thus, we have the following characteristics of uniform flow:

1. The depth and velocity of uniform flow remains constant along the channel.

2. The energy gradient line (EGL), hydraulic gradient line (HGL) (coinciding with free surface) and the channel bed are parallel as shown in Fig. 9-5, that is, the friction slope, the pizeometric slope and the bed slope equals:
$$J = J_P = i \tag{9.7}$$

It is because the HGL coincides with free surface, which is parallel with the channel bed because of the constant depth of uniform flow, $J_P = i$. Meanwhile, the EGL is parallel to the HGL because of the constant velocity of uniform flow, $J = J_P$. Thus, we have Eq. 9.7.

3. The pressure distribution over cross section is same as the hydrostatic pressure distribution
$$z + \frac{p}{\rho g} = C$$
and varies hydrostatically in vertical.

4. The component of gravity force in flow direction is balanced by the frictional force due to the constant average velocity along the channel, as shown in Fig. 9-5:
$$F_f = G\sin\theta \tag{9.8}$$

Thus we can conclude the *restriction conditions* for forming the uniform flow in an open channel:

- Downhill slope of channel to make $i = J$ possible;
- Long straight runs of prismatic channel to get the constant velocity;
- Constant surface roughness to make the flowrate be constant along the channel;
- No flowrate taken from the channel to make sure the flowrate be constant along the channel;
- No obstructions in the flow.

9.2.2 The Chézy Formula for Flow Rate

For uniform flow, Eq. 9.6 presents that all the elevation potential energy is consumed by the friction, that is
$$h_f = z_1 - z_2 = il \tag{9.9}$$

The head loss thus balances the loss in height of the channel. The flow is essentially fully developed, so that the Darcy-Weisbach relation, Eq. 6.3, holds

$$h_f = \lambda \frac{l}{4R_h} \frac{v^2}{2g} \tag{9.10}$$

with $R_h = A/\chi$, as defined in Eq. 6.4, is used to accommodate noncircular channels.

Combining Eqs. 9.9 and 9.10 gives the flow velocity in uniform channel flow

$$v = \sqrt{\frac{8g}{\lambda}} \sqrt{R_h i} \tag{9.11}$$

For a given channel shape and bottom roughness, the quantity $\sqrt{8g/\lambda}$ is a constant and can be denoted by C. For uniform flow, the friction slope is equal to the bed slope, $J = i$ as indicated in Eq. 9.7. Thus, Eq. 9.11 becomes

$$v = C\sqrt{R_h i} = C\sqrt{R_h J} \tag{9.12}$$

$$Q = AC\sqrt{R_h i} = K\sqrt{i} = K\sqrt{J} \tag{9.13}$$

where K is the *modulus of discharge*, $K = AC\sqrt{R_h}$, m³/s; C is the *Chézy coefficient*, varies from about $30\sqrt{m}/s$ for small rough channels to $90\sqrt{m}/s$ for large smooth channels and can be determined by *Manning formula* Eq. 6.68

$$C = \frac{1}{n} R_h^{1/6} \tag{9.14}$$

where n is *Manning coefficient* for a given surface condition for the walls and bottom of the channel, the experimental values of n are listed in Table 6-4 for various channel surfaces.

Equations 9.12 and 9.13 are called the *Chézy formulas* (also see Section 6.6.2), first developed by the French engineer Antoine Chézy in conjunction with his experiments on the Seine River and the Courpalet Canal in 1769.

By considering the Manning formula we can have the *Chézy formulas* for uniform-flow velocity and volume flow rate as thus

$$v = \frac{1}{n} R_h^{2/3} i^{1/2} \tag{9.15}$$

$$Q = \frac{1}{n} A R_h^{2/3} i^{1/2} \tag{9.16}$$

In many man-made channels and in most natural channels, the surface roughness (and hence the Manning coefficient) varies along the wetted perimeter of the channel. A drainage ditch, for example, may have a rocky bottom surface with concrete side walls to prevent erosion. Thus, the effective n will be different for shallow depths than for deep depths of flow. Similarly, a river channel may have one value of n appropriate for its normal channel and another very different value of n during its flood stage when a portion of the flow occurs across fields or through floodplain woods. To determine the flowrate in those channels, a reasonable approximation should be used. It begins to divide the channel cross section into N subsections, each with its own wetted perimeter χ_i, area A_i, and Manning coefficient n_i. Then the total flowrate in channel is assumed to be the sum of the flowrates through each section

$$Q = \Sigma Q_i = \Sigma \frac{1}{n_i} A_i R_{hi}^{2/3} i_i^{1/2} \tag{9.17}$$

It should be noted that the values of χ_i do not include the imaginary boundaries between the different subsections.

EXAMPLE 9-1 Uniform flow

Water is flowing in a weedy excavated earth channel of trapezoidal cross section with a bottom width of 0.8m, trapezoid angle of 60°, and a bottom slope angle of 0.3°, as shown in Fig. E9.1. If the flow depth is measured to be 0.52m, determine the flow rate of water through the channel. What would your answer be if the bottom angle were 1°?

Solution:

The Manning coefficient for an open channel with weedy surfaces is $n = 0.030$ according to Table 6-4. The side slope is $m = \cot 60° = 0.5773$, the bottom slope is $i = \tan 0.3° = 0.005236$.

The cross-sectional area, perimeter, and hydraulic radius of the channel are

$$A = h(b + mh) = (0.52\text{m})[0.8\text{m} + (0.5773)(0.52\text{m})] = 0.572\text{ m}^2$$

$$\chi = b + 2h\sqrt{1 + m^2} = 0.8\text{m} + (2)(0.52\text{m})\sqrt{1 + 0.5773^2} = 2.001\text{m}$$

$$R_h = \frac{A}{\chi} = \frac{0.572\text{ m}^2}{2.001\text{m}} = 0.286\text{m}$$

Thus, the flowrate through the channel is determined by Eq. 9.16

$$Q = \frac{1}{n}AR_h^{2/3}i^{1/2} = \frac{1\text{ m}^{1/3}/\text{s}}{0.030}(0.572\text{ m}^2)(0.286\text{m})^{2/3}(0.005236^{1/2}) = 0.60\text{ m}^3/\text{s}$$

The flow rate for a bottom angle of 1° can be determined by using $i = \tan 1° = 0.01746$ in the last relation. It gives $Q = 1.10\text{ m}^3/\text{s}$.

DISCUSSION The results indicate that the flow rate is a strong function of the bottom slope. Also, there is considerable uncertainty in the value of the Manning coefficient, and thus in the flow rate calculated. A 10 percent uncertainty in n results in a 10 percent uncertainty in the flow rate. Final answers are therefore given to only two significant digits.

EXAMPLE 9-2 Channels with nonuniform roughness

Water flows along the drainage canal having the properties shown in Fig. E9-2. If the bottom slope is $i = 0.002$, estimate the flowrate when the depth is $h = 0.18\text{m} + 0.24\text{m} = 0.42\text{m}$.

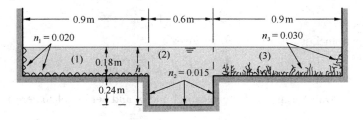

Fig. E9-2

Solution:

We divide the cross section into three subsections as is indicated in Fig. E9-2 and write the flowrate as $Q = Q_1 + Q_2 + Q_3$, where for each section

$$Q_i = \frac{1}{n_i} A_i R_{hi}^{2/3} i_i^{1/2}$$

The appropriate values of A_i, χ_i, R_{hi}, and n_i are listed in the following table.

No.	$A_i (m^2)$	$\chi_i (m)$	$R_{hi} (m)$	n_i
1	0.162	1.08	0.150	0.020
2	0.252	1.08	0.233	0.015
3	0.162	1.08	0.150	0.030

Thus, the total flowrate is

$$Q = Q_1 + Q_2 + Q_3$$
$$= (0.002^{1/2}) \left[\frac{(0.162)(0.15^{2/3})}{0.020} + \frac{(0.252)(0.233^{2/3})}{0.015} + \frac{(0.162)(0.15^{2/3})}{0.030} \right] m^3/s$$
$$= 0.455 \, m^3/s$$

DISCUSSION Note that the imaginary portions of the perimeters between sections (denoted by the dashed lines in Fig. E9-2) are not included in the χ_i. That is, for section (2), $\chi_2 = 0.6m + (2)(0.24m) = 1.08m$, not $\chi_2 = 0.6m + (2)(0.42m) = 1.44m$.

9.2.3 Best Hydraulic Cross Section

One type of problem often encountered in open-channel flows is that of determining the *best hydraulic cross section* defined as the section of the maximum flowrate, Q, or maximum hydraulic radius, R_h, for a given area A, slope i, and roughness coefficient n. It is the most efficient low-resistance sections for given conditions. Since $R_h = A/\chi$, maximizing R_h for given A is the same as minimizing the wetted perimeter χ. The best hydraulic cross section possible is that of a semicircular channel. No other shape has smaller a wetted perimeter for a given area. However, it is often desired to determine the best shape for a class of cross sections such as rectangles, trapezoids, or triangles. The following analysis of the trapezoid section will illustrate the basic results.

Consider the generalized trapezoid in Fig. 9-1b, in which the side slope $m = \cot\theta$. For a given side slope m, the flow area and the wetted perimeter are

$$A = bh + mh^2 \tag{9.18}$$

$$\chi = b + 2h\sqrt{1 + m^2} \tag{9.19}$$

Eliminating b from Eqs. 9.18 and 9.19 gives

$$\chi = \frac{A}{h} - mh + 2h\sqrt{1 + m^2} \tag{9.20}$$

To minimize χ, evaluate $d\chi/dh$ for constant A and m and set equal to zero to give

$$\frac{d\chi}{dh} = -\frac{A}{h^2} - m + \sqrt{1 + m^2} = 0 \tag{9.21}$$

Substituting Eq. 9.18 into Eq. 9.21 yields

$$\beta_m = \left(\frac{b}{h}\right)_m = 2(\sqrt{1+m^2} - m) \tag{9.22}$$

where β_m is the *aspect ratio* of the channel bottom width to flow depth in a best hydraulic trapezoidal cross section, which is the function of the side slope m.

By considering Eq. 9.22 we can have the flow area, the wetted perimeter and the hydraulic radius for best hydraulic trapezoidal cross section

$$A = h^2[2\sqrt{1+m^2} - m] \qquad \chi = 4h\sqrt{1+m^2} - 2mh \tag{9.23}$$

$$R_h = \frac{A}{\chi} = \frac{h^2[2\sqrt{1+m^2} - m]}{4h\sqrt{1+m^2} - 2mh} = \frac{h}{2} \tag{9.24}$$

Equation 9.24 is very interesting: For any *side slope m (trapezoid angle θ)*, the best hydraulic cross section for uniform flow occurs when the hydraulic radius is half the flow depth.

Equations 9.23 and 9.24 are valid for any value of θ. To obtain the best value of m for a given depth and area, evaluate $d\chi/dm$ from Eq. 9.23 with A and h held constant to give

$$2m = \sqrt{1+m^2}$$

$$m = \cot\theta = \frac{1}{3^{1/2}} \qquad \text{or} \qquad \theta = 60° \tag{9.25}$$

Thus the very best hydraulic trapezoid section is half a hexagon.

Similar calculations with a circular channel section running partially full show best efficiency for a semicircle, $h = d/2$. In fact, the semicircle is the best of all possible channel sections (minimum wetted perimeter for a given flow area).

The best hydraulic cross section can be calculated for other shapes in a similar fashion. The results (given here without proof) for circular, rectangular, trapezoidal (with 60° sides), and triangular shapes are shown in Fig. 9-6.

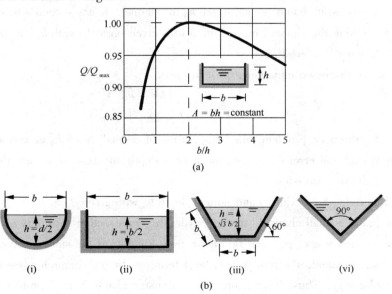

Fig. 9-6 The best hydraulic cross sections of common shapes of channel

In a new channel design, the velocity of flow through channel should be in the ranges from the non-scouring velocity to the non-silting velocity, that is,

$$[v]_{max} > v > [v]_{min} \quad (9.26)$$

where $[v]_{max}$, the *non-couring permissive velocity*, or *maximum permissive velocity*, is the largest velocity permitted in channel to prevent the scouring of the channel surface, its value is based on the surface lined materials of channel; $[v]_{min}$, the *non-silting permissive velocity*, or the *minimum permissive velocity*, is the lowest velocity permitted in channel to prevent the sedimentation and vegetative growth (crude estimates: 0.6 – 0.9m/s for sedimentation and 0.75m/s for vegetation).

9.2.4 Hydraulic Problems of Uniform Flow

In the design and analysis of uniform problem in open channel, we usually encounter three fundamental hydraulic problems as follows.

1. Determining the *flow rate* in a given channel with given shape, sizes (b, h, m), surface lining (and thus the roughness coefficient n) and bottom slope i. The area, wetted perimeter and Chézy coefficient can be directly computed by the above discussed equations, then the flowrate by Eq. 9.16.

2. Determining the *bed slope* of a channel with given shape, sizes (b, h, m), surface lining and the flowrate in the channel. First the modulus of discharge, $K = AC\sqrt{R_h}$, can be directly computed by using the given sectional parameters of channel, then the bottom slope is determined by the discharge formula of uniform flow, Eq. 9.13: $i = Q^2/K^2$.

3. Design the *sectional shape* of channel with the given flow rate in the channel, the surface lining and the bottom slope i of channel. For a trapezoidal section with given side slope m, determine the channel base width b and normal depth h_0 in channel. It has several situations.

Type 1: Determine the channel base width b with given normal depth h_0, or determine normal depth h_0 with given base width b.

From the above discussed equations, we can have

$$Q = \frac{1}{n}\frac{A^{5/3}}{\chi^{2/3}}i^{1/2} = \frac{i^{1/2}}{n}\frac{(bh + mh^2)^{5/3}}{(b + 2h\sqrt{1 + m^2})^{2/3}} \quad (9.27)$$

Therefore, for this type problem, the computation of normal depth h_0 or bottom width b requires iteration or trial and error since both the normal depth and base width are the inplict flow parameters in the above expression.

Type 2: Determine base width b and normal depth h_0 with given β_m.

For the design of a small channel, the best hydraulic cross section is usually used. Thus, the ratio of common section shape, β_m, can be known from Fig. 9-6 or the similar manner to get Eq. 9.22. For the large channel, the ratio β_m can be determined by the comprehensive comparison of economic and technology. Thus, Type 2 problem is transferred into Type 1 problem.

For trapezoidal section, substituting $b = \beta_m h$ into Eq. 9.27 yields

$$Q = \frac{i^{1/2}}{n} \frac{[h^2(\beta_m + m)]^{5/3}}{[h(\beta_m + 2\sqrt{1+m^2})]^{2/3}} \tag{9.28}$$

which gives the explicit expression of bottom width and flow depth

$$\left. \begin{array}{l} h = \left(\dfrac{nQ}{i^{1/2}}\right)^{0.375} \dfrac{(\beta_m + 2\sqrt{1+m^2})^{0.25}}{(\beta_m + m)^{0.625}} \\ b = \beta_m h \end{array} \right\} \tag{9.29}$$

Type 3: Determine bottom width b and normal depth h_0 with given non-scouring permissible velocity $[v]_{max}$.

Once the non-scouring permissible velocity $[v]_{max}$ is given, the flow area and the hydraulic radius are

$$A = \frac{Q}{[v]_{max}} \qquad R_h = \left(\frac{n[v]_{max}}{i^{1/2}}\right)^{3/2} \tag{9.30}$$

Thus, the wetted perimeter

$$\chi = \frac{A}{R_h} = \frac{Qi^{3/4}}{n^{3/2}[v]_{max}^{5/2}} \tag{9.31}$$

Then, combining the geometric relation of channel cross section, we can get the bottom width b and normal depth h_0.

For trapezoidal section, the flow area and the wetted perimeter are computed by Eqs. 9.18 and 9.19. Combining Eqs. 9.18, 9.19 and 9.30, 9.31 to directly give

$$\left. \begin{array}{l} h = \dfrac{\chi \pm \sqrt{\chi^2 - 4A(2\sqrt{1+m^2} - m)}}{2(2\sqrt{1+m^2} - m)} \\ b = \chi - 2h\sqrt{1+m^2} \end{array} \right\} \tag{9.32}$$

EXAMPLE 9-3 Determine normal depth

The asphalt-lined trapezoidal channel carries 8.1 m³/s of water under uniform-flow conditions with a bottom width of 1.8m, bottom slope of 0.0015 and side slope of 0.839. Determine the normal depth of this uniform flow.

Solution:

From Table 6-4, for asphalt-lined channel, $n \approx 0.016$. The area and hydraulic radius are functions of normal depth h_0, which is unknown:

$$A = h_0(b + m h_0) = [h_0(1.8 + 0.839 h_0)] \text{ m}^2$$

$$\chi = b + 2h_0\sqrt{1+m^2} = (1.8 + 2h_0\sqrt{1+0.839^2}) \text{ m} = (1.8 + 2.611 h_0) \text{ m}$$

Substituting into Eq. 9.16 with given flow rate $Q = 8.1$ m³/s gives

$$8.1 = \frac{0.0015^{1/2}}{0.016}[h_0(1.8 + 0.839 h_0)]\left[\frac{h_0(1.8 + 0.839 h_0)}{1.8 + 2.611 h_0}\right]^{2/3}$$

or

$$3.346(1.8 + 2.611 h_0)^{2/3} = [h_0(1.8 + 0.839 h_0)]^{5/3}$$

It has to be solved by iterate. One can eventually find $h_0 = 1.372$m.

EXAMPLE 9-4 Determine sectional sizes

The fine-sand covered natural trapezoidal drainage channel ($n = 0.025$) carries 3.5 m³/s of water under uniform-flow conditions. The cannel has a bed slope of $i = 0.005$ and a side slope of $m = 1.5$. The non-scouring permissible velocity of channel is $[v]_{max} = 0.5$ m/s. Determine the channel bottom width and the flow depth. Discuss whether the channel should be reinforced.

Solution:

In this problem, the section sizes determined by both the best hydraulic section and the non-scouring permissible velocity are proposed and compared.

(1) Design by the non-scouring permissible velocity

The flow area and the hydraulic radius are computed by Eq. 9.30

$$A = \frac{Q}{[v]_{max}} = \frac{3.5 \text{ m}^3/\text{s}}{0.5 \text{m/s}} = 7 \text{ m}^2$$

$$R_h = \left(\frac{n[v]_{max}}{i^{1/2}}\right)^{3/2} = \left[\frac{(0.025)(0.5)}{0.005^{1/2}}\right]^{3/2} \text{m} = 0.07433 \text{m}$$

$$\chi = \frac{A}{R_h} = \frac{7 \text{ m}^2}{0.07433 \text{m}} = 94.17 \text{m}$$

From Eq. 9.32 the normal depth is

$$h_0 = \frac{\chi \pm \sqrt{\chi^2 - 4A(2\sqrt{1+m^2} - m)}}{2(2\sqrt{1+m^2} - m)}$$

$$= \frac{(94.17\text{m}) \pm \sqrt{(94.17\text{m})^2 - (4)(7\text{m}^2)(2\sqrt{1+1.5^2} - 1.5)}}{2(2\sqrt{1+1.5^2} - 1.5)}$$

$$= \begin{matrix} 0.07446\text{m} \\ 44.65\text{m} \end{matrix}$$

For those two flow depths, one is too small and one is too large. Both depths are meaningless in the engineering practice, which indicates it is impossible to design the section shape by non-scouring permissible velocity.

(2) Design by the best hydraulic section

From Eq. 9.22 we have the aspect ratio for trapezoidal channel

$$\beta_m = 2(\sqrt{1+m^2} - m) = 2(\sqrt{1+1.5^2} - 1.5) = 0.6056$$

From Eq. 9.29 to give the flow depth

$$h_0 = \left(\frac{nQ}{i^{1/2}}\right)^{0.375} \frac{(\beta_m + 2\sqrt{1+m^2})^{0.25}}{(\beta_m + m)^{0.625}}$$

$$= \left[\frac{(0.025)(3.5)}{0.005^{1/2}}\right]^{0.375} \frac{(0.6056 + 2\sqrt{1+1.5^2})^{0.25}}{(0.6056 + 1.5)^{0.625}} \text{m} = 0.97\text{m}$$

$$b = \beta_m h_0 = (0.6056)(0.97\text{m}) = 0.59\text{m}$$

Now we have the section sizes and they make sense in the engineering practice. However, we need to check the flow velocity through the channel:

$$v = \frac{1}{n}R_h^{2/3}i^{1/2} = \frac{1}{n}\left(\frac{h}{2}\right)^{2/3}i^{1/2} = \frac{1}{0.025}\left(\frac{0.97}{2}\right)^{2/3}(0.005^{1/2}) \text{ m/s} = 1.75 \text{m/s}$$

DISCUSSION This velocity is much larger than the non-scouring permissible velocity $[v]_{max}$. Therefore, this channel must be reinforced. For example, using the dry masonry block stones as the channel surface lining, the non-scouring permissible velocity $[v]_{max}$ can be raised to 2.0m/s. However, the Manning coefficient of dry masonry block stones is different from the one of fine-sand surface. Thus, the section size has to be recomputed by the exact same manner above.

9.3 Uniform Flow in a Partially Full Circular Pipe

The flow in closed conduits with free surface is open channel flow. It is used in most storm and sanitary sewer systems, sewage treatment plants, many industrial waste applications, and some water treatment plants.

The uniform flow in a partly full circular pipe is almost same as the uniform flow in a channel with a shape other than circular discussed in Section 9.2. The form conditions of flow and the fundamental formulas of velocity and flow rate are the same. However, the uniform flow in a partly full circular pipe actually has one more hydraulic feature: The maximum velocity and the flow rate actually occur before the pipe is completely filled by the transporting liquid.

9.3.1 Geometric Properties

Consider the uniform flow in a partially full pipe of Fig. 9-7. In terms of the pipe diameter d and the arc angle θ up to the free surface, the geometric properties are

$$\left.\begin{aligned} A &= \frac{d^2}{8}(\theta - \sin\theta) \\ \chi &= \frac{d}{2}\theta \\ R_h &= \frac{d}{4}\left(1 - \frac{\sin\theta}{\theta}\right) \\ \alpha &= \frac{h}{d} = \sin^2\frac{\theta}{4} \end{aligned}\right\} \quad (9.33)$$

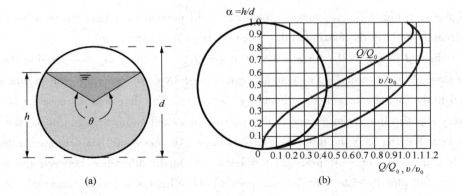

Fig. 9-7 Uniform flow in a partialy full circular channel: (a) geometry; (b) velocity and flow rate versus pipe fullness

where α is the ratio of the depth in pipe divided by the full depth (pipe diameter), termed the *percent depth of flow*, or *fullness of pipe*.

9.3.2 Capacity of Pipe

For a uniform flow in a partially filled pipe with given pipe diameter, bottom slope and Manning coefficient, the flow rate Q and velocity changes with the flow depth h as indicated in Eqs. 9.15 and 9.16

$$v = \frac{1}{n} R_h^{2/3} i^{1/2}$$

$$Q = \frac{1}{n} A R_h^{2/3} i^{1/2} = \frac{i^{1/2}}{n} \frac{A^{5/3}}{\chi^{2/3}}$$

Substituting the geometric properties of partly filled circular pipe into the above formulas to give

$$v = \frac{i^{1/2}}{n} \left[\frac{d}{4} \left(1 - \frac{\sin\theta}{\theta}\right) \right]^{2/3} = 0.397 \frac{i^{1/2}}{n} d^{2/3} \left(\frac{\theta - \sin\theta}{\theta}\right)^{2/3}$$

$$Q = \frac{i^{1/2}}{n} \frac{\left[\frac{d^2}{8}(\theta - \sin\theta)\right]^{5/3}}{\left[\frac{d}{2}\theta\right]^{2/3}} = 0.0496 \frac{i^{1/2}}{n} d^{8/3} \frac{(\theta - \sin\theta)^{5/3}}{\theta^{2/3}}$$

For a given n and slope i, we may plot these two relations versus θ. In particular, the maximum flowrate, Q_{max}, and the maximum flow velocity, v_{max}, do not occur when the pipe is full. The flow rate and velocity for full pipe Q_0 and v_0 are about $Q_0 = 0.929 Q_{max}$ and $v_0 = 0.862 v_{max}$. Therefore, there are two different maxima. By applying $\frac{dQ}{d\theta} = 0$ and $\frac{dv}{d\theta} = 0$ yields

$$v_{max} = 0.452 \frac{1}{n} d^{2/3} i^{1/2} \quad \text{at } \theta \approx 257.5° \quad \text{and } \alpha = 0.813 \quad (9.34)$$

$$Q_{max} = 0.335 \frac{1}{n} d^{8/3} i^{1/2} \quad \text{at } \theta \approx 302.4° \quad \text{and } \alpha = 0.938 \quad (9.35)$$

As shown in Fig. 9-7b, the maximum velocity is 14 percent more than the velocity when running full, and similarly the maximum discharge is about 7 percent more.

As indicated in Fig. 9-7b, for any $0.929 < Q/Q_{max} < 1$ there are two possible depths that give the same Q. The reason for this behavior can be seen by considering the gain in flow area, A, compared to the increase in wetted perimeter, χ, for $h \approx d$. The flow area increase for an increase in h is very slight in this region, whereas the increase in wetted perimeter, and hence the increase in shear force holding back the fluid, is relatively large. The net result is a decrease in flowrate as the depth increases. For most practical problems, the slight difference between the maximum flowrate and full pipe flowrate is negligible, particularly in light of the usual inaccuracy of the value of n.

Since real pipes running nearly full tend to have somewhat unstable flow, the sanitary waste system is usually designed with flowing half full or little fuller. In China, based on 'Code for De-

sign of Outdoor Waste Water Engineering (GB 50014 − 2006)', $\alpha(=h/d)$ is allowed to be 0.55 for $d = 200 - 300$mm, 0.65 for $d = 350 - 450$mm, 0.70 for $d = 500 - 900$mm, and 0.75 for $d \geqslant 1000$mm. For stormwater piping system, the full flow ($\alpha = 1$), or pressure flow, is temporally permissible.

For the design of uniform flow in a partially full pipe for engineering practice, the non-scouring and non-silting permissible velocity, the minimum pipe diameter and the minimum bottom slope also have their limited values which can be checked up in the design manuals.

9.4 Wave Speed and Froude Number

Suppose a rock is thrown into a river, a wave with a *wave velocity c* produced on the surface of a moving stream with velocity v can be observed. If water in river is not flowing ($v = 0$), the wave spreads equally in all directions (Fig. 9-8a). If water in river is near to stationary or moving in a tranquil manner, the wave moves upstream and downstream (Fig. 9-8b, Fig. 9-9a). Upstream locations are said to be in hydraulic communication with the downstream locations. That is, an observer upstream of a disturbance can tell that there has been a disturbance on the surface because that disturbance can propagate upstream to the observer. Viscous effects, which have been neglected in this discussion, will eventually damp out such waves far upstream. Such flow condition ($v < c$) is termed *tranquil* or *subcritical*.

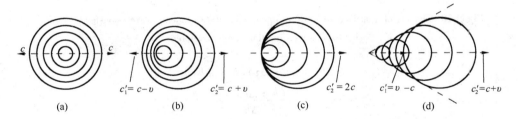

Fig. 9-8 Propagation of wave: wave speed (plane view). (a) Stationary water, (b) Subcritical flow, (c) Critical flow, and (d) Supercritical flow

On the other hand, if water is moving rapidly so that the flow velocity is greater than the wave speed c, no upstream communication with downstream locations is possible. Any disturbance on the surface downstream from the observer will be washed farther downstream (Fig. 9-8d, Fig. 9-9b). Such flow condition ($v > c$) is termed *rapid* or *supercritical*.

For the special case of $v = c$, the upstream propagating wave remains stationary (Fig. 9-8c) and the flow is termed *critical*.

Now, we are going to derive the wave speed with reference to a rectangular channel shown in Fig. 9-10a, which shows a wave of height Δh propagating at speed c into still liquid. To achieve a steady-flow inertial frame of reference, we fix the coordinates on the wave as in Fig. 9-10b, so that the still water moves to the right at velocity c.

For the control volume in Fig. 9-10b, the one-dimensional continuity relation is

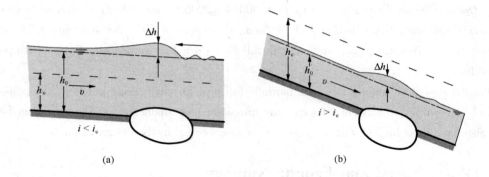

Fig. 9-9 Propagation of wave (side view). (a) Subcritical flow and (b) Supercritical flow

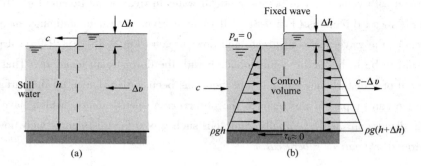

Fig. 9-10 Analysis of a surface wave in an open channel; (a) moving wave, nonsteady frame; (b) fixed wave, inertial frame of reference

$$chb = (c - \Delta v)(h + \Delta h)b$$

or

$$c = \frac{(h + \Delta h)\Delta v}{\Delta h}$$

where b is the channel width.

In the limit of small amplitude waves with $\Delta h \ll h$ gives

$$c = h\frac{\Delta v}{\Delta h} \tag{9.36}$$

If we neglect bottom friction in the short distance across the wave in Fig. 9-10b, namely assume $\tau_0 \approx 0$, the momentum relation is a balance between the net hydrostatic pressure force and momentum

$$\frac{1}{2}\rho gbh^2 - \frac{1}{2}\rho gb(h + \Delta h)^2 = \rho chb[(c - \Delta v) - c]$$

We again impose the assumption of small amplitude waves, i.e., $(\Delta h)^2 \ll h\Delta h$, the momentum equation reduces to

$$\frac{\Delta v}{\Delta h} = \frac{g}{c} \tag{9.37}$$

By combining Eqs. 9.36 and 9.37 we obtain the desired expression for wave propagation speed

$$c^2 = gh \qquad (9.38)$$

or

$$c = \sqrt{gh} \qquad (9.39)$$

For a disturbance wave in a flowing stream with velocity v, the absolute propagation speed of wave are $v + c$ towards downstream, and $v - c$ towards upstream as indicated in Fig. 9-8. Based on the above discussion about *tranquil*, *critical* and *rapid* flow, the flow with $v - c > 0$, or $v/\sqrt{gh} > 1$, is *rapid flow*; the flow with $v - c < 0$, or $v/\sqrt{gh} < 1$, is *tranquil flow*; otherwise, $v/\sqrt{gh} = 1$, is *critical flow*. Moreover, $v/\sqrt{gh}$ is the Froude number defined in Section 5.4.2 when the characteristic length L in Eq. 5.20 is taken to be the flow depth for wide rectangular channel. To make it more generally, the Froude number in open channel flow with any section shape is redefined as

$$\mathrm{Fr} = \frac{v}{\sqrt{g\bar{h}}} \qquad (9.40)$$

where $\bar{h}$ is the averaged flow depth in a channel with any section shape and taken to be the characteristic length in Froude number, $\bar{h} = A/B$, in which A is the flow area and B is the width of free surface of flow in a cross section.

Thus, the flow can be classified as

Fr < 1.0, $v < c$ Tranquil, or subcritical flow
Fr = 1.0, $v = c$ Critical flow
Fr > 1.0, $v > c$ Rapid, or supercritical flow

The character of an open-channel flow may depend strongly on whether the flow is subcritical or supercritical. For supercritical flows it is possible to produce steplike discontinuities in the fluid depth (called a *hydraulic jump*; see Section 9.6). Fig. 9-11 illustrates a channel flow accelerates from subcritical to critical to supercritical flow and then returns to subcritical flow through a *hydraulic jump*. For subcritical flows, however, changes in depth must be smooth and continuous (see Section 9.7). Certain open-channel flows, such as the broad-crested weir (Chapter 10), depend on the existence of critical flow conditions for their operation.

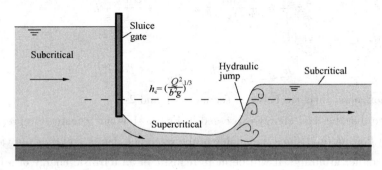

Fig. 9-11 Subcritical, critical and supercritical flow in an open channel

9.5 Fundamentals of Nonuniform Flow

If the channel slope or cross section changes or there is an obstruction in the flow, i. e. a gate, then the flow depth varies, thus the flow become *nonuniform* or *varied*. It could be *rapidly varied flow* (RVF) or *gradually varied flow* (GVF).

9.5.1 Specific Energy

Consider the flow of a liquid in a channel at a cross section where the flow depth is h, the average flow velocity is v, and the elevation of the channel bottom at that location relative to reference level 0-0 is z as shown in Fig. 9-12. Thus, the total mechanical energy of this liquid relative to reference level 0-0 and in terms of head is

$$E = z + h + \frac{\alpha v^2}{2g} \tag{9.41}$$

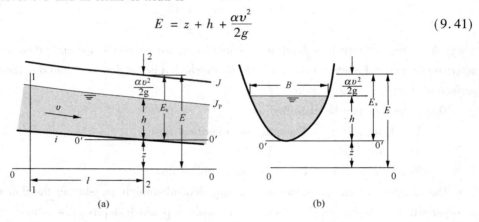

Fig. 9-12 Specific energy and total energy of flow in an open channel: (a) Side view and (b) profile of section 2-2 in (a)

Assume the sectional parameters between sections 1-1 to 2-2 to be constant, thus, the energy equation between sections 1-1 and 2-2

$$z_1 + h_1 + \frac{\alpha_1 v_1^2}{2g} = z_2 + h_2 + \frac{\alpha_2 v_2^2}{2g} + h_w \tag{9.42}$$

with $z_1 - z_2 = il$, $h_w = Jl$ yields

$$h_1 + \frac{\alpha_1 v_1^2}{2g} = h_2 + \frac{\alpha_2 v_2^2}{2g} + (J - i)l \tag{9.43}$$

where l is the horizontal distance between sections 1-1 and 2-2.

We now see that the total energy expressed in Eq. 9.41 is not a realistic representation of the true energy of a flowing fluid since the choice of the reference datum and thus the value of the elevation head z is rather arbitrary. It will be more realistic of an intrinsic energy of a fluid at a cross section if the reference datum is taken to be the bottom of the channel so that $z = 0$ there. Thus, *specific energy*, E_s, suggested by Bakhmeteff in 1911, is defined as the sum of the pressure and dynamic heads of a liquid in an open channel above the reference level 0'-0'

$$E_s = h + \frac{\alpha v^2}{2g} \qquad (9.44)$$

Specific energy is often useful in open-channel flow considerations. Thus, the energy equation, Eq. 9.43, can be written in terms of E_s as

$$E_{s1} = E_{s2} + (J - i) l \qquad (9.45)$$

If head losses are negligible, then $J = 0$, so that $(J - i) l = - il = z_2 - z_1$, thus Eq. 9.45 becomes

$$E_{s1} + z_1 = E_{s2} + z_2 \qquad (9.46)$$

Equation 9.46 indicates that the sum of the specific energy and the elevation head of channel bottom remains constant, which is another statement of the Bernoulli equation.

Specific energy E_s defined in Eq. 9.44 can also be interpreted as the mechanical energy per unit weight liquid in channel when the datum is located at the lowest point of bottom in a section. It differs from the total mechanical energy E. As shown in Fig. 9-12, the total mechanical energy E of fluid in channel has only one reference level and always decrease in the direction of flow, while the specific energy E_s is based on the lowest point of bottom in a section, which may increase, decrease or remain constant (i. e., uniform flow).

SPECIFIC ENERGY DIAGRAM

Consider flow in an open prismatic channel $[A = f(h)]$ with a given flow rate Q, the channel width varies with flow depth h, the specific energy must be in the form

$$E_s = h + \frac{\alpha v^2}{2g} = h + \frac{\alpha Q^2}{2g A^2} = f(h) \qquad (9.47)$$

This equation is very instructive as it shows the variation of the specific energy with flow depth. To gain insight into the flow processes involved, we consider the *specific energy diagram*, a graph of $E_s = f(h)$ with per width discharge q fixed drawn in a coordinate system by letting flow depth h to be the vertical (y) coordinate and specific energy E_s to be the horizontal (x) coordinate, as shown in Fig. 9-13.

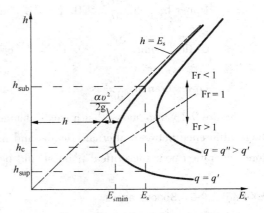

Fig. 9-13 Variations of specific energies E_s with depth h for specified flowrates

For given q and E_s, Eq. 9.47 is a cubic equation with three solutions, h_{sup}, h_{sub}, and h_{neg}. Two of the solutions, h_{sup} and h_{sub}, are positive and the other, h_{neg}, is negative, which has no physical meaning and can be ignored. Thus, for a given flowrate and specific energy there are two possible depths, termed *alternate depths*, unless the vertical line from the E_s axis does not intersect the specific energy curve corresponding to the value of q given (i. e., $E_s \leq E_{smin}$).

When the flow depth is very small, $h \to 0$, then we have $A \to 0$, and from Eq. 9.47 to give

$E_s = h + \dfrac{Q^2}{2gA^2} \approx \dfrac{Q^2}{2gA^2} \to \infty$, the specify energy curve tends to infinity and approaches to x-coordinate; when the flow depth grows large to be infinity, $h \to \infty$, then $A \to \infty$, and $E_s = h + \dfrac{Q^2}{2gA^2} \approx h \to \infty$, the specify energy curve approaches to the line of $h = E_s$, a line through the origin of coordinate system at an angle of 45° with the horizontal coordinate.

The specific energy reaches a minimum value E_{smin} at some intermediate point, characterized by the *critical depth* h_c and *critical velocity* v_c. The minimum specific energy, E_{smin}, is the minimum value of E_s required to support the specified flow rate q. Therefore, E_s cannot be below E_{smin} for a given q. Moreover, the minimum specific energy separates the specific energy curve into two branches: upper branch and lower branch. Thus, for $E_s > E_{smin}$, a vertical line intersects the curve at two points, indicating that a flow can have *alternate* depths corresponding to a fixed value of specific energy. For flow through a sluice gate with negligible frictional losses (and thus $E_s =$ constant), the upper depth corresponds to the upstream flow, and the lower depth to the downstream flow.

The differential of E_s with respect to h gives

$$\frac{dE_s}{dh} = 1 - \frac{\alpha Q^2}{g A^3}\frac{dA}{dh} = 1 - \frac{\alpha Q^2}{gA^3}B = 1 - \frac{\alpha v^2}{g\dfrac{A}{B}} = 1 - \mathrm{Fr}^2 \qquad (9.48)$$

where $\dfrac{dA}{dh} \approx B$, and B is the channel width at the free surface.

Thus, we have

Upper branch: $\dfrac{dE_s}{dh} > 0$, Fr < 1, subcritical flow

Critical point: $\dfrac{dE_s}{dh} = 0$, Fr = 1, critical flow

Lower branch: $\dfrac{dE_s}{dh} < 0$, Fr > 1, supercritical flow

Figure 9-13 also shows that a small change in specific energy near the critical point causes a large difference between alternate depths and may cause violent fluctuations in flow level. Therefore, operation near the critical point should be avoided in the design of open channels.

EXAMPLE 9-5 Specify energy

Water flows up a 0.15-m-tall ramp in a constant width rectangular channel at a rate of $q = 0.5175$ m²/s as in Fig. E9-5a. If the upstream depth is 0.69m, determine the elevation of the water surface downstream of the ramp, $h_2 + z_2$. Neglect viscous effects.

Solution:

Take the kinematic correction factor to be $\alpha = 1.0$.

With $\Delta h = z_1 - z_2 = il$, and $h_w = 0$, conservation of energy (Eq. 9.42) requires that

$$z_1 + h_1 + \frac{\alpha_1 v_1^2}{2g} = z_2 + h_2 + \frac{\alpha_2 v_2^2}{2g} \qquad (1)$$

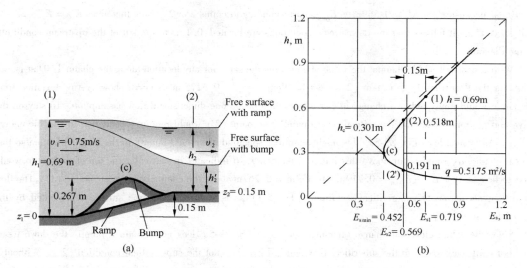

Fig. E9-5

The continuity equation provides the second equation
$$v_1 h_1 = v_2 h_2 \tag{2}$$

For the conditions given ($z_1 = 0$, $z_2 = 0.15$m, $h_1 = 0.69$m, $q = 0.5175$ m²/s, and $v_1 = q/h_1 = 0.75$m/s, $v_2 = q/h_2$), combining Eqs. 1 and 2 gives
$$h_2^3 - 0.5687 h_2^2 + 0.01365 = 0$$

which has solutions
$$h_2 = 0.518\text{m}, \qquad h_2 = 0.190\text{m} \quad \text{or} \quad h_2 = -0.139\text{m}$$

Note that two of these solutions are physically realistic, but the negative solution is meaningless. This is consistent with the previous discussions concerning the specific energy (recall the two roots indicated in Fig. 9-13). The corresponding elevations of the free surface downstream are either
$$h_2 + z_2 = 0.518\text{m} + 0.15\text{m} = 0.668\text{m}$$
or
$$h_2 + z_2 = 0.190\text{m} + 0.15\text{m} = 0.340\text{m}$$

The question is which of these two flows is to be expected? This can be answered by use of the specific energy diagram obtained from Eq. 9.47, which for this problem is
$$E_s = h + \frac{\alpha v^2}{2g} = h + \frac{\alpha q^2}{2gh^2} = h + \frac{0.01365}{h^2}$$

For the specified points, the specify energy upstream is
$$E_{s1} = h_1 + \frac{0.01365}{h_1^2} = 0.719\text{m}$$

The minimum specify energy can be obtained from the critical status, thus from Fr = 1 we have
$$h_c = \sqrt[3]{\frac{\alpha q^2}{g}} = 0.301\text{m}$$

and
$$E_{smin} = h_c + \frac{0.01365}{h_c^2} = 0.452\text{m}$$

Thus, the specify energy diagram is shown in Fig. E9-5b. The upstream condition corresponds to subcritical flow since it locates at the upper branch of diagram ($h_1 > h_c$); the downstream condition is either subcritical or supercritical because $E_{s1} - \Delta h = 0.569\text{m} > E_{smin}$, corresponding to points 2 or 2′. Note that since $E_{s1} = E_{s2} + z_2 - z_1 = E_{s2} + 0.15\text{m}$, it follows that the downstream conditions are located 0.15m to the left of the upstream conditions on the diagram.

With a constant width channel the value of q remains the same for any location along the channel. That is, all points for the flow from (1) to (2) or (2′) must lie along the $q = 0.5175\ \text{m}^2/\text{s}$ curve shown. Any deviation from this curve would imply either a change in q or a relaxation of the one-dimensional flow assumption. To stay on the curve and go from (1) around the critical point (point c) to point (2′) would require a reduction in specific energy to E_{smin}. As is seen from Fig. E9-5a, this might require a specified elevation (bump) in the channel bottom so that critical conditions would occur above this bump. Equation 9.46 indicates in particular the top of this bump would need to be $z_c - z_1 = E_{s1} - E_{smin} = 0.719\text{m} - 0.452\text{m} = 0.267\text{m}$ above the channel bottom at section (1). The flow could then accelerate to supercritical conditions ($\text{Fr}_{2'} > 1$) as is shown by the free surface represented by the dashed line in Fig. E9-5a.

Since the actual elevation change (a ramp) shown in Fig. E9-5a does not contain a bump, the downstream conditions will correspond to the subcritical flow denoted by (2), not the supercritical condition (2′). Without a bump on the channel bottom, the state (2′) is inaccessible from the upstream condition state (1). Such considerations are often termed the *accessibility of flow regimes*. Thus, the surface elevation is

$$h_2 + z_2 = 0.518\text{m} + 0.15\text{m} = 0.668\text{m}$$

DISCUSSION It should be noted the free surface has dropped by $(z_1 + h_1) - (h_2 + z_2) = 0.69\text{m} - 0.668\text{m} = 0.022\text{m}$, which indicates that the elevation of the free surface decreases as it goes across the ramp.

9.5.2 Critical Depth

As discussed in last section, the flow depth at which the specific energy is minimum is the *critical depth*, designated as h_c as shown in Fig. 9-13. It occurs where $dE_s/dh = 0$ at constant Q. Thus, Eq. 9.48 yields

$$\frac{dE_s}{dh} = 1 - \frac{\alpha Q^2}{gA^3}B = 0 \tag{9.49}$$

which is equivalent to

$$\frac{\alpha Q^2}{g} = \frac{A_c^3}{B_c} \tag{9.50}$$

where subscript c presents the corresponding values of parameters when the flow is critical flow ($\text{Fr} = 1$), A_c and B_c are the flow area and the channel width at the free surface when $h = h_c$.

The *critical velocity* can be directly given from Eq. 9.50

$$v_c = \frac{Q}{A_c} = \sqrt{\frac{gA_c}{B_c}} \tag{9.51}$$

for taking $\alpha = 1.0$.

For a given channel shape $A(h)$ and $B(h)$ and a given Q, Eq. 9.50 has to be solved by trial and error to find the critical depth h_c, and then v_c computed from Eq. 9.51.

RECTANGULAR CHANNEL

For the rectangular channel, the channel width at the free surface, B, equals to the channel bottom width, b. Let $q = Q/b = vh$ be the discharge per unit width of a rectangular channel.

Thus, with constant q Eq. 9.50 becomes

$$\frac{\alpha Q^2}{g} = \frac{(bh_c)^3}{b} = b^2 h_c^3$$

Thus

$$h_c = \sqrt[3]{\frac{\alpha Q^2}{gb^2}} = \sqrt[3]{\frac{\alpha q^2}{g}} \qquad (9.52)$$

The associated minimum energy is

$$E_{smin} = E_s(h_c) = \frac{3}{2} h_c \qquad (9.53)$$

Actually, the critical depth h_c for an open channel flow equal to the shallow-water wave propagation speed c from Eq. 9.39. To see this, take the kinematic factor α to be unity and rewrite Eq. 9.52 as

$$q^2 = gh_c^3 = (gh_c) h_c^2 \qquad (9.54)$$

By comparison $q = v_c h_c$ it follows that the critical velocity in open channel is

$$v_c = \sqrt{gh_c} = c$$
$$Fr = 1 \qquad (9.55)$$

By comparing the actual depth and velocity with the critical values, we can determine the local flow condition

$h > h_c, \; v < v_c$ Subcritical flow
$h = h_c, \; v = v_c$ Critical flow
$h < h_c, \; v > v_c$ Supercritical flow

EXAMPLE 9-6 Critical depth

A wide rectangular clean-earth channel has a flow rate $q = 4.5 \, m^3/(s \cdot m)$. Determine: (a) the critical depth; (b) the local flow condition if $h = 0.9m$?

Solution:

From Eq. 9.52 yields critical depth for a rectangular channel

$$h_c = \sqrt[3]{\frac{\alpha q^2}{g}} = \sqrt[3]{\frac{(1.0)[4.5 \, m^3/(s \cdot m)]^2}{9.81 m/s^2}} = 1.27m$$

If the actual depth is $h = 0.9m$, thus $h < h_c$. This flow must be supercritical flow.

9.5.3 Critical Slope

For a uniform flow in a prismatic channel with a constant shape, size and surface roughness, its normal depth is only a function of bottom slope according to Eq. 9.13, $Q = AC\sqrt{R_h i}$. For a given flow rate, the normal depth decreases with the increasing bottom slope.

If the normal depth of a uniform flow in a channel is equal to the critical depth corresponding to the flow rate of this uniform flow, the bottom slope of this channel is called the *critical slope*, designated as i_c. That is when $h_0 = h_c$, $i = i_c$. This condition is analyzed by equating Eq. 9.50 to the Chézy formula Eq. 9.13:

$$Q = A_c C_c \sqrt{R_{hc} i_c} \left.\begin{array}{c} \\ \\ \end{array}\right\}$$
$$\frac{\alpha Q^2}{g} = \frac{A_c^3}{B_c}$$

which yields

$$i_c = \frac{g}{\alpha C_c^2} \frac{\chi_c}{B_c} \tag{9.56}$$

where C_c, χ_c, B_c are the Chézy coefficient, wetted perimeter and channel width at the free surface when the channel flow is a critical flow with $h_0 = h_c$.

Equation 9.56 is valid for any channel shape. For a wide rectangular channel, $B_c \gg h_c$, $\chi_c \approx B_c$, with Eq. 6.62 the formula reduces to

$$i_c \approx \frac{g}{\alpha C_c^2} = \frac{\lambda}{8\alpha} \approx \frac{\lambda}{8} \tag{9.57}$$

This is a special case, a reference point. In most channel flows $h_0 \neq h_c$. For fully rough turbulent flow, the critical slope varies between 0.002 and 0.008.

It should be noted that the critical slope is a particular value of the bed slope for a channel corresponding to a certain discharge and a given shape and dimension. It is nothing to do with the actual bed slope of channel.

By comparing the actual positive bed slope with the critical slope, we can determine the type of channel slope and also the uniform flow regime in this long straight prismatic channel

$i < i_c, h > h_c$ Mild slope, subcritical flow
$i = i_c, h = h_c$ Critical slope, critical flow
$i > i_c, h < h_c$ Steep slope, supercritical flow

It should keep in mind that the actually local uniform flow condition in a channel with a given bottom slope, i, changes with the change of flow rate because the value of critical slope changes with the changing flow rate in the channel as indicated by Eq. 9.56. It could be a subcritical flow when the uniform flow rate is small, and then to be a supercritical flow if the flow rate in the same channel becomes larger.

EXAMPLE 9-7 Judgment of flow regime in a channel

The normal depth of a uniform flow in a long straight rectangular channel is $h_0 = 0.6$m. The bottom width, bottom slope and surface Manning coefficient of the channel are $b = 1$m, $i = 0.0004$ and $n = 0.014$ respectively. Determine the flow regime of this channel flow.

Solution:

(1) By wave speed

The hydraulic radius of this channel flow is

$$R_h = \frac{A}{\chi} = \frac{bh_0}{b + 2h_0} = \frac{(1\text{m})(0.6\text{m})}{1\text{m} + (2)(0.6\text{m})} = 0.273\text{m}$$

Thus, the flow velocity can be determined by Eq. 9.12

$$v = C\sqrt{R_h i} = \frac{1}{n} R_h^{2/3} i^{1/2} = \frac{1}{0.014}(0.273^{2/3})(0.0004^{1/2})\text{ m/s} = 0.601\text{ m/s}$$

The wave speed is
$$c = \sqrt{gh_0} = \sqrt{(9.81\,\text{m/s}^2)(0.6\,\text{m})} = 2.43\,\text{m/s}$$
Thus, we have $c > v$, the flow is subcritical flow.

(2) By Froude number

The Froude number is
$$\text{Fr} = \frac{v}{\sqrt{gh_0}} = \frac{0.601\,\text{m/s}}{\sqrt{(9.81\,\text{m/s}^2)(0.6\,\text{m})}} = 0.25 < 1$$
which indicate the flow is a subcritical flow.

(3) By critical depth

The critical depth in a rectangular channel can be determined by Eq. 9.52
$$h_c = \sqrt[3]{\frac{\alpha q^2}{g}} = \sqrt[3]{\frac{\alpha(vh_0)^2}{g}} = \sqrt[3]{\frac{(1.0)[(0.601\,\text{m/s})(0.6\,\text{m})]^2}{9.81\,\text{m/s}^2}} = 0.237\,\text{m} < h_0 = 0.6\,\text{m}$$
which indicate the flow is a subcritical flow.

(4) By critical slope

Compute the parameters based on the critical depth we have computed
$$B_c = b = 1\,\text{m}$$
$$\chi_c = b + 2h_c = 1\,\text{m} + (2)(0.237\,\text{m}) = 1.474\,\text{m}$$
$$R_{hc} = \frac{A_c}{\chi_c} = \frac{bh_c}{\chi_c} = \frac{(1\,\text{m})(0.237\,\text{m})}{1.474\,\text{m}} = 0.161\,\text{m}$$
$$C_c = \frac{1}{n}R_{hc}^{1/6} = \frac{1}{0.014}(0.161)^{1/6}\sqrt{\text{m}}/\text{s} = 52.68\,\sqrt{\text{m}}/\text{s}$$

Thus, the critical slope can be determined by Eq. 9.46
$$i_c = \frac{g}{\alpha C_c^2}\frac{\chi_c}{B_c} = \frac{9.81\,\text{m/s}^2}{(1.0)(52.68\sqrt{\text{m}}/\text{s})^2}\frac{1.474\,\text{m}}{1\,\text{m}} = 0.0052 > i$$
which indicates the channel is a mild channel, the flow in it is a subcritical flow.

DISCUSSION This example shows that for the flow regime, subcritical, critical and supercritical flow in a channel can be determined by the wave speed, the Froude number, the critical depth, the critical velocity and specify energy diagram for both the uniform or non-uniform flow, while the critical slope can only be used for the uniform flow in the channel.

9.6 Rapidly Varied Flow; The Hydraulic Jump

In many open channels, flow depth changes occur markedly over a relatively short distance in the flow direction so that $dh/ds \sim 1$. This flow in open channels is called *rapidly varied flow* (RVF) as introduced in Section 9.1.2. Such flows occur in sluice gates, broad- or sharp-crested weirs, waterfalls, and the transition sections of channels for expansion and contraction. A change in the cross section of the channel is an important reason for the occurrence of rapidly varied flow. But some rapidly varied flows, such as flow through a sluice gate, occur even in regions where the channel cross section is constant.

The flow in steep channels can be supercritical flow, and can change quickly to a subcritical flow by passing through a hydraulic jump, as indicated in Fig. 9-14. The upstream flow is fast and shallow, and the downstream flow is slow and deep. The hydraulic jump is quite thick, ranging in length from 4 to 6 times the downstream depth h''.

A *hydraulic jump* is a local rapidly varied flow phenomenon that occurs whenever flow changes from supercritical to subcritical in an open channel, which has a rather abrupt change in depth and a steep upwards slope of the water surface profile. A hydraulic jump involves considerable mixing and agitation due to the surface rollers formed in a standing wave like the one shown in Fig. 9-14, and thus a significant amount of mechanical energy dissipation.

Fig. 9-14 Hydraulic jump formed in an inclined open channel (*Courtesy of the Environmental Hydraulic Laboratory, Zhejiang University*)

9.6.1 Momentum Function of Hydraulic Jump in Horizontal Channel

For a flow with a given flow rate Q in a horizontal channel, consider steady flow through a control volume between pre-jump and post-jump sections that encloses the hydraulic jump as shown in Fig. 9-15, in which the depth of supercritical flow (*pre-jump depth*), h', ' jumps' up to its subcritical *conjugate depth*, h'', over the *length of jump*, L_j. To make a simple analysis possible, we make the following assumptions:

1. The wall shear stress and the associated losses are negligible relative to the losses that occur during the hydraulic jump due to the intense agitation.

2. The pre-jump and post-jump sections are gradually varied sections, the pressure in the liquid varies hydrostatically in vertical, and the pressure in the surface rollers also varies hydrostatically.

3. The momentum-flux correction factors for pre-jump and post-jump sections are taken to be unity, namely $\beta_1 = \beta_2 = 1.0$.

4. There are no external or body forces other than gravity.

Noting that the only forces acting on the control volume in the horizontal x-direction are the pressure forces because of negligible wall shear stress, the momentum equation in the x-direction becomes a balance between hydrostatic pressure forces and momentum transfer

$$p_{c1}A_1 - p_{c2}A_2 = \rho Q(\beta_2 v_2 - \beta_1 v_1) \tag{9.58}$$

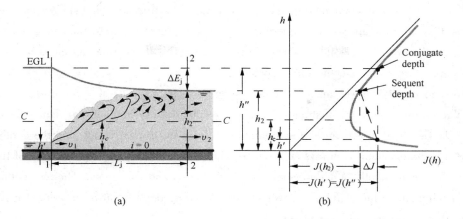

Fig. 9-15 Schematic and $M - h$ curve for a hydraulic jump

where p_{c1} and p_{c2} are the dynamic pressures at the centroid of pre-jump and post-jump sections, respectively, $p_{c1} = \rho g h_{c1}$ and $p_{c2} = \rho g h_{c2}$; h_{c1} and h_{c2} are the water depths of the centroids of pre-jump and post-jump sections, respectively; A_1, v_1 and A_2, v_2 are the flow areas and average velocities at pre-jump and post-jump sections, respectively.

Introducing the continuity equation to give $Q = v_1 A_1 = v_2 A_2$, with $\beta_1 = \beta_2 = 1.0$ and $p_{c1} = \rho g h_{c1}$, $p_{c2} = \rho g h_{c2}$ Eq. 9.58 becomes

$$\rho g h_{c1} A_1 - \rho g h_{c2} A_2 = \rho Q \left(\frac{Q}{A_2} - \frac{Q}{A_1} \right)$$

or

$$\frac{Q^2}{gA_1} + h_{c1} A_1 = \frac{Q^2}{gA_2} + h_{c2} A_2 \qquad (9.59)$$

Equation 9.59 is the fundamental formula of hydraulic jump in a horizontal channel. It indicates that in the zone of hydraulic jump, the sum of the momentum flowing into the control volume through pre-jump section per unit time and the dynamic pressure force acting on the pre-jump section equals to the sum of the momentum flowing out of the control volume through post-jump section per unit time and the dynamic pressure force acting on the post-jump section.

For a given flow rate in a given open channel, the flow areas, A_1, A_2, and the water depths of the centroids of pre-jump and post-jump sections, h_{c1}, h_{c2}, are the function of flow depth, thus the both sides of Eq. 9.59 are the function of flow depth which can be simplified as

$$J(h') = J(h'') \qquad (9.60)$$

where $J(h)$ is called the *function of hydraulic jump*, also the *momentum function*, defined as

$$J(h) = \frac{Q^2}{gA} + h_c A \qquad (9.61)$$

Let the flow depth to be the vertical coordinate, the function of hydraulic jump J to be the horizontal coordinate, we can have the *momentum-depth ($M - h$) diagram* for a hydraulic jump as indicated in Fig. 9-15b, which is similar to the specific energy curve.

The $M - h$ Diagram is a graphical representation of the conservation of momentum and can be applied over a hydraulic jump to find the upstream and downstream depths. We can see from Fig. 9-15b that the flow approaches supercritically at a depth of h', and jump to the subcritical conjugate depth of h' which is labeled as h'' in the figure. The $M - h$ Diagram helps in visualizing how two depths can exist with the same momentum.

Also from Fig. 9-15b we can see conjugate depths refer to the depth (h') upstream and the depth (h'') downstream of the hydraulic jump whose momentum functions are equal for a given flow rate Q. The depth upstream of a hydraulic jump is always supercritical, and the depth downstream of a hydraulic jump is always subcritical. The deeper the depth (h') upstream of the jump is, the shallower the depth (h'') downstream of the jump is, and vice versa. The momentum function has a minimum value, $[J(h)]_{min}$, whose corresponding depth is also the *critical depth*, h_c, similar to that of the specific energy curve.

It is important to note that the *conjugate depths* differ from the *alternate depths* of specific energy of channel flow that conjugate depths are used in momentum conservation calculations as follows.

9.6.2 Conjugate Depths in Horizontal Rectangular Channel

The calculation of conjugate depths is the fundamental of the calculation of hydraulic jump. When one of the conjugate depths (h', or h'') is known, we can get the momentum function $J(h')$, or $J(h'')$, then calculate the another conjugate depth by the formula of hydraulic jump, Eq. 9.59, which is usually calculated by the graphical method.

For the hydraulic jump in a horizontal rectangular channel, $A_1 = bh'$, $A_2 = bh''$, $h_{c1} = h'/2$, $h_{c2} = h''/2$, $q = Q/b$, substituting into Eq. 9.59 and cancelling the width b yields

$$\frac{q^2}{gh'} + \frac{h'^2}{2} = \frac{q^2}{gh''} + \frac{h''^2}{2}$$

Rearrange the above equations to give

$$h'h''(h' + h'') = \frac{2q^2}{g} \tag{9.62}$$

Solving it yields

$$h' = \frac{h''}{2}\left(\sqrt{1 + \frac{8q^2}{gh''^3}} - 1\right) \tag{9.63}$$

$$h'' = \frac{h'}{2}\left(\sqrt{1 + \frac{8q^2}{gh'^3}} - 1\right) \tag{9.64}$$

where

$$\frac{q^2}{gh'^3} = \frac{v_1^2}{gh'} = \mathrm{Fr}_1^2$$

$$\frac{q^2}{gh''^3} = \frac{v_2^2}{gh''} = \mathrm{Fr}_2^2$$

Thus Eqs. 9.63 and 9.64 become

$$h' = \frac{h''}{2}(\sqrt{1 + 8\text{Fr}_2^2} - 1) \qquad (9.65)$$

$$h'' = \frac{h'}{2}(\sqrt{1 + 8\text{Fr}_1^2} - 1) \qquad (9.66)$$

where Fr_1 and Fr_2 are Froude numbers at pre-jump and post-jump sections of the hydraulic jump respectively.

From Eq. 9.66 we can have the *depth ratio*

$$\eta = \frac{h''}{h'} = 0.5(\sqrt{1 + 8\text{Fr}_1^2} - 1) \qquad (9.67)$$

Equation 9.67 indicates that the depth ratio increases with the increasing Froude number upstream of the jump.

9.6.3 Length and Height of Hydraulic Jump

The *length* of a hydraulic jump is often hard to measure in the field and during laboratory investigations due to the sudden changes in surface turbulence, in addition to the formation of roller and eddies. The length of a hydraulic jump is often an important factor to know when considering the design of structures like settling basins. The equation derived for length is based on experimental data, and relates the length to the upstream Froude number.

The *length of a hydraulic jump* in a horizontal rectangular channel can be calculated by the following equations:

$$L_j = 6.1 h'' \qquad \text{for } 4.5 < \text{Fr}_1 < 10 \qquad (9.68)$$

$$L_j = 6.9(h'' - h') \qquad (9.69)$$

$$L_j = 9.4(\text{Fr}_1 - 1)h' \qquad (9.70)$$

The *height* of the hydraulic jump, h_j, similar to length, is useful to know when designing waterway structures like settling basins or spillways. The height of the hydraulic jump is simply the difference in flow depths prior to and after the hydraulic jump, $h_j = h'' - h'$. The height can be determined using the Froude number and upstream energy.

9.6.4 Energy Loss in Hydraulic Jump

Although momentum discussed above is conserved throughout the hydraulic jump, the energy is not. There is an initial loss of energy when the flow jumps from supercritical to subcritical depths. The resulting loss of energy is equal to the change in specific energy across the jump and can be derived from the steady-flow energy equation

$$\Delta E_j = h_w = \left(h' + \frac{\alpha_1 v_1^2}{2g}\right) - \left(h'' + \frac{\alpha_2 v_2^2}{2g}\right) \qquad (9.71)$$

With $\alpha_1 = \alpha_2 = 1.0$ and the continuity equation,

$$q = v_1 h' = v_2 h'' \qquad (9.72)$$

Eq. 9.62 gives

$$\frac{\alpha_1 v_1^2}{2g} = \frac{q^2}{2gh'^2} = \frac{1}{4}\frac{h''}{h'}(h' + h'')$$

$$\frac{\alpha_2 v_2^2}{2g} = \frac{q^2}{2gh''^2} = \frac{1}{4}\frac{h'}{h''}(h' + h'')$$

Substituting back to Eq. 9.71 gives the dissipation head loss across the jump

$$\Delta E_j = \frac{(h'' - h')^3}{4h'h''} \qquad (9.73)$$

For a given Fr_1 and h', Eq. 9.73 shows that the larger the height of hydraulic jump is, the larger the dissipation loss is. It also indicates that the actual depth downstream of the hydraulic jump should be less than the theoretical conjugate depth of h', h'', because of this existing dissipation loss in the jump. This actual depth downstream of the hydraulic jump is called the *sequent depth* as in Fig. 9-15b.

Equation 9.73 shows that the dissipation loss is positive only if $h'' > h'$, thus Eq. 9.67 requires $Fr_1^2 > 1.25$, that is, the upstream flow must be supercritical flow; while Eq. 9.65 gives $Fr_2^2 < 1$, the downstream flow is subcritical.

Being extremely turbulent and agitated, the hydraulic jump is a very effective energy dissipator and is a feature of stilling-basin and spillway applications. When the jump forms at the bottom of a dam spillway, it is very important that such jumps be located on specially designed aprons; otherwise the channel bottom will be badly scoured by the agitation. Jumps also mix fluids very effectively and have application to sewage and water treatment designs.

The specific energy of the liquid before the hydraulic jump is $E_{s1} = h' + \frac{\alpha_1 v_1^2}{2g}$. Then the *energy dissipation ratio* can be expressed as

$$\text{Dissipation ratio} = \frac{\Delta E_j}{E_{s1}} = \frac{\Delta E_j}{h' + \frac{\alpha_1 v_1^2}{2g}} = \frac{\Delta E_j}{h'\left(1 + \frac{\alpha_1 Fr_1^2}{2}\right)} \qquad (9.74)$$

The fraction of energy dissipation ranges from just a few percent for weak hydraulic jumps ($Fr_1 < 2$) to 85 percent for strong jumps ($Fr_1 > 9$) as shown in Table 9-1.

9.6.5 Classification of Hydraulic jumps

Experimental studies indicate that hydraulic jumps can be considered in five categories as shown in Table 9-1, depending primarily on the value of the pre-jump Froude number, $Fr_1 = v_1/\sqrt{gh'}$, the principal parameter affecting hydraulic jump performance.

Table 9-1 Hydraulic jump characteristics

Pre-jump F_{r1}	Depth ratio $\eta = h''/h'$	Classification	Description	Dissipation ratio	Sketch
≤ 1.0	1.0	Jump impossible	Flow must be supercritical for jump to occur	none	
$1.0 \sim 1.7$	$1.0 \sim 2.0$	Undular jump (or standing wave)	Standing or undulating wave about $4h''$ long	< 5%	

continued

Pre-jump F_{r1}	Depth ratio $\eta = h''/h'$	Classification	Description	Dissipation ratio	Sketch
1.7~2.5	2.0~3.1	Weak jump	Series of small rollers	5%~15%	
2.5~4.5	3.1~5.9	Oscillating jump	Each irregular pulsation creates a large wave which can travel downstream for kilometers, damaging earth banks and other structures. Not recommended for design conditions	15%~45%	
4.5~9.0	5.9~12.0	Steady jump	Stable and well-balanced; best performance and action, insensitive to downstream conditions. Best design range	45%~70%	
>9.0	>12.0	Strong jump	Clearly defined, turbulent, somewhat intermittent; Very effective energy dissipation	70%~85%	

1. Data source: U. S. Bureau of Reclamation (1955). 2. The above classification is very rough. Undular hydraulic jumps have been observed with inflow/prejump Froude numbers up to 3.5 to 4.

EXAMPLE 9-8 Hydraulic jump

Water discharging into a 10-m-wide rectangular horizontal channel from a sluice gate is observed to have undergone a hydraulic jump as in Fig. E9-8. The flow depth and velocity before the jump are 0.8m and 7m/s, respectively. Determine (a) the flow depth and the Froude number after the jump, (b) the head loss and the dissipation ratio, and (c) the wasted power production potential due to the hydraulic jump.

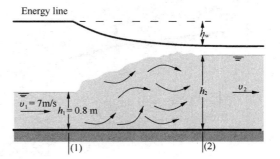

Fig. E9-8

Solution:

(a) The Froude number prior to the hydraulic jump

$$Fr_1 = \frac{v_1}{\sqrt{gh'}} = \frac{7 \text{m/s}}{\sqrt{(9.81 \text{m/s}^2)(0.8 \text{m})}} = 2.50$$

is greater than 1. Therefore, the flow is indeed supercritical prior to the jump. The flow depth, velocity, and Froude number after the jump are

$$h'' = \frac{h'}{2}(\sqrt{1 + 8Fr_1^2} - 1) = \frac{0.8\text{m}}{2}(\sqrt{1 + (8)(2.5^2)} - 1) = 2.46\text{m}$$

$$v_2 = \frac{h'}{h''}v_1 = \frac{0.8\text{m}}{2.46\text{m}}(7\text{m/s}) = 2.28\text{m/s}$$

$$Fr_2 = \frac{v_2}{\sqrt{g\,h''}} = \frac{2.28\,\text{m/s}}{\sqrt{(9.81\,\text{m}^2/\text{s})(2.46\,\text{m})}} = 0.464$$

Note that the flow depth triples and the Froude number reduces to about one-fifth after the jump.

(b) The head loss is determined from the energy equation to be

$$\Delta E_j = \frac{(h'' - h')^3}{4\,h'h''} = \frac{(2.46\,\text{m} - 0.8\,\text{m})^3}{(4)(2.46\,\text{m})(0.8\,\text{m})} = 0.581\,\text{m}$$

The specific energy of water before the jump and the dissipation ratio are

$$E_{s1} = h' + \frac{\alpha_1 v_1^2}{2g} = 0.8\,\text{m} + \frac{(1.0)(7\,\text{m/s})^2}{(2)(9.81\,\text{m/s}^2)} = 3.30\,\text{m}$$

$$\text{Dissipation ratio} = \frac{\Delta E_j}{E_{s1}} = \frac{0.581\,\text{m}}{3.30\,\text{m}} = 0.176$$

Therefore, 17.6 percent of the available head (or mechanical energy) of the liquid is wasted (converted to thermal energy) as a result of frictional effects during this hydraulic jump.

(c) The flow rate of water is

$$Q = bh'v_1 = (10\,\text{m})(0.8\,\text{m})(7\,\text{m/s}) = 56\,\text{m}^3/\text{s}$$

The power dissipation corresponding to a head loss of 0.581m becomes

$$\dot{E}_{\text{dissipated}} = \rho g Q \Delta E_j = (1000\,\text{kg/m}^3)(9.81\,\text{m/s}^2)(56\,\text{m}^3/\text{s})(0.581\,\text{m}) = 319.2\,\text{kW}$$

DISCUSSION The results show that the hydraulic jump is a highly dissipative process, wasting 319.2kW of power production potential in this case. That is, if the water is routed to a hydraulic turbine instead of being released from the sluice gate, up to 319.2kW of power could be generated. But this potential is converted to useless thermal energy instead of useful power, causing a temperature rise of

$$\Delta T = \frac{\dot{E}_{\text{dissipatd}}}{\rho Q c_p} = \frac{319.2\,\text{kJ/s}}{(1000\,\text{kg/m}^3)(56\,\text{m}^3/\text{s})(4.18\,\text{kJ/kg}\cdot\text{°C})} = 0.00136\,\text{°C}$$

for water. Note that a 319.2kW resistance heater would cause the same temperature rise for water flowing at a rate of 56 m³/s.

9.6.6 Hydraulic drop

A *hydraulic drop* is a local hydraulic phenomenon in the open channel as indicated in Fig. 9-16. It is a transition for the flow varying from the subcritical flow to the supercritical flow with the water surface dropping sharply. It is often caused by an abrupt change in the channel slope from the mild to steep slopes or the sudden change of the sectional shape of the downstream channel.

Along with a natural change in the channel slope, a hydraulic drop may be created by controlling the discharge of water into the channel. This can be managed using a series of locks to cause

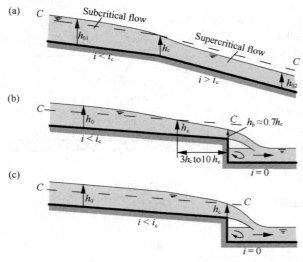

Fig. 9-16 Hydraulic drop. (a) Change in flow from subcritical to supercritical at a break in slope, the critical depth occurs at the break section; (b) Free fall in practice. Critical depth occurs prior to the brink; (c) Free fall in theory. Critical depth is assumed to occur exactly at the brink

the water level to lower when needed.

Theoretically, for a free outfall with subcritical flow in the channel prior to the outfall, critical flow occurs at the brink as indicated in Fig. 9-16c. However, due to the curvature of the streamlines (caused by gravity at the outfall), results in the depth at the brink less than critical, about $0.7\ h_c$ as indicated in Fig. 9-16b.

9.7 Water Surface Profiles of GVF in Prismatic Channel

In this section we consider the gradually varied flows, which is a form of steady nonuniform flow characterized by gradual variations in flow depth and velocity (small slopes and no abrupt changes) and a free surface that always remains smooth (no discontinuities or zigzags).

In gradually varied flow, the bed slope and the friction slope are not equal, the free surface (flow depth) will vary along the flow direction in open channel, $h = h(s)$, either increasing or decreasing in the flow direction. However, the flow depth and velocity vary slowly, and the free surface is stable. This makes it possible to formulate the variation of flow depth along the channel on the basis of the conservation of mass and energy principles and to obtain relations for the profile of the free surface.

9.7.1 Differential Equation of Depth

Consider steady flow in a prismatic open channel, and assume any variation in the bottom slope and water depth to be rather gradual. We again write the equations in terms of average velocity and assume the pressure vertical distribution to be hydrostatic. From Fig. 9-17, the steady-flow energy equation between two sections with a distance of ds and related to the reference level 0-0 is

$$z + h + \frac{\alpha v^2}{2g} = (z + dz) + (h + dh)$$
$$+ \frac{\alpha(v + dv)^2}{2g} + dh_w$$

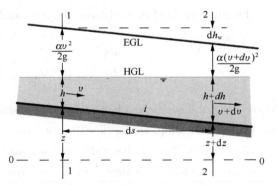

Fig. 9-17 Variation of properties between two sections in a gradually varied flow

Expanding $(v + dv)^2$ and neglecting $(dv)^2$ give

$$dz + dh + d\left(\frac{\alpha v^2}{2g}\right) + dh_w = 0$$

Thus, differentiating with respect to the flow direction in channel s gives

$$\frac{dz}{ds} + \frac{dh}{ds} + \frac{d}{ds}\left(\frac{\alpha v^2}{2g}\right) + \frac{dh_w}{ds} = 0 \qquad (9.75)$$

Noting the terms in the above equation can be expressed as follows.

(1) $\dfrac{dz}{ds} = \dfrac{-[z-(z+dz)]}{ds} = -i$

(2) $\dfrac{d}{ds}\left(\dfrac{\alpha v^2}{2g}\right) = \dfrac{d}{ds}\left(\dfrac{\alpha Q^2}{2gA^2}\right) = -\dfrac{\alpha Q^2}{gA^3}\dfrac{dA}{ds}$

In a prismatic open channel, the flow area is only the function of the flow depth, $A = f(h)$, thus

$$\dfrac{dA}{ds} = \dfrac{\partial A}{\partial h}\dfrac{dh}{ds} = B\dfrac{dh}{ds}$$

where B is the channel width at the free surface. Thus we have

$$\dfrac{d}{ds}\left(\dfrac{\alpha v^2}{2g}\right) = -\dfrac{\alpha Q^2}{gA^3}B\dfrac{dh}{ds}$$

(3) In gradually varied flow, the flow depth and velocity vary slowly, the minor losses are negligible, the head loss only includes the frictional loss, that is $dh_w = dh_f$, thus

$$\dfrac{dh_w}{ds} = J$$

Substituting all of the above relevant equations in (1), (2), (3) into Eq. 9.75 gives

$$-i + \dfrac{dh}{ds} - \dfrac{\alpha Q^2}{gA^3}B\dfrac{dh}{ds} + J = 0$$

in which J can be approximately determined by the flow rate formula of uniform flow

$$J = \dfrac{Q^2}{A^2 C^2 R_h} = \dfrac{Q^2}{K^2}$$

Finally, the final desired form of the gradually varied flow equation is

$$\dfrac{dh}{ds} = \dfrac{i-J}{1-\dfrac{\alpha Q^2}{gA^3}B} = \dfrac{i-J}{1-\mathrm{Fr}^2} = \dfrac{i-\dfrac{Q^2}{K^2}}{1-\mathrm{Fr}^2} \qquad (9.76)$$

where i is the slope of the channel bottom (positive as shown in Fig. 9-17) and J is the slope of the EGL (which drops in the flow direction due to wall friction losses).

Equation 9.76 is the basic differential equation of depth of gradually varied flow in a prismatic channel. A plot of flow depths gives the *surface profile* of the flow. Thus, it can be used to determine the *water surface profile* of gradually varied flow in an open channel.

9.7.2 Curves for Water Surface Profiles

The general characteristics of surface profiles for gradually varied flow depend on the bottom slope and flow depth relative to the critical and normal depths.

BOTTOM SLOPES

As discussed in Section 9.1.1, there are three kinds of bottom slope: positive, negative and horizontal slope classified by the magnitude of the slope i. However, the character of a gradually varied flow is often classified in terms of the actual channel slope, i, compared with the slope required to produce uniform critical flow, i_c, as indicated in Section 9.5.3. Thus, positive slope can be again subclassified into mild, critical and steep slopes, which leads to five types of slopes in the analysis of water surface profiles:

    Mild slope $i < i_c$, the flow would be subcritical if it were uniform depth

    Critical slope $i = i_c$, the flow would be critical if it were uniform depth

    Steep slope $i > i_c$, the flow would be supercritical if it were uniform depth

    Horizontal slope $i = 0$

    Adverse slope $i < 0$, flow uphill

PROFILES OF LIQUID SURFACE

Since we are considering nonuniform depth flows, the flows within these categories may have Fr < 1 or Fr > 1, depending on whether $h > h_c$ or $h < h_c$ respectively, or $J > i$, $J = i$, or $J < i$ depending on $h < h_0$, $h = h_0$ or $h > h_0$ respectively. Thus, substitution of the combinations of those values into Eq. 9.76 can give positive, zero, or negative values of dh/ds as follows and indicated in Fig. 9-18:

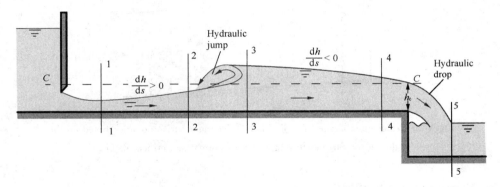

Fig. 9-18 Depth variation of flow in an open channel

(1) $\dfrac{dh}{ds} > 0$, the depth increases along the channel, surface profile is a *backwater curve*, i.e. the free surface variation between sections 1-1 to 2-2 in Fig. 9-18;

(2) $\dfrac{dh}{ds} < 0$, the depth decreases along the channel, surface profile is a *dropdown curve*, i.e. the free surface variation between sections 3-3 to 4-4 in Fig. 9-18;

(3) $\dfrac{dh}{ds} = 0$, the water surface asymptotically approaches the normal depth line of uniform flow;

(4) $\dfrac{dh}{ds} = i$, that is $\dfrac{dh}{ds} = -\dfrac{dz}{ds}$, the surface asymptotically approaches the horizontal line as indicated in Fig. 9-19;

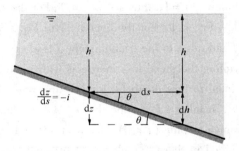

Fig. 9-19 Open channel flow with horizontal free surface

(5) $\dfrac{dh}{ds} \to \infty$, a *hydraulic jump* occurs with a discontinuity of the water surface as indicated in Fig. 9-18;

(6) $\dfrac{dh}{ds} \to -\infty$, a *hydraulic drop* occurs with a discontinuity of the water surface as indicated in Fig. 9-18.

FLOW ZONES

In the analysis of water surface profile, the *normal depth line* $N-N$ ($h = h_0$) and the *critical depth line* $C-C$ ($h = h_c$) are introduced to analyze the water surface profile qualitatively. They separate the flow region into three flow zones as shown in Fig. 9-20:

    Zone I : $h > h_0$ and $h > h_c$
    Zone II : $h_c > h > h_0$ (subcritical flow) or $h_c < h < h_0$ (supercritical flow)
    Zone III: $h < h_0$ and $h < h_c$

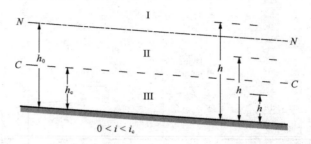

Fig. 9-20 Designation of zone I, II, and III for water surface profiles based on the value of the flow depth relative to the normal and critical depths

9.7.3 Water Surface Profile

The increasing or decreasing of the flow depth of gradually varied flow along the prismatic channel, dh/ds, can be identified by the sign of both the numerator and denominator of Eq. 9.76. For a given Q, n, i, and channel geometry, the variation of dh/ds is determined by the relative magnitudes of J and Fr at a flow depth h. If the flow depth equals to the critical depth, $h = h_c$, it follows $Fr = 1$, thus $1 - Fr^2 = 0$. For the subcritical flow, $1 - Fr_2 > 0$, and for supercritical flow, $1 - Fr_2 < 0$. For the nominator in Eq. 9.76, $i - J$, mainly depends on the relative magnitude of J and i due to the friction slope J is always positive. For the uniform flow, $i = J$, thus $i - J = 0$. Noting that head loss increases with increasing velocity, and that the velocity is inversely proportional to flow depth for a given flow rate. Thus, if $h < h_0$, it follows $J > i$ and $i - J < 0$; if $h > h_0$, it follows $J < i$ and thus $i - J > 0$; and $i - J$ is always negative for horizontal ($i = 0$) and advance-sloping ($i < 0$) channels. Therefore, twelve distinct types of solution curves can be obtained in those five classes of prismatic open channel as in Table 9-2, so do some representative surface profiles in the engineering practices.

Table 9-2 Summary of surface profiles in gradually varied flow

Channel Slope	Slope Notation	Flow depth	Profile designation	Water profile	Representative surface profiles in the engineering practices
$i < i_c$	Mild slope	$h > h_0 > h_c$ $h_0 > h > h_c$ $h_0 > h_c > h$	M_1 M_2 M_3		
$i = i_c$	Critical slope	$h > h_0\,(h_c)$ $h < h_0\,(h_c)$	C_1 C_3		
$i > i_c$	Steep slope	$h > h_c > h_0$ $h_c > h > h_0$ $h_c > h_0 > h$	S_1 S_2 S_3		
$i = 0$	Horizontal slope	$h > h_c$ $h < h_c$	H_2 H_3		
$i < 0$	Adverse slope	$h > h_c$ $h < h_c$	A_2 A_3		

MILD SLOPE ($0 < i < i_c$)

For the flow in a mild slope channel, $h_0 > h_c$. The normal depth line, $N-N$, and the critical depth line, $C-C$, divide the flow field into three depth ranges, as shown in Table 9-2.

In zone I ($h > h_0 > h_c$), $h > h_0$, it follows $i > J$, $i - J > 0$; $h > h_c$, thus Fr < 1, $1 - \text{Fr}^2 > 0$. From Eq. 9.76 we have $\frac{dh}{ds} > 0$, which means the water surface gradually rises along the channel and the water surface profile is a backwater curve, designated M_1. As $h \to h_0$ at the far away upstream, $J \to i$, that is $i - J \to 0$, $\frac{dh}{ds} \to 0$, or in other words, the water surface asymptotically approaches the normal depth line $N - N$, namely the normal depth line is the upstream asymptote of curve M_1. As $h \to +\infty$ at the far away downstream, $J \to 0$, $i - J \to i$, thus $\frac{dh}{ds} \to i$, the water surface approaches a horizontal line.

In zone II ($h_0 > h > h_c$), $h < h_0$, it follows $i < J$, $i - J < 0$; $h > h_c$, thus Fr < 1, $1 - \text{Fr}^2 > 0$. From Eq. 9.76 we have $\frac{dh}{ds} < 0$, which means the water surface gradually drops along the channel and the water surface profile is a dropdown profile, designated M_2. As $h \to h_0$ at the far away upstream, $J \to i$, that is $i - J \to 0$, $\frac{dh}{ds} \to 0$, the water surface asymptotically approaches the normal depth line $N - N$, namely the normal depth line is the upstream asymptote of curve M_2. As $h \to h_c$ at the far away downstream, Fr $\to 1$, thus $\frac{dh}{ds} \to -\infty$, that is, the water surface tends to intersect the critical depth line $C - C$ with a right angle, or hydraulic drop occurs.

In zone III ($h_0 > h_c > h$), $h < h_0$, $i < J$, $i - J < 0$; $h < h_c$, thus Fr > 1, $1 - \text{Fr}^2 < 0$. From Eq. 9.76 we have $\frac{dh}{ds} > 0$, which indicates the water surface profile is a backwater curves, designated M_3. At the far away upstream, the water surface profile is determined by the upstream condition. As $h \to h_c$ at the far away downstream, Fr $\to 1$, thus $\frac{dh}{ds} \to +\infty$, that is, the water surface tends to intersect the critical depth line $C - C$ with a right angle, or hydraulic jump occurs.

STEEP SLOPE ($i > i_c > 0$)

For the flow in steep slope channel, $h_0 < h_c$. The normal depth line, $N - N$, and the critical depth line, $C - C$, also divide the flow field into three depth ranges:

In zone I ($h > h_c > h_0$), $h > h_0$, $i > J$, $i - J > 0$; $h > h_c$, thus Fr < 1, $1 - \text{Fr}^2 > 0$. From Eq. 9.76 we have $\frac{dh}{ds} > 0$, which indicates the water surface profile is a backwater curve, designated S_1. As $h \to h_c$ at the far away upstream, Fr $\to 1$, $\frac{dh}{ds} \to \infty$, the water surface tends to intersect the critical depth line $C - C$ with a right angle, or hydraulic jump occurs. As $h \to +\infty$ at the far away downstream, $J \to 0$, $i - J \to i$, thus $\dfrac{dh}{ds} \to i$, the water surface approaches a horizontal line.

In zone II ($h_c > h > h_0$), $h > h_0$, thus $i > J$, $i - J > 0$; $h < h_c$, thus $\mathrm{Fr} > 1$, $1 - \mathrm{Fr}^2 < 0$. From Eq. 9.76 we have $\dfrac{dh}{ds} < 0$, which indicates the water surface profile is a dropdown profile, designated S_2. As $h \to h_c$ at the far away upstream, $\mathrm{Fr} \to 1$, $\dfrac{dh}{ds} \to -\infty$, the water surface tends to intersect the critical depth line $C - C$ with a right angle, or hydraulic drop occurs. As $h \to h_0$ at the far away downstream, $J \to i$, that is $i - J \to 0$, $\dfrac{dh}{ds} \to 0$, the water surface profile asymptotically approaches the normal depth line $N - N$.

In zone III ($h_c > h_0 > h$), $h < h_0$, $i < J$, $i - J < 0$; $h < h_c$, thus $\mathrm{Fr} > 1$, $1 - \mathrm{Fr}^2 < 0$. From Eq. 9.76 we have $\dfrac{dh}{ds} > 0$, which indicates the water surface profile is a backwater curves, designated S_3. At the far away upstream, the water surface profile is determined by the upstream condition. As $h \to h_0$ at the far away downstream, $J \to i$, that is $i - J \to 0$, $\dfrac{dh}{ds} \to 0$, the water surface profile asymptotically approaches the normal depth line $N - N$.

CRITIC SLOPE ($i = i_c$)

For the flow in a critical slope channel, $h_0 = h_c$. The normal depth line, $N - N$, and the critical depth line, $C - C$, coincide with each other, which divide the flow field into two depth zones: zone I [$h > h_0(h_c)$] and zone III [$h < h_0(h_c)$]. A backwater curve, C_1 and C_3, forms in zone I and zone III, respectively. As $h \to h_0(h_c)$ for both the profiles C_1 and C_3, both the profiles of C_1 and C_3 asymptotically approach horizontal line.

HORIZONTAL SLOPE ($i = 0$)

For the channel with horizontal slope, uniform flow cannot form in the channel, that is, $h_0 = \infty$. Thus, there are only zone II and zone III in the flow region separated by the critical depth line, $C - C$. With $i = 0$ Eq. 9.76 becomes

$$\frac{dh}{ds} = \frac{-J}{1 - \mathrm{Fr}^2}$$

In zone II, $h > h_c$, $\mathrm{Fr} < 1$, $1 - \mathrm{Fr}^2 > 0$, $\dfrac{dh}{ds} < 0$, the water surface profile is a dropdown profile, designated H_2; In zone III, $h < h_c$, $\mathrm{Fr} > 1$, $1 - \mathrm{Fr}_2 < 0$, $\dfrac{dh}{ds} > 0$, the water surface profile is a backwater profile, designated H_3.

ADVERSE SLOPE ($i < 0$)

For the channel with adverse slope, uniform flow cannot form in the channel, $h_0 =$

imaginary. Thus, there are only zone II and zone III in the flow region separated by the critical depth line, $C - C$. With $i = -|i|$, Eq. 9.76 becomes

$$\frac{dh}{ds} = \frac{-|i| - J}{1 - \mathrm{Fr}^2}$$

In zone II, $h > h_c$, $\mathrm{Fr} < 1$, $1 - \mathrm{Fr}^2 > 0$, $\frac{dh}{ds} < 0$, the water surface profile is a dropdown profile, designated A_2; In zone III, $h < h_c$, $\mathrm{Fr} > 1$, $1 - \mathrm{Fr}^2 < 0$, $\frac{dh}{ds} > 0$, the water surface profile is a backwater profile, designated A_3.

SUMMARY OF PROFILES

According to the above discussion, those twelve distinct types of profiles in the prismatic open channel have the following features:

(1) Flows in zone I, the flow depth increases in the flow direction (backwater curve) and the surface profile approaches the horizontal plane asymptotically; Flows in zone II, the flow depth decreases (dropdown curve) and the surface profile approaches the lower of h_c or h_0; Flows in zone III, the flow depth increases (backwater curve) and the surface profile tends the lower of h_c or h_0. These trends in surface profiles continue as long as there is no change in bottom slope or roughness.

(2) Except profiles C_1 and C_3 in the open channel with critical slope, all the profiles in channel with other slopes asymptotically approach the normal depth line $N - N$ if $h \to h_0$, and tend to intersect the critical depth line $C - C$ with a right angle, hydraulic jump or hydraulic drop occur if $h \to h_c$.

(3) Control section is a section where a unique relationships between the discharge and the depth of flow. The disturbance wave in supercritical flow propagates only downstream, thus the control section for the water surface profile in supercritical flow (M_3, S_2, S_3, C_3, H_3, A_3) must be located at the upstream. The disturbance wave in subcritical flow can propagate both downstream and upstream, thus the control section for the water surface profile in subcritical flow (M_1, M_2, S_1, C_1, H_2, A_2) must be located at the downstream. In general, gates, weirs, and sudden falls and critical depths are some examples of control sections as indicated in Table 9-2 or Fig. 9-21. The normal depth h_0, which is the flow depth reached when uniform flow is established, also serves as a control point.

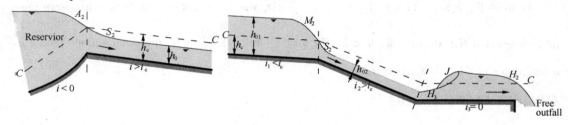

Fig. 9-21 Common control sections

EXAMPLE 9-9 Water surface profiles

Sketch the water surface profiles for two reaches of the open channel given in Fig. E9-9 below. There is no change in bottom slopes or roughness of the long, straight upstream and downstream prismatic channels.

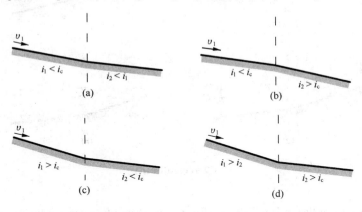

Fig. E9-9

Solution:

Because the upstream and downstream channels are long, straight channels with a constant bottom slope and roughness, both the free surfaces far upstream and downstream coincide with the normal depth lines. Thus, sketch normal depth line $N - N$ and critical depth line $C - C$ to separate the flow region into three zones, then the water surface profile as follows.

1. The open-channel section in Fig. E9-9a involves a slope change from mild to much mild ($i_2 < i_1 < i_c$). The flow velocity in the milder part is lower, and thus the flow depth is higher when uniform flow is established again (see Fig. E9-9e). Noting that uniform flow with much mild slope must be subcritical ($h > h_c$), its control section is located at upstream end, the intersected section, $h = h_{02}$, thus the flow depth in mild channel increases from the initial to the new uniform level smoothly through an M_1 profile.

2. The open-channel section in Fig. E9-9b involves a slope change from mild to steep ($i_1 < i_c$, $i_2 > i_c$). The flow depth in mild channel (subcritical flow) is higher than the flow depth in steep channel (supercritical flow) when uniform flow is established again (see Fig. E9-9f). It indicates a hydraulic drop could occur, and the control section is located at the intersected section, whose flow depth is $h = h_c$. Thus, the flow depth in mild channel decreases from the initial depth, h_{01}, to the critical depth, h_c, at the intersected section smoothly through an M_2 profile, then to h_{02} in steep channel through an S_2 profile.

3. The open-channel section in Fig. E9-9c involves a slope change from steep to mild ($i_1 > i_c$, $i_2 < i_c$). The flow depth in steep channel (supercritical flow) is much lower than the flow depth in mild channel (subcritical flow) when uniform flow is established again (see Fig. E9-9g). It indicates a hydraulic jump will occurs. If the depth h_{01} and h_{02} are conjugate depths, the hydraulic jump (J_2 in the figure) occurs at the intersected section. If h_{02} is greater than the conjugate depth of h_{01}, the hydraulic jump (J_1 in the figure) occurs prior to the intersected section, then flow depth in steep channel increases from the conjugate depth of h_{01} to h_{02} at the intersected section smoothly through an S_1 profile. If h_{02} is less than the conjugate depth of h_{01}, the hydraulic jump (J_3 in the figure) occurs after the intersected section, depth at the intersected section is $h = h_{01}$, then the flow depth in mild channel increases from h_{01} to the pre-jump conjugate depth of h_{02} smoothly through an M_3 profile, then hydraulic jump occurs.

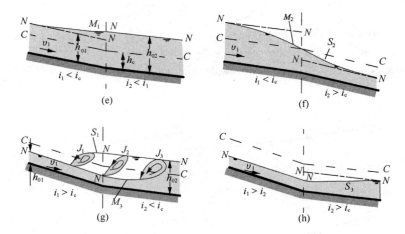

Fig. E9-9(continued)

4. The open-channel section in Fig. E9-9d involves a slope change from steep to less steep ($i_1 > i_2 > i_c$). The flow velocity in the less steep part is lower (a smaller elevation drop to drive the flow), and thus the flow depth is higher when uniform flow is established again (Fig. E9-9h). Noting that uniform flow with steep slope must be supercritical ($h < h_c$), its control section is located at its downstream end, the intersected section, $h = h_{01}$, thus the flow depth in less steep channel increases from the initial to the new uniform level smoothly through an S_3 profile.

9.8 Numerical Solution of Surface Profile

The prediction of the surface profile $h(s)$ is an important part of the design of open-channel systems. A good starting point for the determination of the surface profile is the identification of the control section along the channel, at which the flow depth can be calculated from the knowledge of flow rate as discussed above. Once flow depths at control sections are available, the surface profile upstream or downstream can be determined usually by the nonlinear differential equation, Eq. 9.76

$$\frac{dh}{ds} = \frac{i - J}{1 - \mathrm{Fr}^2} = \frac{i - J}{1 - \dfrac{\alpha Q^2}{gA^3}B}$$

and the specific energy relation, Eq. 9.47

$$\frac{dE_s}{dh} = 1 - \frac{\alpha Q^2}{gA^3}B$$

Combining those two relations yields

$$ds = \frac{dE_s}{i - J}$$

in form of differential gives

$$\Delta l = \Delta s = \frac{\Delta E_s}{i - \bar{J}} = \frac{E_{sd} - E_{su}}{i - \bar{J}} \qquad (9.77)$$

where Δl is the length of differential reach of channel; ΔE_s is the difference of specific energies between the downstream and upstream ends, $\Delta E_s = E_{sd} - E_{su} = \left(h_d + \frac{\alpha_d v_d^2}{2g}\right) - \left(h_u + \frac{\alpha_u v_u^2}{2g}\right)$; $\bar{J}$ is the average friction slope of flow in the differential reach, $\bar{J} = \frac{\bar{v}^2}{\bar{C}^2 \bar{R}_h}$, where $\bar{v} = \frac{v_u + v_d}{2}$, $\bar{C} = \frac{C_u + C_d}{2}$, $\bar{R}_h = \frac{R_{hu} + R_{hd}}{2}$; subscripts 'u' and 'd' present the upstream and downstream ends of the river reach respectively.

Thus, the total length in the flow direction is

$$l = \Sigma l = \Sigma \frac{\Delta E_s}{i - \bar{J}} \qquad (9.78)$$

Equations 9.77 and 9.78 can be used to calculate the free surface profile of one dimension flow in open channel, which can be used for both the non-prismatic and prismatic channel. The procedure is:

1. Starting from the control section, let its depth to be the initial depth of h_u (or h_d) of first differential reach;

2. Assuming the depth at the adjacent section to be the second depth of h_d (or h_u) of the first differential reach;

3. Calculating ΔE_s and $\bar{J}$, then substituting into Eq. 9.77 to get the length, Δl_1, of the first differential reach;

4. Let the second depth of Δl_1 to be the initial depth of h_u (or h_d) of the second differential reach, assume the second depth of h_d (or h_u) of the second differential reach, then calculate the length, Δl_2, of the second differential reach, and so on;

5. When the sum of lengths of every reaches, $\Sigma \Delta l$, equals to the total length of channel, l, the calculated length and depth then can be used to sketch the surface profile.

EXAMPLE 9-10 Numerical calculation of water surface profile

Given is a 3386-m-long trapezoidal drainage channel connecting a lake and a sluice gate with a bottom slope $i = 1/3000$. The channel bottom width $b = 45$m, surface roughness $n = 0.025$, and the side slope of trapezoidal section $m = 2.0$. If the depth prior to the gate is 8.95m at a flow rate of $Q = 500$ m³/s, determine the surface profile along the channel from lake to gate.

Solution:

Assume the flow is steady and incompressible.

1. Determine the type of water surface profile.

From Eq. 9.13 gives the normal depth $h_0 = 4.92$ m. From Eq. 9.50 gives the critical depth $h_c = 2.25$m. Thus, $h_0 > h_c$, the flow in channel is a subcritical flow. Meanwhile, the depth prior to the gate, $h = 8.95$m, gives $h > h_0 > h_c$, which indicates the surface profile is located at zone I of subcritical flow, it is a M_1 profile.

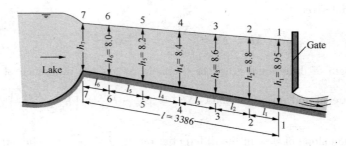

Fig. E9-10

2. Calculation of surface profile

Separate the whole length of channel into six reaches. Let the section prior to the gate to be the control section and start the calculation from this initial section to upstream.

Here is the procedure: The depth at the section 1-1 is $h_1 = 8.95$m. Assume the depth at other sections from 2-2 to 6-6 are 8.8m, 8.6m, 8.4m, 8.2m and 8.0m, respectively, calculate the length of each reach, Δl_{1-2}, ..., Δl_{5-6} by Eq. 9.77. Then let Δl_{6-7} to be $l - \sum \Delta l$ and get the depth at the entrance of channel h_7 from Eq. 9.78. Finally, draw the surface profile in the figure based on the calculated lengths of reaches and depths at the sections.

(1) Calculate the lengths of reaches

The length of reach 1 between section 1-1 to 2-2, Δl_{1-2}:

For section 1-1, the initial depth at section 1-1 is $h_1 = 8.95$ m, thus,

$$A_1 = h_1(b + mh_1) = (8.95\text{m})[45\text{m} + (2)(8.95\text{m})] = 562.96 \text{ m}^2$$

$$\chi_1 = b + 2h_1\sqrt{1 + m^2} = 45\text{m} + (2)(8.95\text{m})\sqrt{1 + 2.0^2} = 85.02\text{m}$$

$$R_{h1} = \frac{A}{\chi} = \frac{562.96 \text{ m}^2}{85.02\text{m}} = 6.62\text{m}$$

$$C_1 = \frac{1}{n} R_{h1}^{1/6} = \frac{1}{0.025} 6.62^{1/6} = 54.81 \text{ m}^{1/2}/\text{s}$$

$$v_1 = \frac{Q}{A_1} = \frac{500 \text{ m}^3/\text{s}}{562.96 \text{ m}^2} = 0.89\text{m/s}$$

$$E_{s1} = h_1 + \frac{\alpha_1 v_1^2}{2g} = 8.95\text{m} + \frac{(1.0)(0.89\text{m/s})^2}{(2)(9.81\text{m/s}^2)} = 8.99\text{m}$$

For section 2-2, assume $h_2 = 8.8$m, with the exact same manner gives

$A_2 = 550.88 \text{ m}^2$, $\chi_2 = 84.35$m, $R_{h2} = 6.53$m, $C_2 = 54.69 \text{ m}^{1/2}/\text{s}$, $v_2 = 0.907$m/s, $E_{s2} = 8.84$m

Thus we have

$$\Delta E_{s1-2} = E_{s1} - E_{s2} = 8.99\text{m} - 8.84\text{m} = 0.15\text{m}$$

$$\bar{v}_{1-2} = \frac{v_1 + v_2}{2} = \frac{0.89\text{m/s} + 0.907\text{m/s}}{2} = 0.899\text{m/s}$$

$$\bar{C}_{1-2} = \frac{C_1 + C_2}{2} = \frac{54.81 \text{ m}^{1/2}/\text{s} + 54.69 \text{ m}^{1/2}/\text{s}}{2} = 54.75 \text{ m}^{1/2}/\text{s}$$

$$\bar{R}_{h1-2} = \frac{R_{h1} + R_{h2}}{2} = \frac{6.62\text{m} + 6.53\text{m}}{2} = 6.58\text{m}$$

$$\bar{J}_{1-2} = \frac{\bar{v}_{1-2}^2}{\bar{C}_{1-2}^2 \bar{R}_{h1-2}} = \frac{(0.899\text{m/s})^2}{(54.75 \text{ m}^{1/2}/\text{s})^2(6.58\text{m})} = 0.0000409$$

Thus the length of reach between section 1-1 to section 2-2 can be obtained by Eq. 9.77:

$$\Delta l_{1\text{-}2} = \frac{\Delta E_{s1\text{-}2}}{i - J_{1\text{-}2}} = \frac{0.15\text{m}}{\frac{1}{3000} - 0.0000409} = 513\text{m}$$

With the assumption of $h_3 = 8.6$m, $h_4 = 8.4$m, $h_5 = 8.2$m and $h_6 = 8.0$m at the sections from 3-3 to 6-6, we can get the lengths of reaches as indicated in the following table.

Depths and lengths	Sections						
	1-1	2-2	3-3	4-4	5-5	6-6	7-7
h(m)	8.95	8.8	8.6	8.4	8.2	8.0	7.97
Δl(m)		513	677	690	700	710	95
$\sum \Delta l$(m)		513	1190	1880	2580	3290	3385 ≈ 3386

(2) Determine the depth at the entrance of channel

The length of last reach $\Delta l_{6\text{-}7} = l - \sum_{i=2}^{6} \Delta l_i = 3386\text{m} - 3290\text{m} = 96\text{m}$ as indicated in the table. Thus, the depth at the entrance of channel can be obtained by Eq. 9.77 by the method of trail and error

$$h_7 = 7.97\text{m}$$

Finally, draw the surface profile in the figure based on the calculated lengths of reaches and depths at the sections and connect them to be the surface profile as indicated in Fig. E9-10.

DISCUSSION If the computed value of $\Delta l_{6\text{-}7}$ is less than 0 it means the assumption of the depths h_2 to h_6 at section 2 to 6 is overestimated. It should decrease the depths at these sections and restart the calculation from section 2 over again.

9.9 Flow Measurements in Open Channel

For flow measurement in open channel, weirs and flumes applications are widespread and typically employ a non-contact ultrasonic level controller to provide level-to-flow calculations. For small open channels, sewers, culverts, streams and rivers, measurement technologies include Doppler sensor, submerged probe, bubbler, area-velocity and laser-based flow meters. For large channels and rivers, ADCP flow meters (Accoustic Doppler Current Profiler) can be used. For shallow applications it is possible to use multiple area-velocity flow meters in series to provide flow monitoring for one larger cross-sectional area, e.g. the braided rivers or low-flow rivers. For summary, the methods for measurement in open channel can be approximately divided into four categories: the hydraulic structures, the area-velocity method, the slope-area method and the tracer method.

9.9.1 Hydraulic Structures

The most common method of measuring open channel flow is the *hydraulic structures method*. A calibrated restriction inserted into the channel controls the shape and velocity of the flow. The flow rate is then determined by measuring the liquid level in or near the restriction. The hydraulic structures have two broad categories: weirs and flumes.

A *weir* is an obstruction or dam built across an open channel over which the liquid flows through a specially shaped opening: triangular (or V-notch), rectangular, or trapezoidal. The flowrate over a weir is determined by measuring the liquid depth in the three to four times the liquid depth upstream from the weir, which is discussed in detailed in Chapter 10.

A *flume* is a specially shaped open channel flow section providing a restriction in channel area and/or a change in channel slope. The flowrate in the channel is determined by measuring the liquid depth at a specified point in the flume. There are three typical flumes: Venturi, Parshall and Palmer-Bowlus flume (Fig. 9-22). In China, most commonly used flumes are Parshall flumes and Palmer-Bowlus flumes.

Fig. 9-22 (a) Venturi flume[1], (b) Parshall flume[2] and (c) Palmer-Bowlus flume[3] in a manhole

PARSHALL FLUME

The *Parshall flume* was designed in the later 1920s to measure the flow of irrigation water. Today, this type of flume is often used to measure the flow of wastewater, for permanent or temporary installations. Unlike the Venturi flume with a flat bottom as in Fig. 9-22a, Parshall flume consists of a converging upstream section, a throat, and a diverging downstream section. The walls are vertical and the floor of the throat is inclined *downward*, as shown in Fig. 9-22b. The flow rate through a Parshall flume is determined by measuring the liquid level one third of the way into the converging section. Parshall flumes are designed to handle moderate flowrates, which make the device self-cleaning.

The free-flow discharge of Parshall flume can be summarized as

$$Q = CH^n$$

where Q is flow rate; C is the free-flow coefficient for the flume, which is a function of the dimension of the constriction; H is the head measured at the required point; n is the constant of the exponent, the value of which is a function of the dimension of the constriction, e.g. 1.55 for a 25.4-mm flume.

Parshall flumes are designated by the width of the throat, and are standardized into 22 sizes

---

[1] http://www.aquamonitoring.cz/produkty/products/show/?categoryId=49&itemId=33

[2] http://www.panoramio.com

[3] http://www.openchannelflow.com/blog/article/using-ultrasonic-flow-meters-with-flumes

ranging from 25.4mm to 15.24m and covering flow ranges from 0.1416 l/s to 92,890 l/s. The throat width and all other dimensions must be strictly followed so that standard discharge tables can be used. Parshall flumes are available in a variety of materials, including aluminum, fiberglass, galvanized steel, lexan (small sizes), PVC (small sizes) and stainless steel.

It should be noted that the drop in the floor of the Parshall flume makes it difficult to install a Parshall flume in an existing channel.

PALMER-BOWLUS FLUME

In contrast, a *Palmer-Bowlus flume* was designed in the 1930s for use as a flume that can be inserted into an existing channel with a slope of less than 2%. The Palmer-Bowlus is a Venturi type flume with a uniform throat over a flat bottom, which is usually installed in the sewerage system with a circular cross section. Thus, they are designated by the size of the pipe into which they fit, that is, the length of the throat is equal to the diameter of the corresponding flume. Standard sizes of a Palmer-Bowlus flume are 101.6mm to 1066.8mm. Its dimensional configuration is not rigidly established for each flume size. However, a Palmer-Bowlus flume with a trapezoidal throat with a flat bottom has emerged as the standard design for circular pipes. Palmer-Bowlus flumes are usually made of prefabricated fiberglass that is reinforced with plastic. The flow rate through it is determined by measuring the liquid depth at a point one-half pipe diameter upstream from the flume throat.

Flumes are more expensive and more difficult to install than weirs. However, flumes result in a lower head loss and are self-cleaning, requiring less maintenance than a weir. Preferably, weirs and flumes are installed so that *non-submerged flow* exists, that is, so that downstream conditions do not affect the flow rate through the device. If *submerged flow* occurs, submerged flow tables should be referred.

When the weirs and flumes are used to measure the flowrate in the open channel, the technologies to measure the level in the channel are needed. The most common technologies are *ultrasonic sensors*, *bubblers* and *submerged pressure transducers*. Once the liquid level at the point in the channel is measured, it can be converted into flowrate based on the known *level-to-flowrate* relationship of the weir or flume. For most modern open channel flow meters, it uses software to convert the measured level into flowrate.

9.9.2 Area-velocity Method

The method of determining the area and velocity of flow is perhaps the most common method of measuring the flowrate of a river. The *area-velocity method* calculates flowrate by multiplying the cross sectional area A of the flow by its average velocity v: $Q = Av$.

The flowmeters measuring the average velocity can approximately be divided into four categories: anemometer-propeller velocity meter; electromagnetic velocity meters; Doppler velocity meters and optical strobe velocity meters, which are briefly described below. Depth measurements are obtained by either a separate ultrasonic sensor or an integrated pressure transducer. Some of the meters utilize both of these technologies. Flowrate is then calculated using a *stage-area relationship* that is pre-programmed into each of the meters using manufacturer specific software.

ANEMOMETER-PROPELLER CURRENT METERS

Anemometer-propeller current meters are the most common type used for irrigation and watershed measurements. These meters use anemometer conical cup wheels with vertical axis of rotation or propellers with horizontal axis to sense velocity. These meters are rated by dragging them through tanks of still water at known speeds. The reliability and accuracy of measurement with these meters are easily assessed by checking mechanical parts for damage and using spin-time tests for excess change of bearing friction. This type current meter does not sense direction of velocity, which may cause problems in complicated flow where backflow might not be readily apparent. For irrigation needs, this problem can be avoided by proper gage station or single measurement site selection.

ELECTROMAGNETIC CURRENT METERS

Electromagnetic current meters produce voltage (Faraday's law) directly proportional to the velocity of the water when water moving through this magnetic field. The higher the velocity, the greater voltage created. The measured voltage is used to determine the velocity at the sensor, which is used to estimate the average channel velocity based on theoretical velocity profiles. One advantage of these current meters is direct analog reading of velocity; counting of revolutions is not necessary. These current meters can also measure crossflow and are directional. The use of electromagnetic velocity meters is very popular among water districts.

DOPPLER TYPE CURRENT METERS

Doppler type current meters determine velocity by measuring the change of source light or sound frequency from the frequency of reflections from moving particles such as small sediment and air bubbles. Laser light is used with laser Doppler velocimeters (LDV) (see Section 8.7.1), and sound is used with acoustic Doppler velocimeters (ADV). Acoustic Doppler current profilers (ADCP) have also been developed. Most of the meters in this class are multidimensional or can simultaneously measure more than a single directional component of velocity at a time.

Acoustic Doppler velocimetry (ADV) is designed to record instantaneous velocity components in x-, y- and z-direction at a *single-point* with a relatively high frequency (Fig. 9-23). It transmits acoustic signals and then receives their echoes reflected by sound scatterers in the water. From the measured frequency shift between emissions and echoes, flow velocity is de-

Fig. 9-23 ADV measurement in a channel (*Courtesy of the Environmental Hydraulic Laboratory, Zhejiang University*)

duced based on the Doppler shift effect.

Acoustic Doppler Current Profiler (ADCP), or Acoustic Doppler Profiler (ADP) measures the speed and direction of currents moving across the entire water column by using the principle of 'Doppler shift'. To make a discharge measurement, the ADCP is mounted onto a boat or into a small watercraft with its acoustic beams directed into the water from the water surface (Fig. 9-24). The ADCP is then guided across the surface of the river to obtain measurements of velocity and depth across the channel. The river-bottom tracking capability of the ADCP acoustic beams or a Global Positioning System (GPS) is used to track the progress of the ADCP across the channel and provide channel-width measurements. Because the emitted sound extends from the ship down to the channel bottom, the ADCP measures the current at many different depths simultaneously. This way, it is possible to determine the speed and direction of the current from the surface of the channel to the bottom, and measure average velocities of cells of selected size in a vertical series. ADCP measures *vertical current profiles*, and is more frequently used in deep flow in reservoirs, oceans, and large rivers now.

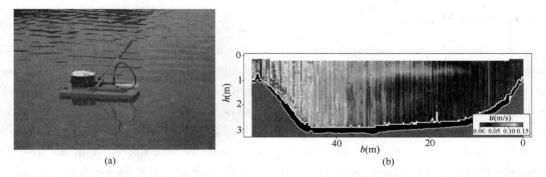

Fig. 9-24 (a) ADCP measurement of Huaicheng river in Jiaxing, Zhejiang (*Courtesy of Huiping, Zhou at Zhejiang University*); (b) Velocity profile.

OPTICAL STROBE VELOCITY METERS

Optical strobe velocity meters use optical methods to determine surface velocities of streams. This meter uses the strobe effect. Mirrors are mounted around a polygon drum that can be rotated at precisely controlled speeds. Light coming from the water surface is reflected by the mirrors into a lens system and an eyepiece. The rate of rotation of the mirror drum is varied while viewing the reflected images in the eyepiece. At the proper rotational speed, images become steady and appear as if the surface of the water is still. By reading the rate of rotation of the drum and knowing the distance from the mirrors to the water surface, the velocity of the surface can be determined. The discharge rate of the stream may be estimated by applying the proper coefficient to this surface velocity and multiplying by the cross-sectional area of the flow section.

This class of meter has several advantages. No parts are immersed in the flowing stream. Moreover, it can be used for high-velocity flows and for flows carrying debris and heavy sediment. The meter can measure large floodflows from bridges. However, the meter measures only the water surface velocity and is very dependent upon the selection of the proper coefficient.

9.9.3 Slope-area Method

Various resistance equations are used to estimate flow rate based on measurements of the water surface slope, cross-sectional area, and wetted perimeter over a length of uniform channel. The most popular of these equations is the *Chézy formula*, Eq. 9-16, as discussed in Section 9.2. Given the size, shape, slope and roughness of the channel, an open channel flow meter can calculate flow rate using the Chézy formula based on a measurement of the liquid depth.

Because the proper selection of the roughness factor, n, for many streams is difficult and is, at best, an estimate, the discharge determined by the *slope-area method* is only approximate. Care must be taken to determine the slope and areas simultaneously if the water levels are changing. Nevertheless, the slope-area method is not as accurate as the hydraulic structures and area-velocity methods, but it can provide sufficient accuracy in some applications. In addition, no weir or flume is required.

9.9.4 Tracer Methods

Tracer methods can be used to determine discharge with accuracies that can vary considerably from about ±1% to over ±30%, depending on the equipment used and the care in applying the techniques. Basically, a tracer is considered anything that mixes with or travels with the flow and is detectable. Some tracers that have been used are: dyes of various colors; other chemicals such as fertilizer, salt, and gases; radioisotopes; neutrally buoyant beads or floats. For irrigation measurements, salts and dyes are the most convenient and commonly used tracers. They are used to determine discharge in two basic ways: (1) the *tracer-velocity-area method*, in which time of tracer travel through a known channel length and average cross-sectional area determine discharge; and (2) the *tracer-dilution method*, in which discharge is determined by the downstream concentration of fully mixed tracer, which has been added upstream at a constant rate, and by accounting for the amount of tracer solids. Tracer methods are rarely used in China.

Summary

This chapter is an introduction to the steady, one dimensional *open-channel flow*, a flow driven by the component of gravity in the direction of flow and open to the atmosphere or in partially filled conduits.

Flow at constant slope and depth (and thus the average velocity) is said to be *uniform flow*, which only occurs in the long and straight *prismatic* downhill channel with constant surface roughness and flow rate along the flow direction. The uniform flow represents a balance between weight and friction forces, and satisfies the classical *Chézy equation*, a relationship among the flowrate of uniform flow, the slope of the channel, the geometry of the channel, and the roughness of the channel surfaces and expressed as

$$v = C\sqrt{R_h i} = C\sqrt{R_h J}$$
$$Q = AC\sqrt{R_h i} = K\sqrt{i} = K\sqrt{J}$$

with the *Chézy coefficient C* determined by *Manning formula*

$$C = \frac{1}{n} R_h^{1/6}$$

The uniform flow in a partly full *circular* pipe differs from the one in an open channel that its maximum velocity and flow rate occur before the pipe is completely filled by the transporting liquid.

Straight prismatic channels can be optimized to find the *best hydraulic cross section* with the maximum hydraulic radius, or equivalently, the one with the minimum wetted perimeter for a specified cross section. For a best cross section for trapezoidal channel, the *aspect ratio* of width to depth is

$$\beta_m = \left(\frac{b}{h}\right)_m = 2(\sqrt{1+m^2} - m)$$

which gives the best cross section for trapezoidal channel is half of a hexagon, for rectangular channel is $h = b/2$, and for circular channel is semicircular.

With variable flow rate, roughness or slope, *nonuniform flow* occurs. The *wave speed* ($c = \sqrt{gh}$), *specify energy* ($E_s = h + \frac{\alpha v^2}{2g}$), *critical depth* ($\frac{\alpha Q^2}{g} = \frac{A_c^3}{B_c}$ for general shape and $h_c = \sqrt[3]{\frac{\alpha q^2}{g}}$ for rectangular shape) and *critical slope* ($i_c = \frac{g}{\alpha C_c^2} \frac{\chi_c}{B_c}$) are presented to describe the nonuniform flow. For the *critical* condition of Froude number unity ($\text{Fr} = \frac{v}{\sqrt{gh}} = 1.0$), the flow has a minimum specific energy and gets a velocity equaling the speed of a small-amplitude surface wave in the channel. If the flow becomes *supercritical* ($\text{Fr} > 1$), it may undergo a *hydraulic jump* to a greater depth and lower velocity (*subcritical*).

In *rapidly varied flow* (RVF), the flow depth changes markedly over a relatively short distance in the flow direction. Any change from supercritical to subcritical flow occurs through a *hydraulic jump*, a highly dissipative process; and from subcritical to supercritical flow occurs through a *hydraulic drop*. The *depth ratio*, *energy loss*, and *energy dissipation ratio* during hydraulic jump are expressed as

$$\frac{h''}{h'} = 0.5(\sqrt{1 + 8\text{Fr}_1^2} - 1)$$

$$\Delta E_j = \frac{(h'' - h')^3}{4h'h''}$$

$$\text{Dissipation ratio} = \frac{\Delta E_j}{E_{s1}} = \frac{\Delta E_j}{h' + \frac{\alpha_1 v_1^2}{2g}} = \frac{\Delta E_j}{h'\left(1 + \frac{\alpha_1 \text{Fr}_1^2}{2}\right)}$$

In *gradually varied flow* (GVF), the application of the *momentum equation* on a control volume leads to a differential equation for the analysis of depth variation

$$\frac{dh}{ds} = \frac{i - J}{1 - Fr^2}$$

which can be used in both the qualitatively analysis or quantitatively calculation of the *water surface profiles* in an open channel.

For the five classes of channel *bed slope i* in a prismatic open channel- *mild, critical, steep, horizontal* and *adverse slope*, the qualitatively analysis of the differential equation gives rise to twelve distinct types of water surface profiles. All profiles in zone Ⅰ and Ⅲ are backwater curves and in zone Ⅱ are dropdown curves. Except profiles C_1 and C_3 in the open channel with critical slope, all profiles asymptotically approach the normal depth line $N - N$ if $h \rightarrow h_0$, and tend to intersect the critical depth line $C - C$ with a right angle if $h \rightarrow h_c$.

In the end, the methods of flow measurement in open channels, the flumes, the area-velocity method, the slope-area method and the traced methods, are briefly introduced.

Word Problems

W9-1 How does open-channel flow differ from pipe flow?

W9-2 What is the driving force for flow in an open channel? How is the flow rate in an open channel established?

W9-3 How does the pressure change along the free surface in an open-channel flow?

W9-4 Consider steady fully developed flow in an open channel of rectangular cross section with a constant slope of 5° for the channel bottom. Will the slope of the free surface also be 5°? Explain.

W9-5 What causes the flow in an open channel to be varied (or nonuniform)? How does rapidly varied flow differ from gradually varied flow?

W9-6 Define the hydraulic radius in an open channels. How does it differ from the one for the circular pipe?

W9-7 Given the average flow velocity and the flow depth, explain how you would determine if the flow in open channels is tranquil, critical, or rapid.

W9-8 What is the Froude number? How is it defined? What is its physical significance?

W9-9 In what kind of channels is uniform flow observed? Under what conditions is the uniform flow established?

W9-10 What are the hydraulic features of the uniform flow?

W9-11 Uniform flow can only be observed in the downhill-slope channel, and cannot be established in a horizontal or adverse-slope channel. Explain.

W9-12 What is normal depth? Explain how it is established in open channels.

W9-13 How is the friction slope defined? Under what conditions is it equal to the bottom slope of an open channel?

W9-14 Consider steady flow of water through a wide rectangular channel. Is the energy line of flow parallel to the channel bottom when the frictional losses are negligible? Explain.

W9-15 Consider steady one-dimensional flow through a wide rectangular channel. Someone claims that the total mechanical energy of the fluid at the free surface of a cross section is equal to that of the fluid at the channel bottom of the same cross section. Do you agree? Explain.

W9-16 Consider two uniform flows passing through two different rectangular channels. For the given parameters below, determine the relation between the normal depths of those two flows:

(a) $Q_1 > Q_2$ if $n_1 = n_2$, $i_1 = i_2$, $b_1 = b_2$;
(b) $b_1 > b_2$ if $n_1 = n_2$, $i_1 = i_2$, $Q_1 = Q_2$;
(c) $i_1 > i_2$ if $n_1 = n_2$, $b_1 = b_2$, $Q_1 = Q_2$;
(d) $n_1 > n_2$ if $i_1 = i_2$, $b_1 = b_2$, $Q_1 = Q_2$.

W9-17 In uniform open-channel flow, what is the balance of forces? Can you use such a force balance to derive the Chézy equation, Eq. 9.13?

W9-18 What is the best hydraulic cross section for an open channel? Is it the best cross section of the open channel in the engineering practice design? Explain.

W9-19 Someone claims that the maximum flow rate through a circular cross section of an open channel usually occurs before the circular section is fully filled by the transporting liquid. Do you agree? Explain.

W9-20 Which is a best hydraulic cross section for each condition of open channel below?

(1) an open channel with a small, or a large hydraulic radius;
(2) an open channel with a cross section of (a) circular, (b) rectangular, (c) trapezoidal, or (d) triangular;
(3) a rectangular open channel whose fluid height is (a) half, (b) twice, (c) equal to, or (d) one-third the channel width;
(4) a trapezoidal channel of base width b is one for which the length of the side edge of the flow section is (a) b, (b) $b/2$, (c) $2b$, or (d) $\sqrt{3}b$.

W9-21 Consider uniform flow through an open channel lined with bricks with a Manning coefficient of $n = 0.015$. If the Manning coefficient doubles ($n = 0.030$) as a result of some algae growth on surfaces while the flow cross section remains constant, the flow rate will (a) double, (b) decrease by a factor of $\sqrt{2}$, (c) remain unchanged, (d) decrease by half, or (e) decrease by a factor of $2^{1/3}$.

W9-22 How does nonuniform or varied flow differ from uniform flow?

W9-23 Define the specific energy of a fluid flowing in an open channel in terms of head.

W9-24 In a downhill prismatic open channel, if the specific energy of the flow remains constant along the channel, is the flow a uniform flow? If the specific energy of the flow increases along the channel, does the friction slope equal the bed slope?

W9-25 Consider steady flow of water through two identical open rectangular channels at identical flow rates. If the flow in one channel is subcritical and in the other supercritical, can the specific energies of water in these two channels be identical? Explain.

W9-26 For a given flow rate through an open channel, which one is true: (a) the specific energy of the fluid will be minimum when the flow is critical, or (b) the specific energy will be minimum when the flow is subcritical?

W9-27 For a steady supercritical flow of water through an open rectangular channel at a constant flow rate, the larger is the flow depth, the larger the specific energy of water. Is it true? Explain.

W9-28 For a steady uniform flow through an open channel of rectangular cross section, which one is true: (a) the specific energy of the fluid remains constant, or (b) the specific energy decreases along the flow because of the frictional effects and thus head loss? Explain.

W9-29 What is critical depth in open-channel flow? For a given average flow velocity, how is it determined?

W9-30 What is critical slope of an open channel flow? For a given bed slope, can it be determined to be a mild or steep slope? Why?

W9-31 For a given flowrate and cross sectional shape of an open channel, how do the normal depth and critical depth vary with the increasing bed slope?

W9-32 Consider steady uniform flows through two long straight prismatic open channel A and B with identical sectional shape and sizes, under the conditions below, determine the relation between the normal depth and the critical depth for each pair of channels.
(a) $Q_A > Q_B$ if $n_A = n_B$, $i_A = i_B$;
(b) $i_A > i_B$ if $n_A = n_B$, $Q_A = Q_B$;
(c) $n_A > n_B$ if $i_A = i_B$, $Q_A = Q_B$.

W9-33 Define subcritical, critical and supercritical flow. For uniform or nonuniform flow, how to determine the flow is subcritical, critical or supercritical?

W9-34 How does gradually varied flow (GVF) differ from rapidly varied flow (RVF)?

W9-35 Someone claims that frictional losses associated with wall shear on surfaces can be neglected in the analysis of rapidly varied flow, but should be considered in the analysis of gradually varied flow. Do you agree with this claim? Justify your answer.

W9-36 For a steady subcritical flow of water in a horizontal channel of rectangular cross section, the flow depth will (a) increase, (b) remain constant, or (c) decrease in the flow direction. What happens to the flow depth if the flow is supercritical?

W9-37 For a steady flow of water in a wide and shallow channel of rectangular cross section, with the increasing of flowrate, the critical slope will (a) increase, (b) remain constant, (c) decrease or (d) uncertain.

W9-38 Consider steady flow of water in a downward sloped channel of rectangular cross section. If the flow is subcritical and the flow depth is greater than the normal depth, the flow depth will (a) increase, (b) remain constant, or (c) decrease in the flow direction. What happen to the flow depth if the flow depth is less than the normal depth? How about the case of the supercritical flow?

W9-39 Consider steady flow of water in an adverse-sloped channel of rectangular cross section. If the flow is supercritical, the flow depth will (a) increase, (b) remain constant, or (c) decrease in the flow direction. What happen to the flow depth if the flow is subcritical.

W9-40 The flow in an open channel is observed to have undergone a hydraulic jump. Is the flow upstream from the jump necessarily supercritical? Is the flow downstream from the jump necessarily subcritical?

W9-41 Is it possible for subcritical flow to undergo a hydraulic jump? Explain.

W9-42 Why is the hydraulic jump sometimes used to dissipate mechanical energy? How is the energy dissipation ratio for a hydraulic jump defined?

Problems

Introduction

9-1 Consider a steady flow through an open channel at a water depth of 2m, determine the cross-sectional area A, the wetted perimeter χ and hydraulic radius R_h in the following channels with the cross section of (a) rectangular with a base width of 3m; (b) trapezoidal with base width of 3m and side slope of 1.5; (c) triangle with right angle; and (d) semi-circular.

9-2 Water at 20℃ is flowing uniformly in a wide rectangular channel at an average velocity of 2m/s. If the water depth is 0.2m, determine (a) whether the flow is laminar or turbulent and (b) whether the flow is subcritical or supercritical.

9-3 Water at 15℃ is flowing uniformly in a 2-m-wide rectangular channel at an average velocity of 4m/s. If the water depth is 8 cm, determine whether the flow is subcritical or supercritical.

9-4 Water at 10℃ is flowing uniformly in a 3-m-diameter circular channel half-full at an average velocity of 2.5m/s. Determine the hydraulic radius, the Reynolds number, and the flow regime (laminar or turbulent).

Uniform flow

9-5 The finished-concrete channel of Fig. P9-5 with no barrier is designed for a flow rate of 6 m³/s at a normal depth of 1m. Determine (a) the design slope of the channel and (b) the

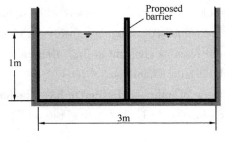

Fig. P9-5

percentage of reduction in flow if the surface is asphalt.

9-6 In Prob. 9-5, for finished concrete, determine the percentage of reduction in flow if the channel is divided in the center by the proposed barrier in Fig. P9-5. How does your estimate change if all surfaces are clay tile? Assume the Manning coefficient unchanged.

9-7 A rectangular channel with bottom width of 5m and a bed slope of 0.00136 carries water at a flow rate of $40 \text{m}^3/\text{s}$ at a normal depth of 3.05m. Determine the Manning coefficient of the channel.

9-8 A rectangular channel ($n = 0.013$) with a bed slope of 0.005 carries water at a flow rate of $11 \text{ m}^3/\text{s}$. If the aspect ratio $\beta = 1.5$, determine the base width of channel and the normal depth.

9-9 A trapezoidal aqueduct has a base width of 5m and the angle of side edge with respect to the horizontal line of 40° and carries a normal flow of 60 m^3/s of water when the normal depth is 3.2m. For clay tile surfaces, estimate the required elevation drop in m/km.

9-10 Two canals join to form a larger canal as shown in Fig. P9-10. Each of the three rectangular canals is lined with the same material and has the same bottom slope. The water depth in each is to be 2m. Determine the width of the merged canal, b. Explain physically (i.e., without using any equations) why it is expected that the width of the merged canal is less than the combined widths of the two original canals (i.e., $b < 4\text{m} + 8\text{m} = 12\text{m}$).

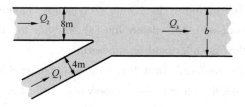

Fig. P9-10

9-11 A clean-earth trapezoidal channel with a bottom width of 1.5m and a side slope of 1:1 is to drain water uniformly at a rate of 8 m^3/s to a distance of 1km. If the flow depth is not to exceed 1m, determine the required elevation drop.

9-12 A trapezoidal channel with a base width of 5m, free surface width of 10m, and flow depth of 2.2m discharges water at a rate of 120 m^3/s. If the surfaces of the channel are lined with asphalt, determine the elevation drop of the channel per km.

9-13 Consider water flow through two identical channels with square flow sections of 3m × 3m. Now the two channels are combined, forming a 6-m-wide channel. The flow rate is adjusted so that the flow depth remains constant at 3m. Determine the percent increase in flow rate as a result of combining the channels.

9-14 A trapezoidal channel made of unfinished concrete has a bottom slope of 1°, base width of 5m, and a side surface slope of 1:1. For a flow rate of 25 m^3/s, determine the normal depth.

9-15 The trapezoidal channel of Fig. P9-15 is made of brickwork and slopes at 1 : 500. Determine the flow rate if the normal depth is 80cm.

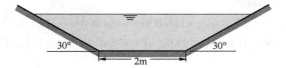

Fig. P9-15

9-16 Let the surface be clean earth for Prob. 9-15, which erodes if the flow velocity exceeds 1.5 m/s. What is the maximum depth to avoid erosion?

9-17 Water flows in a channel with a bed slope of 0.002 and a cross section as in Fig. P9-17. The dimensions and the Manning coefficients for the surfaces of different subsections are also given on the figure. Determine the flow rate through the channel.

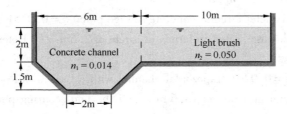

Fig. P9-17

9-18 Water flows in a channel with a bed slope of 0.5° and a cross section as in Fig. P9-18. The dimensions and the Manning coefficients for the surfaces of different subsections are also given on the figure. Determine the flow rate through the channel.

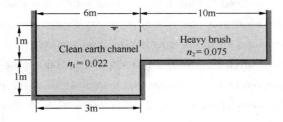

Fig. P9-18

9-19 An old, rough-surfaced, 2-m-diameter concrete pipe with a Manning coefficient of 0.025 carries water at a rate of 5.0 m^3/s when it is half full. It is to be replaced by a new pipe with a Manning coefficient of 0.012 that is also to flow half full at the same flowrate. Determine the diameter of the new pipe.

9-20 Water flows in a 2-m-diameter finished concrete pipe so that it is completely full and the pressure is constant all along the pipe. If the bed slope is 0.005, determine the flowrate by using open-channel flow methods. Compare this result with that obtained by using pipe flow methods of Chapter 8.

9-21 Consider a 1-m-internal-diameter water channel made of finished concrete. The channel bed slope is 0.002. For a flow depth of 0.25m at the center, determine the flow rate of water through the channel.

9-22 A 2-m-internal-diameter circular welded steel storm drain is to discharge water uniformly at a rate of 12 m^3/s to a distance of 1km. If the flow depth is to be 1.5m, determine the required elevation drop.

9-23 A water draining system with a constant slope of 0.0015 is to be built of three asphalt-lined circular channels. Two of the channels have a diameter of 1.2m and drain into the third channel. If all channels are to run half-full and the losses at the junction are negligible, determine the diameter of the third channel.

9-24 A circular pipe with a constant slope of 1/5000 and a Chézy coefficient of 50 $\sqrt{m}$/s carries sewage water at a maximum flow rate of 3.0 m^3/s. If the fullness of pipe is 0.8, determine the required diameter of pipe.

9-25 A 2-m-internal-diameter circular concrete sewage pipe ($n = 0.015$) with a constant slope of 0.00047 is to discharge water uniformly at a rate of 2.6 m^3/s. Determine the normal depth and pipe fullness.

9-26 A clean-earth ($n = 0.020$) trapezoidal channel with a base width of 10m and a side slope of 1.5 carries water at a rate of 15.6 m^3/s. If the non-scouring permissive velocity of the earth is 0.85m/s, determine: (a) the normal depth, and (b) the bed slope of channel.

9-27 Water is to be transported in an open channel whose surfaces are asphalt lined at a rate of 4 m^3/s in uniform flow. The bed slope is 0.0015. Determine the dimensions of the best cross section if the shape of the channel is (a) circular of diameter D, (b) rectangular of base width b, and (c) trapezoidal of base width b.

9-28 A trapezoidal concrete-lined (unfinished) channel with a bed slope of 0.0002 and a side slope of 2.0 carries water at a flow rate of 30 m^3/s. Design the base width of the channel based on the aspect ratio for the best hydraulic cross section.

9-29 A brickwork rectangular channel with a bed slope of 1/5000 carries water at a flow rate of 160 m^3/s to a water power plant. Design the bottom width of the channel and the normal depth based on the theory of the best hydraulic cross section. Is the design possible used in engineering practice? Explain.

9-30 A clean-earth trapezoidal channel ($n = 0.0225$) with a bed slope of 0.0004 and a side slope of 1.5 carries water at a flow rate of 12 m^3/s. If the aspect ratio is $\beta = 5$, determine the base width of channel and the normal depth.

Nonuniform flow

9-31 Water flows steadily in a 0.8-m-wide rectangular channel at a rate of 0.7 m^3/s. If the flow depth is 0.25m, determine the flow velocity and if the flow is subcritical or supercritical. Also determine the alternate flow depth if the character of flow were to change.

9-32 Water at 10°C flows in a 6-m-wide rectangular channel at a depth of 0.55m and a flow rate of 12 m³/s. Determine (a) the critical depth, (b) whether the flow is subcritical or supercritical, and (c) the alternate depth.

9-33 Water flows through a 4-m-wide rectangular channel with an average velocity of 5m/s. If the flow is critical, determine the flow rate of water.

9-34 A 3-m-wide rectangular channel carries flow at a flowrate of 10 m³/s at a depth of 2m. Is the flow subcritical or supercritical? For the same flowrate, what depth will give critical flow?

9-35 Determine the minimum depth in a 3-m-wide rectangular channel if the flow is to be subcritical with a flowrate of 60 m³/s.

9-36 The depth downstream of a sluice gate in a rectangular wooden channel of width 5m is 0.60m. If the flowrate is 18 m³/s, determine the channel slope needed to maintain this depth. Will the depth increase or decrease in the flow direction if the slope is (a) 0.02; (b) 0.01?

9-37 Determine the critical depth for a flow of 200 m³/s through a rectangular channel of 10m width. If the water flows 3.8m deep, is the flow supercritical? Explain.

9-38 When the channel of triangular cross section shown in Fig. P9-38 was new, a flowrate of Q caused the water to reach $L = 2$m up the side as indicated. After considerable use, the walls of the channel became rougher and the Manning coefficient, n, doubled. Determine the new value of L if the flowrate stayed the same.

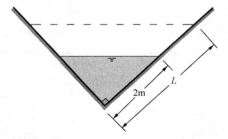

Fig. P9-38

9-39 Water flows uniformly in a rectangular channel with finished-concrete surfaces. The channel width is 3m, the flow depth is 1.2m, and the bottom slope is 0.002. Determine if the channel should be classified as mild, critical, or steep for this flow.

9-40 A smooth masonry-lined rectangular channel with a base width of 4m and a bed slope of 0.0009 carries water at a flow rate of 8 m³/s. Take the kinematic factor to be $\alpha = 1.1$. Determine whether the flow is subcritical or supercritical by the critical depth, wave speed, Froude number, specific energy and critical slope.

9-41 A clean-earth trapezoidal channel ($n = 0.0225$) with a base width of 10m, a side slope of 1.5 and a bed slope of 0.0004 carries water at a flow rate of 20 m³/s. Take the kinematic factor to be $\alpha = 1.1$. Try to determine the flow to be subcritical or supercritical by the critical depth, wave speed, Froude number, specific energy and critical slope.

Hydraulic Jump

9-42 Under appropriate conditions, water flowing from a faucet, onto a flat plate, and over the edge of the plate can produce a circular hydraulic jump as shown in Fig. P9-39. Consider a situation where a jump forms 75mm from the center of the plate with depths upstream and downstream of the jump of 1.74mm and 5.08mm, respectively. Determine the flowrate from the faucet.

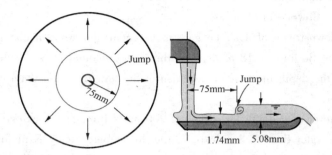

Fig. P9-42

9-43 Water flowing in a wide horizontal channel at a flow depth of 35cm and an average velocity of 12m/s undergoes a hydraulic jump. Determine the head loss associated with hydraulic jump.

9-44 During a hydraulic jump in a wide channel, the flow depth increases from 0.6 to 3m. Determine the velocities and Froude numbers before and after the jump, and the energy dissipation ratio.

9-45 A 4-m-wide rectangular channel flow in Fig. P9-45 changes from a steep slope to mild slope channel with identical bottom widths. When the carried flow rate is 16m³/s, the normal depth in the steep channel is $h_{01} = 0.6$m, in the mild channel is $h_{02} = 1.6$m. (a) Sketch the water-surface profiles expected for gradually varied flow in two channels and the hydraulic jump between profiles. (b) Assume the pre-jump section is located at the end of the steep channel, what is the normal depth in mild channel h_{02}? (c) What happens to the hydraulic jump if $h_{02} = 3.5$m. The bed slope for mild-slope channel can be assumed to be nearly horizontal when the theory of hydraulic jump is employed.

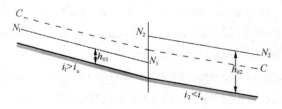

Fig. P9-45

9-46 A horizontal trapezoidal channel with a bottom width of 3m and a side slope of 1.5 carries water at a flow rate of 12 m³/s. If a hydraulic jump occurs, determine: (a) the after-jump depth h'' if the prejump depth $h' = 0.4$m; (b) the pre-jump depth h' if the after jump depth $h'' = 1.5$m.

Gradually Varied Flows and Water Surface Profiles

9-47 A 4-m-wide brickwork rectangular channel carries water at a flowrate of 8.0 m³/s on a slope of 0.1°. Is this a mild, critical, or steep slope? What type of gradually varied solution curve are we on if the local water depth is (a) 1m, (b) 1.5m, and (c) 2m?

9-48 A gravelly earth wide channel is flowing at 10 m³/s per meter of width on a slope of 0.75°. Is this a mild, critical, or steep slope? What type of water surface profiles are we on if the local water depth is (a) 1m, (b) 2m, and (c) 3m?

9-49 A clean-earth wide-channel flow is flowing up an adverse slope of 0.002. If the flow rate per meter width is 4.5 m³/s·m, use gradually varied theory to compute the distance for the depth to drop from 3.0m to 2.0m.

9-50 The wide-channel flows in Fig. P9-50 changes from one profile to another profile gradually. Sketch and label the water-surface profiles which are expected for those gradually varied flows.

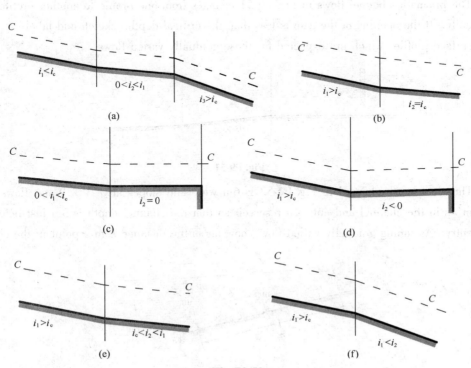

Fig. P9-50

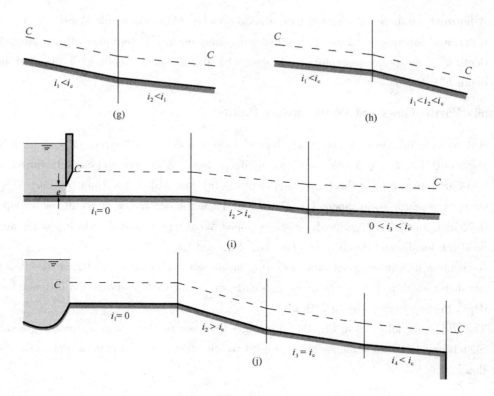

Fig. P9-50(continued)

9-51 The prismatic-channel flows in Fig. P9-51 changes from one profile to another profile gradually. If the opening of the gate is less than the critical depth, sketch and label the water-surface profiles which are expected for those gradually varied flows.

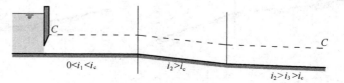

Fig. P9-51

9-52 The clean-earth channel in Fig. P9-52 is 6m wide and slopes at 0.3°. Water flows at 30 m^3/s in the channel and enters a reservoir so that the channel depth is 3m just before the entry. Assuming gradually varied flow, how far is the distance L to a point in the channel

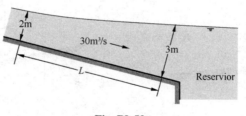

Fig. P9-52

426

where the water depth is 2m? What type of curve is the water surface?

9-53 A concrete rectangular channel ($n = 0.015$) with a base wide of 8m and a bed slope of 0.0002 carries water at a flow rate of 40 m³/s. A sluice gate is located at the downstream end of the channel, which raises the water depth in channel to 5m just before the gate. Take the kinematic correction factor to be 1.0, determine: (a) the distance away from the gate to the point in the channel where the depth is $h = 1.1h_0$; (b) the water depth where is 5000m upstream from the gate.

9-54 A masonry-lined trapezoidal channel ($n = 0.018$) with a base wide of 3.5m, a side slope of 1.0 and a bed slope of 0.0003 carries water at a flow rate of 35 m³/s. A sluice gate is located at the downstream end of the channel, which raises the water depth in channel to 5m just before the gate. Take the kinematic correction factor to be 1.0, determine the distance away from the gate where the depth in the channel is 4.2m.

9-55 A gradually contracted channel is used to connect two rectangular channels, which have identical bed slope of 0.005, identical Manning coefficient of 0.0225 and different base widths, as in Fig P9-55. The contraction channel is contracted linearly. If flowrate in the channel is 18.6 m³/s, determine the flow depths at the two ends of the contraction channel.

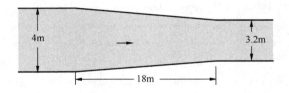

Fig. P9-55

9-56 An unfinished concrete-lined rectangular channel with a base wide of 1m and a bed slope of 0.0004 drains water to a lower river through a outfall as shown in Fig. P9-56. The normal depth in the channel is 0.5m. (a) Determine the flow depth at the brink of the outfall at the end of channel; (b) sketch the water surface profile in the channel.

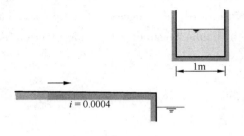

Fig. P9-56

427

10 Weir Flow

CHAPTER OPENING PHOTO: A weir is a common obstruction in a river to backup water behind it and allow water to flow over its top. It can be used to alter the flow of rivers to prevent flooding, measure discharge, and help render rivers navigable. The principle of discharge measurement by weir is discussed in the present chapter. (*Courtesy of Paul Fenwick at Wikimedia*)

A *weir* is an obstruction on a channel bottom over which the fluid must flow. Weirs are usually used as the open-channel flow-measuring devices, such as sharp-crested weirs, broad-crested weirs and sluice gates, or the obstruction to backup the water as a dam, such as the ogee-crested weir. Besides, water flowing through the span under a small bridge is also considered as the broad-crested weir flow.

This chapter begins with the definition and classification of the weir and weir flows. It follows the brief discussion of the fundamental formula of weir flow, the empirical expressions of the discharge coefficients of sharped-crested weir, ogee-crested weir and broad-crested weir, and the flow under gate. Finally the hydraulic design of small bridge aperture is briefly presented based on the theory of broad-crested weir.

10.1 Introduction

10.1.1 Weir and Weir Flow

A *weir*, of which the ordinary dam is an example, is a channel obstruction over which the flow must deflect. The flow over a weir is termed *weir flow*. For simple geometries the channel discharge Q correlates with gravity and with the blockage height H to which the upstream flow is backed up above the weir elevation (see Fig. 10-1). Thus a weir is a simple but effective open-channel flowmeter.

Figure 10-1 also shows the main parameters concerned in weir flow: *weir head* over the weir crest, H; weir width, b, and channel width, B; upstream weir height, P_w, and downstream weir height P'_w; weir crest thickness, δ (parallel to the flow); approach velocity, v_0, prior to the weir; water depth downstream of weir, h; the drowned depth of the weir crest, h_s.

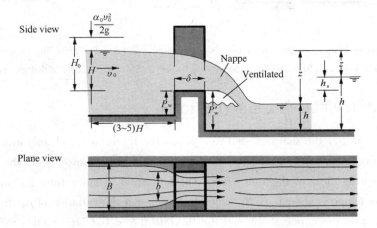

Fig. 10-1 Weir flow over a polygonal ogee-crested weir.

If the downstream water surface is higher than the weir crest, that is $h_s > 0$, the nappe is then drowned in the downstream liquid body as shown in Fig. 10-2. Under this condition, the changes of the downstream flow depth will affect the flow rate through weir. Such a weir is termed *drowned weir*, and such a weir flow is termed *drowned outflow*. Otherwise, if the downstream water surface is low enough that its change will not affect the flow rate through the weir, the nappe is then well ventilated and surrounded by the air, such a weir flow is termed *free outflow*.

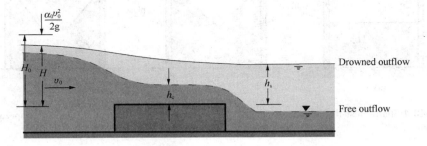

Fig. 10-2 Free outflow and drowned outflow through a broad-crested weir

10.1.2 Classification of Weirs

In Fig. 10-3 there are three common weirs, sharp-crested, ogee-crested and broad-crested weirs, which is termed based on the ratio of the thickness of weir δ and the weir head H. In all cases the flow upstream is subcritical, then accelerates to critical near the top of the weir, and finally spills over into a supercritical *nappe*.

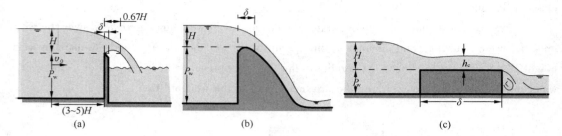

Fig. 10-3 Flow over wide, well ventilated (a) sharp-crested weir ($\delta < 0.67H$), (b) ogee-crested weir ($0.67H < \delta < 2.5H$) and (c) broad-crested weir ($2.5H < \delta < 10H$) with free outflow

SHARP-CRESTED WEIR, $\dfrac{\delta}{H} < 0.67$ (Fig. 10-3a)

When $\delta/H < 0.67$, the sharp-crested weir allows the liquid to fall cleanly away from the weir to be a supercritical nappe, even though sometimes it should be *ventilated* under the nappe to make sure the nappe spring clear of the weir crest. The horizontal distance from the weir crest to the lower edge of the nappe is about $0.67H$, which indicates that the variation of the thickness of weir crest has no effect on the nappe shape and the flowrate if $\delta < 0.67H$. Such a weir is termed as *sharp-crested weir*, which is usually used as the open-channel flow-measuring device.

Sharp-crested weirs come in many different shapes and styles, such as *rectangular* (with and without end contractions), *triangular* (*V-notch*) and *trapezoidal weir*s (see Fig. 10-4). Under controlled conditions, sharp crested weirs can exhibit accuracies as good as $\pm 2\%$, although under field conditions accuracies greater than $\pm 5\%$ should not be expected.

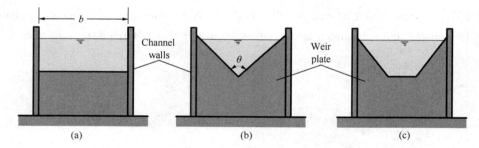

Fig. 10-4 Sharp-crested weir plate geometry: (a) rectangular, (b) triangular, and (c) trapezoidal

OGEE-CRESTED WEIR, $0.67 < \dfrac{\delta}{H} < 2.5$ (Fig. 10-3b)

When weir is suppressed ($0.67 < \delta/H < 2.5$) and its height is large, the nappe emerging out may be subjected to the problems of ventilation. Hence, in such cases the weir profile downstream is constructed conforming to the shape of the lower side of the nappe. Such a weir is known as a *spillway* or *ogee weir*. For the ogee weir, its crest thickness can affect the flowrate through it, and the free surface of flow prior to the weir drops continuously over the weir crest. In general, *curved ogee-crested weir* (Fig. 10-3b), i.e. WES weir, is used in the engineering practice of medium to large hydraulic structures, whereas the simple *polygonal weir* (Fig. 10-1) is used for the small engineering practice.

BROAD-CRESTED WEIR, $2.5 < \dfrac{\delta}{H} < 10$ (Fig. 10.3c)

A *broad-crested weir is* a flat-crested structure, with a long crest compared to the flow depth ($2.5 < \delta/H < 10$). When the crest is 'broad', the free surface drops around the entrance, then becomes near a critical flow parallel to the crest invert and a hydrostatic pressure distribution above the crest, and finally drops again around the exit of the crest. For example, the undergate flow with fully opened gate, the flow through a small bridge, the flow through a culvert under road, railroad, trail, or similar obstruction can be considered as a broad-crested weir flow. Practical experience showed that the weir overflow is affected by the upstream flow conditions and the weir.

But if the weir thickness is too long ($\delta/H > 10$), wall shear effects dominate over the weir crest and cause the flow over the weir to be subcritical, the friction loss cannot be neglected any more and it should be considered as an open channel flow as discussed in Chapter 9.

For all three weirs discussed above, the weir width must be equal to or less than the channel width. If the weir opening is narrower than the channel width, $b < B$ as shown in Fig. 10-1b, the flow contracts once it pasts the weir and then expands downstream of weir. Such a weir is known as *contracted weir*, of which the change of the weir opening affects the flow rate through the weir. Otherwise, if the weir opening across the entire channel width, $b = B$, there is no contraction effect on the flow rate, it is termed *suppressed weir*.

10.2 Sharp-crested Weirs

A *sharp-crested weir* is a vertical plate placed in a channel that forces the liquid to flow through an opening to measure the flow rate. The type of the weir is characterized by the shape of the opening. A vertical thin plate with a straight top edge is referred to as *rectangular weir* since the cross section of the flow over it is rectangular (see Fig. 10-4); a weir with a triangular opening is referred to as a *triangular weir*; etc. They are used to meter flow of water over the weir (and through the open channel) by measuring the head of water over the weir crest.

The flow-rate correlations given below are based on the free overfall of liquid, *nappes*, being clear from the weir as in Fig. 10-1. Since the overfall spans the entire channel, artificial ventilation, i.e. the holes in the channel walls, to the space under the nappe may be needed to assure atmospheric pressure underneath. Empirical relations for drowned weirs are also available in the following section.

10.2.1 General Form of Flowrate over Weir

Consider the free flow of a liquid over a sharp-crested rectangular weir placed in a horizontal channel as shown in Fig. 10-5. The flow over the weir is not one-dimensional since the liquid undergoes large changes in velocity and direction over the weir. But the pressure within the nappe is atmospheric. For simplicity, the velocities upstream of the weir at section 1-1 and in nappe are averaged cross the section.

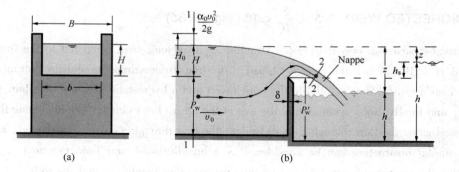

Fig. 10-5 Flow over a sharp-crested rectangular weir with a free outflow: (a) View from upstream to downstream; (b) Side view

A simple relation for the variation of liquid velocity in the nappe can be obtained by writing the energy equation between section 1-1, about $(3 \sim 5) H$ upstream of the weir, and the section 2-2 at the nappe, of which the center is located at the horizontal plane passing the weir crest. Let $H_0 = H + \dfrac{\alpha_0 v_0^2}{2g}$, $\alpha_2 = \alpha$, $v_2 = v$, and we have $p_2 = p_{at}$ because the pressure within the nappe is atmospheric. Thus, the average velocity at the cross section of the nappe is determined to be

$$H + \frac{p_{at}}{\rho g} + \frac{\alpha_0 v_0^2}{2g} = \frac{p_2}{\rho g} + \frac{\alpha_2 v_2^2}{2g} + \zeta \frac{v_2^2}{2g}$$

$$v = \frac{1}{\sqrt{\alpha + \zeta}} \sqrt{2gH_0} = \varphi \sqrt{2gH_0}$$

and then the flowrate flowing over the weir crest is

$$Q = vA = \varphi be \sqrt{2gH_0} \tag{10.1}$$

where φ is the *velocity coefficient*, $\varphi = \dfrac{1}{\sqrt{\alpha + \zeta}}$, which is determined by the velocity distribution at section 2-2 (α) in nappe and the minor loss when liquid flows over the weir; A is the area of the cross section 2-2 in nappe, $A = be$; b is the weir opening width, $b = B$ for the rectangular suppressed weir without contraction; e is the thickness of nappe at section 2-2.

Let $e = kH_0$, Eq. 10.1 becomes

$$Q = \varphi bk H_0 \sqrt{2gH_0} = mb\sqrt{2g}\, H_0^{3/2} \tag{10.2}$$

where m is the *weir discharge coefficient*, $m = k\varphi$, which depends on the weir type, the shape of the leading-edge nose and the relative height of weir, P_w/H.

Consider the approaching velocity v_0 in the discharge coefficient, Eq. 10.2 becomes

$$Q = m_0 b \sqrt{2g}\, H^{3/2} \tag{10.3}$$

where m_0 is the *weir discharge coefficient* considering the approaching velocity, and can be determined by the experiments.

Equations 10.2 or 10.3 can not only be used to determine the flow rate over the sharp-edged rectangular weir, but also be applicable for flows over the ogee-created weir and the broad-crested weir. They are the *general form* of the flowrate of the weir flow.

For the flow over a contracted weir ($b < B$) with a drowned outflow ($h_s > 0$ see Fig. 10-5b), Eqs. 10.2 and 10.3 should be changed into the following forms:

$$Q = \sigma_s \varepsilon m b \sqrt{2g}\, H_0^{3/2} \tag{10.4}$$

$$Q = \sigma_s \varepsilon m_0 b \sqrt{2g}\, H^{3/2} \tag{10.5}$$

where σ_s is the *drowned coefficient*, $\sigma_s \leq 1$; ε is the *contraction coefficient*, $\varepsilon \leq 1$. Both σ_s and ε are determined by the experiments or the empirical formula.

10.2.2 Rectangular Weir

For the free weir flow over a sharp-edged, well ventilated, compressed *rectangular weir*, its flow rate can be determined by Eq. 10.3, in which the weir discharge coefficient is determined by *Bazin formula*, proposed by Bazin, a France engineer in 1898:

$$m_0 = \left(0.405 + \frac{0.0027}{H}\right)\left[1 + 0.55 \left(\frac{H}{H + P_w}\right)^2\right] \tag{10.6}$$

where the units of H and P_w are in meters.

Equation 10.6 is available for $H \leq 1.24\text{m}$, $b \leq 2.0\text{m}$ and $P_w \leq 1.13\text{m}$.

If the weir opening width is less than the channel width, $b < B$, which will cause the sides of the overfall to contract inward and reduce the flow rate. Thus, the discharge coefficient m_0 in Eq. 10.6 must be corrected as

$$m_0 = \left(0.405 + \frac{0.0027}{H} - 0.03\frac{B - b}{B}\right)\left[1 + 0.55 \left(\frac{H}{H + P_w}\right)^2 \left(\frac{b}{B}\right)^2\right] \tag{10.7}$$

where the unit of B is in meters.

If the downstream free surface is higher than the weir crest, $h_s > 0$, and $z/P'_w < 0.7$, this flow is termed as *drowned outflow*. The flow capacity decreases, and the downstream free surface is full of undular, oscillating waves. Thus, the sharp-crested weir is not applicable under the condition of submerged outflow.

10.2.3 Triangular Weir

Another type of sharp-crested weir commonly used for flow measurement is the *triangular weir* (also called the *V-notch weir*) shown in Fig. 10-4b and Fig. 10-6. The triangular weir has the advantage that it maintains a high weir head H even for small flow rates because of the decreasing flow area with decreasing H, and thus it can be used to measure a wide range of flow rates accurately.

For a *triangular weir* with a *V-notch angle* θ, the flow rate pass a differential weir width db

with a weir head h can be determined by Eq. 10.3 as

$$dQ = m_0 \sqrt{2g}\, h^{3/2} db$$

From geometric consideration, the notch width can be expressed as $b = (H - h)\tan(\theta/2)$. Thus, $db = -dh\tan(\theta/2)$. Substituting into the above equation

$$dQ = - m_0 \tan\frac{\theta}{2} \sqrt{2g}\, h^{3/2} dh$$

and performing the integration gives the flow rate for a triangular weir to be

$$Q = -2m_0 \tan\frac{\theta}{2}\sqrt{2g}\int_H^0 h^{3/2}dh = \frac{4}{5} m_0 \tan\frac{\theta}{2}\sqrt{2g}\, H^{5/2} \tag{10.8}$$

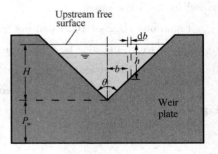

Fig. 10-6 A triangular (or V-notch) sharp-crested weir plate geometry. The view is from downstream looking upstream

When the V-notch angle is $\theta = 90°$, weir head is within the range of $H = 0.05 \sim 0.25$m, the value of discharge coefficient obtained from experiments is $m_0 = 0.395$. Thus, Eq. 10.8 becomes

$$Q = 1.4 H^{5/2} \tag{10.9}$$

where H is the *weir head* over the lowest point of the weir crest, the unit is in meters; the flow rate, Q, is in the unit of m³/s.

When $\theta = 90°$, $H = 0.25 \sim 0.55$m, the flow rate is determined by the empirical formula as

$$Q = 1.343 H^{2.47} \tag{10.10}$$

10.3 Ogee-crested Weirs

10.3.1 Effect of Weir Head

Ogee-crested weir is a block structure across the open channel in such a way that free surface of liquid begins to drop near the upstream of structure, and then liquid flows over the structure surface till to the downstream pool, such as dam for storing water. The most common ogee-crested weir is *curved weir* for large structures (Fig. 10-3b) and *polygonal weir* for small structures (Fig. 10-1a) in engineering practice. The curved surface of the weir is usually designed as the lower interface of nappe of the free flow passing sharp-edged weir.

The formula for flow rate over the ogee-crested weir with a free outflow is same as general form of the flowrate of the weir flow, Eq. 10.3

$$Q = mb\sqrt{2g}\, H_0^{3/2} \tag{10.11}$$

where m is the *discharge coefficient* for ogee-crested weir with a wide range, which depends on the weir geometric shape, weir head and the inlet shape of weir. For the first estimation of flow rate passing the ogee-crested weir, the following value can be used: $m = 0.45$ for a curved ogee-crested weir and $m = 0.35 \sim 0.42$ for a polygonal ogee-crested weir.

For a standard WES weir composed by a vertical wall AB and three curves as shown in Fig. 10-7, the weir discharge coefficient is determined by

$$m = -0.024 \frac{H_0}{H_d} + 0.18503 \sqrt{\frac{H_0}{H_d}} + 0.3406 \tag{10.12}$$

where H_0 is the total head prior to the weir, $H_0 = H + \frac{\alpha_0 v_0^2}{2g}$; v_0 is the approaching velocity prior to the weir; H_d is the *designing weir head* for this standard WES weir.

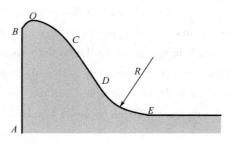

Fig. 10-7 Profile of standard WES weir

Equation 10.12 is applicable for $P_w/H_d \geq 1.33$m, $H_0/H_d < 1.5$. When $H_0/H_d = 1.0$, from Eq. 10.12 gives the discharge coefficient, $m = m_d = 0.502$, which is the supposed value of the discharge coefficient for the standard WES weir at $H_0 = H_d$.

In general, the pressure on the weir surface for $H_0/H_d \leq 1.0$ is equal to or larger than the atmospheric pressure. However, the vacuum occurs when $H_0/H_d > 1.0$. The vacuum height will be up to $0.5H_d$ when $H_0/H_d = 1.33$.

10.3.2 Effect of Drowned Flow

When the downstream free surface is higher than the weir crest, $h_s > 0$ (see Fig. 10-1a), the outflow over the weir is drowned in the downstream liquid, the flow rate decreases and Eq. 10.11 should be corrected as

$$Q = \sigma_s mb \sqrt{2g}\, H_0^{3/2} \tag{10.13}$$

where σ_s is the *drowned coefficient*, which decreases with the increasing of h_s/H as listed in Table 10-1.

Table 10-1 Drowned coefficient for flow over ogee-crested weir

h_s/H	0.05	0.20	0.30	0.40	0.50	0.60	0.70	0.80	0.90	0.95	0.975	0.995	1.00
σ_s	0.997	0.985	0.972	0.957	0.935	0.906	0.856	0.776	0.621	0.470	0.319	0.100	0

10.3.3 Effect of Contraction

When the weir width is less than the channel width, $b < B$ (see Fig. 10-1b), the sides of the overfall contract inward. It leads to a reduction of the flow rate through the weir opening. This effect on flow rate is presented by applying the contraction coefficient to Eq. 10.11

$$Q = \varepsilon mb\sqrt{2g}H_0^{3/2} \tag{10.14}$$

where ε is the *contraction coefficient*, $\varepsilon = 0.85 \sim 0.95$ for the first estimation of flow rate.

10.4 Broad-crested Weirs

A *broad-crested weir* is a rectangular block of height P_w and length δ that has a horizontal crest over which a nearly critical flow occurs. When fluid is approaching the weir entrance, the kinetic energy increases due to smaller flow area over the crest than the sections prior to the weir, and the potential energy decrease due to the minor energy loss around the entrance, which cause a drop of surface around the entrance and leads to a supercritical flow over the crest with a depth close to the critical depth. For a free outflow, the water surface drops again around the exit and connects to the downstream free surface (Fig. 10-8).

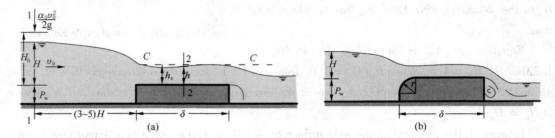

Fig. 10-8 Flow over a broad-crested weir with (a) right-angle leading-edge nose and (b) rounded leading-edge nose with a free outflow

10.4.1 Ideal Flow Rate

The broad-crested weir of Fig. 10-8 can be analyzed more accurately because it creates a short run of nearly one-dimensional critical flow over the weir crest, as shown. Take the weir crest to be the reference level and disregard the energy losses for flow passing weir, the energy equation from upstream section 1-1 to the section 2-2 over the weir crest yields

$$H_0 = H + \frac{\alpha_0 v_0^2}{2g} = h + \frac{\alpha v^2}{2g}$$

Take the kinematic correction coefficient to be unity, that is $\alpha = 1.0$, the average velocity on the section over the weir crest is

$$v = \sqrt{2g(H_0 - h)} \qquad (10.15)$$

and the flow rate per unit weir width is

$$q = h\sqrt{2g(H_0 - h)} \qquad (10.16)$$

Because there is no energy loss for an ideal liquid passing the broad-crested weir, the flow rate obtained from Eq. 10.16 should be maximum if

$$\frac{dq}{dh} = \frac{d[h\sqrt{2g(H_0 - h)}]}{dh} = 0$$

or

$$h = \frac{2}{3}H_0 \qquad (10.17)$$

Combining Eqs. 10.16 and 10.17 yields

$$h = \sqrt[3]{q^2/g} = h_c \qquad (10.18)$$

Equation 10.18 indicates that the depth over the broad-created weir is the critical depth when the energy losses are negligible, which agrees with the experimental results.

Finally, the flow rate for an ideal liquid flowing through the broad-crested weir with a free outflow can be obtained by substituting Eq. 10.17 into Eq. 10.16

$$Q = mb\sqrt{2g}\,H_0^{3/2} \tag{10.19}$$

where $m = \dfrac{2}{3}\sqrt{\dfrac{1}{3}} = 0.3849$, the maximum discharge coefficient for flow past a broad-crested weir.

10.4.2 Actual Flow Rate

For a real liquid, the discharge coefficient in the theoretical weir-flow formula, Eq. 10.19, must be modified experimentally. It is considerably sensitive to geometric parameters, including the surface roughness of the crest. Based on the experiments, А. Р. Березинский proposed the following empirical formulas of the discharge coefficient:

For the broad-crested weir with right-angle leading-edge noses (Fig. 10-8a)

$$m = 0.32 + 0.01\,\frac{3 - P_w/H}{0.46 + 0.75 P_w/H} \quad \text{for } 0 \leqslant \frac{P_w}{H} \leqslant 3.0 \tag{10.20}$$

or

$$m = 0.32 \quad \text{for } \frac{P_w}{H} > 3.0 \tag{10.21}$$

and for the broad-crested weir with rounded leading-edge noses as shown in Fig. 10.8b ($r/H \geqslant 0.2$, where r is the radius of the round edge)

$$m = 0.36 + 0.01\,\frac{3 - P_w/H}{1.2 + 1.5 P_w/H} \quad \text{for } 0 \leqslant \frac{P_w}{H} \leqslant 3.0 \tag{10.22}$$

or

$$m = 0.36 \quad \text{for } \frac{P_w}{H} > 3.0 \tag{10.23}$$

10.4.3 Effect of Drowned Flow

When the downstream free surface increases till to be higher than the weir crest, the depth over the crest changes from the depth less than the critical depth to a depth larger than critical depth, flow statute changes from supercritical to subcritical flow, such a flow is a *drowned flow* as shown in Fig. 10-2.

According to the experimental results, the *drowned depth* (see Fig. 10-2) $h_s > 0$ is the prerequisite condition for a drowned weir, and $h_s/H_0 > 0.8$ is a sufficient condition for a drowned weir to make sure the supercritical flow downstream of weir to become subcritical.

For a drowned broad-crested weir, the flowrate is corrected by introducing a *drowned coefficient* as in Eq. 10.13

$$Q = \sigma_s mb\sqrt{2g}\,H_0^{3/2}$$

in which the *drowned coefficient*, σ_s, decreases with the increasing h_s/H_0 as shown in Table 10-2 or the following correlated empirical equation:

$$\sigma_s = -96.01\left(\frac{h_s}{H_0}\right)^3 + 235.32\left(\frac{h_s}{H_0}\right)^2 - 193.23\frac{h_s}{H_0} + 54.14 \quad \text{for} \quad 0.8 \leq \frac{h_s}{H_0} \leq 0.98$$

(10.24)

Table 10-2　Drowned coefficient for flow over broad-crested weir

h_s/H_0	0.80	0.81	0.82	0.83	0.84	0.85	0.86	0.87	0.88	0.89	0.9	0.91	0.92	0.93	0.94	0.95	0.96	0.97	0.98
σ_s	1.00	0.995	0.99	0.98	0.97	0.96	0.95	0.93	0.90	0.87	0.84	0.82	0.78	0.74	0.70	0.65	0.59	0.50	0.40

10.4.4　Effect of Contraction

When the weir width is less than the channel width, $b < B$, the sides of the overfall contract inward, which reduces the flow rate. This effect on flowrate is corrected by applying the contraction coefficient as in Eq. 10.14

$$Q = \varepsilon m b \sqrt{2g}\, H_0^{3/2}$$

where ε is the *contraction coefficient*, which depends on the relative weir height, P_w/H, relative weir width, b/B, and shapes of the leading-edges. For a broad-crested weir with a single slot, it can be determined by the following empirical formula

$$\varepsilon = 1 - \frac{a}{\sqrt[3]{0.2 + \frac{P_w}{H}}}\sqrt[4]{\frac{b}{B}\left(1 - \frac{b}{B}\right)}$$

(10.25)

where a is the shape coefficient of piers, which strongly depends on the shape of the leading-edge of abutment pier, $a = 0.19$ for a abutment pier with a right-angle leading-edge and $a = 0.1$ for a abutment pier with a rounded leading-edge.

EXAMPLE 10-1　Sharp-crested and broad-crested weirs

Water flows in a rectangular channel of width $b = 2$m with flowrates between $Q_{min} = 0.02$ m³/s and $Q_{max} = 0.6$ m³/s. This flowrate is to be measured by using either (a) a rectangular sharp-crested weir, (b) a triangular sharp-crested weir with $\theta = 90°$, or (c) a broad-crested weir. In all cases the invert of weir crest above the channel bottom is $P_w = 1$m. Plot a graph of $Q = Q(H)$ for each weir and comment on which weir would be the best for this practices.

Solution:

For the rectangular weir with $P_w = 1$m, Eqs. 10.3 and 10.6 give

$$Q = \left(0.405 + \frac{0.0027}{H}\right)\left[1 + 0.55\left(\frac{H}{H + P_w}\right)^2\right] b\sqrt{2g}\, H^{3/2}$$

$$= \left(0.405 + \frac{0.0027}{H}\right)\left[1 + 0.55\left(\frac{H}{H + 1}\right)^2\right](2)\sqrt{(2)(9.81)}\, H^{3/2} \quad (1)$$

For a triangular sharp-crested weir with $\theta = 90°$ and $P_w = 1$m, Eqs. 10.9 and 10.10 give

$$Q = 1.4 H^{5/2} \quad \text{for } H = 0.05 \sim 0.25\text{m}$$
$$Q = 1.343 H^{2.47} \quad \text{for } H = 0.25 \sim 0.55\text{m} \quad (2)$$

For a broad-crested weir with right-angle leading-edge noses with $P_w = 1$m and negligible approaching velocity, Eqs. 10.19 and 10.20 give

$$Q = \left(0.32 + 0.01 \frac{3 - P_w/H}{0.46 + 0.75 P_w/H}\right) b \sqrt{2g}\, H^{3/2}$$

$$= \left(0.32 + 0.01 \frac{3 - 1/H}{0.46 + 0.75/H}\right)(2)\sqrt{(2)(9.81)}\, H^{3/2} \tag{3}$$

where H and Q in Eqs. 1 to 3 are in meters and m³/s, respectively. The results from Eqs. 1 to 3 are plotted in Fig. E10-1.

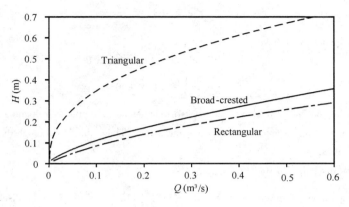

Fig. E10-1

Although it appears as though any of the three weirs would work well for the upper portion of the flowrate range, neither the rectangular nor the broad-crested weir would be very accurate for small flowrates near $Q = Q_{min}$ because of the small head, H, at these conditions. The triangular weir, however, would allow reasonably large values of H at the lowest flowrates. The corresponding heads with $Q = Q_{min} = 0.02$ m³/s for rectangular, triangular, and broad-crested weirs are 0.0272m, 0.1828m, and 0.0378m, respectively.

In addition, as is discussed in this section, for proper operation the broad-crested weir geometry is restricted to $2.5 < \delta/H < 10$, where δ is the weir block length. From Eq. 3 with $Q_{max} = 0.6$ m³/s, we obtain $H_{max} = 0.3547$m. Thus, we must have $\delta > 2.5 H_{max} = 0.887$m to maintain proper critical flow conditions at the largest flowrate in the channel. On the other hand, with $Q = Q_{min} = 0.02$ m³/s, we obtain $H_{min} = 0.0378$m. Thus, we must have $\delta < 10 H_{min} = 0.378$m to ensure that frictional effects are not important. Clearly, these two constraints on the geometry of the weir block, δ, are incompatible.

A broad-crested weir will not function properly under the wide range of flowrates considered in this example. The sharp-crested triangular weir would be the best of the three types considered, provided the channel can handle the $H_{max} = 0.7217$m head.

DISCUSSION This example indicates that the triangle weir would be the better choose for the flow rate measuring if the weir head varies at a larger magnitude starting from a small minimum flowrate.

EXAMPLE 10-2 Broad-crested weir

A 2-m-wide broad-crested weir with a single slot was constructed in a 3-m wide rectangular channel to raise water for irrigation as shown in Fig. E10-2. The abutment pier of weir has a right-angle leading-edge nose. The height of weir is $P_w = P'_w = 1$m. Assume the weir head to be $H = 2$m and the downstream water depth to be $h_t = 2$m. Determine the flow rate through the weir.

Solution:

(1) Determine the flow type through the weir

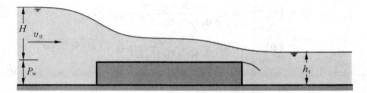

Fig. E10-2

From the given parameters we have
$$h_s = h_t - P'_w = 2\text{m} - 1\text{m} = 1\text{m} > 0$$
$$\frac{h_s}{H_0} < \frac{h_s}{H} = \frac{1\text{m}}{2\text{m}} = 0.5 < 0.8$$

which indicates it is a free outflow through a broad-crested weir. Meanwhile, the weir opening width $b = 2\text{m}$ is less than the channel width $B = 3\text{m}$, which leads to a flow contraction. Thus, the flow through this broad-crested weir is a contracted free outflow.

(2) Determine the discharge coefficient

The weir has a right-angle leading-edge nose, and $\frac{P_w}{H} = \frac{1\text{m}}{2\text{m}} = 0.5 < 3.0$. Thus the discharge coefficient can be calculated by Eq. 10.20
$$m = 0.32 + 0.01 \frac{3 - P_w/H}{0.46 + 0.75 P_w/H} = 0.32 + 0.01 \frac{3 - 0.5}{0.46 + (0.75)(0.5)} = 0.35$$

(3) Determine the contraction coefficient

For the right-angle abutment pier, we have $a = 0.19$. Thus, the contraction coefficient for the broad-crested weir with a single slot is determined by Eq. 10.25
$$\varepsilon = 1 - \frac{a}{\sqrt[3]{0.2 + \frac{P_w}{H}}} \sqrt[4]{\frac{b}{B}} \left(1 - \frac{b}{B}\right) = 1 - \frac{0.19}{\sqrt[3]{0.2 + \frac{1}{2}}} \sqrt[4]{\frac{2}{3}} \left(1 - \frac{2}{3}\right) = 0.936$$

(4) Determine the flow rate through the weir

For the contracted free flow through the broad-crested weir, the flowrate is determined by Eq. 10.14
$$Q = \varepsilon m b \sqrt{2g}\, H_0^{3/2} = (0.936)(0.35)(2)\sqrt{(2)(9.81)}\, H_0^{3/2}\ \text{m}^3/\text{s} = 2.9 H_0^{3/2}\ \text{m}^3/\text{s} \quad (1)$$

where
$$H_0 = H + \frac{\alpha_0 v_0^2}{2g} = \left[2 + \frac{(1.0) v_0^2}{(2)(9.81)}\right]\text{m} = \left[2 + \frac{v_0^2}{19.62}\right]\text{m} \quad (2)$$

$$v_0 = \frac{Q}{B(H + P_w)} = \frac{Q}{(3)(2+1)}\text{m/s} = \frac{Q}{9}\text{m/s} \quad (3)$$

Thus, Eq. 1 has to be solved by the literation. First time we assume $H_{0(1)} \approx H = 2\text{m}$, from Eq. 1 we have $Q_{(1)} = 8.202\ \text{m}^3/\text{s}$, substituting into Eq. 3 gives $v_{0(1)} = 0.911\text{m/s}$, thus to Eq. 2 we have $H_{0(2)} = 2.042\text{m}$, and the relative error
$$\frac{H_{0(2)} - H_{0(1)}}{H_{0(2)}} \times 100\% = 2.05\%$$

Thus, we need to do the second literation. Substitute $H_{0(2)} = 2.042\text{m}$ into Eq. 1, again we can have $Q_{(2)} = 8.465\ \text{m}^3/\text{s}$. Repeat the above steps we can finally obtain the flow rate $Q = Q_{(2)} = 8.48\ \text{m}^3/\text{s}$, $v_0 = 0.94\text{m/s}$ and $H_0 = 2.04\text{m}$.

(5) Check the flow regime prior to the weir

$$\text{Fr} = \frac{v_0}{\sqrt{g(H+P_w)}} = \frac{0.94 \text{m/s}}{\sqrt{(9.81\text{m/s}^2)(2\text{m}+1\text{m})}} = 0.17 < 1$$

Thus, the flow prior to the weir is a subcritical flow. When it approaches to the block, a weir flow occurs, which indicates the above computation is applicable.

10.5 Flow under a Sluice Gate

An obstruction with an adjustable opening at the bottom that allows the liquid to flow underneath it is called an *underflow gate*. A variety of *underflow gate* structures is available for flowrate control at the crest of an overflow spillway, or at the entrance of an irrigation canal or river from a lake, as shown in Fig. 10-9a. If the flow is allowed free discharge through the gap, as in Fig. 10-9a, the flow smoothly accelerates from subcritical (upstream) to critical (near the gap) to supercritical (downstream).

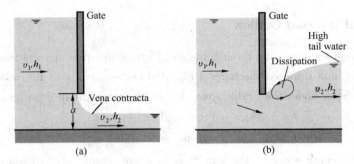

Fig. 10-9 Flow under a sluice gate passes through critical flow:
(a) free outflow with vena contracta; (b) drowned dissipative flow under a drowned gate

The free discharge, Fig. 10-9a, contracts to a depth h_2 about 40 percent less than the gate's gap height, as shown. This is similar to a free *orifice* discharge, as in Fig. 7-1. Take the height of the gate gap to be a and the gap width into the paper be b, we can approximate the flow rate by orifice theory:

$$Q = C_d ab \sqrt{2gh_1} \qquad (10.26)$$

where C_d is the *discharge coefficient* of the flow under a sluice gate, which is a function of the contraction coefficient h_2/a and the depth ratio h_1/a.

From Eq. 10.26 we know that a continuous variation in flow rate is accomplished by raising the gate. However, if the gate is raised too much, the flow changes to weir flow, of which the flow rate is determined by the formulas in section 10.3 or 10.4. Thus, before we go further to determine the flow rate, the flow type must be determined first. If the underflow gate is located at the crest of a broad-crested weir, the flow would be *weir flow* if $a/h_1 \geqslant 0.65$, otherwise it is a *undergate flow*; If the underflow gate is located at the highest point of the crest of a ogee-crested weir, the flow is a *weir flow* when $a/h_1 \geqslant 0.75$, otherwise it is a *undergate flow*.

If the tail water is high, as in Fig. 10-9b, free discharge is not possible. The sluice gate is said to be *drowned* or partially drowned. There will be energy dissipation in the exit flow, probably in the form of a *drowned hydraulic jump*, and the downstream flow will return to be subcritical. Equation 10.26 does not apply to this situation, and experimental discharge correlations are necessary.

10.6 Hydraulic Design of Small Bridge Aperture

Water flowing through the slot of small bridge will contract inward with a surface drop due to the effect of bridge abutment and piers. This hydraulic phenomenon is similar to the one when water flows over the broad-crested weir with a zero weir height ($P_w = P'_w = 0$), and is common in engineering practice. For example, the undergate flow with fully opened gate above a flat channel bed, the flow through a culvert under a road, railroad, trail, or similar obstruction are broad-crested weir flow with zero weir height.

10.6.1 Free and Drowned Outflow

The discharge of water passing through the small bridge aperture appears two kinds, free discharge (Fig. 10-10) and drowned discharge (Fig. 10-11), according to the relative relation between the critical depth under the bridge span, h_c and the water depth downstream of bridge, h

$$h < 1.3h_c \quad \text{Free outflow,} \quad h_b = \psi h_c \quad (10.27)$$

$$h \geq 1.3h_c \quad \text{Drowned outflow,} \quad h_b = h \quad (10.28)$$

where h_b is the water depth under bridge span; ψ is the *depth-contracted coefficient* depending on the shape of bridge abutment, $\psi < 1$. In general, $\psi = 0.80 \sim 0.85$ for smooth entrance and $\psi = 0.75 \sim 0.80$ for other entrances.

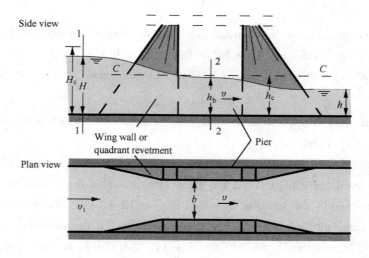

Fig. 10-10 Flow through a small bridge aperture with a free outflow

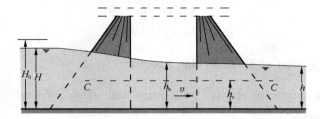

Fig. 10-11 Flow through a small bridge aperture with a drowned outflow (Side view)

FREE OUTFLOW

For free outflow, water depth downstream of the bridge does not affect the flow passing through the bridge. The water surface of flow drops around the entrance and the exit of bridge aperture, supercritical flow occurs under the bridge span ($h_b < h_c$) as shown in Fig. 10-10.

The energy equation between section 1-1 upstream of bridge and section 2-2 under the bridge span referred to the river bed invert gives

$$H + \frac{p_{at}}{\rho g} + \frac{\alpha_0 v_0^2}{2g} = h_b + \frac{p_{at}}{\rho g} + \frac{\alpha v^2}{2g} + \zeta \frac{v^2}{2g}$$

Taking $H_0 = H + \frac{\alpha_0 v_0^2}{2g}$ and $h_b = \psi h_c$ yields

$$v = \varphi \sqrt{2g(H_0 - \psi h_c)} \qquad (10.29)$$

$$Q = vA = \varepsilon b \psi h_c \varphi \sqrt{2g(H_0 - \psi h_c)} \qquad (10.30)$$

where b is the *clear span width* of bridge; ε is the *contraction coefficient* (width-contracted); $\varphi = \frac{1}{\sqrt{\alpha + \zeta}}$ is the *velocity coefficient* of bridge. The empirical values of ε and φ are summarized in Table 10-3.

Table 10-3 Empirical values of ε and φ for small bridge aperture

Shape of bridge abutment	Velocity coefficient φ	Width-contracted coefficient ε
Single aperture, quadrant revetment	0.90	0.90
Single aperture, convergence wing wall	0.90	0.85
Multi-apertures, or multi-apertures without quadrant revetment, or bridge abutment stretching out of quadrant revetment	0.85	0.80
Arch bridge with arch leg dipping in water	0.80	0.75

DROWNED OUTFLOW

For drowned outflow, water depth downstream of the bridge $h \geqslant 1.3 h_c$, which drowns the discharge of flow. The water surface of flow drops only around the entrance of bridge aperture. Flow under the bridge span comes into a subcritical flow ($h_b = h_c$) as shown in Fig. 10-11. Thus, the velocity and flow rate through the bridge aperture can be determined by

$$v = \varphi\sqrt{2g(H_0 - h)} \qquad (10.31)$$

$$Q = \varepsilon b h \varphi \sqrt{2g(H_0 - h)} \qquad (10.32)$$

10.6.2 Hydraulic Design of Small Bridge Aperture

For the safety and economic of small bridge, hydraulic design of small bridge should satisfy the following requirements:

(1) The *clear bridge span b* should be wide enough to pass the design flowrate determined by the hydrological method. Then a standard bridge span B close to the computed span is taken to be the required clear bridge span.

(2) When the flow rate reaches its design value, the average velocity v under the bridge should not exceed the non-scouring velocity, v_{max}, of the soil of river bed or lining materials of bank, that is the bridge foundation should not be scoured.

(3) The *backward water level H* referred to the bed invert upstream of the bridge should not exceed the permissive level H' determined by the elevation of road shoulder and the surplus height of the lowest point of bridge to the water surface.

Thus, the hydraulic design of small bridge can start with the computation of clear bridge span b from the non-scouring velocity v_{max}, then check the backward water level H, or start with the computation of b from H', then check the velocity v under the bridge. The following shows the steps designing the small bridge start from the non-scouring velocity v_{max}.

(1) Determine the critical depth under the bridge span

Take the cross section of the small bridge aperture to be rectangular with a clear width of b, thus the effective width of bridge is εb, the critical depth under the bridge span is

$$h_c = \sqrt[3]{\frac{\alpha Q^2}{g(\varepsilon b)^2}} \qquad (10.33)$$

Take the velocity under the bridge to be the non-scouring velocity v_{max}. Assume water is free discharged out of the bridge aperture, $h_b = \psi h_c$, thus the flowrate pass the bridge aperture is

$$Q = \varepsilon b h_b v_{max} = \varepsilon b \psi h_c v_{max} \qquad (10.34)$$

Substituting back into Eq. 10.33 gives

$$h_c = \frac{\alpha \psi^2 v_{max}^2}{g} \qquad (10.35)$$

(2) Determine the clear bridge span b

Compare the water depth downstream of bridge and the critical depth from Eq. 10.35. If $h < 1.3h_c$, water is freely discharged, we have $h_b = \psi h_c$, thus

$$b = \frac{Q}{\varepsilon \psi h_c v_{max}} \qquad (10.36)$$

If $h \geq 1.3h_c$, water discharge is drowned, we have $h_b = h$, thus

$$b = \frac{Q}{\varepsilon h v_{max}} \qquad (10.37)$$

Choose the standard bridge span B according to the calculated bridge span and make sure B

⩾ b. In engineering practice, the standard spans of small bridge for railway and highway are 4m, 5m, 6m, 8m, 10m, 12m, 16m, 20m, etc.

(3) Check the discharge type and scouting condition

Once the standard bridge span B is chosen, recalculate the critical depth by

$$h'_c = \sqrt[3]{\frac{\alpha Q^2}{g(\varepsilon B)^2}} \qquad (10.38)$$

If $h < 1.3h'_c$, compute the actual velocity under bridge by $v = \dfrac{Q}{\varepsilon B \psi h'_c}$; and if $h \geqslant 1.3h'_c$, $v = \dfrac{Q}{\varepsilon B h}$. Check whether v is less than v_{max}.

If water is discharged freely in step 2, but in this step it is discharged drowned after the standard bridge span is used, the critical depth in step 1 should be recalculated by

$$h_c = \sqrt[3]{\frac{\alpha (hv_{max})^2}{g}} \qquad (10.39)$$

and thus again the bridge span b and the standard span B in step 2. Then check again as listed in step 3.

(4) Check backward water level H

The backward water level H can be determined from Eq. 10.29 to Eq. 10.32

$$H_0 = h_b + \frac{v^2}{2g\varphi^2} = h_b + \frac{Q^2}{2g\varphi^2 (\varepsilon B h_b)^2} \qquad (10.40)$$

where $h_b = \psi h'_c$ for the free discharge and $h_b = h$ for the drowned discharge.

Generally, $H < H_0$. Thus, if H is taken to be H_0, it would be safer for the engineering practice. Thus, we just need to check the backward water level H by comparing H_0 with the permissive level H'.

EXAMPLE 10-3 Design small bridge aperture

A small bridge with a single aperture conducted by smoothly convergent wing walls will be constructed above a river at a design flow rate of $Q = 25$ m³/s. The corresponding water depth downstream of bridge is $h = 0.9$m. From the design manual of small bridge, the permissive level H' referred to the river bed is $H' = 1.6$m, the non-scouring velocity under the bridge is $v_{max} = 3.5$m/s. Design the clear bridge span.

Solution:

Because water flows into the bridge aperture will be conducted smoothly by convergent wing walls, thus from Table 10-3 we have $\varepsilon = 0.85$, $\varphi = 0.90$. And we take $\psi = 0.85$.

1. Determine the critical depth under the bridge span

From Eq. 10.35 gives

$$h_c = \frac{\alpha \psi^2 v_{max}^2}{g} = \frac{(1.0)(0.85^2)(3.5\text{m/s})^2}{9.81 \text{ m/s}^2} = 0.90\text{m}$$

Thus, $1.3h_c = (1.3)(0.90\text{m}) = 1.17\text{m} > h = 0.9\text{m}$, water will be discharged freely.

2. Determine the clear bridge span b

For free discharge, we can have the bridge span b from Eq. 10.36

$$b = \frac{Q}{\varepsilon \psi h_c v_{max}} = \frac{25 \text{ m}^3/\text{s}}{(0.85)(0.85)(0.90\text{m})(3.5\text{m/s})} = 10.98\text{m}$$

Take the standard bridge span to be $B = 12\text{m}$, which is greater than $b = 10.98\text{m}$.

3. Check the discharge type and scouting condition

From Eq. 10.38 gives the critical depth

$$h'_c = \sqrt[3]{\frac{\alpha Q^2}{g(\varepsilon B)^2}} = \sqrt[3]{\frac{(1.0)(25 \text{ m}^3/\text{s})^2}{(9.81\text{m/s}^2)(0.85^2)(12\text{m})^2}} = 0.85\text{m}$$

Thus, $1.3 h'_c = (1.3)(0.85\text{m}) = 1.11\text{m} > h = 0.9\text{m}$, water will still be discharged freely. Thus, the actual velocity under the bridge is

$$v = \frac{Q}{\varepsilon B \psi h'_c} = \frac{25 \text{ m}^3/\text{s}}{(0.85)(12\text{m})(0.85)(0.85\text{m})} = 3.39\text{m/s}$$

Thus, $v < v_{\max} = 3.5\text{m/s}$, which indicates no scouring occurring under the bridge.

4. Check the backward water level H

Take H to be H_0, from Eq. 10.40 we have

$$H \approx H_0 = \psi h'_c + \frac{v^2}{2g\varphi^2} = (0.85)(0.85\text{m}) + \frac{(3.39\text{m/s})^2}{(2)(9.81\text{m/s}^2)(0.90^2)} = 1.45\text{m}$$

which gives $H < H' = 1.6\text{m}$. Thus, $B = 12\text{m}$ is the required clear bridge span.

Summary

An obstruction that allows the liquid to flow over it is called a *weir*, which has three common types, *sharp-crested*, *ogee-crested* and *broad-crested weirs*, characterized by the ratio of the weir thickness δ to the weir head H. The flow rates past these weirs can be determined by the identical fundamental formula

$$Q = \sigma \varepsilon m b \sqrt{2g} \, H_0^{3/2}$$

or

$$Q = \sigma \varepsilon m_0 b \sqrt{2g} \, H^{3/2}$$

but with different empirical expressions of the discharge coefficients:

For *rectangular sharp-crested weir*,

$$m_0 = \left(0.405 + \frac{0.0027}{H} - 0.03 \frac{B-b}{B}\right)\left[1 + 0.55 \left(\frac{H}{H+P_w}\right)^2 \left(\frac{b}{B}\right)^2\right]$$

For *ogee-crested weir*,

$$m = -0.024 \frac{H_0}{H_d} + 0.18503 \sqrt{\frac{H_0}{H_d}} + 0.3406$$

For the *broad-crested weir* with right-angle leading-edge noses $m = 0.32$ for $\frac{P_w}{H} > 3.0$ and

$$m = 0.32 + 0.01 \frac{3 - P_w/H}{0.46 + 0.75 P_w/H} \quad \text{for } 0 \leq \frac{P_w}{H} \leq 3.0$$

and for the broad-crested weir with rounded leading-edge noses $m = 0.36$ for $\frac{P_w}{H} > 3.0$ and

$$m = 0.36 + 0.01 \frac{3 - P_w/H}{1.2 + 1.5 P_w/H} \quad \text{for } 0 \leq \frac{P_w}{H} \leq 3.0$$

However, for the first estimated, the discharge coefficients of rectangular sharped-crested, ogee-crested and broad-crested weir can be taken as 0.42, 0.502 and 0.3849 respectively.

For the *triangle sharp-crested weir*, the flow rate is determined by
$$Q = 1.4H^{5/2} \quad \text{for } H = 0.05 \sim 0.25\text{m}$$
$$Q = 1.343H^{2.47} \quad \text{for } H = 0.25 \sim 0.55\text{m}$$

The obstruction with an adjustable opening at the bottom that allows the liquid to flow underneath it is called an *underflow gate*. The flow would be a *weir flow* if the gate opening is large enough.

The hydraulic design of *small bridge aperture* is based on the theory of broad-crested weir, whose main purpose is to determine the clear bridge span b, check the backward water level H and make sure the bridge foundation will not be scoured. The flow past the bridge is a *free outflow* with a shrink water depth $h_b = \psi h_c$ under the bridge if $h < 1.3h_c$ and a *drowned flow* with $h_b = h$ if $h \geq 1.3h_c$.

Word Problems

W10-1 The weir flow is (a) a subcritical flow over an obstacle, (b) a supercritical flow over an obstacle, (c) an uniform flow with free surface exposed to ambient pressure or (d) an uniform pressure flow.

W10-2 On what basis are the sharp-crested weir, ogee-crested weir and broad-crested weir classified?

W10-3 Define sharp-crested weir. On what basis are the sharp-crested weirs classified?

W10-4 What is the basic principle of operation of a broad-crested weir used to measure flow rate through an open channel?

W10-5 For the outflows past broad-crested weir and small bridge aperture, on what basis the free or drowned outflow classified?

W10-6 The flow depth, h, over the crest of broad-crested weir flow with a free outflow is (a) $h < h_c$ (where h_c is the corresponding critical depth of this flow), (b) $h = h_c$, (c) $h > h_c$ or (d) Uncertain.

W10-7 The drown coefficient of weir σ_s is (a) $\sigma_s < 1$, (b) $\sigma_s = 1$, (c) $\sigma_s > 1$ or (d) All of above is possible.

W10-8 Water is flowing through the small bridge aperture with a free outflow, thus the flow depth under the bridge span h_b is (a) $h_b < h_c$ (where h_c is the corresponding critical depth of the flow), (b) $h_b = h_c$, (c) $h_b > h_c$ or (d) Uncertain.

Problems

10-1 The flow rate of water in a 4-m-wide horizontal channel is being measured by using a 0.75-m-high sharp-crested rectangular weir that spans across the channel. If the water depth upstream referred to the bed invert is 2.2m, determine the flow rate of water.

10-2 Water flows over a 2-m-high sharp-crested rectangular weir. The flow depth upstream of the weir is 3m, and water is discharged from the weir into an unfinished-concrete channel of equal width where uniform-flow conditions are established. If no hydraulic jump is to occur in the downstream flow, determine the maximum slope of the downstream channel.

10-3 The flow rate of water in a 6-m-wide rectangular channel is to be measured using a 1.1-m-high sharp-crested rectangular weir that spans across the channel. If the head above the weir crest is 0.60m upstream from the weir, determine the flow rate of water.

10-4 A full-width rectangwar sharp-crested weir is to be used to measure the flow rate of water in a 3-m-wide rectangular channel. The maximum flow rate through the channel is 4.25m³/s, and the flow depth upstream from the weir is not to exceed 1.5m. Determine the appropriate height of the weir.

10-5 Water flows over a 1.5-m-wide rectangular sharp-crested weir with a weir height of 1.35m. If the depth upstream is 1.5m, determine the flowrate.

10-6 A rectangular sharp-crested weir is used to measure the flowrate in a channel of width 3.0m. It is desired to have the channel flow depth be 1.8m when the flowrate is 1.41m³/s. Determine the height, P_w, of the weir plate.

10-7 Water flows from a storage tank, over two triangular weirs, and into two irrigation channels as shown in Fig. P10-7. The head for each weir is 0.12m, and the flowrate in the channel fed by the 90°-V-notch weir is to be twice the flowrate in the other channel. Determine the angle θ for the second weir. Take the discharge coefficient for the second V-notch weir to be 0.396.

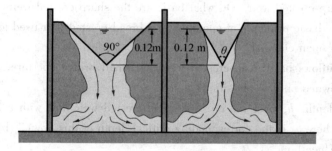

Fig. P10-7

10-8 The flow rate of water flowing in a 3-m-wide channel is to be measured with a sharp-crested triangular weir 0.5m above the channel bottom with a notch angle of 60°. If the flow depth upstream from the weir is 1.5m, determine the flow rate of water through the channel. Take the weir discharge coefficient to be 0.60.

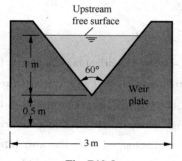

Fig. P10-8

10-9 A sharp-crested triangular weir with a notch angle of 100° is used to measure the discharge rate of water from a large lake into a spillway. If a weir with half the notch angle ($\theta = 50°$) is used instead, determine the percent reduction in the flow rate. Assume the water depth in the lake and the weir discharge coefficient remain unchanged.

10-10 (a) The rectangular sharp-crested weir shown in Fig. P10-10a is used to maintain a relatively constant depth in the channel upstream of the weir. How much deeper will the water be upstream of the weir during a flood when the flowrate is $1.27 \text{m}^3/\text{s}$ compared to normal conditions when the flowrate is $0.85 \text{m}^3/\text{s}$? Assume the weir coefficient remains constant at $m = 0.62$. (b) Repeat the calculations if the weir of part (a) is replaced by a rectangular sharp-crested 'duck bill' weir which is oriented at an angle of 30° relative to the channel centerline as shown in Fig. P10-10b. The weir coefficient remains the same.

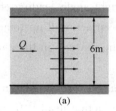
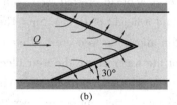

Fig. P10-10

10-11 Water flows over a triangular weir as shown in Fig. P10-11a. It is proposed that in order to increase the flowrate, Q, for a given head, H, the triangular weir should be changed to a trapezoidal weir as shown in Fig. P10-11b. (a) Derive an equation for the flowrate as a function of the head for the trapezoidal weir. Neglect the upstream velocity head and assume the weir coefficient is 0.60, independent of H. (b) Use the equation obtained in part (a) to show that when $b \ll H$ the trapezoidal weir functions as if it were a triangular weir. Similarly, show that when $b \gg H$ the trapezoidal weir functions as if it were a rectangular weir.

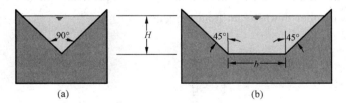

Fig. P10-11

10-12 A 20-m-wide, 60-m-high ogee-crested weir with a section of standard WES weir has a design weir head of $H_d = 6$m. It spans across the whole width of the channel. Water over the weir is discharged freely. (a) Determine the design flow rate. (b) If the flowrate is 1000 m^3/s, determine the corresponding weir head above the weir.

10-13 A 1-m-high broad-crested weir with the right-angle leading-edge nose is used to measure the flow rate of water in a 5-m-wide rectangular channel. The flow depth well upstream from the weir is 1.6m. Determine the flow rate through the channel and the minimum flow depth above the weir.

10-14 Consider water flows over a 0.80-m-high sufficiently long broad-crested weir with the right-angle leading-edge nose. If the minimum flow depth above the weir is measured to be 0.50m, determine the flow rate per meter width of channel and the flow depth upstream of the weir.

10-15 Consider two identical 3.6-m-wide rectangular channels each equipped with a 0.6-m-high full-width weir, except that the weir is sharp-crested in one channel and broad-crested with the right-angle leading-edge nose in the other. For a flow depth of 1.5-m in both channels, determine the flow rate through each channel.

10-16 Water flows over a broad-crested weir with the right-angle leading-edge nose that has a width of 4m and a height of $P_w = 1.5$m. The free-surface well upstream of the weir is at a height of 0.5m above the weir crest. Determine the flowrate in the channel and the minimum depth of the water above the weir block.

10-17 A 2-m-high broad-crested weir with the rounded-angle leading-edge nose is placed across a 6-m-wide channel. Determine the flowrate if the acting head is 0.30m.

10-18 Water flows over a broad-crested weir with right-angle leading-edge noses has a width of 4m and a height of $P_w = 0.6$m and is placed across a 4.0-m-wide horizontal channel. The free-surface well upstream of the weir is at a height of 1.2m above the weir crest. Determine the flowrate if the downstream water depth is (a) 0.8m and (b) 1.7m.

10-19 A broad-crested weir with rounded leading-edge noses has a height of $P_w = 1.0$m and is placed across a 5.0-m-wide horizontal channel. If the flow rate is 15.0m^3/s and downstream water depth is 1.8m, determine the flow depth well upstream.

10-20 The flow rate of water through a 5-m-wide (into the paper) channel is controlled by a sluice gate as shown in Fig. P10-20. If the flow depths are measured to be 1.1 and 0.45m upstream and downstream from the gates, respectively, determine the flowrate and the Froude number downstream from the gate.

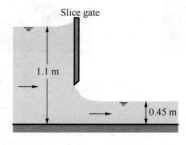

Fig. P10-20

10-21 A small bridge with a single aperture conducted by smoothly convergent wing walls (ε = 0.85, φ = 0.90, ψ = 0.85) will be built above a river at a design flow rate of 15 m^3/s.

The natural water depth downstream of bridge is 1.3m. The river bed under bridge is enforced by the pebble, which has a non-scouring velocity v_{max} = 3.5m/s. Design the clear bridge span and the backward water level upstream of bridge.

10-22 Experiments show that the discharge coefficient for water passing through the small bridge aperture is about 0.34. Thus, the drowned outflow condition for flow through the small bridge aperture, $h \geq 1.3h_c$, is basically same as the one for flow past a broad-crested weir, $h_s \geq 0.8H_0$. Verify this conclusion.

11 Seepage Flow

CHAPTER OPENING PHOTO: A piping is an underseepage in a levee and resurfaces on the landside in the form of a volcano-like cone of sand, as in the figure, which may induce the erosion of the levee toe or foundation and result in the levee breach. The phreatic surface and location of piping can be determined by the principle of seepage flow in the present chapter. (ⓒ巴彦网 ❶)

Seepage flow is a flow that pore fluid flows in the porous medium, which has a great application in the field of water conservancy, civil, petroleum and mining engineering.

In the practice of civil engineering, it predominantly works on problems that involve water and flowing in earth (in earth's gravity) or rock under the ground surface. Such a flow is termed *groundwater flow*. Its theory is widely used in the practice of dewatering of foundation pit, excavation and waterproof of tunnel, urban water supply and drainage system and the design of hydraulic structure.

In this chapter, only the simple case, the steady groundwater flow in a homogeneous and isotropic soil is discussed. It begins with the simplification of the seepage flow and the Darcy's law, follows the brief introduction of the gradually varied seepage flow and its phreatic surfaces, ends with the well yields and the shapes of the drawdown cones of the different types of wells.

---

❶ http://www.bayan.ccoo.cn/forum/thread-7635896-1-1.html

11.1 Introduction

11.1.1 Permeability of a Soil

Permeability is the measure of the soil's ability to permit water to flow through its pores or voids. It is one of the most important soil properties of interest to geotechnical engineers.

Permeability plays an important role in the engineering practices. It greatly influences the rate of settlement of a saturated soil under load, the stability of slopes and retaining structures. The design of earth dams and the filters made of soils are very much based upon the permeability of the soils used.

The *hydraulic conductivity*, or *coefficient of permeability*, of a soil is a measure of the soil's permeability when submitted to a hydraulic gradient. It is dependent on the grain size and the useful pore space. If the hydraulic conductivity of a soil layer is such low that its permeability can be neglected relative to the adjacent soil layers, such a soil layer is termed *impermeable stratum*; otherwise, *permeable stratum*. In less permeable soils the seepage water can be stored temporarily. If the seepage water encounters an impermeable soil layer or impermeable rock, seepage will no longer take place and the seepage water accumulates permanently. Such underground water accumulations are known as *groundwater*.

If a soil has an identical hydraulic conductivity at any point, such a soil is termed *homogeneous soil*; otherwise, it is *heterogeneous soil*. If a soil has an identical hydraulic conductivity in any direction of a point, such a soil is termed *isotropic soil*; otherwise, *anisotropic soil*.

In this chapter, only the steady groundwater flow in a homogeneous and isotropic soil is briefly presented.

11.1.2 Conceptual Model of Seepage Flow

In the porous soil, groundwater has five statuses: *gaseous water, hydroscopic water, pellicular water, capillary water* and *gravitational water*. Gravitational water is the water which occupies the pores of the soil and moves freely in the pores under the effect of gravity, namely the *groundwater flow* defined above. However, other four states of water do not flow in the pores, excepting the capillary water may be rising in the very fine pores in the soil.

In engineering, the grain sizes and the pore spaces are random. They may vary from a soil to another soil, even for the identical soil they also vary from one site to another site. Neither theoretical analysis nor experiment methods can provide a real simulation of fluid flow in the pore spaces. Therefore, a hypothetical simplified *conceptual model* of seepage flow has been developed to replace the real one. In this model, the seepage flow is simplified as an open channel flow (see Fig. 11-1): the soil grains are ignored so that the seepage flow region is assumed to occupy the whole flow field, including the volume supposed to be occupied by the soil grains in real case, and the boundary conditions, such as the seepage flux, pressure and resistance of the seepage flow, are identical to the real case.

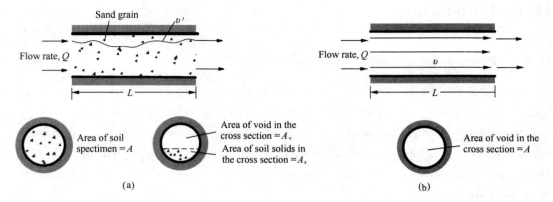

Fig. 11-1 Seepage flow. (a) The seepage flow in practice; (b) the conceptual model of seepage flow

Thus, the permeating velocity in the seepage flow model can be determined by

$$v = \frac{Q}{A}$$

where A is the area of the cross section in the model; Q is the flow rate.

In the practice case of seepage flow, the area of cross section, A, does not include the area occupied by the soil grains, A_s. Thus, the actual velocity in the prototype is bigger than that in the model because they have the identical flow rates. The relation is

$$v' = \frac{Q}{A_v} = \frac{vA}{A_v} = \frac{v}{n}$$

where v' is the *permeating velocity* in the prototype; n is the *actual porosity* in the cross section in the prototype, which is the ratio of area of actual pore spaces to the total area of the soil of interest, $n = A_v/A$.

The application of the seepage flow model in the theory of the seepage flow not only keeps the macroscopical quantities, i. e. the seepage flowrate, the pressure and the resistance of seepage flow, identical to the actual one, but also makes the methods and concepts of the continuum assumption available in the seepage flow model.

11.1.3 Classification of Seepage Flow

Based on the conceptual model of seepage flow, the seepage flow also can be classified as one-, two- or three dimensional flow on basis of the velocity distribution; steady or unsteady flow according to the temporal variation of properties; uniform and nonuniform flow (gradually or rapid varied seepages for sub-classification) on the spatial variation of properties; seepage flow with free surface in unconfined aquifer and seepage flow in confined aquifer.

11.2 Darcy Law

11.2.1 Darcy Law

Henry P. G. Darcy (1803-1858), a French hydraulic engineer interested in purifying water supplies using sand filters, conducted experiments to determine the flow rate of water through the filters. Published in 1856, his conclusions, Darcy's law known nowadays, have served as the basis for all modern analysis of groundwater flow.

Darcy performed a quantifiable filtration tests by an instrument as in Fig. 11-2. The main part is a straight identical diameter column with an open top and filled with homogeneous sand. Water is continuously supplied and overflowed to make sure the water level in the column steady. Two piezometer tubes are installed apart from each other with a vertical distance of l on the side wall of cylinder. A porous plate near the bottom of the cylinder is used to prevent the sand grain from passing. Thus, the flow in Darcy's experiment is a pressurized, steady and uniform seepage flow.

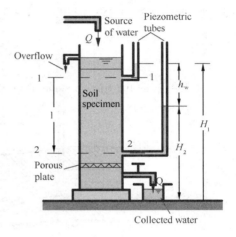

Fig. 11-2 Apparatus of Darcy experiment (constant-head permeameter)

For seepage flow, the velocity head in the energy equation is usually negligible due to its very slow velocity. Therefore, the energy between the two measured sections is completely dissipated with an energy grade

$$J = \frac{h_w}{l} = \frac{H_1 - H_2}{l}$$

where H_1, H_2 are the piezometric head in section 1-1 and 2-2 respectively, J is the friction slope and l is the distance between two sections.

The experiment results show that the volume of flow of the pore fluid through a porous medium per unit time is proportional to the rate of change of excess fluid pressure with distance.

$$Q \propto A \frac{H_1 - H_2}{l} = AJ \qquad (11.1)$$

Introducing the coefficient of proportionality into the above equation yields

$$v = kJ \qquad (11.2)$$

where v is the average sectional seepage velocity in the conceptual model of seepage flow; k is a comprehensive coefficient for the permeability properties of a soil, termed *coefficient of permeability*, which has the dimension of velocity and is used to describe the hydraulic conductivity of a soil.

Equation 11.2 is known as *Darcy Law* for seepage flow, which shows that the friction slope is directly proportional to the velocity of the seepage flow. So it is also called *linear law of seepage flow*.

11.2.2 Darcy Law with Point Velocity

In Eq. 11.2, the velocity is sectional-averaged. It is because the actual fluid velocity varies throughout the pore space, due to the connectivity and geometric complexity of that space. This variable velocity is thus characterized by its mean or average value. However, for a flow in a streamtube in a *confined aquifer* (*artesian aquifer*) as shown in Fig. 11-3, the velocity at any point, u, can also be expressed as the form of Darcy law

$$u = kJ = -k\frac{dH}{ds} \tag{11.3}$$

where $J = -\dfrac{dH}{ds}$ is the friction slope along the streamtube length ds; dH is the energy drop along ds.

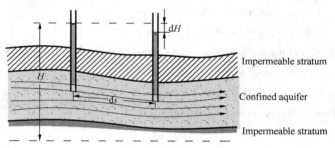

Fig. 11-3 Seepage flow in a confined aquifer

11.2.3 Applicability of Darcy's Law

Darcy's Law makes some assumptions, which can limit its applicability:

(1) We assume the kinetic energy can be ignored, which indicates Darcy law only can be used for laminar flow in which viscous forces dominates, i.e. seepage flow has very low Reynolds number, $Re < 1$, or may be up to 10 as indicated in Fig. 11-4. In seepage flow, Reynolds number has a definition as

$$Re = \frac{vd_{10}}{\nu} \tag{11.4}$$

where v is the cross sectional average velocity of the seepage flow; d_{10} is the *effective grain size diameter*, defined as the sieve pore diameter of the grading screen through which 10 percent of the soil (in weight) can pass; ν is the kinematic viscosity of water.

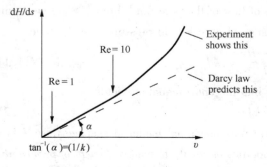

Fig. 11-4 Relation of velocity to the head gradient (Todd and Mays, 2004)

However, the seepage flow in coarse sand soils such as gravels, cobbles are no longer linear (Re > 1 ~ 10), while the friction slope is no longer proportional to the velocity of the seepage flow. So Darcy Law is no longer valid. In this chapter, only the seepage flow meeting Darcy Law will be discussed.

(2) We assume the average properties control discharge, which indicates that Darcy law is applicable only on macroscopic scale, for which areas are not less than 5 to 10 times the average pore cross section, as illustrated in Fig. 11-5.

(3) We assume the fluid properties are constant, which indicates that Darcy law is applicable only for constant temperature or salinity settings, or convert the head H_1, H_2 of water with variable temperature and salinity into fresh-water equivalent head, then to friction slope J.

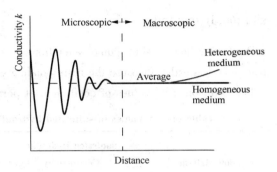

Fig. 11-5 Variation of rock properties with scale. Note heterogeneous media have average properties that vary positively (shown) or negatively (not shown) with distance.

11.2.4 Coefficient of Permeability

As was discussed in the preceding section, *coefficient of permeability*, k, is the velocity of the seepage flow per unit friction slope, and gets a dimension as LT^{-1}. In SI the units for engineering are cm/s or m/d. It reflects the effect of the interaction between porous soil and water to the permeability. It relates the amount of water which will flow through a unit cross-sectional area of aquifer under a unit gradient of hydraulic head.

The value of coefficient of permeability is affected by many factors, the main factors are the shape, size of the soil granule, the average grading of the soil, and the water temperature. It is rather difficult to determine its value accurately. Generally, it can be determined by the following three methods.

Empirical formula

In engineering, some empirical formulas can be used to estimate the value of k. For example, for nearly identical grain size, efficient diameter of the soil granule can be used to estimate k

$$k = Cd_{10}^2 \tag{11.5}$$

where C is the *Hazen's empirical coefficient*, which takes a value between 1.0 and 1.5 with d_{10} in mm and k in cm/s.

Laboratory methods

In laboratory, there are two methods to get the value of permeability coefficient: *constant-head method* and *falling-head method*. The theory of constant-head method is exact the theory of Darcy law with an apparatus similar to the one in Fig. 11-2, then calculate k by using Eq. 11.3.

This method is typically used on granular soil. The falling-head method is totally different than the constant head methods in its initial setup: the soil sample is first saturated under a specific head condition; the water is then allowed to flow through the soil without maintaining a constant pressure head. However, the advantage to the falling-head method is that it can be used for both fine-grained and coarse-grained soils.

The laboratory methods are simple and easy, but the experimental result may have large error due to the disturbance to the soil samples.

In-situ (field) method

Dig testing hole in situ, pump water from it or into it, measure some necessary data when the seepage flow is steady, then use the corresponding equations (Eq. 11.3) to calculate the value of k.

The common used values of coefficient of permeability are tabulated in Table 11-1.

Table 11-1 Values of saturated hydraulic conductivities for sands and gravels

Soil Materials	Saturated Hydraulic Conductivity k (m/d)	Soil Materials	Saturated Hydraulic Conductivity k(m/d)
Clay	<0.005	Coarse sand	20~50
Silt clay	0.005~0.1	Homogeneous coarse sand	60~75
Silt	0.1~0.5	Cobble	50~100
Loess	0.25~0.5	Gravel	100~500
Very fine sand	0.5~1.0	Unfilled cobble	500~1000
Fine sand	1.0~5.0	Rock with few fractures	20~60
Medium sand	5.0~20.0	Rock with much fractures	>60
Homogeneous medium sand	35~50		

Source: Data are from 《工程地质手册(第四版)》(2007).

11.3 Unconfined Steady Gradually Varied Seepage Flow

11.3.1 General Formula

For the unconfined *gradually varied groundwater flow* as illustrated in Fig. 11-6, assume:

1. The water table or free surface is only slightly inclined. In 1863, Dupuit observed that in most groundwater flows, the slope of the water table is very small, which is commonly in the range 1/1000 to 10/1000.

2. Streamlines are assumed to be horizontal and equipotential lines be vertical (see Fig. 11-6b). In steady flow without acceleration, in a vertical two-dimensional xz-plane, as shown in Fig. 11-6, the water table is a streamline.

3. Slopes of energy gradient line and hydraulic gradient line are equal due to the very small seepage velocity as indicated in Table 11-1.

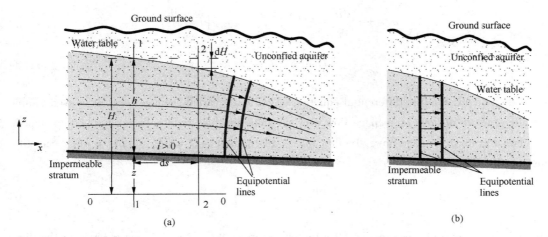

Fig. 11-6 (a) Gradually varied groundwater flow in an unconfined aquifer. (b) Gradually varied groundwater flow in an unconfined aquifer under the Dupuit assumption (equipotentials are vertical, flow is horizontal, uniform velocity along the vertical). Not to scale.

Take horizontal line 0-0 in Fig. 11-6a to be the reference datum, and two sections 1-1 and 2-2 with a distance of ds. We also take the piezometric head at section 1-1 to be $H(= z + h)$ and $H + dH$ at section 2-2. Thus, the head loss between sections 1-1 and 2-2 along any streamtube is dH. When the distance between these two sections approaches zero, the friction slope of a point in the gradually varied seepage flow can be determined by

$$J = -\frac{dH}{ds} = \text{constant}$$

Thus, for the gradually varied seepage flow in a homogeneous soil, the velocity at any point in the cross section, u, is equal to the average velocity of the cross section, v, that is

$$v = u = kJ = -k\frac{dH}{ds} \tag{11.6}$$

where J is the average friction slope of the cross sections. For non-uniform seepage flow, J is different for different cross sections, as well as velocity v.

Equation 11.6 is the general equation for gradually varied seepage flow, known as *Dupuit formula*, named after Arsene Jules Emile Juvenal Dupuit (1804-1866), a French hydraulic engineer, Darcy's associate and successor.

Equation 11.6 is the extension of Darcy Law in gradually varied seepage flow, which relates the average velocity in a cross section to the frictional slope of seepage flow.

11.3.2 Differential Equation

Take the water depth at section 1-1 of a gradually varied seepage flow over the first impermeable stratum as shown in Fig 11-6 to be h, the invert elevation of impermeable stratum be z, and the slope of the impermeable stratum be $i = -dz/ds$. From Dupuit formula, Eq. 11.6, we can have the average velocity in this section

$$v = -k\frac{dH}{ds} = -k\frac{d(z+h)}{ds} = k\left(i - \frac{dh}{ds}\right) \tag{11.7}$$

or
$$Q = kA\left(i - \frac{dh}{ds}\right) \tag{11.8}$$

Equation 11.8 is the differential equation of gradually varied seepage flow. It is used to analyze the *phreatic surfaces* of seepage flow over the impermeable strata with various bottom slopes.

For a uniform seepage flow, the flow rate is determined as $Q = vA_0 = kJA_0 = kiA_0$. Substituting into Eq. 11.8 gives

$$\frac{dh}{ds} = i\left(1 - \frac{A_0}{A}\right) \tag{11.9}$$

where A_0 is the area of cross section for a uniform seepage flow, A is the area of cross section for a actual seepage flow.

11.3.3 Phreatic Surfaces

For the gradually varied groundwater flow in an unconfined aquifer, the water table of the groundwater flow along the flow direction is known as the *phreatic surface*, which is defined as the surface at every point of which the water pressure is atmospheric. It indicates the location where the pore water pressure is under atmospheric conditions (i.e. the pressure head is zero). The slope of the phreatic surface is assumed to indicate the direction of ground water movement in an unconfined aquifer. It is because the phreatic surface, coinciding with the hydraulic gradient line, also coincides with the energy gradient line which continuously drops in the flow direction due to the energy loss.

Phreatic surface is similar to the water surface profiles of the gradually varied flow in the open-channel flow (see Section 9.7). However, the velocity head of the seepage flow is negligible due to its small velocity, the specific energy at its cross section is equal to the water depth, so the critical depth no longer exists, neither the concepts such as subcritical flow and supercritical flow, mild slope and steep slope. Therefore, when we analyze the phreatic surfaces of the groundwater flow, we only need to consider the impermeable stratum with a slope of downhill, horizontal and adverse, and the uniform seepage flow depth, h_0.

SEEPAGE FLOW OVER DOWNHILL IMPERMEABLE STRATUM ($i > 0$)

Similar to the uniform flow in an open channel, uniform seepage flow can also form in an unconfined aquifer over a downhill impermeable stratum as shown in Fig. 11-7. Thus, the normal depth N-N of uniform seepage flow separates the flow region into two regimes: zone I with $h > h_0$ and zone II with $h < h_0$ (see Fig. 11-7).

In zone I, $h > h_0$, so $A > A_0$, from Eq. 11.9 we have $dh/ds > 0$, which means the phreatic surface is a *backwater curve*, designated D_1. As $h \to h_0$ at the far away upstream, $A \to A_0$, thus $dh/ds \to 0$, the water surface asymptotically approaches the normal depth line $N - N$, namely the normal depth line $N - N$ is the upstream asymptote of curve D_1. As $h \to \infty$ at the far away downstream, $A \to \infty$, thus $dh/ds \to i$, the water surface approaches a horizontal asymptote.

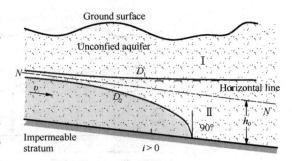

Fig. 11-7 Phreatic surface of seepage flow in an unconfined aquifer over a downhill impermeable stratum

In zone II, $h < h_0$, so $A < A_0$, from Eq. 11.9 we have $dh/ds < 0$, which means the water surface profile is a *dropdown curve*, designated D_2. As $h \to h_0$ at the far away upstream, $A \to A_0$, thus $dh/ds \to 0$, the water surface asymptotically approaches the normal depth line $N - N$. As $h \to 0$ at the far away downstream, $A \to 0$, thus $dh/ds \to -\infty$, the water surface tends to intersect the impermeable stratum with a right angle. Actually, the flow is not a gradually varied flow any more, Eq. 11.9 is not applicable in this region. Thus, the curve should be determine by the actual conditions.

Assume the cross section of the groundwater flow is a rectangular with a very larger width b.

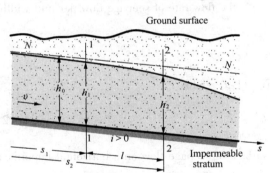

Fig. 11-8 Parameters of the phreatic surface ($i > 0$)

Thus, we have $A = bh$, $A_0 = bh_0$. Substituting into Eq. 11.9 and taking $\eta = h/h_0$ gives

$$ds = \frac{h_0}{i}\left(1 + \frac{1}{\eta - 1}\right)d\eta$$

Integrating it from section 1-1 to section 2-2 (see Fig. 11-8) gives

$$l = \frac{h_0}{i}\left(\eta_2 - \eta_1 + \ln\frac{\eta_2 - 1}{\eta_1 - 1}\right) \quad (11.10)$$

where l is the distance between section 1-1 to section 2-2, $\eta_1 = \frac{h_1}{h_0}$, $\eta_2 = \frac{h_2}{h_0}$.

Equation 11.10 is used to determine the phreatic surface of seepage flow in an unconfined aquifer over a downhill impermeable stratum.

SEEPAGE FLOW OVER HORIZONTAL IMPERMEABLE STRATUM ($i = 0$)

The seepage flow in an unconfined aquifer over a horizontal impermeable stratum is shown in Fig. 11-9. Take the slope in Eq. 11.8 to be zero, that is $i = 0$, we have the equation of phreatic surface of seepage flow over a horizontal impermeable stratum

$$\frac{dh}{ds} = -\frac{Q}{kA} \quad (11.11)$$

No uniform seepage flow over a horizontal impermeable stratum can come into form. Thus,

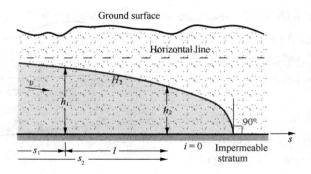

Fig. 11-9 Phreatic surface of seepage flow in an unconfined aquifer over a horizontal impermeable stratum ($i=0$)

From Eq. 11.11 we have $dh/ds < 0$ because all the values of Q, k and A are greater than zero. Thus, there is only one phreatic surface, namely a dropdown curve, H_2, in this situation as shown in Fig. 11-9. At the far end of upstream, $h \to \infty$, $A \to \infty$, thus $dh/ds \to 0$, the water surface approaches horizontal asymptote; at the far end of downstream, $h \to 0$, $A \to 0$, thus $dh/ds \to -\infty$, the water surface tends to intersect the impermeable stratum with a right angle in the downstream.

Again, assume the cross section of the groundwater flow is a rectangular with a very larger width b. Thus, we have $A = bh$, $Q/b = q$, where q is the flow rate of seepage flow per unit width. Substituting into Eq. 11.11 gives

$$\frac{q}{k}ds = -hdh$$

Integrating it from section 1-1 to 2-2 (see Fig. 11-9) yields

$$\frac{2q}{k}l = h_1^2 - h_2^2 \qquad (11.12)$$

Equation 11.12 indicates that the phreatic surface of seepage flow on a horizontal impermeable stratum is a parabola. It is used to determine the phreatic surface of seepage flow in an unconfined aquifer over a horizontal impermeable stratum.

SEEPAGE FLOW OVER ADVERSE IMPERMEABLE STRATUM ($i<0$)

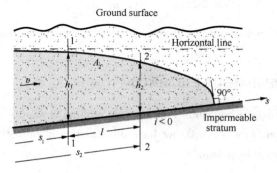

Fig. 11-10 Phreatic surface of seepage flow in an unconfined aquifer over an adverse impermeable stratum ($i<0$)

Similar to the seepage flow in an unconfined aquifer over a horizontal impermeable stratum, there is only one dropdown curve, A_2, of the seepage flow in an unconfined aquifer over an adverse impermeable stratum as shown in Fig. 11-10, which approaches horizontal asymptote in the upstream and tends to intersect the impermeable stratum with a right angle in the downstream.

11.3.4 Seepage Flow of Collecting Gallery

Collecting gallery is a hydraulic structure used to lower the static water table of groundwater by collecting groundwater in a gallery and then pumping it out. Fig. 11-11 shows a schematic cross section of a collecting gallery with an impermeable bottom and permeable side walls. When the pumping flow rate is steady, a *drawdown cone* comes into form, and the water depth, h, in the gallery over the invert of impermeable stratum will not vary any more. Thus, we have $z = h$ at $x = 0$, and $z = H$ at $x = L_0$, where H is the static water table of groundwater above the invert of impermeable stratum; L_0 is the *maximum influencing distance* of the gallery, which is measured from the side wall of gallery to the first vertical section at which the drawdown of static water table is zero.

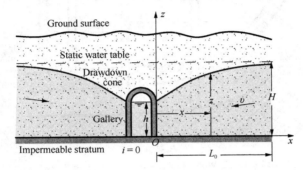

Fig. 11-11 Schematic of flow around a storage gallery

Substituting $l = L_0$, $h_1 = H$, $h_2 = h$ into Eq. 11. 12 gives

$$q = \frac{k(H^2 - h^2)}{2L_0} \qquad (11.13)$$

where q is the per width flow rate of infiltration for each side wall of gallery.

The phreatic surface around the gallery due to the pumping can also be determined from Eq. 11. 12

$$z^2 - h^2 = \frac{2q}{k}x \qquad (11.14)$$

where x is the horizontal distance from the side wall of gallery to the point of interest; z is the elevation of phreatic surface above the invert of impermeable layer.

11.4 Seepage Flow around Wells and Clustered-Wells

In the above section, the seepage flow in an unconfined aquifer bounded from above a phreatic surface and from below an impervious stratum is discussed. It has a water table with the atmospheric pressure. However, the *confined aquifer* is one that is bounded both from above and from below by impervious stratum. The water level in a well (or a piezometer) that is open (i. e., screened) in such an aquifer is higher than the impervious surface that bounds the aquifer from above. If the *piezometric surface*, the surface which at every point indicates the piezometric head, may be anywhere above the ceiling of the confined aquifer as shown in Fig. 11-12, such a confined aquifer is referred to as an *artesian aquifer*. A well in an artesian aquifer is an *artesian well*. Water may even reach the ground surface without pumping if the natural pressure is high enough, in which case the well is called a *flowing artesian well*. A well in an unconfined aquifer is a *water table well*. If the bottom of water table well reaches to the impervious surface of the impermeable stratum, such a well is called *complete water table well*.

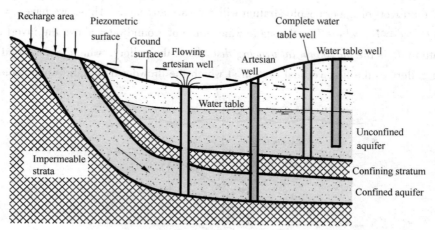

Fig. 11-12 Water table wells and artesian wells

11.4.1 Complete Water Table Well

In Fig. 11-13 shows a complete water table well in a steady statue under pumping. When the well is pumping, the static water level is drawn down most near the well and becomes shallower when moves away from the well. At some distance from the well at any given time there is a point at which the pumping does not change the water table and the drawdown is zero. Such a distance is called the *influence radius of well*, R. The depression in the static water table near the well caused by pumping is called *drawdown cone*.

Set cylinder coordinates system as shown in Fig. 11-13, the pumping flowrate (also the *well yield*) in a steady state equals to the *well flow rate*, the rate that water flows into the well itself from the surrounding soils. Thus, the well flow rate through a cylinder surface with a radius r and a thickness of drawdown cone z can be obtained by using Dupuit formula (Eq. 11.6)

$$Q = Av = 2\pi rzk\frac{dz}{dr} \qquad (11.15)$$

Separate variables

$$\int_h^z 2zdz = \frac{Q}{\pi k}\int_{r_0}^r \frac{dr}{r}$$

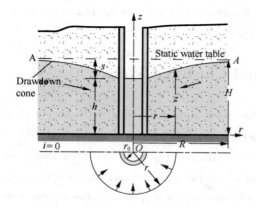

Fig. 11-13 Complete water table well

and integrate it to be

$$z^2 - h^2 = \frac{Q}{\pi k}\ln\frac{r}{r_0} \qquad (11.16)$$

where r_0 is the radius of the well; h is the water level in the well over the well invert; k is the coefficient of permeability.

Equation 11.16 can be used to determine the shape of the drawdown cone of the complete water table well.

When $r = R$, the static water table is not disturbed by the pumping and the drawdown is zero, that is, $z = H$. Thus, substituting $r = R$, $z = H$ into Eq. 11.16 gives the formula of the yield of the complete water table well

$$Q = \pi \frac{k(H^2 - h^2)}{\ln(R/r_0)} \qquad (11.17)$$

where H is static water table level above the well invert.

The influence radius of well, R, strongly depend on the permeability of the soil. Based on the practical experience, for fine sand $R = 100 \sim 200$m, for moderate sand $R = 250 \sim 500$m, for coarse sand $R = 700 \sim 1000$m. R can also be calculated by the following empirical formula:

$$R = 575 s \sqrt{Hk} \quad \text{or} \quad R = 3000 s \sqrt{k} \qquad (11.18)$$

where s is the *maximum drawdown* of static water table in the well, $s = H - h$ as in Fig. 11-13. Both s and R are in meters, and k is in m/s.

Use $s(= H - h)$ replacing h in Eq. 11.17 to give

$$Q = 2\pi \frac{kHs}{\ln(R/r_0)}\left(1 - \frac{s}{2H}\right) \qquad (11.19)$$

When $s/H \ll 1$, Eq. 11.19 is simplified as

$$Q = 2\pi \frac{kHs}{\ln(R/r_0)} \qquad (11.20)$$

In engineering practice, the measurement of s is much easier than that of h, therefore Eq. 11.19 is more practical than Eq. 11.17 in the engineering practice.

EXAMPLE 11-1 Recharging well

In Fig. E11-1 water is recharged into the soil through a 0.2-m-diameter complete water table recharging well. When the steady infiltration cone comes to form, the recharged flowrate into the well is 2×10^{-4} m^3/s, the water depth in the well over the well invert is $h = 5$m and the influence radius of well is $R = 150$m. If the static water table is $H = 3.5$m over the invert of the impermeable stratum, determine the permeability coefficient of the soil.

Solution:

When the infiltration cone is steady, the re-

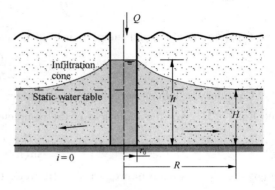

Fig. E11-1

charged flowrate into the well is equal to the flowrate of seepage flow out of the well. Revise Eq. 11.17 (because $h > H$) into

$$Q = \pi \frac{k(h^2 - H^2)}{\ln(R/r_0)}$$

to give

$$k = \frac{Q\ln(R/r_0)}{\pi(h^2 - H^2)} = \frac{(2 \times 10^{-4} \text{m}^3/\text{s})\left[\ln\left(\frac{150\text{m}}{0.1\text{m}}\right)\right]}{(3.14)[(5\text{m})^2 - (3.5\text{m})^2]} = 3.65 \times 10^{-5} \text{m/s}$$

11.4.2 Complete artesian well

As what discussed above, an *artesian well* is a well in which water rises under pressure from a permeable stratum overlaid by impermeable rock. *Complete artesian well* is an artesian well of which the well bottom reaches the invert of the lower artesian aquifer surface, as shown in Fig. 11-14.

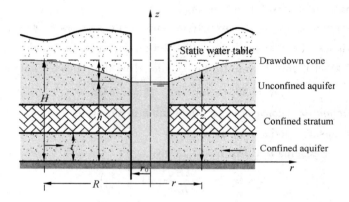

Fig. 11-14 Complete artesian well

Once the flow rate of pumping from the complete artesian well is steady after a period of pumping, a drawdown cone comes into form in the unconfined aquifer. Set cylinder coordinate system as shown in Fig. 11-14, the pumping flow rate (also *well yield*) equals to the well flow rate which water flows into the well through the well wall in artesian aquifer. Thus, the well yield can be obtained by using Dupuit formula (Eq. 11.6) at a cylinder surface with a radius r and a height t

$$Q = Av = 2\pi rtk \frac{dz}{dr}$$

Separate variables and integrate it to give

$$z - h = \frac{Q}{2\pi tk} \ln \frac{r}{r_0} \quad (11.21)$$

Consider the influence radius of well R here, thus, we have $z = H$ when $r = R$. Substituting into Eq. 11.21 gives the well yield of the complete artesian well

$$Q = 2\pi tk \frac{H - h}{\ln(R/r_0)} = \frac{2\pi tks}{\ln(R/r_0)} \quad (11.22)$$

where s is the maximum drawdown of static water table in the well, $s = H - h$; r_0 is the radius of well; h is the water depth in the well over the well invert; k is the coefficient of permeability; H is the piezometric surface level (static water table) of confined artesian aquifer over the well invert; R is the influence radius of well, which can be computed by Eq. 11. 18.

11.4.3 Multiple Well System

If there are more than one wells working together, and the distance between them is far smaller than the common influence radius of all wells, the groundwater flows among wells will affect each other. Thus, all wells as a unity is termed *multiple-well* as shown in Fig. 11-15. In the civil engineering, multiple-well is usually used to dewater the foundation pit, or draw the groundwater from the aquifers in a water supply system.

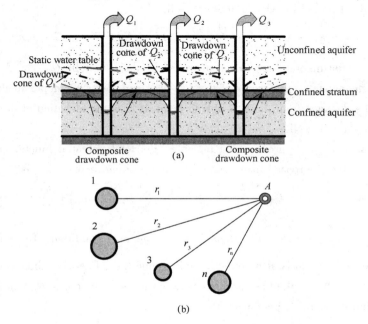

Fig. 11-15 Multiplewell system: (a) Side view; (b) top view

In a well field, when cone of depression of one well overlaps with the cone of depression of other wells, then the actual drawdown will be more than the drawdown calculated for every individual well (Fig. 11-15). In this case, the actual drawdown can be calculated using the principle of superposition of linear system.

According to the potential theory in Chapter 3, the seepage flow of a single complete well can be approximately described as a sink in the potential flow, while its potential function can be written as

$$\varphi = \frac{Q}{2\pi}\ln r + C$$

Thus, we have

$$d\varphi = \frac{Q}{2\pi}\frac{dr}{r}$$

Compared to the differential equation of the complete water table well, Eq. 11. 15, we have

467

$$d\varphi = kz\,dz$$

Integrating it gives the potential function for a complete water table well

$$\varphi = \frac{1}{2}kz^2 + C \tag{11.23}$$

where C is the integral constant; z is the elevation of drawdown cone of well at point A away from well at a distance of r (see Fig. 11-15b); k is the coefficient of permeability of soil.

Assume the multiple well system is composed of n complete water table wells on a horizontal impermeable stratum. Take the impermeable stratum to be the common datum plane of the multiple well system and the single wells. Substituting Eq. 11.16 into Eq. 11.23 gives the potential function at point A attributed by each individual well

$$\varphi_i = \frac{1}{2}kz_i^2 + C_i = \frac{1}{2}k\left(\frac{Q_i}{\pi k}\ln\frac{r_i}{r_{0i}} + h_i^2\right) + C_i \quad (i = 1,2,\cdots,n) \tag{11.24}$$

where φ_i is the potential function at point A attributed by the i-th well; Q_i, r_{0i}, h_i are the steady well yield in the steady state, the radius of the i-th well and the water depth in the i-th well respectively; r_i is the distance from the i-th well to point A; z_i is the elevation of drawdown cone at point A caused by the i-th well.

When all individual wells are working together, the total potential function of the multiple wells at point A can be superposed using potential superposition theory, that is

$$\varphi = \frac{1}{2}kz^2 + C = \sum_{i=1}^{n}\varphi_i = \frac{1}{2}k\sum_{i=1}^{n}\left(\frac{Q_i}{\pi k}\ln\frac{r_i}{r_{0i}} + h_i^2\right) + \sum_{i=1}^{n}C_i \tag{11.25}$$

Let $C' = \sum_{i=1}^{n}C_i - C$, which is determined by the boundary condition. The influence radius, R, of the multiple well system is also applied, and assume the influence radius of each individual well equals to R due to the negligible distances between wells compared to R, that is $r_i = R$, $z = H$. Substituting into the above equation yields

$$C' = \frac{k}{2}H^2 - \frac{k}{2}\sum_{i=1}^{n}\left(\frac{Q_i}{\pi k}\ln\frac{R}{r_{0i}} + h_i^2\right)$$

Substituting C' back into Eq. 11.25 yields

$$z^2 = H^2 - \sum_{i=1}^{n}\left(\frac{Q_i}{\pi k}\ln\frac{R}{r_i}\right) \tag{11.26}$$

If the pumping flow rates from each individual wells, Q_i, are the same, that is $Q_i = Q_0/n$, then Eq. 11.26 becomes

$$z^2 = H^2 - \frac{Q_0}{\pi k}\left[\ln R - \frac{1}{n}\ln(r_1 r_2 \cdots r_n)\right] \tag{11.27}$$

where $Q_0 = nQ$ is the pumped flowrate of the multiple well system; H is the static water table over the invert of impermeable stratum; R is the influence radius of the multiple well system, $R = 575s\sqrt{kH}$; and s is the drawdown of static water table at the center point of the multiple well system when the seepage flow is steady.

Equation 11.26 can be used to determine the depression cone of the multiple well system composed of complete water table wells.

EXAMPLE 11-2 Multiple well system

Figure E11-2 shows a dewatering system for the foundation pit of a building. For the safety of construction, the static water table at the center point of foundation pit should be lowered at least 5m by a multiple well system, which is composed of eight complete water table wells with identical pumping flowrate. The static water table above the invert of impermeable stratum is $H = 15$m. The coefficient of permeability of soil is $k = 0.005$cm/s. If the radius of each individual well is $r = 0.1$m and the total pumping flowrate is $Q_0 = 7.6 \times 10^{-3}$m^3/s, check whether the design of the multi-well system meets the requirement.

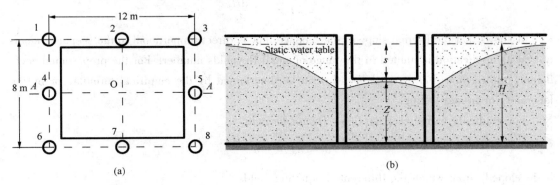

Fig. E11-2 (a) plane view of the foundation pit; (b) side view of section A-A

Solution:

The influence radius of the multiple well system can be obtained from the known parameters

$$R = 575s\sqrt{kH} = (575)(5)\sqrt{(0.00005)(15)}\text{m} = 79\text{m}$$

The horizontal distance from each well to the center point O can be calculated from Fig. E11-2

$$r_1 = r_3 = r_6 = r_8 = \sqrt{(6\text{m})^2 + (4\text{m})^2} = 7.2\text{m}$$

$$r_2 = r_7 = 4\text{m}$$

$$r_4 = r_5 = 6\text{m}$$

Thus, the maximum distance between two wells in this case is

$$r_{max} = \sqrt{(12\text{m})^2 + (8\text{m})^2} = 14.4\text{m} \ll R = 79\text{m}$$

Thus, the distances between wells are negligible compared to R.

Substitute the known parameters and $r_1, r_2, \ldots, r_8$ into Eq. 11.27 to give

$$z^2 = H^2 - \frac{Q_0}{\pi k}\left[\ln R - \frac{1}{n}\ln(r_1 r_2 \cdots r_n)\right]$$

$$= (15\text{m})^2 - \frac{7.6 \times 10^{-3}\text{m}^3/\text{s}}{(3.14)(0.00005\text{m/s})}\left[\ln(79) - \frac{1}{8}\ln[(6^2)(4^2)(7.2^4)]\right] = 99.8\text{m}^2$$

Thus,

$$z = 9.99\text{m}$$

The drop of static water table at the center point of the foundation pit is

$$s = H - z = 15\text{m} - 9.99\text{m} = 5.01\text{m} > 5\text{m}$$

Therefore, the design of this multiple well system meets the requirement.

Summary

This chapter briefly discussed the *seepage flow* in the porous soil, especially the steady flow of the *gravitational water* in a *homogeneous* and *isotropic* soil. Before we start, the *conceptual model of seepage flow* is introduced so the seepage flow can be regarded as a continuum. It assumes the whole flow field is occupied by the flow and the volume supposed to be occupied by the soil grains in real case is ignored. Thus, the *seepage velocity* in the model is less than the actual velocity in the real *groundwater flow*.

Darcy law is the basic principle of the seepage flow and expressed as

$$u = kJ = -k\frac{\mathrm{d}H}{\mathrm{d}s}$$

It shows that the friction slope of the seepage flow is direct proportional to the seepage velocity. So its application is limited in the range of small Reynolds number. For the proportional coefficient, the *coefficient of permeability*, k, can be determined by the empirical formula, or by the laboratory or in-situ testing.

For gradually varied seepage flow, *Dupuit formula*

$$v = kJ = -k\frac{\mathrm{d}H}{\mathrm{d}s}$$

is developed, from which the differential equation yields

$$\frac{\mathrm{d}h}{\mathrm{d}s} = i\left(1 - \frac{A_0}{A}\right)$$

for determining the profiles of the *phreatic surfaces* of gradually varied seepage flow in the unconfined aquifer over impermeable stratum with different slopes.

Finally, the seepage flows around *water table wells*, *artesian wells*, *multiple well system* and *storage gallery* are discussed, the *well yields* and the *drawdown cones* are determined by the series of equations.

Word Problems

W11-1 Define the seepage flow and the phreatic surface in the flow.

W11-2 What is the conceptual model of seepage flow? What conditions should the conceptual model of seepage flow satisfy?

W11-3 What is the relation between the seepage flow velocities in the conceptual model of seepage flow and in the actual seepage flow in soil?

W11-4 What is Darcy's law? What limits does it have when it is employed?

W11-5 In Darcy's experiment, do the liquid drops falling from the constant head tanks have constant velocity?

W11-6 In Darcy's experiment, what happens if the friction gradient is too steep?

W11-7 How does the Darcy's law differ from the Dupuit equation?

W11-8 How can the collecting galleries and wells be used for the engineering practice?

Problems

11-1 In Fig. 11-1 the Darcy experiment apparatus has a 0.2-m-diameter cylinder connected by two piezometer tubes at a vertical distance of 0.3 m. At a constant head, the water level difference in this two piezometer tubes is 0.4 m, the volume of seepage water is 15×10^{-3} m^3 over 24 hours. Determine the coefficient of permeability of the measured soil.

11-2 Two reservoirs are separated by a hill as in Fig. P11-2. A confined aquifer with a size of 4-m-high, 500-m-wide (normal into paper) and 2000-m-long is found located at the lower part of hill and connecting reservoirs A and B. The soil in first half of the aquifer is fine sand and the second half medium sand, whose permeability coefficient are 0.001 cm/s and 0.01 cm/s respectively. Determine the seepage flowrate infiltrating from reservoir A to B.

11-3 The pipe connects two tanks shown in Fig. P11-3 has a square cross section of 0.2m by 0.2m and a length of 2m. Determine the seepage flowrate in the pipe if the pipe is filled with: (a) coarse sand with a permeability coefficient of 0.05cm/s; (b) fine sand with a permeability coefficient of 0.002cm/s; and (c) coarse sand in the first half and fine sand in the second half of the pipe.

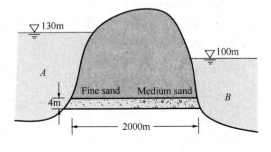

Fig. P11-2

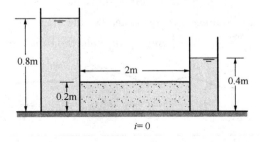

Fig. P11-3

11-4 In Fig. P11-4 a 180-m-long unconfined aquifer connecting the channel and the river is located over an impermeable stratum with a bed slope of 0.02. The coefficient of permeability of soil in the aquifer is 0.005cm/s. (a) Determine the seepage flow per unit width (normal into paper); (b) sketch the phreatic surface of the seepage flow.

11-5 An unconfined aquifer with fine sand is located on a horizontal impermeable stratum. Two wells are dug for the flowrate inspection as in Fig. P11-5. Take the coefficient of permeability of the fine sand to be 7.5m/d. Determine: (a) the seepage flowrate per unit width (normal into paper); (b) again, seepage flowrate if the width is 150m (normal into paper); (c) the elevation of water table at section (3), h, which is 100m apart from well (1).

471

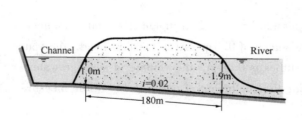
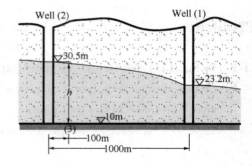

Fig. P11-4 **Fig. P11-5**

11-6 A 100-m-long (normal into paper) storage gallery is located on an impermeable stratum as in Fig. P11-6. The static water table is 4 m above the invert of impermeable stratum. The coefficient of permeability of soil is 0.001 cm/s. When the pumping flowrate from the storage gallery is steady, the water depth in gallery is 2m, the radius of influence is 140m. Determine: (a) the pumped flowrate from the storage gallery; and (b) the elevation of drawdown cone over the invert of impermeable stratum at point C which is 100 m away from the storage gallery wall.

11-7 A 100-m-long (normal into paper) storage gallery is located on an impermeable stratum as in Fig. P11-7, The static water table is 4m above the invert of impermeable stratum. The coefficient of permeability of the soil is 5×10^{-3} cm/s. When the pumping flowrate from storage gallery is steady, the slope of drawdown cone of the seepage flow is 0.03. Determine the pumping flowrate from the storage gallery.

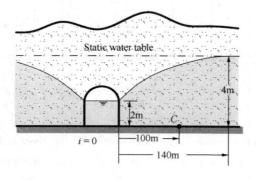
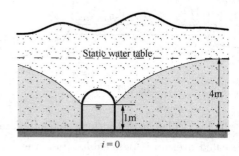

Fig. P11-6 **Fig. P11-7**

11-8 A complete water table well with a radius of 0.1m is located on a horizontal impermeable stratum. The static water table is 8m over the invert of impermeable stratum, the coefficient of permeability of the soil is 0.001cm/s. When the pumping flowrate from the well is steady, the water depth in the well is 3m. Determine the pumping flowrate from the well.

11-9 A complete water table well has a radius of 0.15m located on a horizontal impermeable stratum. An inspection well is dug 60m away from the center of well as in Fig. P11-9. Assume the steady pumping flowrate from the well is $0.00025 \text{m}^3/\text{s}$, determine the coefficient of permeability of the soil.

11-10 A complete water table well with a radius of 0.1m is located on a horizontal impermeable stratum. The static water table is 10m over the invert of impermeable stratum. The coefficient of permeability of soil is 0.0015 cm/s. When the pumping flowrate from the well is steady, the maximum drop of the water table in the well is 4.5 m. If the radius of influence of well is 100m. determine the pumping flowrate.

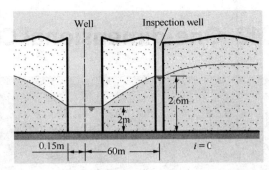

Fig. P11-9

11-11 A complete recharging well with a radius of 0.1m is located on a horizontal impermeable stratum as in Fig. P11-11. If the steady recharging flow rate is 0.25×10^{-3} m³/s, determine the coefficient of permeability of soil.

11-12 In Fig. P11-12 the complete artesian well has a radius of 0.1m, and the inspection hole is 10m away from the well. The static water table is 12m over the invert of confined aquifer. When the steady pumping flowrate from the well is 36m³/h, the drawdown of static water table in the well is 2m and in the inspected hole 1 m. Determine: (a) the coefficient of permeability of soil in the artesian aquifer; (b) the radius of influence of the well.

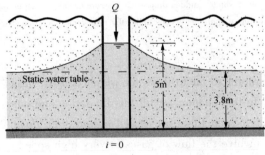

Fig. P11-11

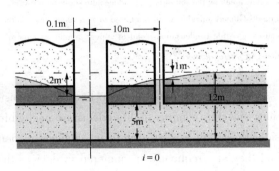

Fig. P11-12

11-13 A complete artesian well has a diameter of 0.304m. The thickness of the artesian aquifer is 14m and the coefficient of permeability of soil in it is 10m/d. When the pumping flow rate from the well is steady, the drawdown of the static water level in the well is 4m. Determine the pumping flow rate from it.

12 One-dimensional Compressible Gas Flow

CHAPTER OPENING PHOTO: F-18 travling at 600 mph, as fast as the speed of sound as discussed in the present chapter, breaking the sound barrier induces the super hornet vapor cone at an air show in San Diego, California, USA in 2011. The Prandtl-Glauert singularity, also referred to as a vapor cone, is the point at which a sudden drop in air pressure occurs, which may causes a visible condensation cloud to appear around an aircraft traveling at transonic speeds under the right atmospheric conditions. (©*Steve Skinner* ❶)

For the most part, we have limited our consideration so far to flows for which density variations and thus compressibility effects are negligible. In this chapter we lift this limitation and consider flows that involve significant changes in density. Such flows are called *compressible flows*, and they are frequently encountered in devices that involve the flow of gases at very high speeds. Compressible flow combines fluid dynamics and thermodynamics in that both are absolutely necessary to the development of the required theoretical background.

For simplicity, in this introductory study of compressibility effects we mainly consider the steady, one-dimensional, constant (including zero) viscosity, compressible flow of an ideal gas. In this chapter, one-dimensional flow refers to flow involving uniform distributions of fluid properties over any flow cross-sectional area. Both frictionless ($\mu = 0$) and frictional ($\mu \neq 0$) compressible flows are considered.

---

❶ svsimagery.smugmug.com

12.1 Fundamental of Compressible Flow

12.1.1 Equation of State for an Ideal Gas

An *ideal gas* is a theoretical inviscid gas composed of a set of randomly moving, non-interacting point particles, which obeys the *ideal gas law* and is amenable to analysis under statistical mechanics. At normal conditions such as standard temperature and pressure, most real gases behave qualitatively like an ideal gas within reasonable tolerances, such as air, nitrogen, oxygen, hydrogen, noble gases, and some heavier gases like carbon dioxide.

Gases are highly compressible in comparison to liquids. The state of an amount of gas is determined by its pressure, volume, and temperature. The changes in gas density directly related to changes in pressure and temperature through the equation

$$\frac{p}{\rho} = RT \tag{12.1}$$

where p is the absolute pressure, Pa; ρ the density, kg/m^3; T is the absolute temperature, K; R is a *gas constant*, which depends on the particular gas and is related to the molecular weight of the gas, $R = 287 \text{J}/(\text{kg} \cdot \text{K})$ for air at standard condition.

Equation 12.1 is commonly termed the *ideal or perfect gas law*, or the *equation of state* for an ideal gas. It is an extension of experimentally discovered gas laws, and known to closely approximate the behavior of real gases under normal conditions when the gases are not approaching liquefaction. Generally, a gas behaves more like an ideal gas at higher temperature and lower pressure, and the ideal gas model tends to fail at lower temperature or higher pressure.

12.1.2 Speed of Sound

An important parameter in the study of compressible flow is the *speed of sound* (or the *sonic speed*), which is the speed at which an infinitesimally small pressure wave (sound) travels through a medium. The pressure wave may be caused by a small disturbance, which creates a slight rise in local pressure.

To obtain a relation for the speed of sound in a medium, consider a duct that is filled with a fluid at rest, as shown in Fig. 12-1. A piston fitted in the duct is now moved to the right with a constant incremental velocity dv, creating a sonic wave. The wave front moves to the right through the fluid at the speed of sound c and separates the moving fluid adjacent to the piston from the fluid still at rest.

Ahead of the pressure pulse, the fluid velocity is zero and the fluid pressure and density are p and ρ. Behind the pressure pulse, the fluid velocity has changed by an amount dv, and the pressure and density of the fluid have also changed by amounts dp and $d\rho$. Thus, an infinitesimally thin control volume $AA'B'B$ that moves with the pressure pulse is selected as sketched in

Fig. 12-1a. The speed of the weak pressure pulse is considered constant and in one direction only; thus, our control volume is inertial.

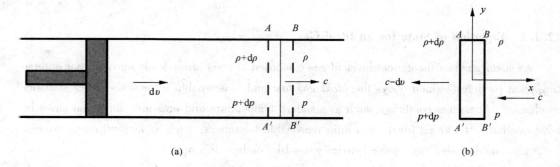

Fig. 12-1 (a) Weak pressure pulse moving through a fluid at rest. (b) The flow relative to a control volume containing a weak pressure pulse.

For an observer moving with this control volume (Fig. 12-1b), it appears as if fluid is entering the control volume through surface area A with speed c at pressure p and density ρ and leaving the control volume through surface area A with speed $c - d\upsilon$, pressure $p + dp$ and density $\rho + d\rho$. When the continuity equation (Eq. 4.5) and the linear momentum equation (Eq. 4.47) are applied to the flow through this control volume, the results are

$$\rho c A = (\rho + d\rho)(c - d\upsilon) A$$
$$pA - (p + dp)A = \rho c A[(c - d\upsilon) - c]$$

Rearranging and neglecting the high order differential terms yield

$$c\,d\rho - \rho\,d\upsilon - \upsilon\,d\rho = 0$$
$$dp = \rho c\,d\upsilon$$

Solving to give $c^2 = \dfrac{dp}{d\rho}\left(1 + \dfrac{d\rho}{\rho}\right)$. Consider the relative density incremental $\dfrac{d\rho}{\rho} \ll 1$ to give

$$c = \sqrt{\dfrac{dp}{d\rho}} \qquad (12.2)$$

This expression for the speed of sound results from application of the conservation of mass and conservation of linear momentum principles to the flow through the control volume of Fig. 12-1b, which can be used for both the liquid flow and gas flow.

For gases, the pressure pluses moves very quickly, there is no heat exchange between the gas and the outside gas, and the incremental amounts of velocity, pressure and density are very small. Thus, the moving process of pressure pluses can be considered as an *isentropic process* (*adiabatic* and frictionless), or *constant entropy process*. For the isentropic flow of an ideal gas (with constant c_p and c_v), we have

$$\dfrac{p}{\rho^k} = C \qquad (12.3)$$

where k is the *specific-heat ratio* of the gas, $k = c_p/c_v$; c_p is the ideal gas *specific heat at constant pressure*; c_v is the ideal gas *specific heat at constant volume*. For the common gases, k decreases slowly with temperature and lies between 1.0 and 1.7; for air $k = 1.40$.

Differentiating Eq. 12.3 and combining the *equation of state* for an ideal gas, Eq. 12.1, give

$$\frac{dp}{d\rho} = Ck\rho^{k-1} = \frac{p}{\rho^k}k\rho^{k-1} = \frac{p}{\rho}k = RTk$$

Substituting back into Eq. 12.2 yields the speed of sound for an ideal gas

$$c = \sqrt{\frac{dp}{d\rho}} = \sqrt{kRT} \tag{12.4}$$

Noting that the gas constant R has a fixed value for a specified ideal gas, and the specific heat ratio k of an ideal gas is, at most, a function of temperature, we see that the speed of sound in a specified ideal gas is a function of temperature alone.

More generally, the bulk modulus of elasticity, K, of any fluid including liquids is defined as (see Section 1.4.2) $K = \rho \frac{dp}{d\rho}$, thus Eq. 12.4 becomes

$$c = \sqrt{\frac{K}{\rho}} \tag{12.5}$$

From experience we know that air is more easily compressed than water, and the speed of sound in air is much less than it is in water. From Eq. 12.5, we can conclude that if a fluid is truly incompressible, its bulk modulus would be infinitely large, as would be the speed of sound in that fluid. Thus, an incompressible flow must be considered as an idealized approximation of reality.

12.1.3 Mach Number

The *Mach number*, Ma, introduced in section 5.4.2 as a dimensionless measure of compressibility in a fluid flow, is defined as the ratio of the value of the local flow velocity, v, to the local *speed of sound*, c

$$\text{Ma} = \frac{v}{c} \tag{12.6}$$

Note that the Mach number depends on the speed of sound, which depends on the state of the fluid. Therefore, the Mach number of an aircraft cruising at constant velocity in still air may be different at different locations.

Fluid flow regimes are often described in terms of the Mach number of flow. The flow is called *sonic* if Ma = 1, *subsonic* if Ma < 1, *supersonic* if Ma > 1, *hypersonic* if Ma $\gg$ 1, and *transonic* if Ma $\cong$ 1 ; *incompressible flow* if Ma < 0.3, *compressible flow* if Ma > 0.3.

EXAMPLE 12-1 Mach number of air entering a diffuser

Air enters a diffuser shown in Fig. E12-1 with a velocity of 200m/s. Determine (a) the speed of sound and

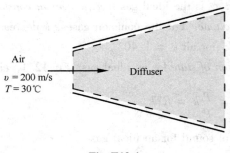

Fig. E12-1

Air
$v = 200$ m/s
$T = 30\ ℃$

(b) the Mach number at the diffuser inlet when the air temperature is 30℃. Assuming air at specified conditions behaves as an ideal gas.

Solution:

Air enters a diffuser with a high velocity. The speed of sound and the Mach number are to be determined at the diffuser inlet. The gas constant of air is $R = 0.287 \text{kJ}/(\text{kg·K})$, and its specific heat ratio at 30℃ is $k = 1.4$.

(a) The speed of sound in air at 30℃ can be determined by Eq. 12.4

$$c = \sqrt{kRT} = \sqrt{(1.4)[0.287\text{kJ}/(\text{kg·K})][(273+30)\text{K}]} = 349 \text{m/s}$$

(b) Then the Mach number becomes

$$\text{Ma} = \frac{v}{c} = \frac{200\text{m/s}}{349\text{m/s}} = 0.573$$

DISCUSSION   The flow at the diffuser inlet is subsonic since Ma < 1.

12.2   One-Dimensional Adiabatic and Isentropic Steady Flow

During fluid flow through many devices such as nozzles, diffusers, and turbine blade passages, flow quantities vary primarily in the flow direction only, thus the flow can be approximated as one-dimensional isentropic flow with good accuracy.

12.2.1   Fundamental Formulas

CONTINUITY EQUATION

When fluid flows steadily through a conduit with a flow cross-sectional area that varies along the axial distance, the conservation of mass (continuity) equation is

$$\dot{m} = \rho c A = \text{constant} \tag{12.7}$$

Take a logarithm on both sides of the continuity equation and then differentiate it to give

$$\frac{d\rho}{\rho} + \frac{dc}{c} + \frac{dA}{A} = 0 \tag{12.8}$$

For compressible flow, density, cross-sectional area, and flow velocity may vary from section to section. Thus, Eq. 12.8 is used to relate the flow rates at different sections and to determine how fluid density and flow velocity change with axial location in a variable area duct when the fluid is an ideal gas and the flow through the duct is steady and isentropic.

DIFFERENTIAL EQUATION OF MOTION

In Chapter 3, Newton's second law was applied to the inviscid (frictionless) and steady flow of a fluid particle. For the streamwise direction, the result (Eq. 3.59) for either compressible or incompressible flows is

$$gdz + \frac{1}{\rho}dp + \frac{1}{2}d(v^2) = 0 \qquad (12.9)$$

The frictionless flow from section to section through a finite control volume is also governed by Eq. 12.9, if the flow is one-dimensional, because every particle of fluid involved will have the same experience. For ideal gas flow, the potential energy difference term, gdz, can be dropped because of its small size in comparison to the other terms, namely, dp and $d(v^2)$. Thus, an appropriate equation of motion in the streamwise direction for the steady, one-dimensional, and frictionless flow of an ideal gas is obtained from Eq. 12.9 as

$$\frac{1}{\rho}dp + d\left(\frac{v^2}{2}\right) = 0 \qquad (12.10)$$

ENERGY EQUATION

Integrate the differential equation of motion, Eq. 12.10, to give the energy equation for the steady, one-dimensional, and frictionless flow of an ideal gas

$$\int \frac{dp}{\rho} + d\left(\frac{v^2}{2}\right) = C \qquad (12.11)$$

where C is the integration constant.

The density of compressible gas in Eq. 12.11 is not a constant, which depends on the pressure and temperature. Therefore, the equation of state for gases, the relation between ρ, p and T must be employed before we go forward to integrate Eq. 12.11.

(1) Isochoric process

An *isochoric process*, also called a *constant-volume process*, an *isovolumetric process*, or an *isometric process*, is a thermodynamic process during which the volume of the closed system undergoing such a process remains constant. That is, the gas density in an isochoric process is a constant (ρ = constant), such a gas is an *incompressible* gas.

Integrating Eq. 12.11 with a constant density gives the energy equation of isochoric process

$$\frac{p}{\rho} + \frac{v^2}{2} = C \qquad (12.12)$$

Equation 12.12 is the energy equation for the steady, one-dimensional, and frictionless flow of an incompressible ideal gas with negligible gravity and constant-volume. It presents the conservation of the sum of the pressure potential energy and kinematic energy per unit mass gas at any section along the axial distance.

(2) Isothermal flow

Isothermal flow is a model of compressible fluid flow whereby the flow remains at the same temperature while flowing in a conduit, that is, T = constant. The equation of state, Eq. 12.1, is to be $\frac{p}{\rho} = RT$ = constant, thus we have $\rho = \frac{p}{RT}$. Substituting it into Eq. 12.11 yields

$$RT\int \frac{dp}{p} + d\left(\frac{v^2}{2}\right) = C$$

Integrate it to give the energy equation for isothermal flow

$$RT\ln p + \frac{v^2}{2} = C \qquad (12.13)$$

Equation 12.13 is the energy equation for the steady, one-dimensional, and frictionless flow of an ideal gas with negligible gravity and constant temperature.

(3) Isentropic process

An *isentropic process* or *isoentropic process* is the one in which one may assume that the process takes place from initiation to completion without an increase or decrease in the entropy of the system, i. e., the entropy of the system remains constant throughout. An isentropic flow is a flow that is both *adiabatic* and *reversible*. That is, no heat is added to the flow, and no energy transformations occur due to friction or dissipative effects. For an isentropic flow of a perfect gas, several relations can be derived to define the pressure, density and temperature along a streamline.

From the state equation of isentropic flow, Eq. 12.3, $\frac{p}{\rho^k} = C$, we have $\rho = p^{1/k} C^{-1/k}$, thus

$$\int \frac{dp}{\rho} = C^{1/k} \int \frac{dp}{p^{1/k}} = \frac{k}{k-1} \frac{p}{\rho}$$

Substituting it into Eq. 12.11 and integrating it yield the energy equation for isentropic process

$$\frac{k}{k-1} \frac{p}{\rho} + \frac{v^2}{2} = C \qquad (12.14a)$$

or

$$\frac{1}{k-1} \frac{p}{\rho} + \frac{p}{\rho} + \frac{v^2}{2} = C \qquad (12.14b)$$

Equation 12.14b is the energy equation for the steady, one-dimensional, and frictionless flow of an ideal gas with negligible gravity and constant entropy. It is not only applicable for the *isentropic process*, but also for the *adiabatic process* with friction, because even if the mechanical energy is converted into the heat due to friction, heat cannot be released from the flow to outside of duct due to the adiabatic process and only can be transfer into the internal energy of flow. Thus, the total energy of gas remains constant.

12.2.2 Stagnation Properties

In dealing with problems involving the compressible flow, many discussions and equations can be simplified by introducing the concept of the isentropic stagnation state and the properties associated with it.

The *stagnation state* is a state a flowing flow undergoing a deceleration to zero velocity and associating with the entropy value that corresponds to the entropy of the flowing fluid. The stagnation state is called the *isentropic stagnation state* when the stagnation process is reversible as well as adiabatic (i. e., isentropic), that is, the *isentropic stagnation state* is a state a flowing flow would attain if it underwent a reversible adiabatic deceleration to zero velocity. The entropy of a fluid remains constant during an isentropic stagnation process.

During a stagnation process, the kinetic energy of a fluid is converted to enthalpy (internal energy + flow energy), which results in an increase in the fluid temperature and pressure. The properties of a fluid at the stagnation state are called *stagnation properties* (stagnation temperature, stagnation pressure, stagnation density, etc.). The subscript 0 is used to designate the stagnation state such as T_0, p_0, ρ_0 and c_0.

For example, if the fluid flowing through the converging-diverging duct were drawn isentropically from the atmosphere, the atmospheric pressure and temperature would represent the stagnation state of the flowing fluid. The stagnation state can also be achieved by isentropically decelerating a flow to zero velocity. This can be accomplished with a diverging duct for subsonic flows or a converging-diverging duct for supersonic flows.

If the fluid were brought to a complete stop as shown in Fig. 12-2, then the velocity at state S would be zero and the energy equation for *adiabatic* process, Eq. 12.14a, becomes

$$\frac{k}{k-1}\frac{p}{\rho} + \frac{v^2}{2} = \frac{k}{k-1}\frac{p_0}{\rho_0} \tag{12.15}$$

or
$$\frac{k}{k-1}RT + \frac{v^2}{2} = \frac{k}{k-1}RT_0 \tag{12.16}$$

or in term of the speed of sound at stagnation state, $c_0 = \sqrt{kRT_0}$ and local speed of sound $c = \sqrt{kRT}$

$$\frac{c^2}{k-1} + \frac{v^2}{2} = \frac{c_0^2}{k-1} \tag{12.17}$$

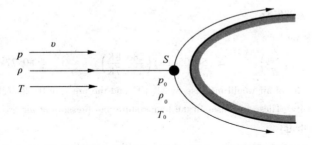

Fig. 12-2 Flow past an obstacle at stagnation state

For the convenience of calculation, the ratios of *stagnation properties* and properties of motion are introduced and expressed as the function of Mach number. Thus, from Eq. 12.16 gives

$$\frac{T_0}{T} = 1 + \frac{k-1}{2}\frac{v^2}{kRT} = 1 + \frac{k-1}{2}\frac{v^2}{c^2} = 1 + \frac{k-1}{2}\text{Ma}^2 \tag{12.18}$$

Then we can have other ratios by combining Eq. 12.3 and Eq. 12.1

$$\frac{p_0}{p} = \left(\frac{T_0}{T}\right)^{\frac{k}{k-1}} = \left(1 + \frac{k-1}{2}\text{Ma}^2\right)^{\frac{k}{k-1}} \tag{12.19}$$

$$\frac{\rho_0}{\rho} = \left(\frac{T_0}{T}\right)^{\frac{1}{k-1}} = \left(1 + \frac{k-1}{2}\text{Ma}^2\right)^{\frac{1}{k-1}} \tag{12.20}$$

$$\frac{c_0}{c} = \left(\frac{T_0}{T}\right)^{1/2} = \left(1 + \frac{k-1}{2}\text{Ma}^2\right)^{1/2} \tag{12.21}$$

EXAMPLE 12-2 Compression of high-speed air in an aircraft

An aircraft is flying at a cruising speed of 250m/s at an altitude of 5000m where the atmospheric pressure is 54.05kPa and the ambient air temperature is 255.7K. The ambient air is first decelerated in a diffuser before it enters the compressor (Fig. E12-2). Assuming both the diffuser and the compressor to be isentropic, determine (a) the stagnation pressure at the compressor inlet and (b) the required compressor work per unit mass if the stagnation pressure ratio of the compressor is 8.

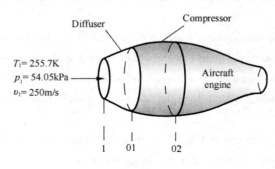

Fig. E12-2

Solution:

Assume air is an ideal gas with constant specific heats at room temperature.

(a) Take the constant-pressure specific heat c_p, the gas constant R, and the specific heat ratio k of air at room temperature to be $c_p = 1.005$ kJ/kg·K, $R = 0.287$ kJ/(kg·K) and $k = 1.40$.

Under isentropic conditions, the stagnation pressure at the compressor inlet (diffuser exit) can be determined from Eq. 12.19. However, first we need to find the stagnation temperature T_{01} at the compressor inlet. Under the stated assumptions, T_{01} can be determined from Eq. 12.18 to be

$$T_{01} = T\left(1 + \frac{k-1}{2}\frac{v^2}{kRT}\right)$$

$$= (255.7\text{K})\left[1 + \frac{1.40 - 1}{2}\frac{(250\text{m/s})^2}{(1.40)[287\text{ J/(kg·K)}](255.7\text{K})}\right]$$

$$= 286.8\text{K}$$

Then from Eq. 12.19 we have

$$p_{01} = p\left(\frac{T_{01}}{T}\right)^{k/(k-1)} = (54.05\text{kPa})\left(\frac{286.8\text{K}}{255.7\text{K}}\right)^{1.40/(1.40-1)} = 80.77\text{kPa}$$

That is, the temperature of air would increase by 31.1°C and the pressure by 26.72kPa as air is decelerated from 250m/s to zero velocity. These increases in the temperature and pressure of air are due to the conversion of the kinetic energy into enthalpy.

(b) To determine the compressor work, we need to know the stagnation temperature of air at the compressor exit T_{02}. The stagnation pressure ratio across the compressor p_{02}/p_{01} is specified to be 8. Since the compression process is assumed to be isentropic, T_{02} can be determined from the ideal gas isentropic relation (Eq. 12.19):

$$T_{02} = T_{01}\left(\frac{p_{02}}{p_{01}}\right)^{(k-1)/k} = (286.8\text{K})[8^{(1.40-1)/1.40}] = 519.5\text{K}$$

Disregarding potential energy changes and heat transfer, the compressor work per unit mass of air can be determined by

$$W_{in} = c_p(T_{02} - T_{01}) = (1.005\text{kJ/kg·K})(519.5\text{K} - 286.8\text{K}) = 233.9\text{kJ/kg}$$

Thus the work supplied to the compressor is 233.9kJ/kg.

DISCUSSION Notice that using stagnation properties automatically accounts for any changes in the kinetic energy of a fluid stream.

12.2.3 Compressibility of Gases

We have taken a brief look to see when we might safely neglect the compressibility inherent in every real fluid. We found that the proper criterion for a nearly incompressible flow is a small Mach number

$$\text{Ma} \ll 1$$

Under those small Mach number conditions, changes in fluid density are everywhere small in the flow field. The energy equation becomes uncoupled, and temperature effects can be either ignored or put aside for later study. The equation of state degenerates into the simple statement that density is nearly constant. This means that an incompressible flow requires only a momentum and continuity analysis, as we showed with many examples in Chapters 6 to 11.

For an incompressible gas, the commonly accepted limit is Ma < 0.3. Thus, from Eq. 12.20 we have the maximum density change for air at standard conditions

$$\frac{\rho_0}{\rho} = 1.0456$$

and also the maximum velocity of air from Eq. 12.6

$$v = c\text{Ma} = (343\text{m/s})(0.3) = 102.9\text{m/s}$$

It indicates that for air at standard conditions, a flow can thus be considered incompressible if the velocity is less than about 100 m/s or the density change is less than 5%. This encompasses a wide variety of airflows: automobile and train motions, light aircraft, landing and takeoff of high-speed aircraft, most pipe flows, and turbomachinery at moderate rotational speeds. Further, it is clear that almost all liquid flows are incompressible, since flow velocities are small and the speed of sound is very large.

The limited change of pressure for an incompressible gas can be derived by computing the stagnation properties from the energy equation for incompressible gas, Eq. 12.12

$$\frac{p}{\rho} + \frac{v^2}{2} = \frac{p_0}{\rho}$$

which can be rearranged as

$$\frac{p - p_0}{\rho v^2 / 2} = 1 \tag{12.22}$$

Expand the ratio expression of stagnation pressure for isentropic compressible flow, Eq. 12.19, by the binomial theorem, and take the first two terms to give

$$\frac{p_0}{p} = 1 + \frac{k}{2}\text{Ma}^2\left(1 + \frac{\text{Ma}^2}{4}\right)$$

where $\frac{k}{2}\text{Ma}^2 = \frac{k}{2}\frac{v^2}{c^2} = \frac{\rho v^2}{2p}$, substituting it back into the above equation yields

$$\frac{p - p_0}{\rho v^2 / 2} = 1 + \frac{\text{Ma}^2}{4} \tag{12.23}$$

Compare Eq. 12.22 to Eq. 12.23 to give the computational error of gas pressure computed by the energy equation for incompressible fluid

$$\delta = \frac{\text{Ma}^2}{4} \tag{12.24}$$

For example, if Ma = 0.3, we have $\delta = 0.0225$, that is the limited pressure error for the incompressible gas should be less than 2.25%.

However, the limited value of pressure error or density change is usually first determined by the requirement in engineering practice, then be used to determine the velocity of gas. For example, if the limited pressure error is taken to be $\delta = 1\%$, thus from Eq. 12.24 to give Ma = 0.2, the velocity for air at standard conditions would be $v = c\text{Ma} = (343\text{m/s})(0.2) = 68.6\text{m/s}$. Or if the limited density change is 1%, thus from Eq. 12.20 to give Ma = 0.141, the velocity for air at standard conditions would be $v = 48.36\text{m/s}$.

12.3 Compressible Flow with Friction in Constant-area Duct

12.3.1 Fundamental Equations of Fanno Flows

Wall friction associated with high-speed flow through short devices with large cross-sectional areas such as large nozzles is often negligible, and flow through such devices can be approximated as being *frictionless*. But wall friction is significant and should be considered when studying flows through long flow sections, such as long ducts, especially when the cross sectional area is small. In this section we consider compressible flow with significant wall friction but negligible heat transfer in ducts of constant cross-sectional area.

The steady, one-dimensional, adiabatic flow of an ideal gas with constant specific heats through a constant-area duct with significant frictional effects is referred to as *Fanno flows*.

As an elementary introduction, this section treats only the effect of friction, neglecting area change and heat transfer. The basic assumptions are

(1) Steady one-dimensional adiabatic flow
(2) Perfect gas with constant specific heats
(3) Constant-area straight duct
(4) Negligible shaft-work and potential-energy changes
(5) Wall shear stress correlated by a Darcy friction factor

In effect, we are studying a Moody-type pipe-friction problem but with large changes in kinetic energy, enthalpy, and pressure in the flow.

Consider the elemental duct control volume of area A and length dx in Fig. 12-3. The area is constant, but other flow properties (p, ρ, T, v) may vary with x. Application of the three conservation laws to this control volume gives three differential equations.

(1) Continuity equation:
$$\rho v = C \quad (12.25)$$

Take alogarithm and differential on both sides of the equation to give

$$\frac{d\rho}{\rho} + \frac{dv}{v} = 0 \quad (12.26)$$

(2) Momentum equation:

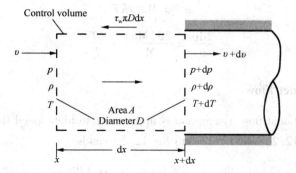

Fig. 12-3 Elemental control volume for flow in a constant-area duct with friction

$$pA - (p + \mathrm{d}p)A - \tau_0 \pi D \mathrm{d}x = \dot{m}(v + \mathrm{d}v - v)$$

With $A = \frac{1}{4}\pi D^2$ and $\dot{m} = \rho v A = (\rho + \mathrm{d}\rho)(v + \mathrm{d}v)A$ for steady flow we have

$$\frac{\mathrm{d}p}{\rho} + v\mathrm{d}v + \frac{4\tau_0 \mathrm{d}x}{\rho D} = 0 \tag{12.27}$$

To eliminate τ_0 as an unknown, it is assumed that wall shear is correlated by a local Darcy friction factor λ by Eq. 6.21 with $v^2 = \mathrm{Ma}^2 kRT$

$$\tau_0 = \frac{\lambda}{8}\rho v^2 = \frac{\lambda}{8} k p \mathrm{Ma}^2 \tag{12.28}$$

Thus Eq. 12.27 will be

$$\frac{\mathrm{d}p}{\rho v^2} + \frac{\mathrm{d}v}{v} + \frac{\lambda}{2D}\mathrm{d}x = 0 \tag{12.29}$$

(3) Perfect-gas law(Eq. 12.1)

$$p = \rho RT$$

Take alogarithm and differential on both sides of the equation to give

$$\frac{\mathrm{d}p}{p} = \frac{\mathrm{d}\rho}{\rho} + \frac{\mathrm{d}T}{T} \tag{12.30}$$

(4) Energy equation for isentropic flow(Eq. 12.14a)

$$\frac{k}{k-1}\frac{p}{\rho} + \frac{v^2}{2} = C$$

Substitute Eq. 12.1 to Eq. 12.14a and differentiate to give

$$\frac{k}{k-1}R\mathrm{d}T + v\mathrm{d}v = 0 \tag{12.31}$$

or

$$\frac{\mathrm{d}T}{T} = -\frac{v^2}{kRT}(k-1)\frac{\mathrm{d}v}{v} = -\mathrm{Ma}^2(k-1)\frac{\mathrm{d}v}{v} \tag{12.32}$$

(5) State equation for isentropic flow(Eq. 12.3)

$$\frac{p}{\rho^k} = C$$

(6) From the definition of Mach number, $\mathrm{Ma} = \dfrac{v}{c} = \dfrac{v}{\sqrt{kRT}}$, we have

$$v^2 = \text{Ma}^2 kRT \tag{12.33}$$

or
$$\frac{2dv}{v} = \frac{2d(\text{Ma})}{\text{Ma}} + \frac{dT}{T} \tag{12.34}$$

12.3.2 Adiabatic Duct Flow

The adiabatic frictional-flow assumption is appropriate to high-speed flow in short ducts. Substituting Eqs. 12.26 and 12.32 into Eq. 12.30 yields

$$\frac{dp}{p} = -\frac{dv}{v} - (k-1)\text{Ma}^2 \frac{dv}{v} \tag{12.35}$$

With Eq. 12.1 and $\text{Ma} = \dfrac{v}{c} = \dfrac{v}{\sqrt{kRT}}$ gives

$$\frac{dp}{\rho v^2} = \frac{dp}{(p/RT)v^2} = \frac{dp}{p}\frac{1}{k\text{Ma}^2} \tag{12.36}$$

Substituting Eqs. 12.35 and 12.36 into Eq. 12.29 yields

$$\frac{dv}{v} = \frac{k\text{Ma}^2}{2(1-\text{Ma}^2)}\lambda\frac{dx}{D} \tag{12.37a}$$

Substituting Eq. 12.37a into Eqs. 12.35, 12.26, 12.30 and 12.34 gives

$$\frac{dp}{p} = -k\text{Ma}^2 \frac{1+(k-1)\text{Ma}^2}{2(1-\text{Ma}^2)}\lambda\frac{dx}{D} \tag{12.37b}$$

$$\frac{d\rho}{\rho} = -\frac{dv}{v} = -\frac{k\text{Ma}^2}{2(1-\text{Ma}^2)}\lambda\frac{dx}{D} \tag{12.37c}$$

$$\frac{dT}{T} = -\frac{k(k-1)\text{Ma}^4}{2(1-\text{Ma}^2)}\lambda\frac{dx}{D} \tag{12.37d}$$

$$\frac{d(\text{Ma}^2)}{\text{Ma}^2} = k\text{Ma}^2 \frac{1+\frac{1}{2}(k-1)\text{Ma}^2}{1-\text{Ma}^2}\lambda\frac{dx}{D} \tag{12.37e}$$

All these have the factor $1 - \text{Ma}^2$ in the denominator, so that, subsonic and supersonic flow have opposite effects as tabulated in Table 12-1.

Table 12-1 The effects of friction on the properties of adiabatic flow

Property	Subsonic	Supersonic
p	Decreases	Increases
ρ	Decreases	Increases
v	Increases	Decreases
T	Decreases	Increases
Ma	Increases (maximum is 1)	Decreases (minimum is 1)

The key parameter above is the Mach number. Whether the inlet flow is subsonic or supersonic, the duct Mach number always tends downstream toward Ma = 1 because this is the path along which the entropy increases.

The second law of thermodynamics states that, based on all past experience, entropy can only remain constant or increase for adiabatic flows. A subsonic Fanno flow is accelerated by friction to a higher Mach number, but never greater than 1.0. A supersonic flow is decelerated by friction to a lower Mach number, but never less than 1.0. That is, the Mach number approaches unity (Ma = 1) in both cases, the gas velocity tends to the speed of sound as flow approaches to the exit of the adiabatic duct.

SONIC LENGTH

Considering that all Fanno flows tend to Ma = 1, it is convenient to use the sonic state as the reference point and to express flow properties relative to the sonic state properties, even if the actual flow never reaches the critical point.

For a given Fanno flow (constant specific heat ratio, duct diameter, and friction factor) the length of duct (from x to $x + l$) required to change the Mach number from Ma_1 to Ma_2 can be determined by integrating Eq. 12.37e. Thus, separate the variables and integrate to give:

$$\int_{x}^{x+l} \lambda \frac{dx}{D} = \int_{Ma_1^2}^{Ma_2^2} \frac{1 - Ma^2}{k Ma^4 \left[1 + \frac{1}{2}(k-1)Ma^2\right]} d(Ma^2)$$

For an approximate solution, we can assume that the friction factor is constant at an average value over the integration length l. We also consider a constant value of k. Thus, the integration of above equation gives

$$\lambda \frac{l}{D} = \frac{1}{k}\left(\frac{1}{Ma_1^2} - \frac{1}{Ma_2^2}\right) + \frac{k+1}{2k}\ln\left[\frac{Ma_1^2}{Ma_2^2} \cdot \frac{2 + (k-1)Ma_2^2}{2 + (k-1)Ma_1^2}\right] \quad (12.38)$$

For noncircular ducts, D is replaced by the *hydraulic diameter* $D_h = (4 \times \text{area})/\text{perimeter}$ as in Eq. 6.69.

It is clear from the previous discussions that friction causes subsonic Fanno flow in a constant-area duct to accelerate toward sonic velocity, and the Mach number becomes exactly unity at the exit for a certain duct length. This duct length is referred to as the *maximum length*, the *sonic length*, or the *critical length*, and is denoted by l_{max}.

Let the upper limit is the sonic point ($Ma_2 = 1$), whether or not it is actually reached in the duct flow. The lower limit is placed at the position $x = 0$, where the Mach number is $Ma_1 = Ma$. Thus the *sonic length* can be obtained from Eq. 12.38

$$\lambda \frac{l_{max}}{D} = \frac{1 - Ma^2}{k Ma^2} + \frac{k+1}{2k}\ln\frac{(k+1)Ma^2}{2 + (k-1)Ma^2} \quad (12.39)$$

In Eq. 12.39, l_{max} is the duct length required for the Mach number to reach unity under the influence of wall friction. Therefore, l_{max} represents the distance between a given section where the

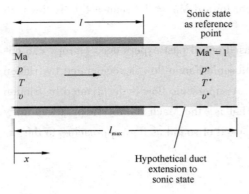

Fig. 12-4 Schematic of the sonic length l_{max}

Mach number is Ma and a section (an imaginary section if the duct is not long enough to reach Ma = 1.0) where sonic conditions occur (Fig. 12-4).

Note that the value of $\lambda l_{max}/D$ is fixed for a given Mach number, and thus values of $\lambda l_{max}/D$ can be tabulated versus Ma for a specified k. Also, the value of sonic length l_{max} needed to reach sonic conditions is inversely proportional to the friction factor. Therefore, for a given Mach number, l_{max} is large for ducts with smooth surfaces and small for ducts with rough surfaces.

EXAMPLE 12-3 Exit conditions of Fanno flow in a duct

Air enters a 27-m-long 5-cm-diameter adiabatic duct at $v_1 = 85$m/s, $T_1 = 450$K, and $p_1 = 220$kPa (Fig. E12-3). The average friction factor for the duct is estimated to be 0.023. Determine the Mach number at the duct exit and the mass flow rate of air. Assume the friction factor is constant along the duct.

Solution:

Take the constant-pressure specific heat c_p, the gas constant R, and the specific heat ratio k of air at room temperature to be $c_p = 1.005$kJ/kg·K, $R = 0.287$kJ/(kg·K) and $k = 1.40$.

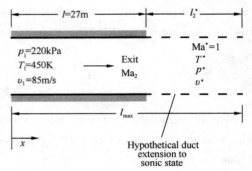

Fig. E12-3

The inlet Mach number and the corresponding value of the function are

$$c_1 = \sqrt{kRT_1} = \sqrt{(1.40)[0.287\text{kJ}/(\text{kg}\cdot\text{K})](450\text{K})} = 425\text{m/s}$$

$$\text{Ma}_1 = \frac{v_1}{c_1} = \frac{85\text{m/s}}{425\text{m/s}} = 0.20$$

From Eq. 12.39 we can get the sonic length

$$l_{max} = \frac{D}{\lambda}\left[\frac{1-\text{Ma}^2}{k\text{Ma}^2} + \frac{k+1}{2k}\ln\frac{(k+1)\text{Ma}^2}{2+(k-1)\text{Ma}^2}\right]$$

$$= \frac{0.05\text{m}}{0.023}\left[\frac{1-0.2^2}{(1.40)(0.2^2)} + \frac{1.40+1}{(2)(1.40)}\ln\frac{(1.40+1)(0.2^2)}{2+(1.40-1)(0.2^2)}\right] = 31.6\text{m}$$

Thus, $l_{max} > l = 27$m, which indicated that the flow is not chocked, the flow at the duct exit has not reached sonic conditions. Thus, we use Eq. 12.38 to obtain the Mach number at the duct exit, $\text{Ma}^2 = 0.42$.

The mass flow rate of air is determined from the inlet conditions to be

$$\rho_1 = \frac{p_1}{RT_1} = \frac{220\text{kPa}}{[0.287 \text{ kJ}/(\text{kg}\cdot\text{K})](450\text{K})} = 1.703\text{kg/m}^3$$

$$\dot{m} = \rho_1 A_1 v_1 = (1.703\text{kg/m}^3)[\pi(0.05)^2/4](85\text{m/s}) = 0.285\text{kg/s}$$

DISCUSSION   Note that it takes a duct length of 27 m for the Mach number to increase from 0.20 to 0.42, but only 31.6m − 27m = 4.6m to increase from 0.42 to 1.0. Therefore, the Mach number rises at a much higher rate as sonic conditions are approached.

12.3.3 Isothermal Flow with Friction

For flow in long ducts, e.g., natural-gas pipelines, the gas state more closely approximates an isothermal flow.

Some of the differential equations for isothermal flow with friction are identical to those for adiabatic flow, Eqs. 12.26 to Eq. 12.30. However, an additional simple relation

$$T = \text{constant} \quad \text{or} \quad dT = 0 \tag{12.40}$$

should be added.

Again it is possible to write all property changes in terms of the Mach number. Substituting Eqs. 12.26, 12.30, 12.36 and 12.40 into Eq. 12.29 gives

$$\frac{dv}{v} = \frac{k\text{Ma}^2}{2(1 - k\text{Ma}^2)}\lambda\frac{dx}{D} \tag{12.41a}$$

Substituting Eq. 12.41a into Eqs. 12.26, 12.30 and 12.34 gives

$$\frac{dp}{p} = \frac{d\rho}{\rho} = -\frac{dv}{v} = -\frac{k\text{Ma}^2}{2(1 - k\text{Ma}^2)}\lambda\frac{dx}{D} \tag{12.41b}$$

$$\frac{d(\text{Ma})}{\text{Ma}} = \frac{dv}{v} = \frac{k\text{Ma}^2}{2(1 - k\text{Ma}^2)}\lambda\frac{dx}{D} \tag{12.41c}$$

Similar to the Fanno flow in an adiabatic duct, all the properties of isothermal flow have the factor $1 - \text{Ma}^2$ in the denominator. So that, the critical point of Mach number, $\text{Ma}_{\text{crit}} = \sqrt{1/k} = 0.845$ if $k = 1.40$, is the critical point for flows in a isothermal duct with friction have opposite effects as tabulated in Table 12-2. Note that the Mach number approaches the critical Mach number ($\text{Ma}_{\text{crit}} = \sqrt{1/k}$) in both cases of subsonic with $\text{Ma} < \sqrt{1/k}$, and supersonic or subsonic with $\text{Ma} > \sqrt{1/k}$. The Mach number at the exit of isothermal duct with constant area should be not greater than $\sqrt{1/k}$, that is, $\text{Ma} \leq \sqrt{1/k}$.

Table 12-2 The effects of friction on the properties of isothermal flow

Property	Subsonic with $\text{Ma} < \sqrt{1/k}$	Supersonic or Subsonic with $\text{Ma} > \sqrt{1/k}$
p	Decreases	Increases
ρ	Decreases	Increases
v	Increases	Decreases
T	Constant	Constant
Ma	Increases (maximum is $\sqrt{1/k}$)	Decreases (minimum is $\sqrt{1/k}$)

MAXIMUM LENGTH

For a given isothermal flow, we also can get the length of duct (from x to $x + l$) required to change the Mach number from Ma_1 to Ma_2 by integrating Eq. 12.41c. Again, for an approximate solution, assume that the friction factor is constant at an average value over the integration length l and a constant value of k. Thus, integration of the Mach-number-friction relation yields

$$\lambda \frac{l}{D} = \frac{1}{k}\left(\frac{1}{Ma_1^2} - \frac{1}{Ma_2^2}\right) + \ln \frac{Ma_1^2}{Ma_2^2} \qquad \text{for } Ma^2 \leq \sqrt{1/k} \qquad (12.42)$$

Let the upper limit be the critical point ($Ma^2 = \sqrt{1/k}$), the lower limit is placed at the position $x = 0$, where the Mach number is $Ma_1 = Ma$. Thus the *maximum length* for isothermal duct flow can be obtained from Eq. 12.42

$$\lambda \frac{l_{max}}{D} = \frac{1 - kMa^2}{kMa^2} + \ln(kMa^2) \qquad (12.43)$$

Equation 12.43 also indicates that this friction relation has the interesting result that l_{max} becomes zero not at the sonic point but at $Ma_{crit} = \sqrt{1/k} = 0.845$ if $k = 1.40$. The inlet flow, whether subsonic or supersonic, tends downstream toward this limiting Mach number $Ma_{crit} = \sqrt{1/k}$. If the tube length l is greater than l_{max} from Eq. 12.43, a subsonic flow will choke back to a smaller Ma_1 and mass flow and a supersonic flow will experience a normal-shock adjustment.

The mass flow rate in an isothermal duct can be obtained by the continuity equation:

$$\dot{m} = \rho v A = \frac{p}{RT} Ma \sqrt{kRT} A = \sqrt{\frac{k}{RT}} \, p Ma A$$

Thus

$$Ma = \frac{\dot{m}}{A}\sqrt{\frac{RT}{k}} \frac{1}{p}$$

Substituting into Eq. 12.43 and rearranging yields the mass flow rate in an isothermal duct with constant-area

$$\dot{m} = A \sqrt{\frac{p_1^2 - p_2^2}{RT\left(\lambda \frac{l}{D} + 2\ln \frac{p_1}{p_2}\right)}} \qquad \text{for } Ma^2 \leq \sqrt{1/k} \qquad (12.44)$$

EXAMPLE 12-4 Mass flow rate for Isothermal flow

Air enters a pipe of 1-cm diameter and 1.2-m length at $T_1 = 300K$, and $p_1 = 220kPa$. If $\lambda = 0.025$ and the exit pressure is $p_2 = 140kPa$, estimate the mass flow rate for isothermal flow.

Solution:

Take the gas constant R, and the specific heat ratio k of air at room temperature to be $R = 0.287$ kJ/(kg·K) and $k = 1.40$.

For isothermal flow, Eq. 12.44 applies without iteration:

$$\dot{m} = \frac{1}{4}\pi D^2 \sqrt{\frac{p_1^2 - p_2^2}{RT\left(\lambda \frac{l}{D} + 2\ln \frac{p_1}{p_2}\right)}}$$

$$= \frac{\pi}{4}(0.01\text{m})^2 \sqrt{\frac{(220\text{kPa})^2 - (140\text{kPa})^2}{[0.287\text{kJ}/(\text{kg}\cdot\text{K})](300\text{K})\left(0.025\frac{1.2\text{m}}{0.01\text{m}} + 2\ln\frac{220\text{kPa}}{140\text{kPa}}\right)}}$$

$$= 0.023 \text{kg/s}$$

Check that the exit Mach number is not choked:

$$\rho_2 = \frac{p_2}{RT} = \frac{140\text{kPa}}{[0.287 \text{ kJ}/(\text{kg}\cdot\text{K})](300\text{K})} = 1.626 \text{kg/m}^3$$

$$\text{Ma}_2 = \frac{v_2}{c_2} = \frac{\frac{4\dot{m}}{\rho_2 \pi D^2}}{\sqrt{kRT}} = \frac{\frac{(4)(0.023\text{kg/s})}{(1.626\text{kg/m}^3)(3.14)(0.01\text{m})^2}}{\sqrt{(1.40)[0.287\text{kJ}/(\text{kg}\cdot\text{K})](300\text{K})}} = 0.519 < 0.845$$

This is well below choking, and the isothermal solution is accurate.

Summary

This chapter is a brief introduction to a very broad subject, compressible flow, sometimes called *gas dynamics*. The primary parameter is the Mach number Ma, which is large and causes the fluid density to vary significantly. This means that the continuity and momentum equations must be coupled to the energy relation and the equation of state to solve for the four unknowns (p, ρ, T, **u**).

The chapter reviews the thermodynamic properties of an ideal gas and derives a formula for the speed of sound of a fluid. The analysis is then simplified to one dimensional steady adiabatic flow without shaft work, for which the stagnation enthalpy of the gas is constant. A further simplification to isentropic flow enables formulas to be derived for high-speed gas flow in a variable-area duct. Finally a Fanno flow, steady, frictional, and adiabatic flow of an ideal gas with constant specific heats through a constant-area duct, is briefly introduced.

Word Problems

W12-1 Define the ideal gas.

W12-2 How does the isentropic process differ from adiabatic process? Show examples.

W12-3 A high-speed aircraft is cruising in still air. How does the temperature of air at the nose of the aircraft differ from the temperature of air at some distance from the aircraft?

W12-4 What is sound? How is it generated? How does it travel? Can sound waves travel in a vacuum?

W12-5 Is it realistic to assume that the propagation of sound waves is an isentropic process? Explain.

W12-6 Is the sonic velocity in a specified medium a fixed quantity, or does it change as the properties of the medium change? Explain.

W12-7 In which medium does a sound wave travel faster: in cool air or in warm air?

W12-8 In which medium will sound travel fastest for a given temperature: air, helium, or argon?

W12-9 In which medium does a sound wave travel faster: in air at 20℃ and 1 atm or in air at 20℃ and 5 atm?

W12-10 Does the Mach number of a gas flowing at a constant velocity remain constant? Explain.

W12-11 What is the limited condition for a real gas treated as an ideal gas? If the limited density change of air is less than 1%, what is the velocity for air at standard conditions?

W12-12 What are the characteristics of adiabatic duct flow and isothermal flow with friction in constant area ducts?

W12-13 What is the characteristic aspect of Fanno flow? What are the main assumptions associated with Fanno flow?

W12-14 What is the effect of friction on flow velocity in subsonic Fanno flow? Answer the same question for supersonic Fanno flow.

W12-15 Consider subsonic Fanno flow accelerated to sonic velocity (Ma = 1) at the duct exit as a result of frictional effects. If the duct length is increased further, will the flow at the duct exit be supersonic, subsonic, or remain sonic? Will the mass flow rate of the fluid increase, decrease, or remain constant as a result of increasing the duct length?

W12-16 Consider supersonic Fanno flow that is decelerated to sonic velocity (Ma = 1) at the duct exit as a result of frictional effects. If the duct length is increased further, will the flow at the duct exit be supersonic, subsonic, or remain sonic? Will the mass flow rate of the fluid increase, decrease, or remain constant as a result of increasing the duct length?

Problems

12-1 Determine the speed of sound in air at (a) 15℃ at the see level and (b) −55℃ in stratosphere.

12-2 When the plane flies at a speed of 1500km/h in the air listed in Prob.12-1, can it reach the supersonic?

12-3 Determine the speed of sound in air at (a) 300K and (b) 1000K. Also determine the Mach number of an aircraft moving in air at a velocity of 240m/s for both cases.

12-4 A duct is used to transport air at a flowrate of $2.25 \times 10^{-3} \text{m}^3/\text{s}$ with the air pressure of 240 kPa and the air temperature of 27℃. If the maximum permissive air speed in the duct is $v_{max} = 5.8\text{m/s}$, determine the required minimum diameter of duct.

12-5 Carbon dioxide flows isentropically at a speed of 14.8m/s along a streamline from a location with a temperature of 60℃ to another location with a temperature of 30℃. Let the gas constant of the Carbon dioxide be $R = 189\text{J}/(\text{kg}\cdot\text{K})$, the specific-heat ratio be $k = 1.29$. Determine the gas speed on the second location.

12-6 Air expands isentropically from 1.5MPa to 0.4MPa at 60℃. Calculate the ratio of the initial speed to the final speed of sound.

12-7 Determine the sound speed at the stagnation point, and the air speed, the sound speed and the pressure at the point where Ma = 0.8 if air flows isentropically and its stagnation temperature and stagnation pressure are T_0 = 20℃ and p_0 = 490kPa respectively.

12-8 Air isothermally flows in a 50-mm-diameter and 1400-m-long duct at a constant temperature of 5℃. The average friction factor of the duct is 0.016. If the pressures at the entrance is 1MPa and at the duct exit is 0.7MPa, determine the mass flow rate in the duct.

12-9 Determine the stagnation temperature and stagnation pressure of air that is flowing at 44kPa, 245.9K, and 470m/s.

12-10 Air at 300K is flowing in a duct at a velocity of (a) 1, (b) 10, (c) 100, and (d) 1000m/s. Determine the temperature that a stationary probe inserted into the duct will read for each case.

12-11 Air enters a 5-cm-diameter adiabatic duct at Ma_1 = 0.2, v_1 = 73m/s, T_1 = 400K, and p_1 = 200kPa. The average friction factor for the duct is estimated to be 0.016. If the Mach number at the duct exit is 0.8, determine the duct length, temperature, pressure, and velocity at the duct exit.

12-12 Air enters a constant diameter L-long adiabatic duct at Ma = 0.4 and leaves at Ma = 0.6. Determine the location in the pipe when Ma = 0.5.

12-13 Air adiabatically flows out the air tank at 300kPa and 15℃ through a 50-mm-diameter and 85-m-long duct. If the pressure at the exit of the duct is 120kPa and the average friction factor of duct is 0.024, determine (a) the velocity at the duct inlet, and (b) the mass flowrate of in the duct.

12-14 Air enters a 15-m-long, 4-cm-diameter adiabatic duct at v_1 = 70m/s, p_1 = 300kPa and T_1 = 500K. The average friction factor for the duct is estimated to be 0.023. Determine (a) the Mach number at the duct exit, and (b) the mass flow rate of air.

12-15 Air in a room at T_0 = 290K and p_0 = 95kPa is drawn steadily by a vacuum pump through a 1-cm-diameter, 50-cm-long adiabatic tube equipped with a converging nozzle at the inlet. The flow in the nozzle section can be assumed to be isentropic, and the average friction factor for the duct can be taken to be 0.018. Determine the maximum mass flow rate of air that can be sucked through this tube and the Mach number at the tube inlet.

12-16 Air enters a 3-cm-diameter pipe 15m long at v_1 = 73m/s, p_1 = 550kPa and T_1 = 60℃. The friction factor is 0.018. Compute v_2, p_2, T_2 at the end of the pipe. How much additional pipe length would cause the exit flow to be sonic?

12-17 Air enters a 5cm by 5cm square duct at v_1 = 900m/s and T_1 = 300K. The friction factor is 0.02. For what length duct will the flow exactly decelerate to Ma = 1.0

12-18 Air flows adiabatically in a 3-cm-diameter duct. The average friction factor is 0.015. If, at the entrance, v = 950m/s and T = 250K, how far down the tube will the Mach number be 1.8.

12-19 How do the compressible-pipe-flow formulas behave for small pressure drops? Let air at

20°C enter a tube of diameter 1 cm and length 3m. If $\lambda = 0.028$ with $p_1 = 102$kPa and $p_2 = 100$kPa, estimate the mass flow in kg/h for (a) isothermal flow, (b) adiabatic flow, and (c) incompressible flow (Chapter 6) at the entrance density.

12-20 Natural gas, with $k = 1.3$ and a gas constant of 520J/(kg·K), is to be pumped through 100km of 81-cm-diameter pipeline. The downstream pressure is 150kPa. If the gas enters at 60°C, the mass flow is 20kg/s, and $\lambda = 0.024$, estimate the required entrance pressure for isothermal flow.

References

1. 毛根海主编. 毛根海, 邵卫云, 张燕编. 应用流体力学. 北京: 高等教育出版社. 2006.
2. 邵卫云改编. 工程流体力学(第五版, 英文改编版). 北京: 电子工业出版社, John Wiley & Sons. 2006.
3. 陈玉璞, 王惠民主编. 流体动力学(第2版). 北京: 清华大学出版社. 2013年.
4. 王惠民编著. 流体力学基础(第3版). 北京: 清华大学出版社. 2013年.
5. 刘鹤年. 流体力学(第二版). 北京: 中国建筑工业出版社. 2004.
6. 刘树红, 吴玉林主编, 周雪漪, 陈庆光副主编. 应用流体力学. 北京: 清华大学出版社. 2006.
7. 李玉柱, 贺五洲. 工程流体力学(上册). 北京: 清华大学出版社. 2006.
8. 王惠民主编. 王惠民, 王泽, 张淑君编. 流体力学. 南京: 河海大学出版社. 2010.
9. 《工程地质手册》编委会. 工程地质手册(第四版). 北京: 中国建筑工业出版社. 2007.
10. 上海市建设和交通委员会. 室外排水设计规范(GB 50014—2006)(2014年版). 北京: 中国计划出版社. 2014.
11. AISI (American Iron and Steel Institute), Modern Sewer Design. 1980.
12. Bear, J. Dynamics of Fluids in Porous Media. Elsevier, New York, NY, 1972.
13. Bear, Jacob. Modeling groundwater flow and contaminant transport. Computer-Mediated Distance Learning Course. Haifa, Israel. 2010. http://www.interpore.org/reference_material/mgfc-course/mgfcclas.html
14. Çengel, Yunus A. and Cimbala, John M. Fluid Mechanics: Fundamentals and Applications. New York: McGraw-Hill, 2006.
15. Çengel, Yunus A, Cimbala, John M. and Turner, Robert H. Fundamentals of Thermal-Fluid Sciences, 4th Edition. McGraw-Hill, 2012.
16. Cengel, Yunus A. and Boles, Michael A. Thermodynamics: An Engineering Approach, 4th Edition, McGraw-Hill, 2002.
17. Forchheimer PE(1902). Discussions of the Bohio Dam by George S, Morison. Transactions, A. S. C. E., 48: 302.
18. Haaland, S. E. (1983). Simple and Explicit for the Friction Factor in Turbulent Pipe Flow. Journal of Fluids Engineering, 105 (1): 89-90.
19. Hauser, Barbara A. Practical Hydraulics Handbook, 2rd Edition. NewYork: CRC Press, 1995.
20. Heiner, Bryan J., Vermeyen, Tracy B. Laboratory Evaluation of Open Channel Area-Velocity Flow Meters. Hydraulic Laboratory Report HL-2012-03. U. S. Department of the Interior Bureau of Reclamation, Denver, Colorado, USA. http://www.usbr.gov/pmts/hydraulics_lab/pubs/HL-2012-03.pdf
21. Howell, R. H., Sauer, H. J., Coad, jr., W. J. Principles of Heating, Ventilating, and

Air Conditioning. ASHRAE, 2005, Atlanta, GA. http://www.engr.ipfw.edu/~renie/ME301%20Page/ch16.pdf

22. Lighthill, Sir James. Fluid Mechanics of Tropical Cyclones. Theoretical and Computatonal Fluid Dynamics. 1998 (10): 3-21.

23. Marstella, Joseph Edward. The Concept, Analysis and Critical Thinking Involved in Flow Net Theory- A Project Based Learning Approach Using Cofferdams [D]. University of Southern Queensland, 2010.

24. McDonough, J. M. Lectures in Elementary Fluid Dynamics: Physics, Mathematics and Applications. University of Kentucky, USA, 2009. (Lecture notes)

25. Metcalf and Eddy, Inc., G. Tchobanoglous, editor. Wastewater Engineering: Collection and Pumping of Wastewater. McGraw-Hill, 1981.

26. Ministère Du Développement Durable, De L' Environnement Et Des Parcs Du Québec, May 2007, Sampling Guide for Environmental Analysis: Booklet 7 - Flow Measurement Methods in Open Channels, Centre d' expertise en analyse environnementale du Québec, 223 p., http://www.ceaeq.gouv.qc.ca/documents/publications/guides_ech.htm

27. Moody, Lewis F. Friction Factors for Pipe Flow. The Transactions of the ASME, 671-684, Nov., 1944.

28. Munson, Bruce R., Young, Donald F. and Okiishi, Theodore H. Fundamentals of Fluid Mechanics, 5th Edition. New York: Wiley, 2006.

29. Potter, Mertle C. and Wiggert, David C. Mechanics of Fluids, 3rd Edition. Brooks/Cole, Wadsworth Group. 2002.

30. Sleigh, Andrew. An Introduction to Fluid Mechanics. University of Leeds, UK, 2001. (Lecture notes)

31. Todd, David Keith, Mays, Larry W. Groundwater Hydrology, John Wiley & Sons, 2004.

32. U.S. Department of the Interior Bureau of Reclamation. Water Measurement Manual. A Water Resources Technical Publication. U. S. Department of the Interior Bureau of Reclamation, Denver, Colorado, USA. http://www.usbr.gov/pmts/hydraulics_lab/pubs/wmm/

33. Ven Te Chow. Open Channel Hydraulics. New York: McGraw-Hill, 1959.

34. White, Frank M. Fluid Mechanics, 4th Edition. New York: McGraw-Hill, 1998.

35. White, Frank M. Fluid Mechanics, 5th Edition. New York: McGraw-Hill, 2003.

36. Young, Donald F., Munson, Bruce R., Okiishi, Theodore H. and Huebsch, Wade W. A Brief Introduction to Fluid Mechanics, 5th Edition. New York: Wiley, 2010.

Appendix A Property Tables

Table A-1 Properties of water at 1 atm pressure

Temperature T, ℃	Density ρ, kg/m^3	Dynamic viscosity μ, 10^{-3} kg/m·s	Kinematic viscosity ν, 10^{-6} m^2/s
0	999.87	1.792	1.792
4	1000.0	—	—
5	999.9	1.519	1.519
10	999.7	1.307	1.307
15	999.1	1.138	1.139
20	998.0	1.002	1.004
25	997.0	0.891	0.894
30	996.0	0.798	0.801
35	994.0	0.720	0.724
40	992.1	0.653	0.658
45	990.1	0.596	0.502
50	988.1	0.547	0.554
60	983.3	0.467	0.475
70	977.5	0.404	0.413
80	971.8	0.355	0.468
90	965.3	0.315	0.326
100	957.9	0.282	0.294
150	916.6	0.183	0.200
200	864.3	0.134	0.155
300	713.8	0.086	0.120
374.14	317.0	0.043	0.136

Source: Density and Dynamic viscosity data are from Yunus A. Çengel and John M. Cimbala(2006), *Fluid Mechanics: Fundamentals and Applications*, Table A-3, The kinematic viscosity is calculated from its definitions, $\nu = \mu/\rho$.

Table A-2 Properties of air at 1 atm pressure

Temperature T, ℃	Density ρ, kg/m^3	Dynamic viscosity μ, 10^{-5} kg/m·s	Kinematic viscosity ν, 10^{-5} m^2/s
−150	2.866	0.8636	0.301
−100	2.038	0.1189	0.584
−50	1.582	1.474	0.932
−20	1.394	1.630	1.169
−10	1.341	1.680	1.252
0	1.292	1.729	1.338
5	1.269	1.754	1.382
10	1.246	1.778	1.426
15	1.225	1.802	1.470
20	1.204	1.825	1.516
25	1.184	1.849	1.562
30	1.164	1.872	1.608
40	1.127	1.918	1.702
50	1.092	1.963	1.798
60	1.059	2.008	1.896
70	1.028	2.052	1.995
80	0.9994	2.096	2.097
90	0.9718	2.139	2.201
100	0.9458	2.181	2.306
120	0.8977	2.264	2.522
140	0.8542	2.345	2.745
160	0.8148	2.420	2.975
180	0.7788	2.504	3.212
200	0.7459	2.577	3.455
250	0.6746	2.760	4.091
300	0.6158	2.934	4.765
500	0.4565	3.563	7.806
1000	0.2772	4.826	17.410
2000	0.1553	6.630	42.692

Source: Data are from Yunus A. Çengel and John M. Cimbala (2006), *Fluid Mechanics: Fundamentals and Applications*, Table A-9.

Table A-3 Properties of some common fluids at 1 atm pressure

Fluid	Temperature T, °C	Density ρ, kg/m^3	Dynamic viscosity μ, kg/m·s	Kinematic viscosity ν, m^2/s
LIQUIDS:				
Carbon tetrachloride	20	1590	9.58×10^{-4}	6.03×10^{-7}
Ethyl alcohol	20	789	1.19×10^{-3}	1.51×10^{-6}
Gasoline[a]	15.6	680	3.10×10^{-4}	4.6×10^{-7}
Glycerin	20	1260	1.50	1.19×10^{-3}
Kerosine	20	804	1.92×10^{-3}	2.388×10^{-6}
Mercury	20	13600	1.57×10^{-3}	1.15×10^{-7}
SAE 30W oil[a]	20	891	0.29	3.25×10^{-4}
Seawater	15.6	1030	1.20×10^{-3}	1.17×10^{-6}
GASES:				
Hydrogen	20	0.0838	8.84×10^{-6}	1.05×10^{-4}
Carbon dioxide	20	1.83	1.47×10^{-5}	8.03×10^{-6}
Nitrogen	20	1.16	1.76×10^{-5}	1.52×10^{-5}
Oxygen	20	1.33	2.04×10^{-5}	1.53×10^{-5}

[a]: Typical values. Properties of petroleum products vary.

Source: Data of kerosine are from Frank M. White (1998), *Fluid Mechanics* (4th Edition), Table A.3; other data are from Bruce R. Munson, Donald F. Young and Theodore H. Okiishi (2005), *Fundamentals of Fluid Mechanics* (5th Edition), Tables 1.6 and 1.8.

Appendix B Bibliography

Archimedes (287-212 B. C.) 1.1.3, 2.7.1 Principles of hydrostatics and flotation; Archimedes principle
Bernoulli, Daniel (1700-1782) 3.6.1 Bernoulli equation
Boussinesq, Joseph (1842-1929) 4.4.2, 6.5.2 Momentum-flux correction factor; turbulent shear stress
Buckingham, Edgar (1867-1940) 5.2 Buckingham pi theorem
Cauchy, Augustin Louis de (1789-1857) 5.4.2 Cauchy number
Chézy, Antoine (1718-1798) 6.6.2, 9.2.2 Chézy formula
Colebrook, Cyril Frank (1910-1997) 6.6.2 Colebrook equation
Cross, Hardy (1885-1959) 8.5.3 Hardy cross method
Darcy, Philibert Caspard Henry (1803-1858) 6.1.2, 11.2.1 Darcy-Weisbash equation; Darcy law
da Vinci, Leonado (1452-1519) 1.1.3 Equation of conservation of mass in 1D steady flow and experimented with waves, jets, hydraulic jumps, eddy formation.
Dupuit, Arsene Jules Emile Juvenal (1804-1866) 11.3.1 Dupuit formula
Euler, Leonhard (1707-1783) 1.1.3, 2.2.1, 3.1.2, 3.5.1, 3.5.3, 3.6.1, 5.4.2 Derivation of Bernoulli equation; Euler number; Euler equations of motion; Eulerian approach; continuity equation
Froude, William (1810-1879) 1.1.3, 5.4.2 Froude number, laws of model testing
Galilei, Galileo (1564-1642) 1.1.3 Marked the beginning of the experimental mechanics
Hagen, Gotthilf Heinrich Ludwig (1797-1884) 6.4.1 Hagen-Poiseuille flow
Lagrange, Joseph Louis (1736-1813) 3.1.1, 3.7.1 Lagrangian description of flow field; stream function; velocity potential function
Mach, Ernst (1838-1916) 5.4.2 Mach number
Manning, Robert(1816-1897) 6.6.2 Manning equation
Mariotte, Edme(1620-1684) 1.1.3 Built the first wind tunnel and tested model in it
Moody, Lewis F. (1880-1953) 6.6.2 Moody chart
Navier, Louis Marie Henri (1785-1836) 1.1.3, 3.5.2 Navier-Stokes equation
Newton, Isaac (1642-1727) 1.1.3, 1.4.1 Laws of motion; law of viscosity; Newtonian fluid
Nikuradse, Johann (1894-1979) 6.6.1 Nikuradse experimental curves on friction factor
Pascal, Blaise (1623-1662) 2.1.1 Pascal's law
Pitot, Henri (1695-1771) 3.6.5 Pitot-static probe
Poiseuille, Jean Louis Marie (1799-1869) 6.4.1 Hagen-Poiseuille flow
Prandtl, Ludwig (1875-1953) 1.1.3, 6.5.2, 6.8.1, 6.8.2 Founder of modern fluid mechanics; inviscid outer layer; Boundary layer; semi-empirical theory of Prandtl
Rankine, William John Macquorn (1820-1872) 3.7.5 Rankine half body
Rayleigh, John William Strutt (1842-1919) 1.1.3, 5.2.1 Proposed dimensional analysis; Rayleigh method
Reynolds, Osborne (1842-1912) 1.1.3, 5.4.2, 6.2.1, 6.2.2, 6.5.2 Reynolds number; Reynolds stress; Reynolds experiment; published classic pipe experiment
Stokes, George Gabriel (1819-1903) 1.1.3, 3.5.2, 6.8.3 Navier-Stokes equation, Stokes law

Strouhal, Vincenz (1850-1922) 5. 4. 2 Strouhal number

Taylor, Geoffrey Ingram (1886-1975) 1. 1. 3 Advanced statistical theory of turbulence and the Taylor microscale

Torricelli, Evangelista (1608-1647) 2. 3. 2 Constructed the first mercury barometer

Venturi, Giovanni Battista (1746-1822) 3. 6. 5 Venturi meter

von Kármán, Theodore (1881-1963) 1. 1. 3, 5. 4. 2 von Kármán vortex street

Weber, Moritz (1871-1951) 5. 4. 2 Formalizing the general use of common dimensional groups as a basis of similitude studies; Weber number

Weisbach, Julius (1806-1871) 6. 1. 2 Published the first textbook on hydrodynamics; Darcy-Weisbach equation

Index of Vocabulary

A

Acceleration 3. 1. 3
    convective, or advective 3. 1. 3
    local 3. 1. 3
Acceleration filed 3. 1. 3
Accumulator, or expansion tank 8. 6. 5
Acting head 7. 2. 1, 7. 3. 1, 8. 2. 1
Adiabatic 4. 3. 4, 12. 1. 2, 12. 2. 1, 12. 2. 2
Alternate depths 9. 5. 1
Aquifer 11
    artesian 11. 2. 2, 11. 4
    confined 11. 2. 2, 11. 4
    unconfined 11. 3, 11. 4
Area moment of inertia 2. 5. 1
Archimedes principle 2. 7. 1
Aspect ratio 9. 2. 3

B

Bazin formula 10. 2. 2
Bed slope, or bottom slope 9. 1. 1
    adverse or negative 9. 1. 1, 9. 7. 2, 9. 7. 3, 11. 3. 3
    downhill, or positive 9. 1. 1, 9. 5. 3, 9. 7. 2, 9. 7. 3, 11. 3. 3
        mild 9. 5. 3, 9. 7. 2, 9. 7. 3
        critical 9. 5. 3, 9. 7. 2, 9. 7. 3
        steep 9. 5. 3, 9. 7. 2, 9. 7. 3
    horizontal 9. 1. 1, 9. 7. 2, 9. 7. 3, 11. 3. 3
Bernoulli equation 1. 1. 3, 3. 6
    along a streamline 3. 6. 1
    generalized 3. 6. 1
    for potential flow 3. 6. 2
Best hydraulic cross section 9. 2. 3
Body 2. 7. 1
    floating 2. 7. 1
    submerged, or immersed 2. 7. 1
    sunk 2. 7. 1
Boiling 1. 4. 3
Borda equation 6. 7. 2
Boundary layer 3. 4. 2, 6. 3. 1, 6. 8
    in internal flow 6. 3. 1
    in external flow 6. 8
        boundary layer separation 6. 8. 2
        laminar boundary layer 6. 8. 1
        turbulent boundary layer 6. 8. 1
        wake flow 6. 8. 2
        wake region 6. 8. 1, 6. 8. 2
Boundary layer region 6. 3. 1
Boundary layer thickness 6. 8. 1
Bulk modulus of compressibility 1. 4. 2, 8. 6. 3
Bulk modulus of elasticity 1. 4. 2
Buoyancy 2. 7. 1
Buoyant 2. 7. 1

C

Capillarity 1. 4. 4
    nonwetting 1. 4. 4
    wetting 1. 4. 4
Capillarity drop 1. 4. 4
Capillarity rise 1. 4. 4
Cavitation 1. 4. 3, 4. 3. 3, 5. 4. 2
Cavitation bubble, or vapor bubble 1. 4. 3
Cavitation number 5. 4. 2
Celerity of pressure wave 8. 6. 2, 8. 6. 3
Center, of
    buoyancy 2. 7
    gravity 2. 7
    pressure 2. 5. 1
Centroid 2. 5. 1

CFD, computational fluid mechanics 1. 1. 3, 3. 3. 1
Chézy coefficient 6. 6. 2, 9. 2. 2
Chézy formula 6. 6. 2, 9. 2. 2
Circulation 3. 7. 4
Coefficient, of
    contraction 7. 1, 7. 2. 1, 7. 2. 2, 10. 2. 1, 10. 3. 3, 10. 4. 4, 10. 6. 1
    depth-contracted 10. 6. 1
    discharge 7. 2. 1, 7. 2. 2, 7. 2. 3, 7. 3. 1, 8. 2. 1, 8. 7. 3, 10. 2, 10. 3. 1, 10. 4. 2, 10. 5
    drag 5. 5. 2, 6. 8. 3
    drowned 10. 2. 1, 10. 3. 2, 10. 4. 3
    lift 6. 8. 3
    minor loss, or resistance, see minor loss coefficient
    permeability, or hydraulic conductivity 11. 1. 1, 11. 2. 1, 11. 2. 4
    surface tension 1. 4. 4
    velocity 7. 2. 1, 7. 2. 2, 7. 3. 1, 10. 2. 1, 10. 6. 1
    volume compressibility 1. 4. 2
    volume expansion, or volume expansivity 1. 4. 2
Colebrook equation 6. 6. 2
Compressibility 1. 4. 2, 12. 2. 3
    bulk modulus of compressibility, or bulk modulus of elasticity 1. 4. 2
    coefficient of volume compressibility, or isothermal compressibility 1. 4. 2
    coefficient of volume expansion, or volume expansivity 1. 4. 2
Compressible flow 12
Conservation law of mass 3. 5. 1
Contact angle 1. 4. 4
Continuity equation 3. 5. 1, 4. 2
    differential 3. 5. 1
    of uniform flow 3. 3. 3
    of total flow 4. 2
Continuum 1. 1. 2, 3. 5. 1
Continuum hypothesis, or continuum assumption 1. 1. 2, 1. 2. 1, 3. 5. 1
Control cross section 3. 3. 2
Control surface 3. 3. 2, 4. 1
Control volume 3. 1. 2, 3. 3. 2, 4. 1
Correction factor
    kinematic 4. 1, 4. 3. 2, 6. 4. 1
    momentum 4. 1, 4. 4. 2, 6. 4. 1
Critical depth 9. 5. 1, 9. 5. 2, 9. 6. 1
Critical slope 9. 5. 3
Critical time, or reflection time 8. 6. 3
Critical velocity, in open channel 9. 5. 1, 9. 5. 2
Critical velocity, in pipe 6. 2. 1
    upper 6. 2. 1
    lower 6. 2. 1
Cross section 3. 3. 2
Contraction, of flow 7. 2. 3, 7. 3
    non-perfect 7. 2. 3
    perfect 7. 2. 3

D

Darcy law 11. 2. 1
Darcy-Weisbach equation 5. 2. 2, 6. 1. 2, 6. 4. 2, 6. 6. 1, 6. 6. 3, 9. 2. 2
Deformation of fluid element 3. 4. 1
Density 1. 1. 2, 1. 3. 1
Depressed sewer, or trap 8. 2. 2
Description approach of flow 3. 1
    Eulerian 3. 1. 2
        Eulerian variable 3. 1. 2
    Lagrangian 3. 1. 1
        Lagrangian variable 3. 1. 1
Dimension 5. 1. 1, 5. 1. 2
    basic 5. 1. 2
    derived, or secondary 5. 1. 2
    reference 5. 2. 2
Dimensional analysis 5. 2
    Rayleighs method 5. 2. 1
    Buckingham pi theorem 5. 2. 2
Dimensionless group, or dimensionless produce, or pi term 5. 1. 2, 5. 4
    Cauchy number 5. 4. 2
    Euler number 5. 4. 1, 5. 4. 2
    Froude number 5. 4. 1, 5. 4. 2, 9. 1. 2, 9. 4
    Mach number 5. 4. 2, 12. 1. 3
    Reynolds number 5. 2. 1, 5. 4. 1, 5. 4. 2, 6. 2. 2, 6. 8. 1, 9. 1. 2, 11. 2. 3

        critical 6. 2. 2, 6. 8. 1
        local 6. 8. 1
    Strouhal number 5. 4. 1, 5. 4. 2
    Weber number 5. 4. 2
Dimensional homogeneity 5. 1. 3, 5. 2
Drag force 5. 5. 2, 6. 8. 3
    friction drag 6. 8. 3
    pressure drag, or form drag 6. 8. 3
Draw-off discharge 8. 4. 3
Drowned depth 10. 4. 3
Dupuit formula 11. 3. 1, 11. 4

E

Efficiency, of
    pump 4. 3. 4
    turbine 4. 3. 4
Effective buoyancy 4. 3. 4
Element flow 3. 3. 2
End flowrate 8. 4. 3
Energy equation 4. 3
    differential form 4. 3. 1
    form for converging flow 4. 3. 4
    form for diverging flow 4. 3. 4
    form for incompressible gas flow 4. 3. 4
    form with shaft work 4. 3. 4
    general form 4. 3. 2
Entrance length 6. 3. 1
Entrance region 6. 3. 1
Equation of state, or perfect gas law, or ideal gas law 1. 4. 2, 12. 1. 1
Equi potential lines 3. 7. 2, 3. 7. 3
Equi pressure surface 2. 2. 2, 2. 3. 1, 2. 4
    horizontal 2. 3. 1
    inclined plane 2. 4. 1
    paraboloid 2. 4. 2
Equivalent discharge 8. 4. 3
Equivalent length 8. 3
Euler equation of motion 2. 2. 1
    differential form 2. 2. 1
    equilibrium form 2. 2. 1
    form for inviscid flow 3. 5. 2

External flow 6. 8
    around bluff 6. 8. 2
    around flat plate 6. 8. 1

F

Fluid 1. 1. 1
    liquid 1. 1. 1
    gas 1. 1. 1, 12
    classified by
        compressibility
            compressible 1. 4. 2
            incompressible 1. 4. 2
        viscosity
            ideal, or inviscid 1. 4. 1
            real, or viscous 1. 4. 1
        Newton's law of viscosity 1. 4
            Newtonian 1. 4. 1
            non-Newtonian 1. 4. 1
                Bingham plastic 1. 4. 1
                shear thickening, or dilatant 1. 4. 1
                shear thinning, or pseudoplastic 1. 4. 1
Fluid dynamics 1. 1. 1, 3 ~ 12
Fluid mechanics 1. 1. 1
Fluid properties 1. 3, 1. 4
    density 1. 1. 2, 1. 3. 1
    specify gravity 1. 3. 3
    specify volume 1. 3. 1
    specific weight 1. 3. 2
    surface tension 1. 4. 4
    vapor pressure 1. 4. 3
    viscosity 1. 4. 1, see also viscosity
Fluid statics 1. 1. 1, 2
Flow layer, in turbulent flow near a wall 6. 5. 3, 6. 5. 4
    buffer zone 6. 5. 4
    outer layer 6. 5. 3, 6. 5. 4
    overlap layer 6. 5. 3, 6. 5. 4
    transition layer 6. 5. 4
    viscous sublayer 6. 5. 3, 6. 5. 4
Flow net 3. 7. 3
Flow, of fluid
    Classified by

compressibility
    compressible 1. 4. 2, 12. 1. 3
    incompressible 1. 4. 2, 12. 1. 3
dimensionality
    one dimensional, or total flow 3. 2. 2, 4
    three dimensional 3. 2. 2
    two dimensional 3. 2. 2
Mach number 12. 1. 3
    hypersonic 12. 1. 3
    sonic 12. 1. 3
    subsonic 12. 1. 3
    supersonic 12. 1. 3
    transonic 12. 1. 3
Reynolds number
    laminar 6. 2. 1, 6. 4
    transitional 6. 2. 1
    turbulent 6. 2. 1, 6. 5, 6. 6
spatial variation
    uniform 3. 2. 4, 9. 1. 2, 9. 2
    non-uniform 3. 2. 4, 9. 1. 2
        gradually varied 3. 2. 4, 9. 1. 2, 9. 7
        rapidly varied 3. 2. 4, 9. 1. 2, 9. 6
temporal variation
    steady 3. 2. 3
    unsteady 3. 2. 3, see also unsteady flow
viscosity
    inviscid 3. 2. 1, 3. 7
    viscous 3. 2. 1
vorticity
    irrotational, or potential 3. 4. 2, 3. 7
    rotational, or vortex 3. 4. 2
in channel 9
in pipe 8
over weir 10
seepage 11
under a small bridge 10. 6
under sluice gate 10. 5
Flow strength 3. 7. 2
Flow rate
    mass 3. 3. 3, 4. 4. 1
    volume, or discharge 3. 3. 3
Flow visualization 3. 3
    pathline 3. 3. 1

streakline 3. 3. 1
streamline 3. 3. 1
Force
    buoyant, or buoyancy force 2. 7. 1
    drag 5. 5. 2, 6. 8. 3, see also drag force
    gravitational 5. 4. 1
    hydrostatic 2
        on a curved surface, resultant 2. 6
            horizontal component 2. 6. 1
            vertical component 2. 6. 1
        on a plane surface 2. 5
        on a surface in layered fluids 2. 6. 3
    inertia 5. 4. 1
    lift 6. 8. 3
    mass, or body 1. 2. 2
        unit 1. 2. 2
    pressure 5. 4. 1
    surface 1. 2. 1
    surface tension 1. 4
    viscous 5. 4. 1
Friction factor, Darcy 5. 2. 2, 6. 1. 2, 6. 4. 2, 6. 6
    Щифринсон formula 6. 6. 2
    Blasius formula 6. 6. 2
    Chézy formula 6. 6. 2
    Colebrook equation 6. 6. 2
    Manning equation 6. 6. 2
    Moody chart 6. 6. 2
    Prandtl equation 6. 6. 2
    von Kármán equation 6. 6. 2
Friction velocity, or shear velocity 6. 3. 3, 6. 5. 4
Frictional slope 4. 3. 3, 6. 3. 2, 9. 2. 1
Free head, or residual head 8. 5. 2
Free jet, or free discharge 7. 1, 7. 2. 1, 7. 2. 3
Frontal area 6. 8. 3
Fullness of pipe, or percent depth of flow 9. 3. 1
Fully developed region(flow) 6. 3. 1
Fully rough flow 6. 6. 2
Fully rough turbulent flow 6. 6. 2

G

Gage fluid, or manometric fluid 2. 3. 3

Gas constant 1. 4. 2, 12. 1. 1
Grade line 4. 3. 3
    energy grade line (EGL) 4. 3. 3, 8, 9. 1. 2, 9. 2
    hydraulic grade line (HGL) 4. 3. 3, 8, 9. 1. 2, 9. 2
Gravity flow 9. 1. 2
Gas flow, compressible 12
    adiabatic process 12. 1. 2
        isentropic, or isoentropic, or constant entropy process 1. 4. 2, 12. 1. 2, 12. 2. 1
        isochoric, or isovolumetric, or isometric process, or constant-volume process 12. 2. 1
        isothermal process 1. 4. 2, 12. 2. 1
        Fanno flow 12. 3
Gas flow, incompressible 4. 3. 4, 12. 2. 1
Groundwater 11. 1. 1
    capillary 11. 1. 2
    gaseous 11. 1. 2
    gravitational 11. 1. 2
    hydroscopic 11. 1. 2
    pellicular 11. 1. 2
Groundwater flow 11. 1. 2, 11. 3

H

Hagen-Poiseuille flow 6. 4. 1
Half-body, Rankine 3. 7. 5
Hardy Cross method 8. 5. 3
    method of balancing heads 8. 5. 3
Hazen's empirical coefficient 11. 2. 4
Head loss 4. 3. 2, 6. 1
    major, or frictional 6. 1. 1, 6. 6
        Darcy friction factor, see Friction factor, Darcy
        Darcy-Weisbach equation, see Darcy-Weisbach equation
    minor, or local 6. 1. 1, 6. 7
        minor loss coefficient, see minor loss coefficient
Head, of fluid 2. 3. 2, 3. 6. 3, 4. 3. 2
    elevation 2. 3. 2, 3. 6. 3, 4. 3. 2
    free, or residual 8. 5. 2

    loss of 4. 3. 2, see also Head loss
    piezometric 2. 3. 2, 3. 6. 3, 4. 3. 2
    pressure 2. 3. 2, 3. 6. 3, 4. 3. 2
    total 3. 6. 3, 4. 3. 2
    velocity 3. 6. 3, 4. 3. 2
Head, of shaft machine 4. 3. 4
    pump 4. 3. 4
    turbine 4. 3. 4
Height, of fluid column 2. 3. 2
    elevation 2. 3. 2
    piezometric 2. 3. 2, 4. 3. 2
    vacuum 2. 3. 2, 7. 3. 2
Hydraulic 1. 1. 3
Hydraulic diameter 6. 6. 3, 12. 3. 2
Hydraulic jump 9. 4, 9. 6. 1-9. 6. 5
    conjugate depth 9. 6. 1, 9. 6. 2
    depth ratio 9. 6. 2
    dissipation ratio 9. 6. 4, 9. 6. 5
    height of 9. 6. 3
    length of 9. 6. 1, 9. 6. 3
    momentum-depth diagram 9. 6. 1
    momentum function 9. 6. 1
    sequent depth 9. 6. 4
Hydraulic conductivity, or coefficient of permeability 11. 1. 1
Hydraulic radius 6. 1. 2, 6. 2. 2, 9. 1. 2
Hydraulic slope 4. 3. 3
Hydrodynamics 1. 1. 3
Hydrometer 2. 7. 1
Hydrostatic pressure distribution 2. 3. 1, 3. 2. 4

I

Ideal gas 12. 1. 1
Ideal gas law, ot perfect gas law 1. 4. 2, 12. 1. 1
Impermeable stratum 11. 1. 1
Influence radius of well 11. 4. 1
Injector 4. 3. 4
Internal flow 6. 8
Invertal siphon 8. 2. 2
Inviscid core flow region 6. 3. 1
Isothermal compressibility 1. 4. 2

J

Joukowsky's formula 8. 6. 3

L

Laminar flow, fully developed 6. 4
    Hagen-Poiseuille flow 6. 4. 1
    Poiseuille's law 6. 4. 1
Laplace equation 3. 7. 1, 3. 7. 2
Left-side convention 3. 7. 1
Linear law of seepage flow 11. 2. 1

M

Manning coefficient 6. 6. 2, 9. 2. 2
Manning formula 6. 6. 2, 9. 2. 2
Manometry 2. 3. 3
Manometric fluid, or gage fluid 2. 3. 3
Measurement, of
    drag force 6. 8. 4
        water tunnel 6. 8. 4
            cavitation tunnel 6. 8. 4
        wind tunnel 6. 8. 4
            open-return, or Eiffel type 6. 8. 4
            return-flow, or Prandtl type 6. 8. 4
    permeability coefficient 11. 2. 4
        empirical method 11. 2. 4
        in-situ (filed) method 11. 2. 4
        laboratory method 11. 2. 4
            constant-head method 11. 2. 4
            falling-head method 11. 2. 4
    pressure 2. 3
        barometer 2. 3. 2
        Bourdon tube 2. 3. 3
        differential U-tube manometer 2. 3. 3
        inclined-tube manometer 2. 3. 3
        piezometer tube 2. 3. 3
        pressure transducer 2. 3. 3
        U-tube manometer 2. 3. 3
    specific gravity
        hydrometer 2. 7. 1
    velocity
        HFA, Hot Film Anemometry 8. 7. 1
        HWA, Hot Wire Anemometry 1. 1. 3, 8. 7. 1
        LDV, Laser-Doppler Velocimetry 1. 1. 3, 8. 7. 1
        Pitot-static probe 3. 6. 5, 8. 7. 1
            Pitot tube 3. 6. 5
            static pressure tap 3. 6. 5
        PIV, Particle Image Velocimetry 1. 1. 3, 6. 8. 4, 8. 7. 2
    volume flow rate, in pipe
        head-loss instruments 8. 7. 3
            obstruction flowmeter, or Bernoulli-type device 8. 7. 3
            nozzle meter 8. 7. 3
            orifice meter 8. 7. 3
            Venturi meter 3. 6. 5, 8. 7. 3
        mechanical instruments 8. 7. 3
    volume flow rate, in open channel 9. 9
        area-velocity method 9. 9. 2
            anemometer-propeller velocity meter 9. 9. 2
            Doppler velocity meters 9. 9. 2
            ADCP, Acoustic Doppler current profiler 9. 9. 2
            ADV, acoustic Doppler velocimeter 9. 9. 2
            LDV, laser Doppler velocimeter 9. 9. 2
            electromagnetic velocity meters 9. 9. 2
        hydraulic structures
            flume 9. 9. 1
                Palmer-Bowlus flume 9. 9. 1
                Parshall flume 9. 9. 1
                Venturi flume 9. 9. 1
            weir 9. 9. 1, 10
        optical strobe velocity meters 9. 9. 2
        slope-area method 9. 9. 3
        tracer methods 9. 9. 4
            tracer-dilution method 9. 9. 4
            tracer-velocity-area method 9. 9. 4
Meniscus 1. 4. 4
Metacenter 2. 7. 4

Metacentric height 2.7.4
Minor loss coefficient, or resistance coefficient, of 4.3.1, 6.1.2, 6.7.1
    gradual contraction 6.7.4
    gradual expansion, or diffuser 6.7.4
    pipe bend 6.7.5
    pipe exit 6.7.3
    pipe inlet 6.7.3
    sudden contraction 6.7.2
    sudden expansion 6.7.2
    through nozzle 7.3.1
    through orifice 7.2.1, 7.2.2
    typical pipe components 6.7.6
Mixing length 6.5.2
Model 5.3.1
    distorted 5.5.1
Modulus of discharge 9.2.2
Momentum-depth diagram 9.6.1
Momentum Equation, linear 4.4
    form for uniform flow 4.4.1
    general form 4.4.2
Momentum function 9.6.1
Motion of fluid element 3.4.1
    linear strain (dilatation) 3.4.1
    rotation 3.4.1
    shear strain 3.4.1
    translation 3.4.1

N

Navier-Stokes equations 1.1.3, 3.5.2
Nappe 10.1.2, 10.2
Newton's law of viscosity 1.4.1, 3.5.2, 6.5.3
Newton's second law 3.5.2, 3.6.1, 4.4.1
Newton's third law 2.6.1
Nikuradse's experiment 6.6.1
Nonrepeating variables 5.2.2
Normal depth 9.1.2, 9.2
No-slip condition 1.4.1, 3.2.2, 6.3.1, 9.1.2
Nozzle 7.1, 7.3
    converging cone 7.1, 7.3.3
    cylindrical outer 7.1, 7.3

    diverging cone 7.1, 7.3.3
    streamlining outer 7.1, 7.3.3
Nozzle flow 7.1, 7.3

O

Open channel 9
    artificial 9.1.1
    natural 9.1.1
    non-prismatic 9.1.1
    prismatic 9.1.1
        circular 9.1.1
        rectangular 9.1.1
        trapezoidal 9.1.1
        triangular 9.1.1
Open channel flow 9
    classified by 9.1.2
    temporal variation 9.1.2
        steady 9.1.2
        unsteady 9.1.2
    Reynolds number 9.1.2
        laminar 9.1.2
        transitional 9.1.2
        turbulent 9.1.2
    depth variation 9.1.2
        nonuniform or varied 9.1.2, 9.5
            gradually varied 9.1.2, 9.7
            rapidly varied 9.1.2, 9.6
                hydraulic drop 9.6.6
                hydraulic jump 9.6.1-9.6.5, see also hydraulic jump
        uniform 9.1.2, 9.2, 9.3
            in partially full circular pipe 9.3
    Froude number 9.1.2
        critical 9.1.2, 9.4
        rapid, or supercritical 9.1.2, 9.4
        tranquil, or subcritical 9.1.2, 9.4
Orifice 7.1
    big 7.1, 7.2.3
    sharp-edged 7.1, 7.2
    small 7.1, 7.2.1, 7.2.2
Orifice flow 7.1, 7.2

P

Parallel axis theorem 2. 5. 1
Pascal's Law 2. 1. 1
Perfect gas law 1. 4. 2, 12. 1. 1
Perfect vacuum 2. 3. 2
Permeability 11. 1. 1
Permeable stratum 11. 1. 1
Permeating velocity 11. 1. 2
Phreatic surface 11. 3. 2, 11. 3. 3
 backwater curve 11. 3. 3
 drawdown cone 11. 3. 4, 11. 4. 1
 dropdown curve 11. 3. 3
Piezometric surface 11. 4
Pipe 8
 long 8. 1, 8. 3, 8. 4, 8. 5
 short 8. 1, 8. 2
  inverted siphon, or sag culvert, or sag line 8. 2. 2
  siphon 8. 2. 2
  suction pipe 8. 2. 2
Pipe discharge modulus 8. 3
Pipe flow 8
 steady 8. 2 ~ 8. 5
 unsteady, or transient 8. 6
  surge 8. 6. 1
  water hammer, see water hammer
Pipeline, or piping system 8. 1
 complex pipeline 8. 1, 8. 4
  parallel piping 8. 4. 2
  serial piping 8. 4. 1
  with uniform draw-off 8. 4. 3
 pipe network 8. 1, 8. 5
  branched system 8. 5. 1, 8. 5. 2
   branch 8. 5. 2
   trunk 8. 5. 2
  looped system 8. 5. 1, 8. 5. 3
  simple pipeline 8. 1, 8. 2, 8. 3
Pipe specific resistance 8. 3
Pitot formula 3. 6. 5
Poiseuille's law 6. 4. 1

Porosity 11. 1. 2
Potential flow 3. 4. 2, 3. 7. 2, 3. 7. 4
 plane 3. 7. 4
  doublet 3. 7. 4
  free vortex 3. 7. 4
  forced vortex 3. 7. 4
  sink flow 3. 3. 1, 3. 7. 2, 3. 7. 4
  source flow 3. 3. 1, 3. 7. 2, 3. 7. 4
  uniform flow 3. 7. 4
 Superpositon of 3. 7. 5
Pressure 1. 2. 1
 absolute 2. 3. 2
 atmospheric 1. 4. 2, 2. 3. 2
 dynamic 2. 1. 2, 3. 6. 3, 3. 6. 5
 elevation 4. 3. 4
 engineering barometric 2. 3. 2
 gage 2. 3. 2
 hydrostatic 3. 6. 3
 kinematic 4. 3. 4
 stagnation 3. 6. 3, 3. 6. 5
 static 3. 6. 3, 3. 6. 5, 4. 3. 4
 suction 2. 3. 2
 total 3. 6. 3
 vacuum 2. 3. 2, 4. 3. 3
Pressure coefficient 5. 3. 4, 5. 4. 2
Pressure distribution, in vertical
 hydrostatically 2. 3. 1, 2. 4. 2
 linearly 2. 4. 1
Pressure flow 8. 1
Pressure force prism 2. 6. 1, 2. 6. 2
 real 2. 6. 2
 virtual 2. 6. 2, 2. 7. 1
Pressure gradient 6. 8. 2
 adverse 6. 8. 2
 favorable 6. 8. 2
Pressure loss 4. 3. 4
Pressure prism 2. 5. 2
Pressure-recovery coefficient 6. 7. 4
Prototype 5. 3. 1
Pure number 5. 1. 2

Q

Quantity 5. 1. 2

secondary 5.1.2
primary 5.1.2

R

Rankine half-body 3.7.5
Rate, of
    linear strain 3.4.1
    rotation (angle deformation) 3.4.1
    shear strain 1.4.1, 3.4.1
    translation (velocity) 3.4.1
Repeating variables 5.2.2
Resistance square zone 6.6.2
Resultant force 2.5, 2.6
Reynolds' experiment 6.2.1
Rigid-body motion 2.4
    linear 2.4.1
    rotation 2.4.2
Roughness height 6.5.4
    equivalent 6.6.2
    relative 6.6.1

S

Saturation pressure 1.4.3
Scale ratio 5.3
    acceleration 5.3.3
    area 5.3.2
    flow rate 5.3.3
    force 5.3.4
    length 5.3.2
    mass 5.3.4
    time 5.3.3
    velocity 5.3.3
    volume 5.3.2
Secondary flow 6.7.5
Seepage flow 11
    conceptual model of 11.1.2
    gradually varied 11.3
        phreatic surface, see also Phreatic surface
            of collecting gallery 11.3.4
            of well 11.4

Semi-empirical theory of Prandtl 6.5.2
Sequent depth 9.6.4
Shear stress 1.2.1, 6.3.2, 6.5
    laminar 1.2.1, 6.5.2, 6.5.3
    turbulent, or apparent 6.5.2, 6.5.3
    Reynolds 6.5.2, 6.5.3
Side slope 9.2.3
Similitude 5.3.1
Similarity 5.3
    dynamic 5.3.4
    geometric 5.3.2
    kinematic 5.3.3
    of initial condition 5.3.5
    of boundary condition 5.3.5
Singularity 3.3.1, 3.7.2, 3.7.4
Siphoning 8.2.2
Soil 11.1
    anisotropic 11.1.1
    heterogeneous 11.1.1
    homogeneous 11.1.1
    impermeable 11.1.1
    isotropic 11.1.1
Solid 1.1.1
Sonic length, or maximum length, or critical length 12.3.2
Specific energy 9.5.1
Specific energy diagram 9.5.1
Specific heat
    at constant pressure 1.4.2, 12.1.2
    at constant volume 12.1.2
Specific-heat ratio 5.3.4, 12.1.2
Speed of sound, or sonic speed 5.4.2, 8.6.3, 12.1.2
Stability, of 2.7
    floating body 2.7.4
    immersed body 2.7.3
Stagnation point 3.3.1, 3.6.5
Stagnation streamline 3.6.5
Stagnation property 12.2.2
Stagnation state 12.2.2
    isentropic 12.2.2
State of equilibrium 2.7.2
    neutral 2.7.2, 2.7.3, 2.7.4
    stable 2.7.2, 2.7.3, 2.7.4

unstable 2. 7. 2, 2. 7. 3, 2. 7. 4
stokes low 6. 8. 3
stratum 11. 1. 1
    impermeable 11. 1. 1
    permeable 11. 1. 1
Stream function 3. 7. 1
    physical interpretations 3. 7. 1
Streamline surface 3. 3. 1
Streamtube 3. 3. 1, 3. 3. 2
Strength of flow 3. 7. 2, 3. 7. 4
Stress tensor 3. 5. 2
    normal stress 2. 1. 2, 3. 5. 2
        mechanical pressure, or dynamic pressure 2. 1. 2, 3. 5. 2
    viscous stress 3. 5. 2
    shear stress 1. 4. 1, 3. 5. 2
Submerged jet, or submerged discharge 7. 1, 7. 2. 2
Superposition 3. 7. 5
Surface tension 1. 4. 4
Surge 8. 6. 1

T

Tornado 3. 7. 2
Total flow 3. 2. 2, 3. 3. 2
Transient flow 3. 2. 3, 8. 6. 1
Transmission of fluid pressure 2. 3. 1
Turbulent flow, fully developed 6. 5
    time-averaged concept 6. 5. 1
        average value 6. 5. 1
        fluctuating component 6. 5. 1
Turbulence intensity 6. 5. 1

U

Unsteady flow 3. 2. 3
    in pipe
        transient flow 8. 6. 1
            water hammer, or hydraulic shock 8. 6
            surge 8. 6. 1
    in channel 9. 4
    through orifice 7. 4

Unit 5. 1. 1
Underflow gate 10. 5

V

Vapor bubble 1. 4. 3
Vapor pressure 1. 4. 3, 5. 4. 2
Velocity
    angular (rate of rotation) 3. 4. 1
    average, or mean 4. 1
    economic 8. 5. 2
    permeating 11. 1. 2
    permissive 8. 5. 2, 9. 2. 3
        maximum 8. 5. 2, 9. 2. 3
        minimum 8. 5. 2, 9. 2. 3
        non-scouring 9. 2. 3, 10. 6. 2
        non-silting 9. 2. 3
    uniform 3. 3. 3
Velocity gradient 1. 4. 1
Velocity potential function 3. 7. 2
Velocity profile 1. 4. 1, 6. 4. 1, 6. 5. 4
    between two parallel plates 1. 4. 1
        linear 1. 4. 1
    of laminar flow, fully developed 6. 4. 1
        parabolic 6. 4. 1
    of turbulent flow, fully developed 6. 5. 4
        in outer region 6. 5. 4
            power-law velocity profile 6. 5. 4
                one-seventh 6. 5. 4
        in wall region 6. 5. 4
            in overlap layer 6. 5. 4
                logarithmic law 6. 5. 4
                universal velocity profile 6. 5. 4
            in viscous layer 6. 5. 4
                law of the wall 6. 5. 4
Vena contracta 6. 7. 2, 6. 7. 3, 7. 1, 7. 3. 1
Venturi discharge coefficient 3. 6. 5
Viscous sublayer thickness 6. 5. 4
Viscosity, of 1. 4. 1
    apparent 1. 4. 1
    dynamic, or absolute 1. 4. 1
    eddy, or turbulent 6. 5. 2

kinematic 1. 4. 1
kinematic eddy, or kinematic turbulent 6. 5. 3
V-notch angle 10. 2. 3
von Kármán vortex street 1. 1. 3, 5. 4. 2
von Kármán equation 6. 6. 2
Vorticity 3. 4. 2
Vortex strength 3. 7. 4

W

Wake region 6. 8. 1
Wall regime 6. 5. 4
    hydraulically rough wall 6. 5. 4, 6. 6. 1
    hydraulically smooth wall 6. 5. 4, 6. 6. 1
    hydraulically transitional wall 6. 5. 4, 6. 6. 1
Wall shear stress 6. 3. 2
Water hammer, or hydraulic shock 1. 4. 2, 8. 6
    gradual closure 8. 6. 3
    sudden closure 8. 6. 3
Water hammer phenomenon 8. 6. 2
Water surface profile 9. 7. 2, 9. 7. 3
    backwater curve 9. 7. 2, 9. 7. 3
    dropdown curve 9. 7. 2, 9. 7. 3
    hydraulic drop 9. 7. 2, 9. 7. 3
    hydraulic jump 9. 7. 2, 9. 7. 3
Wave velocity 9. 4
Weir 9. 9. 1, 10. 1. 1
    classified by
        weir thickness
            broad-crested 10. 1. 2, 10. 4
            sharp-crested 10. 1. 2, 10. 2
                rectangular 10. 1. 2, 10. 2. 2
                trapezoidal 10. 1. 2
                triangular or V-notch 10. 1. 2, 10. 2. 3
            ogee-crested, or spillway 10. 1. 2, 10. 3
            curved 10. 1. 2
            polygonal 10. 1. 2
    weir width
        contracted 10. 1. 2
        suppressed 10. 1. 2
    drowned weir 10. 1. 1
Weir flow 10. 1. 1
    drowned outflow 10. 1. 1, 10. 2. 2, 10. 3. 2, 10. 4. 3, 10. 6. 1
    free outflow 10. 1. 1, 10. 6. 1
Weir head 10. 1. 1, 10. 2
Well 11. 4
    artesian 11. 4
        complete 11. 4. 2
        flowing 11. 4
    multiple-well 11. 4. 3
    water table 11. 4
        complete 11. 4. 1
Well flow rate 11. 4. 1
Well yield 11. 4
Wetted perimeter 6. 1. 2, 9. 1. 2

Y

Yield stress 1. 4. 1

Nomenclature

a	Opening of sluice gate, m; shape coefficient of bridge pier; half distance between sink and source in a doublet, m
$\mathbf{a}, a$	Acceleration and its magnitude, m/s^2
A	Area, m^2; frontal area, m^2
b	Width or other distance, m; clear span width of bridge, m; weir width, m
B	Width, m; channel width, m; center of buoyancy
c	Speed of sound, m/s; speed of wave, m/s
c_0	Speed of sound in water, m/s;
c_p	Constant-pressure specific heat, J/kg·K
c_v	Discharge coefficient of Venturi meter; constant-volume specific heat, J/kg·K
C	Bernoulli constant, m, N/m^2, m^2/s^2 or W depending on the form of Bernoulli equation; Chézy coefficient, m$^{1/2}$/s; Hazen's empirical coefficient; constant; centroid of body; center of gravity
Ca	Cauchy number
C_a	Cavitation number
CCS	Control cross section
C_D	Drag coefficient
C_d	Discharge coefficient of obstruction flowmeter
C_L	Lift coefficient
C_n	Discharge coefficient of nozzle meter
C_o	Discharge coefficient of orifice meter
CP	Center of pressure
C_p	Pressure recovery coefficient
CS	Control surface
CV	Control volume
d_{10}	Effective grain size diameter, m
D or d	Diameter, m (d typically for a smaller diameter than D)
D_h	Hydraulic diameter, m
e	Thickness of nappe, m
E	Total energy, kJ, or m; elastic modulus, N/m^2
$\dot{E}$	Power of mechanical energy, W
EGL	Energy grade line, m
$\dot{E}_L$	Power of mechanical energy loss, W
E_s	Specific energy in open-channel flows, m
Eu	Euler number
$\mathbf{f}, f$	Unit mass force and its magnitude, m/s^2
f	doppler shift
F	Dimension of force; area, m^2
$\mathbf{F}, F$	Force and its magnitude, N
$\mathbf{F}_b, F_b$	Body force and its magnitude, N
F_B	Magnitude of buoyancy force, N
F_D	Magnitude of drag force, N
F_L	Magnitude of lift force, N
F_n	Magnitude of normal force, N
Fr	Froude number
$\mathbf{F}_s, F_s$	Surface force and its magnitude, N
F_τ	Magnitude of tangential force, N
$\mathbf{g}, g$	Gravitational acceleration and its magnitude, m/s^2
G	Center of gravity; weight, N
GVF	Gradually varied flow
h	Depth of fluid, m; height of fluid column, m; head, m; small gap distance, m; capillary rise or drop, m
h', h''	Conjugate depths of hydraulic jump, m
h_0	Normal depth in open-channel flow, m; original depth of rigid-body liquid at rest in a container, m
h_b	Water depth under bridge, m
h_c	Critical depth in open-channel flow, m

h_f	Major loss, or friction loss, m	m	Mass, kg; side slope of trapezodial channel; discharge coefficient of weir flow; strength of sink or source, m²/s
h_j	Minor loss, or local loss, m; height of hydraulic jump, m		
h_p	Pump head, m	m_0	Discharge coefficient of weir flow
h_s	Drowned depth, m; head of shaft machine, m	$\dot{m}$	Mass flowrate, kg/s
		M	Metacentric center; Dimension of mass
h_T	Turbine head, m	Ma	Mach number
h_w	Head loss for total flow, m	n	Number of variables in Buckingham pi theorem; Manning coefficient; porosity of soil
h'_w	Head loss for streamtube flow, m		
H	Height, m; total total head, m head, m; weir head, m	$\mathbf{n}$	Unit normal vector
		n_j	Total number of nodes in a looped pipe network
H_0	Acting head, m		
H_d	Design head for WES weir, m	n_l	Total number of loops in a looped pipe network
H_z	Residual head, m		
HGL	Hydraulic grade line, m	n_p	Total number of pipes in a looped pipe network
i	Bed slope of channel; index of number		
$\mathbf{i}$	Unit vector in x-direction	N	Turbulence intensity
i_c	Critical bed slope in open-channel flow	$\dot{N}$	Power of shaft machine, W
I	Moment of inertia, N·m·s²	p	Pressure, N/m² or Pa
I_x	Second moment of inertia, m⁴	p_{abs}	Absolute pressure, N/m² or Pa
I_{xy}	Product of inertia, m⁴	p_{at}	Atmospheric pressure, N/m² or Pa
$\mathbf{J}$	Unit vector in y-direction	p_{sat}	Saturation pressure, N/m² or Pa
J	Friction slope, momentum-depth diagram	p_v	Vacuum pressure, N/m² or Pa; vapor pressure, N/m² or Pa
J_p	Hydraulic slope of HGL		
κ	Specific heat ratio; number of pi terms in Buckingham pi theorem; permeability coefficient, m/s	P_w, P'_w	Weir height, m
		q	Volume flow rate per unit width, m³/(s·m); physical variables; draw-off discharge per unit length of pipe, m³/(s·m)
$\mathbf{k}$	Unit vector in z-direction		
$\mathbf{K}, K$	Momentum, kg·m/s	Q	Volume flow rate, m³
K	Modulus of discharge, m³/s; bulk modulus of compressibility, N/m²; strength of doublet, m³/s	Q_0	Total pumping flowrate of multi-well, m³/s; flowrate of channel flow in full pipe, m³/s
l	Length or distance, m	Q_c	Equivalent flowrate, m³/s
l'	Equivalent length, m	Q_t	Total draw-off flowrate, m³/s
L_0	Influencing distance of gallery, m	Q_z	End flowrate of a pipe with uniform draw-off, m³/s
l_m	Mixing length, m		
l_{max}	Sonic length, m	$\mathbf{r}, r$	Radial coordinate, m
L	Dimension of length; Length or distance, m; characteristic length, m	r	Radius, m; number of reference dimension in Buckingham pi theorem
L_e	Entrance length in pipe flow, m	r_0	Radius of well, m;
L_j	Length of hydraulic jump, m	$\mathbf{R}, R$	Resultant external force and its magnitude, N

R	Gas constant, J/kg·K; radius, m; influence radius of well, m		β	Coefficient of volume compressibility, m^2/N; momentum-flux correction factor; angle; throat-to-pipe diameter ratio; exponent
Re	Reynolds number		β_m	Aspect ratio
R_h	Hydraulic radius, m		γ	Specific weight, N/m^3; exponent
RVF	Rapidly varied flow		$\dot{\gamma}$	Rate of shear strain, N/m^2
$\mathbf{s}, s$	Vector of differential length along a surface or streamline and its magnitude, m; distance along a surface or streamline, m		δ	Boundary layer thickness, m; weir crest thickness, m; small change in a quantity; error; thickness of pipe wall, m
s	Maximum drawdown of static water table in the well, m		δ_v	Viscous sublayer thickness, m
S	Pipe specific resistance, s^2/m^6		ε	Coefficient of contraction
S_G	Specific gravity		ε_{ij}	Strain rate, s^{-1}
St	Strouhal number		ζ	Minor loss coefficient
t	Time, s		$\boldsymbol{\zeta}, \zeta$	Vorticity vector and its magnitude, s^{-1}
T	Dimension of time; emptying or filling time, s; temperature, °C or K; time period, s		η	Depth ratio
			θ	Angle; trapezoidal angle; contact angle; V-notch angle
T_c	Time of valve closure, s		$\boldsymbol{\theta}, \theta$	Angular coordinate
T_r	Critical time, s		κ	Coefficient of volume expansion, 1/K or 1/°C; log law constant in turbulent boundary layer
$\mathbf{u}, u$	Velocity and its magnitude, m/s			
u_*	Friction velocity in turbulent boundary layer, m/s		λ	Darcy friction factor; scale ratio; wave length, m
U	Free stream flow velocity, m/s; constant velocity, m/s; Uniform-flow velocity, m/s		μ	Viscosity, kg/m·s; discharge coefficient of orifice and nozzle flows, pipe flow
Ψ	Volume, m^3		μ_{ap}	Apparent viscosity, kg/m·s
Ψ_p	Volume of pressure force prism, m^3		ν	Kinematic viscosity, m^2/s; specific volume, m^3/kg
We	Weber number			
x	Cartesian coordinate (usually to the right), m; physical variable in dimensional analysis		π	Nondimensional parameter in dimensional analysis
			ρ	Density, kg/m^3; metacentric height, m
y	Cartesian coordinate (usually up or into the page), m; distance in y-direction, m		ρ_f	Density of body, kg/m^3
			σ_{ii}	Normal stress, N/m^2
z	Cartesian coordinate (usually up), m; elevation, m		σ_{ij}	Stress tensor, N/m^2
			σ_s	Surface tension, N/m; drowned coefficient
z_s	Depth of free surface of rigid-body liquid over the rotated container bottom, m		$\boldsymbol{\tau}, \tau$	Stress vector and its magnitude, N/m^2
			τ_0	Wall shear stress, N/m^2; yield stress, N/m^2

Greek Letters

α	Angle; kinematic correction factor; exponent; fullness of pipe		τ_{ii}	Viscous stress, N/m^2
			τ_{ij}, τ	Shear stress, N/m^2

$\mathbf{v}, v$	Sectional-averaged velocity vector and its magnitude, m/s	m	Property of a model; property of mercury
v_0	Initial velocity, m/s; approaching velocity, m/s; velocity of full pipe channel flow, m/s	max	Maximum value
		mech	Property of mechanical energy
		min	Minimum value
v_c	Economic velocity in pipe, m/s; Critical velocity in channel flow, m/s	n	Property of a nozzle; normal component
		o	Property of oil; property of orifice
φ	Velocity coefficient	out	Property at exit section of flow
ϕ	Velocity potential function, m²/s	P	Property of pump
χ	Wetted perimeter, m	p	Property of a prototype
ψ	Stream function, m²/s; depth-contracted coefficient	R	Resultant force
		se	Property of sudden expansion
$\boldsymbol{\omega}, \omega$	Angular velocity vector and its magnitude, rad/s;	t	Property of a turbulent flow
		T	Property of turbine
		u	Upstream of flow
ω	Angular frequency, rad/s; frequency of oscillating flow	V	Acting vertically
		w	property of water
Γ	Circulation or vortex strength, m²/s	x	Cartesian component of physical variable in x-direction
Δ	Mean surface roughness, m; small change in a quantity	y	Cartesian component of physical variable in y-direction
Θ	Dimension of temperature; temperature in dimensional analysis, °C or K	z	Cartesian component of physical variable in z-direction; Cylindrical component of physical variable in z-direction

Subscripts

0	Stagnation property; property at the originor at a reference point	r	Cylindrical component of physical variable in r-direction
abs	Absolute	θ	Cylindrical component of physical variable in θ-direction
at	Atmospheric		
B	Buoyant force		

Superscripts

c	Acting at the centroid; critical property; property at section of vena contracta	
d	Downstream of flow	¯ (overbar)	Averaged quantity
e	External of a subject	˙ (overdot)	Quantity per unit time; time derivative
E	Acting horizontally	
Fg	Property of mercury	′ (prime)	Fluctuating quantity; modified variable; opposite force
i	Internal of a subject	
in	Property at inlet section of flow	*	Nondimensional property; sonic property
j	Property of hydraulic jump	
i, j, k	Index	+	Law of the wall variable in turbulent boundary layer
lam	Property of a laminar flow	

Answers to Problems

Chapter 1

1-1　$13.4 \times 10^3 \text{kg/m}^3$, 131.4kN/m^3
1-2　899.36kg/m^3
1-3　$5.88 \times 10^{-6} \text{m}^2/\text{s}$
1-4　$4.0 \times 10^{-3} \text{Pa·s}$
1-5　105.5N; 45.8%
1-6　$U = \dfrac{Gh\sin\theta}{\mu A}$
1-7　$F = \left(\dfrac{\mu_1 U}{h_1} + \dfrac{\mu_2 U}{h_2}\right)A$; no necessary relation
1-8　1.0
1-9　795N
1-10　0.0648Pa·s
1-11　0.55N·m
1-12　39.57N·m
1-13　0.0231Pa·s
1-14　0.0163m^3
1-15　$1.98 \times 10^9 \text{Pa}$
1-16　435.4 kPa
1-17　1%
1-18　8660m
1-19　No; No
1-20　0.0232N/m
1-21　Water: 12.25cm, +22% error; Mercury: 15.91cm, −6% error
1-22　1.36mm; 10200m
1-23　971Pa
1-24　4.95mm; 2.47mm
1-25　14.8mm

Chapter 2

2-1　117.62kPa; 19.62kPa
2-2　$-2.17 \text{mH}_2\text{O}$; $-0.43 \text{mH}_2\text{O}$
2-3　1.52m
2-4　0m, 3m, 10m; 10m, 10m, 10m
2-5　160.8kN
2-6　Right: 34.52cm; Left: 10.94cm
2-7　1.37m^2
2-8　−981Pa
2-9　265kPa
2-10　889N
2-11　56.9kPa
2-12　3.39kPa
2-13　22.6cm
2-14　26.1kPa
2-15　$p_3 = p_4 > p_2 > p_1$, $p_4 > p_5$
2-16　235mm
2-17　51.8cm
2-18　-3.80m/s^2; down; 32.2kPa
2-19　$h = al/g$
2-20　1347r/min
2-21　0.16m
2-22　224r/min; 275r/min
2-23　138r/min
2-24　16.5rad/s; 0, $0.4 \text{mH}_2\text{O}$, $1.15 \text{mH}_2\text{O}$, $1.65 \text{mH}_2\text{O}$; 1.51kN, 2.84kN
2-26　0.38m; 0.52m
2-27　31.01kN
2-28　78.5kN; 2.03m below oil free surface
2-29　$1.175 \times 10^9 \text{N}$; $3.13 \times 10^9 \text{N·m}$; No, Maybe
2-30　14.46m; 21.54m; 26.73m
2-31　0.333m; yes.
2-32　1.414m; 2.586m
2-34　58.86kN, 3.111m; 70.04kN, 0.932; 91.49kN, 40°
2-35　1.17R; 0.875R
2-36　1.5; 0.6
2-37　97.9MN, 153.8MN; 10.74m to the right and 3.3m up from point A

2-38 3.68kN
2-39 11.3kN
2-40 179.1kN; 10.27°to the horizontal; yes
2-41 1095.6kN; 57.5°
2-42 2.51kN; 18.82kN; 0.955
2-43 593N·m
2-44 5.05kN, 3.88kN
2-45 0.0522m^3
2-46 1050kg/m^3
2-47 87.6cm
2-48 0.834
2-49 18.9kPa; 0.208m^3

Chapter 3

3-1   20m/s along the x and y axis; $-90°$, $-45°$, $0°$
3-2 x; $2x^2z + x^2yz$; $x^2z^2 + y^2z$
3-3   0.5m/s^2; 1.0m/s^2; positive
3-4 $u = u_{entrance} + \dfrac{u_{exit} - u_{entrance}}{L^2} x^2$
3-5   0; -297m/s^2
3-6   Unsteady; uniform; 7.21m/s^2; $3x - 2y = C$
3-7   13.6m/s^2; 2D; steady; nonuniform
3-8 $x^2 + y^2 = C$
3-9   streamline equation: $y = e^{(\frac{1}{2}x^2 - x)} - 1$. Same
3-10 $y = 4/x^2$; $y = 4/x^2 (x < 0)$
3-11 $y = y_0 \left(\dfrac{x}{x_0}\right)^{\frac{1}{1+2t}}$; $x = x_0 e^{\ln(y/y_0) + \ln^2(y/y_0)}$
3-13 5.89m^3/s; 7.09kg/s; 69.60N/s
3-15 rotational, no; irrotational, yes.
3-16 $\sqrt{3}/2$; $5\sqrt{3}/2$
3-17 Yes. $-\dfrac{U}{2b}\mathbf{k}$, $-\dfrac{U}{b}\mathbf{k}$, $\dfrac{U}{b}$.
3-18 No; Yes; Yes; No
3-19 $a = -3c$
3-20 $u_y = -ay + f(x)$
3-21 $\dfrac{\partial p}{\partial x} = -2\rho xy^2$
3-22 any a and b; $p = -\dfrac{\rho}{2}ab(x^2 + y^2) + C$
3-23 Yes; $p = p_{at} - \dfrac{\rho}{2}(2x^2y^2 + x^4 + y^4)$
3-24 9.82m/s^2

3-25 $\nabla p = -72\mathbf{i} + 288\mathbf{j} - 282.8\mathbf{k}$ kN/m^3
3-26 $Q = 1.56D^2$m^3/s; $Q = 0.0156$m^3/s for any D
3-27 33.8m/s
3-28 1.06×10^{-3}m^3/s; 3.02×10^{-3}m^3/s; 0.118m^3/s
3-29 6.10×10^{-3}m^3/s
3-30 $U_{1,min} = \sqrt{\dfrac{2gh}{1 - (D_1/D_2)^4}}$
3-31 0.168m^3/s
3-32 4.43L/s
3-33 19.7hrs
3-34 $h = H\sin^2\theta + L\sin\theta\cos^2\theta$
3-35 $h, h, 0.5h$
3-36 $U_2 = \sqrt{2gH}$; $p_{3,min} = p_{at} - \rho g(L + H)$
3-38 0.15m/s
3-39 $\dfrac{dp}{dx} = \rho aU^2 \left(\dfrac{1}{x^2} + \dfrac{a}{x^3}\right)$;
      $p = p_0 - \rho aU^2 \left(\dfrac{1}{x} + \dfrac{a}{2x^2}\right)$
3-40 30°
3-41 13.3, 13.7, 14.1, 14.5, 15.1mm
3-42 2×10^{-4}m^3/s; 0.129m
3-43 Yes, no; No, no; Yes, yes; Yes, yes
3-44 $\phi = -4xy + C$
3-45 $\psi = 3x^2y - y^3 + C$
3-46 Yes; Yes, $\phi = 2(x + y) + C$; 0
3-47 $u_x = -2by + dx$, $u_y = -2ax - c - dy$
3-48 $u_x = 5x + 3$, $u_y = -5y + 4$; $\phi = \dfrac{5}{2}(x^2 - y^2) + (3x + 4y) + C$; 10m/s, 57.5kN/m^2
3-49 $\psi = \dfrac{1}{2\mu}\dfrac{dp}{dx}\left(\dfrac{y^3}{3} - h\dfrac{y^2}{2}\right)$;
      $\psi_{top} = -\dfrac{1}{12\mu}\dfrac{dp}{dx}h^3$
3-50 $q = -\dfrac{1}{12\mu}\dfrac{dp}{dx}h^3$; same
3-51 $u_r = 0$, $u_\theta = \dfrac{C}{r}$, $C = \dfrac{u_0(r_1 - r_2)}{\ln(r_1/r_2)}$; $p_r = p_0 + \dfrac{\rho}{2}\left(u_0^2 - \dfrac{C^2}{r^2}\right)$
3-52 2.34L/s, 10m/s
3-53 8.49×10^{-4}m/s
3-54 4.15m/s, 21.02Pa; 5.14m/s, 12.74Pa

Chapter 4

4-1 $v = \dfrac{1}{2}u_{max}$; $\alpha = 2.0$; $\beta = 4/3$;

4-2 $v = \dfrac{162}{190}u_{max}$; $\alpha = 1.037$; $\beta = 1.013$;

4-3 22.2m/s; $5.4 \times 10^7 \text{kg/s}$

4-4 $0.15\text{m}^3/\text{s}$

4-5 $2.0\text{m}^3/\text{s}$

4-6 $\dfrac{dh}{dt} = \dfrac{Q_1 + Q_3 - Q_2}{\pi d^2/4}$; 4.13m/s

4-7 $Q = \dfrac{3}{8}U_0 b\delta$

4-8 0.0604kg/s; 1060m/s

4-9 6.06m/s

4-10 6.02kg

4-11 15.4m

4-12 $0.042\text{m}^3/\text{s}$; less

4-13 16.4m/s; 1.16MW

4-14 3.7cm

4-15 $0.027\text{m}^3/\text{s}$

4-16 $2.04 \times 10^5 \text{N/m}^2$

4-17 $1.5\text{m}^3/\text{s}$

4-18 $0.045\text{m}^3/\text{s}$, 22.66m

4-19 56.4m; 33.7kW

4-20 $6.12 \times 10^{-3} \text{m}^3/\text{s}$; 340kPa

4-21 1151kW, 530kW; \$28725/year

4-22 125kPa; 17.3m

4-23 -64.5Pa; 967.5Pa

4-24 $0.023\text{m}^3/\text{s}$; 73.2kPa; 1.05kN

4-25 right, right, right, left

4-26 0; 7420 N

4-27 $R_x = 8340\text{N}$, $R_y = 0$

4-28 1.18kN

4-29 163N

4-30 $2.66 \times 10^{-4} \text{m}^3/\text{s}$

4-31 167N; 833W

4-32 980N

4-33 0.108kg

4-34 40N

4-35 45.49N

4-36 (c) > (b) > (a); $180°$; 2

4-37 $0.84\text{m}^3/\text{s}$, $0.0603\text{m}^3/\text{s}$; 8100N; 2500N

4-38 358.2N; 517.4N

4-39 733N; 93.1N

4-40 $F = \dfrac{1}{2}\rho g b h_1^2 \left[1 - \left(\dfrac{h_2}{h_1}\right)^2\right] - \rho h_1 b v_1^2 \left[\dfrac{h_1}{h_2} - 1\right]$;

$\dfrac{h_2}{h_1} = \left(\dfrac{v_1^2}{gh_1}\right)^{1/3}$; $h_2 = h_1/\sqrt{2}$

4-41 $v \approx \zeta + (\zeta^2 + 2\zeta v_j)^{1/2}$, $\zeta = \dfrac{\rho Q}{2k}$

4-42 61.6kN

4-43 (a) > (b)

Chapter 5

5-1 $[\nu] = L^2 T^{-1}$; $[\tau] = ML^{-1}T^{-2}$; $[\sigma] = MT^{-2}$;
$[\dot{N}] = ML^2 T^{-3}$

5-2 $ML^{-2}T^{-2}$; MLT^{-2}; $ML^{-3}T^{-2}$; MLT^{-2}

5-4 L

5-5 L^{-1}

5-6 $\dfrac{\tau}{\rho U^2}$; $\dfrac{\Delta p}{\rho U^2}$; $\dfrac{F}{\rho U^2 L^2}$; $\dfrac{\sigma}{\rho U^2 L}$

5-7 $\dot{N} = K\left(\dfrac{Q^2}{gH^5}\right)^{1-c} \rho Q g H$, in which c is the exponent of g

5-8 $Q = Kb^{\frac{5}{2}-c}g^{\frac{1}{2}}H^c$, where c is the exponent of H

5-9 $\dfrac{\Delta p D_1}{v\mu} = f\left(\dfrac{D_2}{D_1}, \dfrac{\rho D_1 v}{\mu}\right)$

5-10 $\pi = \dfrac{Q\mu}{(\Delta p/L)b^4}$; $16Q$

5-11 $\dfrac{F}{\rho U^2 d^2} = f\left(\dfrac{\mu}{\rho dU}\right)$

5-12 $\dfrac{f_k D}{U} = f\left(\dfrac{\mu}{\rho dU}\right)$

5-13 $\dfrac{u}{U} = f\left(\dfrac{\mu}{\rho dU}, \dfrac{y}{h}\right)$

5-14 16

5-15 $Q = \sqrt{g}H^{3/2}f\left(\dfrac{b}{H}, \dfrac{P}{H}\right)$

5-16 10

5-17 10.73; 10.73, 0.429

5-18 74.67Pa; -35.56Pa

5-19 $\dfrac{l_p}{h_p} = \dfrac{l_m}{h_m}$, $\dfrac{b_p}{h_p} = \dfrac{b_m}{h_m}$, $\dfrac{\rho_p h_p U_p}{\mu_p} = \dfrac{\rho_m h_m U_m}{\mu_m}$, $\dfrac{u_p}{U_p}$

$$= \frac{u_m}{U_m}$$

5-20 0.0647 to 0.0971
5-21 0.0914 m³/s
5-22 0.151 m
5-23 3.3 m/s, 246 m³/s; 40.5 kN
5-24 1.86 m/s; 42900; 254000
5-25 5.33 m/s; 11.95 Pa
5-26 37.5 m/s; 3.7Δp

Chapter 6

6-1   Separate; green
6-2   Yes; yes
6-3   0.05 m
6-4   Turbulent; 7.25×10^{-3} m³/s
6-5   Turbulent
6-6   Laminar; 0.0151 m; Turbulent
6-7   0.123 m³/s
6-8   0.392 Pa; 0.196 Pa; 0.8 m
6-9   17.4 m³/s, 0.034 m; 88 Pa
6-10 0.08; 12 Pa
6-11 0.354D
6-12 −37.6 kPa
6-13 67.2 kPa; 50.5 kPa
6-14 188 kPa; 19.2 m; 0.71 W
6-15 8.0 m/s
6-16 2.0 m/s, 0.00251 m³/s
6-17 32.3 Pa
6-18 1.62×10^{-5} m³/s; 1.00×10^{-5} m³/s; 2.44×10^{-5} m³/s
6-19 0.066 m³/h; 0.066 m³/h
6-21 −4000 Pa/m; 50 Pa; 46%
6-22 0.012 mm
6-23 2.38×10^6 Pa
6-24 0.00031 kg/(m·s); gasoline
6-25 4.0 m³/h
6-26 0.02
6-27 0.80 m/s; 5.1 m³/s
6-28 Turbulent; 6.20 m
6-29 3928 m
6-30 0.109 m³/s
6-31 1.06 mm; 66% less
6-32 15 m³/h
6-33 25 m³/h (to left)
6-34 1520 m³/h; −6.5 Pa
6-35 0.089 m/s; 0.0191
6-36 1.34×10^{-3} m³/s; 1.11×10^{-3} m³/s
6-37 322 kPa; 7.8% (25 kPa more)
6-38 0.904D
6-39 0.972D
6-40 2.09 m
6-41 3.46 MPa
6-42 0.3%; 5 times more
6-43 1.17×10^{-3} m³/s; 0.84 cm
6-44 6.4
6-45 2.15×10^{-3} m³/s
6-46 $7.48 \times 10^{-3} \sqrt{x}$ m
6-47 8.28 m/s; 20.5 km/h
6-48 1.06 m/s
6-49 2.44 m/s; 2.13 m/s
6-50 30.8 hp

Chapter 7

7-1   0.64, 0.993, 0.6355, 0.014
7-2   0.040 m³/s
7-3   3.222 m³/s, 3.172 m³/s
7-4   0.00122 m³/s; 0.00161 m³/s, 1.506 m
7-5   1.20 m; 4.30 m
7-6   5.55 m, 1.56 m; 0.015 m³/s
7-7   1.06 m, 1.44 m; 0.00367 m³/s
7-8   393.6 s
7-9   5.11 m/s; 704 s; yes
7-10 225.8 s
7-11 $T = \dfrac{2}{3} \dfrac{lD^{3/2}}{\mu A \sqrt{g}}$

Chapter 8

8-1   1.0 m
8-2   0.071 m³/s; 2.85 mH$_2$O
8-3   0.007 m³/s; 27.5 m
8-4   0.932 m
8-5   9.19 m³/s; −58.7 kPa

8-6 0.211m; 62.03kPa
8-7 75.46m
8-8 53.4m
8-9 734kPa
8-10 2.96kW
8-11 840W
8-12 2.08m³/s or 14.4m³/s; No flow
8-13 379kW
8-14 0.031m × 0.053m; 0.0445m
8-15 37.2kN
8-16 0.0025m³/s, 0.19m
8-17 1.32hp
8-18 0.0053m³/s
8-19 0.007m³/s, 0.039m³/s; 0.107m
8-20 4976m
8-21 0.05m³/s
8-22 5.66
8-23 0.231m³/s
8-24 0.229m³/s
8-25 217W
8-26 0.91m³/s, 2.09m³/s
8-27 0.023m³/s, 0.013m³/s; 2160kPa
8-28 0.0269m³/s
8-29 0.0281m³/s, 0.0111m³/s and 0.0163m³/s; 503kPa
8-30 0.0667m³/s, 0.0333m³/s; 15.18m
8-31 215kPa
8-32 0.31m³/s, 0.31m³/s, 0.38m³/s; 149m
8-33 24.9m; 41.36m; 14.84m
8-34 19.75m
8-35 0.0152m³/s; 0.0174m³/s
8-36 0.300m³/s, 0.510m³/s, 0.809m³/s
8-37 38.14m
8-38 10.09m
8-39 29.30m; 250mm, 200mm, 100mm (nominal diameters)
8-40 AB, AC, BC, CD, BD: 0.0337, 0.0229, 0.0280, 0.0509, 0.0057m³/s; 744.6, 710.2, 514.0kPa
8-41 EA, ED, AD, AB, BC, CD: 5.322, 4.678, 1.501, 3.821, 0.821, 3.179 m³/s
8-42 25.5m/s; 0.109m³/s; 1.23 Pa
8-43 46.65m/s
8-44 58cm; 20m³/h
8-45 67.3kPa; 64.8kPa

Chapter 9

9-1 $6m^2$, 7m, 0.857m; $12m^2$, 10.21m, 1.175m; $4m^2$, 5.657m, 0.707m; $6.28m^2$, 6.28m, 1.0m
9-2 Turbulent; supercritical
9-3 Supercritical
9-4 0.75m; 1.43×10^6; turbulent
9-5 0.00114; 25% less
9-6 20% less; no change
9-7 0.0174
9-8 2.139m, 1.337m
9-9 0.38m/km
9-10 10.66m
9-11 10.3m
9-12 8.52m
9-13 31%
9-14 0.685m
9-15 5.23m³/s;
9-16 1.025m
9-17 116m³/s
9-18 37.2m³/s
9-19 1.52m
9-20 11.7m³/s (open channel flow), 12.2m³/s (pipe flow)
9-21 0.159m³/s
9-22 6.37m
9-23 1.56m
9-24 2.65m
9-25 1.50m, 0.748
9-26 1.50m, 0.00023
9-27 2.42m; $b = 2.21m$, $h = 1.11m$; $b = 1.35m$, $h = 1.17m$;
9-28 1.43m
9-29 12.58m, 6.29m; No
9-30 7.05m, 1.41m
9-31 3.50m/s; supercritical; 0.82m
9-32 0.742m; supercritical; 1.03m
9-33 51.0m³/s
9-34 subcritical; 1.04m

9-35　3.44m
9-36　0.0136; decrease; increase
9-37　3.44m; No
9-38　2.59m
9-39　mild
9-40　Subcritical
9-41　Subcritical
9-42　$2.15 \times 10^{-4} \text{m}^3/\text{s}$
9-43　4.56m
9-44　9.40m/s, 1.88m/s; 3.873, 0.347; 0.376
9-45　(b) 2.05;
9-46　1.93m; 0.60m
9-47　Mild; all are M_1 curves
9-48　Steep; S_3, S_2, S_1
9-49　345m
9-52　214m; M_1
9-53　13212m; 4.44m
9-54　3897m
9-55　2.17m, 2.0m
9-56　0.202m

Chapter 10

10-1　$15.6 \text{m}^3/\text{s}$
10-2　0.002145
10-3　$5.404 \text{m}^3/\text{s}$
10-4　0.725m
10-5　$0.164 \text{m}^3/\text{s}$
10-6　1.402m
10-7　53.02°
10-8　$1.228 \text{m}^3/\text{s}$
10-9　60.9%
10-10　0.043m; 0.028m
10-11　$Q = m_0 \left(\dfrac{2}{3}\sqrt{2g}bH^{3/2} + \dfrac{8}{15}\sqrt{2g}H^{5/2} \right)$,
　　　where $m_0 = 0.60$
10-12　$653.26 \text{m}^3/\text{s}$; 7.75m
10-13　$3.374 \text{m}^3/\text{s}$; 0.359m
10-14　$1.11 \text{m}^3/\text{s}$; 0.821m
10-15　$6.655 \text{m}^3/\text{s}$; $4.688 \text{m}^3/\text{s}$
10-16　$2.0 \text{m}^3/\text{s}$; 0.295m
10-17　$1.572 \text{m}^3/\text{s}$
10-18　$8.97 \text{m}^3/\text{s}$; $8.13 \text{m}^3/\text{s}$
10-19　1.421m
10-20　$8.806 \text{m}^3/\text{s}$; 1.86
10-21　6m; 1.59m

Chapter 11

11-1　4.15×10^{-6} m/s
11-2　5.45×10^{-4} m^3/s
11-3　$4\text{cm}^3/\text{s}$; $0.16\text{cm}^3/\text{s}$; $0.308\text{cm}^3/\text{s}$
11-4　9.5×10^{-7} m^2/s
11-5　$1.068 \times 10^{-5} \text{m}^2/\text{s}$; $0.0016 \text{m}^3/\text{s}$; 29.9m
11-6　$8.6 \times 10^{-5} \text{m}^3/\text{s}$; 3.55m
11-7　$7.5 \times 10^{-4} \text{m}^3/\text{s}$
11-8　$3.11 \times 10^{-4} \text{m}^3/\text{s}$
11-9　0.0173cm/s
11-10　$4.76 \times 10^{-4} \text{m}^3/\text{s}$
11-11　3.27×10^{-5} m/s
11-12　0.00147m/s; 1020m
11-13　$6.05 \times 10^{-3} \text{m}^3/\text{s}$

Chapter 12

12-1　340.2m/s; 296.0m/s
12-2　Yes; yes
12-3　347m/s, 0.692; 634m/s, 0.379
12-4　0.022m
12-5　225m/s
12-6　1.21
12-7　343.1m/s; 258.4m/s; 323.1m/s; 321.45kPa
12-8　0.23kg/s
12-9　356K; 160kPa
12-10　300.0K, 300.1K, 305.0K, 797.5K
12-11　45.2m, 357K, 47.3kPa, 303m/s
12-12　$0.68L$
12-13　40.42m/s; 0.29kg/s
12-14　0.187, 0.184kg/s
12-15　0.0136kg/s, 0.5225
12-16　107m/s, 371kPa, 330K, 9.2m
12-17　1.13m
12-18　0.56m
12-19　0.00187kg/s, 0.00188kg/s, 0.00189kg/s
12-20　892kPa